| DATE | | | |
|---|---|---|---|
| | | | |
| | | | |
| | | | |
| | | | |
| | | | |
| | | | |
| | | | |
| | | | |
| | | | |
| | | | |
| | | | |

BAKER & TAYLOR

# Introduction to Electronic Devices

# Introduction to Electronic Devices

## Michael Shur

### University of Virginia

JOHN WILEY & SONS, INC.

New York   Chichester   Brisbane   Toronto   Singapore

| Acquisitions Editor | Steven M. Elliot |
| Marketing Manager | Lisa Perrone |
| Production Editor | Deborah Herbert |
| Manufacturing Manager | Dorothy Sinclair |

This book was typeset by the author and Printed and bound by Hamilton Printing Co. The cover was printed by New England Book Components, Inc.

*Library of Congress Cataloging in Publication Data:*

Shur, Michael.
      Introduction to electronic devices / Michael Shur.
           p.       cm.
      Includes index.
      ISBN 0-471-10348-9 (cloth : alk. paper)
      1. Electronic apparatus and appliances.    I. Title.
TK7870.S5137    1995
621.3815'2--dc20
                                  95-10567
                                    CIP

ISBN 0-471-10348-9

Printed in the United States of America

10  9  8  7  6  5  4  3  2  1

# Preface

This is a textbook for electrical engineering students taking their first course in electronic devices during their junior or senior year. For many students, such a course may be the only one they will take on electronic devices; for others, it may become a gateway to an exciting career in this important engineering field. In either case, this course must form a solid foundation for understanding the basics of electronic device technology. This technology has penetrated every aspect of our lives. Electronic devices are found in home electronics, computers, cars, appliances, bank machines, copiers, phones, toys, and medical equipment. More and more, these devices determine the competitiveness of our economy and the strength of our defense. Just recall the role of smart munitions and night vision devices in the Gulf war. Both sides used the same explosives and their shells were made from the same metal. However, the Allied forces had superior semiconductor electronics. Or think about the large trade imbalance between the United States and Japan – a relatively small island country with few natural resources – in favor of the Japanese. We come to the conclusion that small semiconductor chips have become mightier than gold, or oil, or troops, or a large territory, or a large work force.

The importance of the course on the fundamentals of solid state devices for all electrical engineering students, irrespective of their future area of interest, cannot be overstated. A technician may treat an electronic device simply as a black box with characteristics described in manufacturer's data sheets but an electrical engineer cannot. She or he must understand how these devices work because the present dominance of solid-state devices will surely pale compared with what is expected to come in the near future. Three-dimensional flat panel room-size TVs integrated with private video phones with teleconference capabilities, cellular phones for everybody, paperless offices, notebook supercomputers, personal digital assistants, interactive university courses widely available on television, smart robots, intelligent electric cars, widespread use of photovoltaic technology, smart energy-efficient controls of heating and air conditioning in offices and homes, new sensors to control and protect environment are but a few technologies that will very soon become a reality. All these technologies will utilize new generations of electronic devices with vastly improved capabilities and at greatly reduced cost.

According to a great economist, Adam Smith, (who published "An Enquiry into Nature and causes of the Wealth of Nations" in 1776) wealth is created by a laissez-faire economy and free trade. According to John Maynard Keynes (who published "The General Theory of Employment Interest, and Money" in 1936), wealth is created by careful government planning and government stimulation of the economy. However new ideas in economics in 1990s (often linked to Paul Romer of Berkeley) emphasize the role of innovation (such as an invention of a computer chip) in creating wealth.

In the 1960s, a single semiconductor device, which was slow and inefficient by modern standards, may have cost a few dollars. Today, high-speed, low-power transistors used in computers or other electronic equipment cost as little as $10^{-3}$ cents.

Modern semiconductor electronics needs faster and faster devices which operate at less and less power. This requires the scaling down of typical device sizes. Advances in fabrication technology have reduced the minimum device feature size from about 20 microns in the early 1960s to submicron dimensions in the 1990s. In shorter devices, electrons take less time to travel across the device, leading to higher speeds and operating frequencies. A smaller device volume also translates into lower operating power and denser circuits. For example, the new Intel® microprocessor – Pentium$^{tm}$ – uses dimensions as short as 0.8 µm. It contains 3.1 million transistors on a one square inch semiconductor chip and runs approximately 15 times faster than microprocessors ran just a decade ago. Another new microprocessor – PowerPC 620 – (developed jointly by Apple, IBM, and Motorola) utilizes 7 million transistors with 0.5 µm feature sizes and operates at 133 MHz clock rate. Still, its power requirements are quite high (up to 30 W power dissipation at 3.3 V power supply). Texas Instruments is using even smaller, 0.35 µm devices in its new T5 microprocessor operating at up to 200 MHz clock rate. In research laboratories, engineers and scientists are experimenting with devices with dimensions less than one tenth of a micron.

Today, most semiconductor devices are made of silicon. However, submicron devices made of compound semiconductors, such as gallium arsenide or indium phosphide, successfully compete for applications in microwave and ultra-fast digital circuits. Other semiconductors, such as mercury cadmium telluride, are utilized in infrared detectors. Silicon carbide, aluminum nitride, and gallium nitride promise to be suitable for power devices operating at elevated temperatures and in harsh environments. As device dimensions shrink and more exotic compound semiconductor materials are used in electronic circuits, the physics involved in understanding the device behavior becomes more complicated and more fascinating. In novel device structures, the dimensions are so small that quantum effects become important or even dominant. Modeling, or even qualitative understanding, of these structures presents a formidable challenge. In this respect, the physics of semiconductor devices differs from more established "classic" engineering courses, such as courses on electromagnetic fields or circuit theory. The material here is not as firmly established and is somewhat in a state of flux. Kirchoff's current law will never change. However, even basic semiconductor equations, used for decades to analyze semiconductor devices, have to be questioned and revised when applied to very small devices.

All this makes teaching the first course on electronic devices a real challenge. Rational choices and compromises have to be made so that the students do not feel completely overwhelmed and can find links between this course and other core electrical engineering courses.

I wrote this book keeping all that in mind. Above all, I wanted to help the students gain a firm grasp of fundamentals. Ten years from now, solid-state device technology will surely change a great deal, but the fundamentals of semiconductor materials and device physics will probably remain the same. These fundamentals will still provide the firm foundation for mastering whatever new technologies will have been developed. Therefore, a considerable fraction of the material included into this book is devoted to physics and basic principles of device operation.

My other goal was to make the acquired knowledge very practical and to teach students how to use it for solving real problems. I tried to link the semiconductor device models described in the book to the models implemented in the popular integrated circuit simulator SPICE, which is used by a vast majority of electrical engineers. At least two free versions of SPICE are readily available for students - the student version of PSpice$^{tm}$ by Microsim, Inc., and the student version of AIM-Spice. The student version of PSpice$^{tm}$ can be ordered from Prentice Hall; the latest student version of AIM-Spice can be downloaded using an electronic mail (see the Instructions given in Appendix A8). I also included an example of a computer simulation of a semiconductor device using one of the most popular device simulators called PISCES (developed at Stanford). At many universities nowadays, PISCES is available even to undergraduate students.

Finally, my objective was to bring the reader to the forefront of modern electronic device technology. This is the reason why I included such topics as heterojunction devices, liquid crystal displays, and single electron electronics. The last chapter, "Novel Devices," gives the student an opportunity to use his or her basic knowledge of semiconductor device physics gained from the first eight chapters for the understanding of emerging device technologies.

After taking this course, students should be able to design a solar panel power supply for a portable computer. They should be able to use the circuit simulator, SPICE, to predict how the speed of an amplifier or a digital electronic circuit increases when the device dimensions are scaled down, and they should gain enough understanding of electronic devices to make an intelligent choice of device technology for a particular application. This is an example of how students can use the knowledge acquired in this course in combination with a very powerful Computer Aided Design (CAD) tool such as SPICE. For those students who will decide to delve into the subject matter more deeply, this course will give a firm foundation for further studies.

I am hoping that this book will appeal to a wide audience and that most readers will be find this book useful, perhaps, for different reasons. Quoting Alexander Pope (from an *Essay on Man*):

> The learned is happy nature to explore,
> The fool is happy that he knows no more;
> The rich is happy in the plenty given,
> The poor contents him with the care of Heaven.

The vast majority of electrical engineering students definitely fall into the first category.

Almost every section of the book ends with a summary of basic equations in a tabular form. This material can be useful for review purposes and exam preparation. Students taking advanced courses on semiconductor and solid state devices can use these tables to make sure that they know the material required as the prerequisite for these courses. These tables may also be useful for students and engineers preparing for qualifying or professional engineer examinations.

Not all of the material included into the book needs to be covered in a one semester lecture course. The section on tunneling in Chapter 1, the section dealing with one-dimensional and two-dimensional electron gas in Chapter 2, the section on the Hall effect in Chapter 3, the sections in Chapter 4 on the tunnel, Gunn, and IMPATT diodes and on computer simulations, Chapters 7 and 9 may be used for reading assignments or even skipped altogether. However, these sections and chapters (marked with asterisks in the table of contents) will be very useful for the students who intend to specialize in the area of the solid-state devices. Some other sections, such as the section on MOSFET modeling, can be better mastered by solving practical problems using SPICE.

For a typical 40-lecture course, two lectures may be devoted to the material in the Introduction, three lectures to quantum mechanical concepts (Sections 1.2 and 1.4), four lectures may cover basic solid-state physics (Chapter 2), and four more lectures may cover electrons and holes in semiconductors (Chapter 4). Approximately six lectures may be devoted to diodes and contacts (Sections 4.1 through 4.9), six lectures to BJTs (Sections 5.1 through 5.5), ten lectures to MOSFETs (Chapter 6), four lectures to photonic devices (Chapter 8), and two lectures to the final review. Chapter 9 could be also used as a starting point for undergraduate research projects.

This book contains many examples, problems, and review questions. The relative difficulty of review questions is rated by points based on my experience of using these questions for quizzes and exams. Most of the problems are fairly traditional and self-contained "engineering science" problems. Some of the problems will require a student to make reasonable guesses. Many problems and review questions (with numbers underlined) are "engineering design" problems that do not have unique solutions. I also included a fair number of problems that require a student to do calculations using a personal computer. I believe that students should be strongly encouraged to use MathCad$^{tm}$ (a student version is really quite sufficient) or Mathematica$^{tm}$ (at the universities where this software is available for undergraduate students), or even spreadsheets such as Lotus$^{tm}$ or Excel$^{tm}$. This will give students an opportunity to gain some rudimentary number crunching experience and, perhaps more importantly, to build a sense of confidence in their ability to perform relatively long calculations and computations correctly, the ability that is crucial for all electrical engineers. In making up the problems, I tried to ensure that after taking the course the students will be comfortable with units and numbers, since I strongly believe that students must be able not just to understand how to solve a problem in principle but also be able to obtain accurate answers.

A few problems (marked with asterisks) are more advanced. They may be suitable for problem-solving sessions or may be worked in class.

For courses based on this book, I recommend that 75% of the credit hours should be allocated to engineering science and 25% to engineering design.

The book includes references to key monographs and textbooks in the field of semiconductor devices. In a way, I tried to make this book "a reference of references".

The Instructor's Manual (available free of charge to instructors) includes the answers to the review questions and problem solutions, basic semiconductor modeling software for the Macintosh computer, proposed lecture outlines, reading assignments, viewgraphs, and two types of handouts (2 slides per page and 6 slides per page) for each lecture for a 40 lecture course. The Instructor's Manual is also available in electronic form in order to simplify the preparation of homework assignments and exams.

We live in an unsentimental age, but I cannot help but admire the ideas and hard work of thousands and thousands of bright people who have contributed and are contributing to the area of electronic devices. The power of human mind, which transcends national boundaries and differences between generations, is behind every single electronic device. A great deal of thought and innovation went into every step along the way – from material growth and device fabrication to simulation and modeling. I feel truly privileged to be able to work in this field and admire these achievements. As the Japanese would say, I bow twice to device scientists and engineers, past, present, and future.

During the writing of this book, my wife, Paulina, provided me with invaluable support and encouragement. My colleagues at the Applied Electrophysics Laboratory here at the University of Virginia, especially Professors Bob Mattauch, Bill Peatman, Tom Crowe, Steve Jones, Elias Towe, and Kiang Lee, Drs. Boris Gelmont, Bjornar Lund, Hyunchang Park, and Alexei Bykhovski shared with me their insights into device physics and created an excellent environment for both research and teaching. I greatly benefited from working with Professors Tor Fjeldly, Kwyro Lee, Drs. Trond Ytterdal and

Michael Hack on the development of device models. I would like to thank Professor Konstantin Likharev for his suggestions regarding the superconductivity and single electronics sections. Dr. Peter Rabkin of Silvaco International helped me with the ATLAS-II simulation discussed in Section 4.13. Professors Tor Fjeldly, Mikhail Dyakonov, Gennadi Gildenblat, Ms. Holly Slade, Ms. Jodi Bowers, Mr. Mark Jacunski, and Mr. Jason Robertson made many useful comments. I am also most appreciative of the useful comments and suggestions of Professors Pritpal Singh, Reginald Perry, and Jasprit Singh who reviewed this manuscript. I am very grateful to my former and present graduate and undergraduate students for their hard work, enthusiasm, friendship, and support.

I will appreciate any comments, corrections, or suggestions which can be sent via electronic mail to *shur@virginia.edu*.


Michael Shur
Charlottesville, Virginia

# Contents

## CHAPTER 4. DIODES AND CONTACTS   188

# CHAPTER 5. BIPOLAR JUNCTION TRANSISTORS  297

Contents

# List of Symbols

| | |
|---|---|
| $\alpha$ | absorption coefficient |
| $\alpha$ | base transport factor |
| $\alpha$ | short-circuit common-base current gain |
| $\mathbf{a}, \mathbf{b}, \mathbf{c}$ | primitive basis vectors |
| $a_B$ | Bohr radius |
| $A_i$ | current gain |
| $\alpha_{ni}$ | electron impact ionization coefficient |
| $A_p$ | power gain |
| $\alpha_{pi}$ | hole impact ionization coefficient |
| $A_v$ | voltage gain |
| $B$ | magnetic field |
| $\beta$ | short-circuit common-emitter current gain |
| $B_c$ | critical magnetic field |
| $C$ | capacitance |
| $c$ | velocity of light |
| $C_d$ | differential capacitance |
| $C_{dep}$ | depletion capacitance |
| $D$ | detectivity |
| $D$ | transmission coefficient |
| $D$ | two dimensional density of states |
| $D_n$ | electron diffusion coefficient |
| $D_p$ | hole diffusion coefficient |
| $E$ | energy |
| $\varepsilon$ | dielectric permittivity |
| $E_B$ | Bohr energy |
| $E_c$ | conduction band edge |
| $E_F$ | Fermi level |

| | |
|---|---|
| $E_{Fn}$ | electron quasi-Fermi level |
| $E_{Fp}$ | hole quasi-Fermi level |
| $E_g$ | energy gap |
| $E_i$ | intrinsic Fermi level |
| $E_t$ | trap level |
| $E_v$ | valence band edge |
| $\Phi$ | wave function (time dependent) |
| $\phi$ | electric potential |
| $f$ | frequency |
| $f(\nu)$ | radiation density for black-body radiation |
| $\mathbf{F}, F$ | electric field |
| $F_{1/2}$ | Fermi integral |
| $F_{br}$ | breakdown field |
| $FF$ | fill factor |
| $\Phi_m$ | metal work function |
| $f_{max}$ | maximum frequency of oscillations |
| $f_n$ | electron distribution function |
| $f_p$ | hole distribution function |
| $\Phi_s$ | semiconductor work function |
| $f_T$ | cutoff frequency |
| $G$ | generation rate |
| $\gamma$ | injection efficiency |
| $g_d$ | drain conductance |
| $g_m$ | transconductance |
| $g_n$ | electron density of states |
| $g_p$ | hole density of states |
| $\eta$ | efficiency |
| $\eta$ | ideality factor |
| $h$ | Planck constant |
| $\hbar$ | reduced Planck constant |
| $I$ | current |
| $i$ | $\sqrt{-1}$ |
| $I_L$ | light-generated current |
| $I_s$ | saturation current |
| $j$ | current density |
| $j$ | quantum number |
| $j_s$ | saturation current density |
| $\mathbf{K}$ | reciprocal lattice vector |
| $\mathbf{k}, k$ | wave vector |
| $k_B$ | Boltzmann constant |
| $l$ | orbital quantum number |

| | |
|---|---|
| $\lambda$ | wavelength |
| $L_{Dn}$ | electron Debye radius (electron Debye length) |
| $L_{Dp}$ | hole Debye radius (hole Debye length) |
| $L_n$ | electron diffusion length |
| $L_p$ | hole diffusion length |
| $m$ | grading coefficient |
| $m$ | magnetic quantum number |
| $m$ | mass |
| $m_e$ | free electron mass |
| $m_n$ | electron effective mass |
| $m_p$ | hole effective mass |
| $m_{ph}$ | heavy hole effective mass |
| $m_{pl}$ | light hole effective mass |
| $n$ | electron concentration |
| $n$ | principal quantum number |
| $N_a$ | acceptor concentration |
| $N_c$ | electron effective density of states |
| $N_d$ | donor concentration |
| $n_G$ | Gummel number |
| $n_i$ | intrinsic carrier concentration |
| $n_{po}$ | equilibrium electron concentration in $p$-type semiconductor |
| $n_r$ | refraction index |
| $n_s$ | surface electron concentration |
| $N_v$ | hole effective density of states |
| $p$ | hole concentration |
| $\mathbf{p}, p$ | momentum |
| $p_{no}$ | equilibrium hole concentration in $n$-type semiconductor |
| $q$ | electronic charge |
| $Q_c$ | collection efficiency |
| $Q_d$ | depletion charge per unit area |
| $Q_e$ | quantum efficiency |
| $q\phi_b$ | Schottky barrier height |
| $Q_s$ | surface charge per unit area |
| $R$ | recombination rate |
| $R$ | reflection coefficient |
| $R$ | resistance |
| $R$ | responsivity |
| $\rho$ | space charge density |
| $\mathbf{r}$ | space vector |
| $R_c$ | contact resistance |

| | |
|---|---|
| $\rho_c$ | specific Ohmic contact resistance |
| $R_d$ | differential resistance |
| $r_H$ | Hall factor |
| $R_s$ | series resistance |
| $S$ | cross section |
| $S$ | spin |
| $\sigma$ | conductivity |
| $S_n$ | electron surface recombination rate |
| $S_p$ | hole surface recombination rate |
| $T$ | temperature |
| $t$ | film thickness |
| t | time |
| $T_c$ | critical temperature |
| $\tau_{gen}$ | generation time |
| $\tau_{md}$ | Maxwell dielectric relaxation time |
| $\tau_{nE}$ | electron energy relaxation time |
| $\tau_{nl}$ | electron lifetime |
| $\tau_{np}$ | electron momentum relaxation time |
| $\tau_{pl}$ | hole lifetime |
| $U$ | potential difference |
| $U$ | potential energy |
| $V$ | voltage |
| $V_{abr}$ | critical voltage of avalanche breakdown |
| $V_{bi}$ | built-in voltage |
| $V_{brt}$ | critical voltage of tunneling breakdown |
| $V_{BS}$ | substrate bias |
| $V_{FB}$ | flat-band voltage |
| $V_H$ | Hall voltage |
| $\mathbf{v_n}, v_n$ | electron drift velocity |
| $V_{oc}$ | open circuit voltage |
| $\mathbf{v_p}, v_p$ | hole drift velocity |
| $V_s$ | surface potential |
| $v_s$ | saturation velocity |
| $v_{sn}$ | electron saturation velocity |
| $V_T$ | threshold voltage |
| $V_{th}$ | thermal voltage |
| $v_{th}$ | thermal velocity |
| $W$ | device width |
| $\Omega$ | Ohm |
| $\Omega$ | two-dimensional density of states |
| $\omega$ | radian frequency |

| | |
|---|---|
| $x$ | mole fraction |
| $x, y, z$ | space coordinates |
| $x_d$ | depletion region width |
| $X_s$ | electron affinity |
| $\psi$ | wave function (time independent) |
| $\mu_{FET}$ | field effect mobility |
| $\mu_H$ | Hall mobility |
| $\mu_n$ | electron mobility |
| $\mu_p$ | hole mobility |
| $\Delta$ | energy gap in superconductor energy spectrum |
| $\Delta E_c$ | conduction band discontinuity |
| $\Delta E_g$ | energy band discontinuity |
| $\Delta E_v$ | valence band discontinuity |

# Introduction

The purpose of this Introduction is to introduce the basic concepts of semiconductor physics in the simplest possible terms. Later chapters will elaborate upon these concepts in more detail. Reading this Introduction may be compared to looking at a map before taking a long trip.

We shall start from a general discussion of electronic materials. These materials include dielectrics, metals, semiconductors, and superconductors. Good quality dielectrics have extremely large specific resistances so that measurable currents in dielectrics flow only in very high electric fields (on the order of millions of V/cm) and/or at elevated temperatures. Specific resistances for metals at room temperature vary from $10^{-4}$ $\Omega$cm for mercury to $1.6 \times 10^{-6}$ $\Omega$cm for silver and $1.75 \times 10^{-6}$ $\Omega$cm for copper. Specific resistances for semiconductors may vary enormously – from $10^{-4}$ $\Omega$cm to $10^{12}$ $\Omega$cm. Superconductors have no resistance whatsoever. However, their superconducting properties occur only at cryogenic temperatures below a certain critical temperature, $T_c$ ($T_c \leq 30$ K or so for conventional superconductors and $T_c \leq 160$ K for recently discovered superconductors called high $T_c$ materials). Also, superconductivity disappears in high magnetic fields. When the current exceeds a certain critical value, the superconducting state is also destroyed. These limitations hinder large-scale practical applications of superconductors.

In 1821, the German physicist Tomas Seebeck first noticed unusual properties of semiconductor materials, such as lead sulfur (PbS). In 1833, the English physicist Michael Faraday reported on the temperature dependence of the conductivity for a new class of materials – semiconductors. This dependence was very unusual. The resistance of these materials decreases as temperature increases whereas in metals, resistance always increases with temperature!

In 1873, the British engineer W. Smith discovered that the resistivity of selenium, a semiconductor material, is very sensitive to light. The first practical applications of semiconductors date to 1875, when Werner von Siemens invented a selenium photometer, and to 1878, when Alexander Graham Bell used this device for a wireless telephone communication system. Still, until the discovery of a Bipolar Junction Transistor by the American scientists Bardeen, Brattain, and Shockley in 1947, electronic circuits utilized large, power hungry vacuum tubes. Nowadays, electronic circuits use tiny integrated circuit chips containing millions of semiconductor devices.

These integrated circuits operate from a power supply where nonelectrical forces (such as chemical or mechanical forces) separate **positive and negative electric charges**. These separated charges create electric fields, and those electric fields make **charge carriers**, such as free electrons, to carry electric current. Such a supply can operate a computer, in which semiconductor devices perform many millions or even billions of operations per second.

The primary reason why semiconductor devices can meet these challenging tasks is that they can quickly change their conducting properties from those approximating insulators to those closer to good conductors. These "off" and "on" states correspond to zeros and ones of the Boolean logic. To understand how this works, we first have to understand why the resistivity of a typical semiconductor material such as silicon may vary by many, many orders of magnitude – from approximately $10^{-2}$ $\Omega$cm to $10^4$ $\Omega$cm at room temperature. (Also, it can vary enormously with temperature.)

The explanation of the differences between semiconductors, metals, and dielectrics involves **quantum mechanical properties of matter**. In an atom, electrons can have only certain discrete values of energy called **energy states** or **energy levels**. The simplest atom, the hydrogen atom, has only one electron. For the hydrogen atom, the allowed energy levels are given by

$$E_n = -\frac{E_B}{n^2} \tag{I-1}$$

where

$$n = 1, 2, 3, 4, \ldots \tag{I-2}$$

is called the **principal quantum number,**

$$E_B = \frac{q^2}{8\pi\varepsilon_o a_B} \tag{I-3}$$

is called the **Bohr energy**, and

$$a_B = \frac{4\pi\varepsilon_o \hbar^2}{m_e q^2} \tag{I-4}$$

is called the **Bohr radius**.

Electron energies are usually measured in electronvolts. The unit of energy called **electronvolt** (eV) is the energy acquired by an electron accelerated by a 1 V potential difference. Since the electronic charge is equal to $-q$ where $q = 1.602 \times 10^{-19}$ C,

$$1 \text{ eV} = 1.602 \times 10^{-19} \text{ C} \times 1\text{V} = 1.602 \times 10^{-19} \text{ joule} \tag{I-5}$$

[As a general rule, we will use the international system of units in this book (SI) with an electron-volt being one of the important exceptions. The SI units are summarized in Appendix A1.]

The Bohr energy, $E_B = 2.18 \times 10^{-18}$ J $= 13.6$ eV, the Bohr radius, $a_B = 0.52917 \times 10^{-10}$ m $= 0.52917$ Å.

The electronic structure of other atoms that contain many electrons is more complicated, but the same general principle applies, and the electrons still occupy only certain discrete energy levels. The energies of these levels are unique for each chemical element. When atoms form solids, these energy levels split into many close energy levels that form allowed **energy bands** (see Fig. I.1). In each energy band, allowed energy levels are very close to each other so that the electron energy can vary continuously. However, these bands are separated by **forbidden energy gaps**. The band structure, that is, the position and extent of allowed and forbidden energy gaps, determines the properties of solids.

The important property of electrons is determined by the rule that is called the **Pauli exclusion principle**. According to this principle, not more than two electrons can occupy each energy state. Electrons occupy the lowest energy levels first. In semiconductors and dielectrics, almost all the states in the lowest energy bands are filled by electrons, whereas the energy states in the higher energy bands are, by and large, empty. The lower energy bands with mostly filled energy states are called the **valence bands**. The higher energy bands with mostly empty energy states are called **conduction bands**. The difference between the highest valence band and the lowest conduction band is called the **energy band gap** or the **energy gap**. An electron in a valence band needs the energy equal to or higher than the energy gap to experience a transition from the valence to the conduction band.

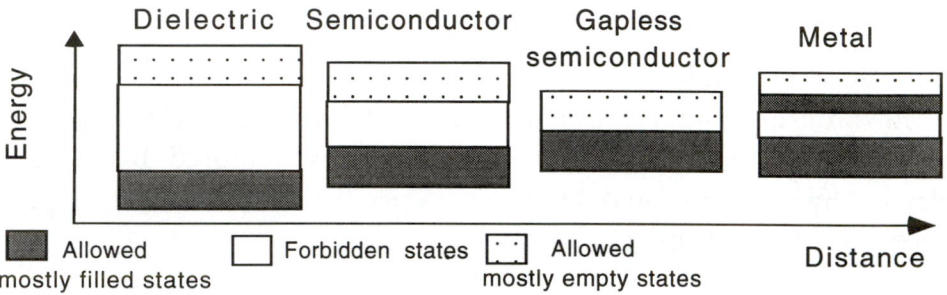

**Fig. I.1.** Band structures of a dielectric, a semiconductor, a gapless semiconductor, and a metal. The shaded regions represent energy levels filled with electrons (two per state, to satisfy the Pauli exclusion principle).

Typically, for semiconductors, energy band gaps vary between, let us say, 0.1 eV and 3.5 eV. In a dielectric, the energy gap, $E_g$, is large so that the valence bands are completely filled and conduction bands are totally devoid of electrons. Typically, for a dielectric $E_g$ is larger then 5 to 6 eV.

In a metal, the lowest conduction band is partially filled with many electrons (usually one electron per atom) and metal's resistance is very small.

Figure I.1 schematically shows the band structures of a metal, a semiconductor, and a dielectric. The rectangles in this figure correspond to the allowed electronic states (the shaded regions corresponding to the filled states).

In equilibrium, an **average thermal electron energy** is equal to $3k_BT/2$ where $k_B = 1.38 \times 10^{-23}$ joule/degree K is called the Boltzmann constant, and $T$ is the semiconductor temperature in degrees Kelvin (see Appendix A2 for more accurate values of $k_B$ and other important physical constants). At room temperature, $T = 27\ ^oC = 300$ K, and

$$k_BT \approx 1.38 \times 10^{-23}\ \frac{J}{K} \times 300\ K = 4.14 \times 10^{-21}\ joule = \frac{4.14 \times 10^{-21} J}{1.602 \times 10^{-19}\ \frac{J}{eV}} \approx 0.0258\ eV \quad (I-6)$$

However, $3k_BT/2$ is just the average energy. This energy corresponds to the average kinetic energy $m_n v_{th}^2/2$ where $m_n$ is the electronic mass and $v_{th}$ is called the electron **thermal velocity**. The actual energies of individual electrons may vary widely. A few electrons may even acquire enough thermal energy to transfer into the conduction band. [The energy gap of **silicon** (Si), which is the most important semiconductor material, is approximately 1.12 eV at room temperature. This is more than 43 times larger than $k_B T$ at room temperature!] In the absence of an electric field, chaotic electron velocities caused by thermal motion compensate each other so that the electric current is equal to zero. An electric field adds to this motion a component of the electron velocity in the direction of the electric field called the drift velocity. This drift velocity results in an electric current.

Of course, the motion in the conduction band electrons is different from the electron motion in free space, since the conduction band electrons are affected by atoms in the semiconductor. The effect of this potential on the conduction band electrons can be approximately described by introducing an electron **effective mass**, $m_n$, which is different from the **free electron mass**, $m_e = 9.11 \times 10^{-31}$ kg. For example, in a semiconductor compound, gallium arsenide, $m_n \approx 0.067\ m_e$.

**Example I-1.**

An electron enters a GaAs sample with nearly zero velocity (see the figure below). The electric field in the sample is 1 kV/cm. The sample length is 0.1 μm. What is the electron transit time across this sample? How does it compare with the transit time in free space under the same conditions?

**Hint:** Use the Second Law of Motion: $m_n dv/dt = qF$ where $v$ is the electron velocity, $q$ is the electronic charge, $F$ is the electric field, $qF$ is the force acting on the electron (please notice that since the electronic charge is equal to $-q$, the force acting on the electron is opposite to the electric field).

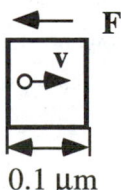

0.1 μm

**Solution:**

Integrating the **equation of motion**, $m_n dv/dt = qF$, with respect to time, we obtain: $v(t) = qFt/m_n + v(0)$. Integrating $v(t)$ with respect to time, we find distance $x$ traveled during time $t$: $x(t)=\dfrac{qFt^2}{2m_n}+v(0)t$. The transit time, $t_{tr}$, is found from $x(t_{tr}) = L$ where $L$ is the sample length. For $v(0) = 0$,

$$t_{tr}=\left(\frac{2m_n L}{qF}\right)^{1/2}=\left(\frac{2\times0.067\times9.11\times10^{-31}\times10^{-7}}{1.602\times10^{-19}\times10^5}\right)^{1/2}=8.73\times10^{-13}(s)$$

Since $t_{tr}$ is proportional to $m_n^{1/2}$, the transit time in free space under similar conditions would be $(1/0.067)^{0.5} = 3.86$, that is, nearly four times longer.

The electrons in the valence band cannot move in the electric field and carry electric current unless they can find vacant spaces in the valence band. A valence band electron can be accelerated by an electric field and can change its energy and velocity if and only if it can be promoted to an empty energy level. The vacant levels in the valence band allow electrons in the valence band to move as shown in Fig. I.2. We can use an analogy of billiard balls on a pool table: if the table is filled with balls, then none of them can move, just like electrons in the valence band.

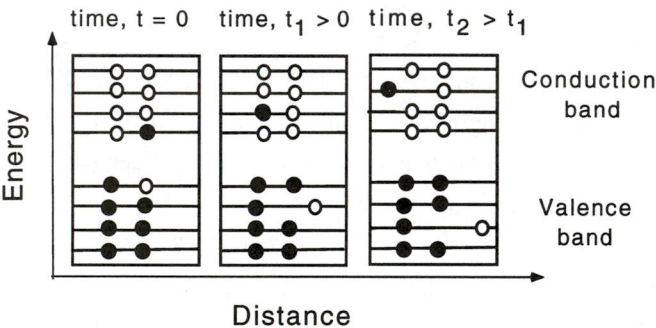

**Fig. I.2.** Motion of electrons in an electric field. These band diagrams are similar to those shown in Fig. I.1. However, the energy scale is increased to show that allowed states consist of very close energy levels that could be occupied by two electrons each. Solid circles represent electrons occupying energy levels; open circles represent energy levels available for electrons.

As can be seen from the figure, the changes in the electron level occupancy in the valence band can be followed more easily by identifying the motion of the vacant energy level. Such a motion can be represented as a fictitious particle (called a **hole**) representing the absence of an electron. This is somewhat similar to the motion of a bubble in a boiling water. A hole is similar to a bubble in the "fluid" of valence electrons.

The absence of a negative electronic charge, $-q$, corresponds to a positive charge $q = -(-q)$. Hence, the hole has a positive charge with a magnitude equal to the electronic charge. Just like conduction band electrons, holes can be considered as free particles with a certain effective mass, $m_p$.

Electrons in a conduction band and holes in a valence band carry along negative and positive electronic charges, respectively, and therefore are often called **charge carriers** or **free charge carriers**.

The occupied states in the conduction band are usually close to the lowest energy in the conduction band, $E_c$, which is called the **conduction band edge** or the **bottom of the conduction band**. The empty states in the valence band are usually close to the highest energy in the valence band, $E_v$. $E_v$ is called the **valence band edge** or the **top of the valence band**. The dependencies of $E_c$ and $E_v$ in a semiconductor device on position are called **band diagrams**. Band diagrams are very useful for illustrating properties and understanding behavior of semiconductor materials and devices. As an example, we show in Fig. I.3 the band diagram of a piece of semiconductor with and without an applied external electric field, $F$.

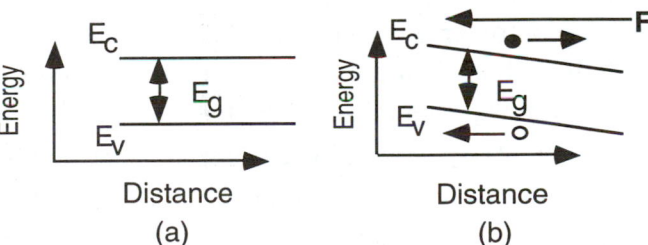

Fig. **I.3.** Energy band diagrams for a uniform semiconductor sample in (a) zero and (b) nonzero electric field.

In the electric field, the bands are tilted, with the slope $-qF$. The arrows in Fig. I.3b represent the directions of forces exerted on the electrons and holes by the electric field. (Since electrons are charged negatively and holes are charged positively, they move in opposite directions.)

In a pure semiconductor, the electron concentration in the conduction band and the hole concentration in the valence band are usually very small compared to the number of energy states with energies within $k_BT$ from the band edges. These concentrations can be changed by many orders of magnitude by **doping**, that is, by adding to a semiconductor impurity atoms called **dopants**. There are two basic types of dopants: **donors** and **acceptors**. A donor atom has more electrons available for bonding with neighboring atoms than is required for an atom in the host semiconductor. For example, a silicon atom has four electrons available for bonding and forms four bonds with the four nearest silicon atoms. A phosphorus atom has five electrons available for bonding. Hence, when a phosphorus atom replaces a silicon atom in a silicon crystal, it can bond with the four nearest neighbors and "donate" one extra electron to the conduction band.

An acceptor atom has fewer electrons than are needed for chemical bonds with neighboring atoms of the host semiconductor. For example, a boron atom has only three electrons available for bonding. Hence, when a boron atom replaces a silicon atom in silicon, it "accepts" one missing electron from the valence band, creating a hole.

A semiconductor doped with donors is called an **n-type** semiconductor. A semiconductor doped with acceptors is called a **p-type** semiconductor. Often, especially at room temperature or elevated temperatures, each donor in an $n$-type semiconductor supplies one electron to the conduction band, and the electron concentration, $n$, in the conduction band is approximately equal to the donor concentration, $N_d$. In a similar way, at room temperature or elevated temperatures, each acceptor creates one hole in the valence band, and the hole concentration, $p$, in the valence band of a $p$-type semiconductor is approximately equal to the acceptor concentration, $N_a$. If both donors and acceptors are added to a semiconductor, they "compensate" each other, since electrons supplied by donors occupy the vacant levels in the valence band created by the acceptor atoms. In this case, the semiconductor is called **compensated**. In a compensated semiconductor, the largest impurity concentration "wins": if $N_a > N_d$, the compensated semiconductor is $p$-type with the effective acceptor concentration, $N_{aeff} = N_a - N_d$; if $N_d > N_a$, the compensated semiconductor is $n$-type with the effective donor concentration, $N_{deff} = N_d - N_a$.

As was mentioned above, electrons in the conduction band and holes in the valence band experience a chaotic thermal motion. In an electric field, the electrons and holes acquire an additional drift velocity caused by the electric field and superimposed on the chaotic thermally induced velocities. In a uniform semiconductor and in a weak electric field, $\mathbf{F}$, the drift velocities, $\mathbf{v}_n$ and $\mathbf{v}_p$, of the electrons and holes are proportional to the electric field:

$$\mathbf{v}_n = -\mu_n \mathbf{F} \tag{I-7}$$

$$\mathbf{v}_p = \mu_p \mathbf{F} \tag{I-8}$$

where $\mu_n$ and $\mu_p$ are called the electron and hole mobility, respectively. (Please notice that the direction of $\mathbf{v}_p$ coincides with the direction of the electric field, and the direction of $\mathbf{v}_n$ is opposite to the direction of the electric field, since holes are positively charged and electrons are negatively charged.) The total charge of the electrons crossing a unit area of a semiconductor in one second is $qnv_n$, where $n$ is the electron concentration in the conduction band. The charge of the holes crossing a unit area of a semiconductor per second is $qpv_p$, where $p$ is the hole concentration in the valence band. Hence, using eqs. (I-3) and (I-4), we obtain the following expressions for the current density, $\mathbf{j}$, and conductivity, $\sigma$, for a semiconductor:

$$\mathbf{j} = \sigma\mathbf{F} \tag{I-9}$$

where

$$\sigma = q\mu_n n + q\mu_p p \tag{I-10}$$

Often, we can neglect the hole contribution to the conductivity of an $n$-type semiconductor because the hole concentration is many orders of magnitude smaller than the electron concentration. Likewise, we can often neglect the electron contribution to the conductivity of a $p$-type semiconductor.

Equation (I-9) is called **Ohm's law**. We can rewrite this equation in a more familiar form:

$$I = V/R \tag{I-11}$$

where $I = jS$ is the total current, $S$ is the sample cross section, $V = FL$ is the potential difference across the sample (here we assume a constant electric field), $L$ is the sample length, and

$$R = \frac{L}{\sigma S} \tag{I-12}$$

is the sample **resistance**. However, we should remember that, in semiconductors, Ohm's law is only valid when an electric field is weak, which is rarely the case in modern semiconductor devices.

All semiconductor devices use junctions between different materials or between regions of the same semiconductor but with different doping. The contact between an *n*-type semiconductor and a *p*-type semiconductor is called a **p-n junction**. In a *p-n* junction, holes diffuse from the *p*-region, where their concentration is high, to the *n*-region where their concentration is low. The deficit of positively charged holes creates a layer of negatively charged acceptors in the *p*-region close to the junction. In a similar way, electrons diffuse from the *n*-region, where their concentration is high, to the *p*-region where their concentration is low. The deficit of negatively charged electrons creates a layer of positively charged donors in the *n*-region near the junction (see Fig. I.4).

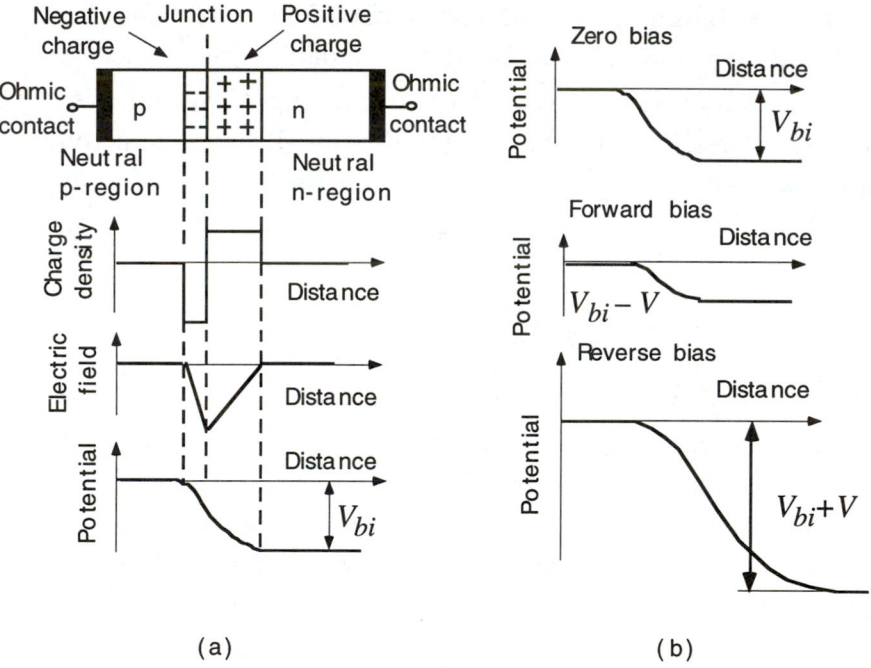

(a)                                                            (b)

**Fig. I.4.** Charge, field, and potential distribution for a *p-n* junction diode (a) under zero bias and (b) comparison of potential distributions under zero, forward, and reverse biases.

This charged region, nearly devoid of holes in the *p*-region and nearly devoid of electrons in the *n*-region, is called a **space charge region** or **depletion region**. The charges in the depletion region create a potential barrier that prevents more electrons from coming into the *p*-region and prevents more holes from coming into the *n*-region. A few important questions may come to mind when you are looking at Fig. I.4. What is the magnitude of this potential barrier, $qV_{bi}$? [The answer to this question will be given in Chapter 4. For now, we just mention without proof that $V_{bi}$ (called the **built-in voltage**) is of the order of $E_g/q$ where $E_g$ is the energy gap and $q$ is the electronic charge.] Why does the charge distribution at the junction have a rectangular shape? Once again, we have to defer the answer to Chapter 4, and for now just take it as given.

   The potential barrier at the $p$-$n$ junction exists without any applied bias. This potential difference is caused by different doping of the $p$ and $n$ regions, that is, by chemical forces that separate electrical charges, somewhat similar to the potential difference in a battery. The magnitude (the height) of this barrier can be increased or decreased by applying an external voltage to such a junction. A positive potential, $V$ (plus with respect to the $n$-region), applied to the $p$-region of the diode shown in Fig. I.4 attracts electrons (which have a negative charge) from the $n$-region and pushes holes into the $n$-region, reducing the potential barrier to $V_{bi} - V$. This voltage polarity corresponds to the **forward bias**. A negative potential (with respect to the $n$-region) applied to the $p$-region of the diode shown in Fig. I.4 pushes electrons away from the $p$-region and pushes holes away from the $n$-region. Such a potential (**reverse bias**) increases the potential barrier to $V_{bi} + |V|$. (An easy way to remember these polarities is to follow the rule stating that charges of opposite polarities are attracted to each other, and charges of the same polarity are repelled.) As a result, the diode resistance under the forward and reverse bias will be dramatically different, and the diode current voltage characteristic will look like that shown in Fig. I.5a.

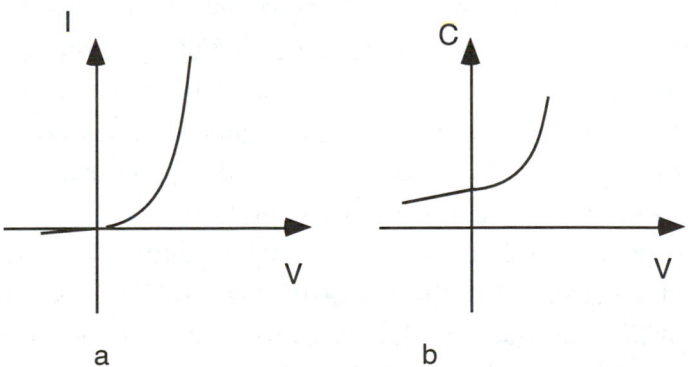

**Fig. I.5.** (a) Qualitative current-voltage and (b) capacitance-voltage characteristics of a $p$-$n$ junction diode.

The space charge region is depleted of free carriers and, under reverse or small forward bias, it isolates the conducting $p$ and $n$ regions from each other. The field distribution is very similar to that in a parallel plate capacitor, where the space charge region plays the role of a dielectric and neutral $p$ and $n$ regions play the role of metal plates. The parallel plate capacitance is given by

$$C = \frac{\varepsilon S}{d} \qquad (I\text{-}13)$$

where $d$ is the separation between the plates, $S$ is the plate area, and $\varepsilon$ is the dielectric permittivity of the material between the plates. In a $p\text{-}n$ junction, the width of the depletion region, $d_{dep}$, is similar to the plate separation in a parallel plate capacitor, and the capacitance of a $p\text{-}n$ junction is given by eq. (I-13) where $d$ is replaced with $d_{dep}$. Under reverse bias, $d_{dep}$ increases, and the capacitance of the $p\text{-}n$ junction diode decreases. Under forward bias, the space charge region shrinks, and the capacitance increases (see Fig. I.5b).

In order to apply an external bias to a $p\text{-}n$ junction, we have to make metal contacts to both the $n$ and $p$ regions connecting these regions to the metal wires that are attached to a power supply. Ideally, these contacts should not present any resistance to an electric current. Since this ideal situation cannot be achieved, at least we require that these contacts present a small constant resistance to the current, $I$, flowing through the device, that is, that they satisfy Ohm's law

$$I = \frac{V_c}{R_c} \qquad (I\text{-}14)$$

where $R_c$ is the contact resistance, and $V_c$ is the voltage drop (usually fairly small) across the contact. Such contacts are called **ohmic contacts**.

A $p\text{-}n$ junction diode can be used as a rectifier, as a nonlinear circuit element, and as a variable capacitance whose value depends on the applied bias (called a **varactor** diode). More details about $p\text{-}n$ junctions diodes and metal-semiconductor contacts will be given in Chapter 4.

Two $p\text{-}n$ junction diodes can be merged into one device with three terminals called a **Bipolar Junction Transistor (BJT)**. As was mentioned in the beginning of this Introduction, the discovery of a Bipolar Junction Transistor by the American scientists Bardeen, Brattain, and Shockley in 1947 started a revolution in electronics. Bipolar Junction Transistors are analyzed in Chapter 5.

An example of another semiconductor device is depicted in Fig. I.6. This device is called a **Junction Field Effect Transistor (JFET)**.

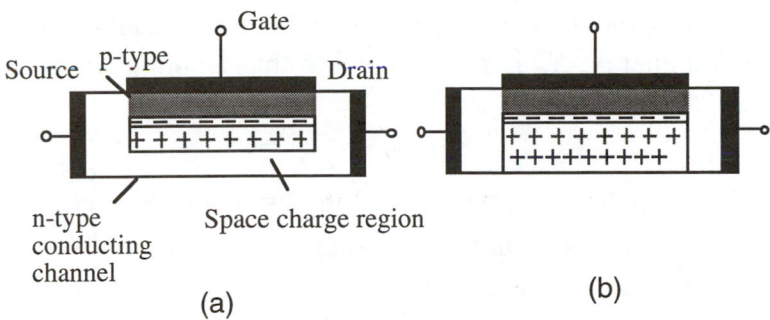

**Fig. I.6.** Schematic diagram of a JFET with (a) zero bias and (b) negative (reverse) bias on the gate. The transistor behaves either as (a) an open switch or (b) a closed switch.

A JFET has three contacts: **source**, **drain**, and **gate**. The source and drain are ohmic contacts to an $n$-type semiconductor material. A positive bias is applied to the drain electrode, and the source contact is grounded. The $p$-type gate layer forms a $p$-$n$ junction with $n$-type channel with a depletion region that is practically devoid of charge carriers and does not conduct. The thickness of the depletion region and, as a consequence, the thickness of the remaining **conducting channel** between the source and the drain depend on the gate bias with respect to the device channel. When the gate contact is forward biased, the depletion region is very narrow and almost the entire region between the source and the drain contributes to the channel conductance. Under strong enough reverse bias, this entire region may become totally depleted, and the channel conductance may become very low. More sophisticated **Field Effect Transistors (FETs)** may operate differently, but most of them, just like practically all semiconductor devices, utilize the same wonderful property of semiconductors – their ability to change their conductance or charge in response to a relatively small variation of external factors, such as applied voltages, temperature, or illumination. Field Effect Transistors are considered in Chapters 6 and 7.

Under forward bias, electrons and holes in a $p$-$n$ junction move into the space charge region from the neutral $n$-type and $p$-type regions, respectively (see Fig. I.4). Hence, both electrons and holes (that is, electron vacancies) wind up in the same region, and electrons may drop into the vacant spots in the valence band. In this process, called **recombination**, electron-hole pairs are annihilated. Their energy (which is approximately equal to the energy gap for each recombining electron–hole pair) may be emitted as light quanta called **photons**. The photon energy, $E$, is proportional to the radiation frequency, $\omega$,

$$E = \hbar\omega$$

where $\hbar = h/2\pi = 1.055 \times 10^{-34}$ Js is called the **reduced Planck constant**, $h = 6.62 \times 10^{-34}$ Js is the **Planck constant**, and $\omega = 2\pi/T$ where $T$ is the period.

The radiative recombination of electrons and holes allows a $p$-$n$ junction diode under a forward bias to operate as a **Light Emitting Diode (LED)** or a **laser diode**.

If we shine light on a $p$-$n$ junction such that the photon energies are larger than the energy gap of the semiconductor material, $E_g$, these photons can create electron–hole pairs that are separated by the electric field in the depletion layer and by other forces. This separation of positive (for holes) and negative (for electrons) electric charges leads to an additional voltage difference across the junction and/or electric current. Hence, such a device can operate as a **detector** of electromagnetic radiation, that is, a **photodetector**. A similar device can be used as a source of electricity. Such a device is called a **solar cell,** since solar radiation is used as a source of light. LEDs, photodetectors, solar cells, and other photonic devices (i. e., devices interacting with visible or invisible light) are considered in Chapter 8.

Modern semiconductor technology allows us to control material composition and/or doping literally within one atomic distance. New semiconductor materials hold promise of faster electronic circuits and may allow us to reach operating temperatures above the temperature of a red hot glowing metal. Many new device concepts have emerged, including an idea that, under certain conditions, even a single electron may control the current flow. Even though these concepts are not yet fully developed, they point to further evolutionary or, in certain cases, perhaps even revolutionary developments in solid–state electronics. Some of these novel device concepts and new semiconductor materials are considered in Chapter 9.

Table I.1 summarizes some of the basic concepts discussed in this Introduction.

| Band structure | Bands of allowed and forbidden energy states |
|---|---|
| Pauli exclusion principle | Not more than two electrons can occupy each energy level |
| Energy gap | The difference between the top of the highest valence band and the bottom of the lowest conduction band |
| Electron-volt (eV) | Energy of an electron accelerated by 1 V: $1\,eV = 1.602 \times 10^{-19}$ joule |
| Average thermal electron energy | $3k_BT/2$ |
| Hole | Positive particle representing propagating vacant state in valence band |
| Band diagram | Dependence of $E_c$ and $E_v$ on distance |
| Donors | Impurities that "donate" electrons to the conduction band |
| Acceptors | Impurities that "accept" electrons from the valence band, creating holes |
| $n$-type semiconductor | Semiconductor doped by donors |
| $p$-type semiconductor | Semiconductor doped by acceptors |
| Compensated semiconductor | Semiconductor doped by both donors and acceptors |
| Effective doping concentrations in compensated semiconductor | If $N_a > N_d$, $N_{aeff} = N_a - N_d$<br>If $N_d > N_a$, $N_{deff} = N_d - N_a$ |
| Electron and hole drift velocities and mobilities | $\mathbf{v}_n = -\mu_n\mathbf{F}, \quad \mathbf{v}_p = \mu_p\mathbf{F}$ |
| Current density in a semiconductor | $j = q\mu_n nF + q\mu_p pF$ |
| Conductivity of a semiconductor | $\sigma = q\mu_n n + q\mu_p p$ |
| Resistance | $R = L/(\sigma S)$ |
| $p$-$n$ junction | Contact between $n$-type semiconductor and $p$-type semiconductor |
| $p$-$n$ junction capacitance | $C = \varepsilon_s S / d_{dep}$ |
| Potential barrier between $n$-type and $p$-type regions in $p$-$n$ junction | $qV_{bi}$ under zero bias<br>$q(V_{bi} - V)$ under forward bias, $V$<br>$q(V_{bi} + |V|)$ under reverse bias, $V$ |
| Ohmic contact | Metal–semiconductor contact with a small constant resistance to electric current (ideally, vanishingly small) |
| Recombination | Annihilation of electron–hole pairs |

**Table I.1.** Basic concepts.

# BIBLIOGRAPHY

M. E. LEVINSHTEIN AND G. S. SIMIN, *Getting to Know Semiconductors*, World Scientific, Singapore (1992)
*Popular introduction to the physics of semiconductor devices.*

B. G. STREETMAN, *Solid State Devices*, Fourth Edition, Prentice Hall, Englewood Cliffs, NJ (1995)
*Basic undergraduate text on solid–state devices.*

E. S. YANG, *Microelectronic Devices*, McGraw Hill, New York (1988)
*Basic undergraduate text on solid–state devices.*

D. A. NEAMEN, *Semiconductor Physics and Devices. Basic Principles*, Irwin, Homewood, IL (1992)
*Basic undergraduate text on solid–state devices.*

# REVIEW QUESTIONS

**1.** Consider a small crystal of solid hydrogen (even though it is impossible to make such a crystal under normal conditions!). The crystal has 1,000,000 hydrogen atoms. Each hydrogen atom has only one electron. When these atoms are brought close together in this crystal, the lowest allowed energy state of the hydrogen atom splits into 1,000,000 levels, which form the lowest energy band.

a. How many electrons may be placed into this energy band and why?

$\square$ 1 point

b. Will this crystal be a dielectric, a semiconductor, or a metal and why?

$\square$ 1 point

**2.** Express the electron thermal energy, $3k_BT/2$, in electronvolts at $T = 77$ K.

$\square$ 1 point

**3.** At room temperature, the electron mobility, $\mu_n$, in Si is approximately 1,000 cm$^2$/Vs.

a. Calculate the resistance of the silicon bar shown in the figure.  The bar is doped by donors at $10^{17}$ cm$^{-3}$.  (Each donor provides one electron to the conduction band.)

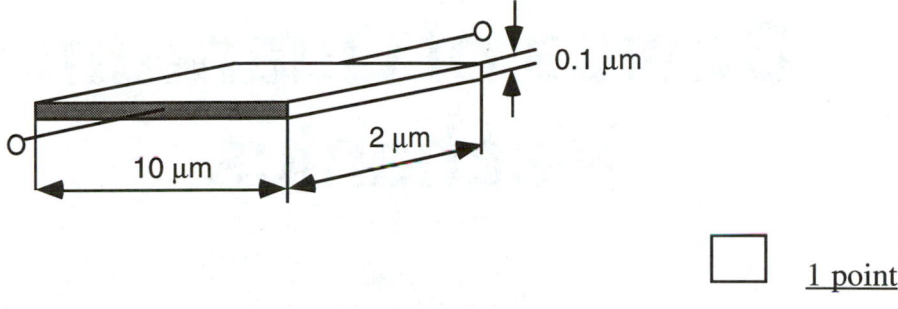

$\square$  1 point

b. How low should **specific ohmic contact resistance**, $\rho_c$, be so that both ohmic contacts add no more than 10% to the resistance of this silicon bar? ($\rho_c = R_c S$ where $R_c$ and $S$ are the resistance and cross section of the ohmic contact, respectively.)

$\square$  1 point

c. What concentration of the compensating acceptors should be added to this silicon sample to increase its resistance by a factor of a hundred?

$\square$  1 point

# 1

# Basics of Quantum Mechanics

## 1-1. INTRODUCTION

Understanding the principle of operation of many important electronic devices requires only a rudimentary knowledge of quantum mechanics, and the subsequent chapters of this book can be understood without mastering the material in this chapter. However, the basics of quantum mechanics not only allow us to achieve a deeper understanding of modern electronic devices but they are also essential even for the philosophy of a contemporary engineer or scientist. One can argue that they simply must be a part of his or her intellectual baggage. In addition, quantum mechanics is essential for understanding material properties, basic chemistry, and many other facets of modern science.

## 1-2. QUANTUM MECHANICAL CONCEPTS

### 1.2.1. Wave particle duality and Uncertainty Principle.

On the basis of our everyday experience, we can easily understand the concept of a particle (which we might visualize as something similar to a golf ball) and the concept of a wave (like waves in shallow harbor water, sound waves, or electromagnetic waves). However, on an atomic scale, these classical concepts are not sufficient to explain numerous experimental facts. Phenomena on this scale, which determine the properties of electronic materials, involve quantum

**1 8**

mechanical concepts. One of the most important concepts of quantum mechanics is the concept of dual particlelike and wavelike properties of matter. For example, electrons in a semiconductor material behave both like particles and like waves. This concept dates back to the pioneering work of Max Planck, who explained the energy distribution of **black body radiation**. A black body is defined as an object that absorbs all incoming radiation at all frequencies.

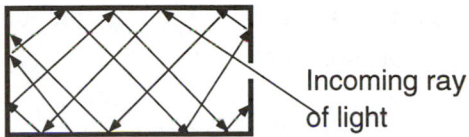

**Fig. 1.2.1.**  Black body model.

A fairly good model of a black body can be obtained by piercing a small hole in a nontransparent sealed empty container. Fig. 1.2.1 shows a cross section of such a container and the propagation of the reflected rays of light inside. If the opening is very small and the light is partially absorbed at every reflection, very little reflected light, if any, will be emitted back through the small hole. It is easy to check that this small hole will look black from the outside even if the inside container walls are not black. Ideally, that is, for a very small hole in a very large container, the spectrum of the radiation emitted through the opening does not depend on the material the container is made of or on the color of the internal walls or on the shape of the container. (In practice, black absorbing walls will make the spectrum of such an opening be even closer to an ideal black body.) This property of the black body spectrum was first established by Kirchhoff (1824–1887). In 1901, Planck showed that the energy distribution of the black body radiation can be explained only by assuming that this radiation (i. e., electromagnetic waves) is emitted and absorbed as discrete energy quanta – **photons**. The photon energy, $E$, is proportional to the radiation radian frequency, $\omega$,

$$E = \hbar\omega \tag{1-2-1}$$

where constant $\hbar = h/2\pi = 1.055 \times 10^{-34}$ Js is called the **reduced Planck constant**, $h = 6.62 \times 10^{-34}$ Js is the **Planck constant**, and $\omega = 2\pi/T$ where $T$ is the period. The peak of the human eye sensitivity corresponds to green light with a wavelength of 0.555 μm , frequency

$$\omega = \frac{2\pi c}{\lambda} \approx \frac{2\pi \times 3.00 \times 10^8 \frac{m}{s}}{0.555 \times 10^{-6} m} = 3.40 \times 10^{15} \frac{1}{s}$$

and a photon energy $E = \hbar\omega = 3.58 \times 10^{-19}$ J.  Here $c$ is the velocity of light in vacuum (see Appendix A2).  A more convenient measure of energy for such small values is an **electron volt** (eV), which is the energy acquired by an electron accelerated by a 1 volt potential:

$$1 \text{ eV} = 1.602 \times 10^{-19} C \times 1 \text{ volt} = 1.602 \times 10^{-19} \text{ joule}$$

The energy of the green photon is $3.58 \times 10^{-19}$ J/$1.602 \times 10^{-19}$ J/eV = 2.23 eV.

Albert Einstein was the first to show that each photon has a **momentum**

$$p = E / c \tag{1-2-2}$$

where $c$ is the velocity of light.  Hence, Planck and Einstein demonstrated that light has not only wavelike but also particlelike properties.

In 1924, de Broglie suggested that this quantum mechanical duality described by eqs. (1-2-1) and (1-2-2) applies not only to electromagnetic radiation but also to particles, such as electrons.  He introduced a wave associated with an electron (called the **de Broglie wave**).  Later, Erwin Schrödinger and Max Born introduced the **wave function** $\Phi(x,y,z,t)$, such that the probability, $dP$, of finding a particle in an incremental volume $dxdydz$ is equal to $|\Phi(x,y,z,t)|^2 dxdydz$.  Here $x$, $y$, $z$ are space coordinates and $t$ is time.  The wave function, $\Phi(x, y, z, t)$, can be interpreted as the amplitude of the probability density of finding the particle in a given point of space at a given time.  For a particle with a given momentum $p$ in free space, the wave function, $\Phi(x,y,z,t)$, is proportional to $\exp[i(k_x x + k_y y + k_z z)] \exp(-i\omega t)$ where $\omega$ is frequency and $k_x$, $k_y$, and $k_z$ are components of the **wave vector, k** (which is related to the wavelength by $|\mathbf{k}| = 2\pi/\lambda$):

$$\Phi(x, y, z, t) = A \exp[i(k_x x + k_y y + k_z z)] \exp(-i\omega t) \tag{1-2-3}$$

Here $A$ is a constant, and

$$\lambda = h / p = 2\pi\hbar / p \tag{1-2-4}$$

is called the **de Broglie wavelength**.  The momentum $p = mv$  where $m$ is the mass of a particle and $v$ is the particle velocity.  Since $\lambda = 2\pi/k$, we obtain from eq. (1-2-4)

$$p = \hbar k \qquad (1-2-5)$$

In 1927, Davisson and Germer experimentally demonstrated that electrons may behave like waves.  They observed the scattering of an electron beam from the surface of nickel and studied the intensity of the reflected beam as a function of angle.  The metal surface represents a periodic array of atoms (see Fig. 1.2.2).

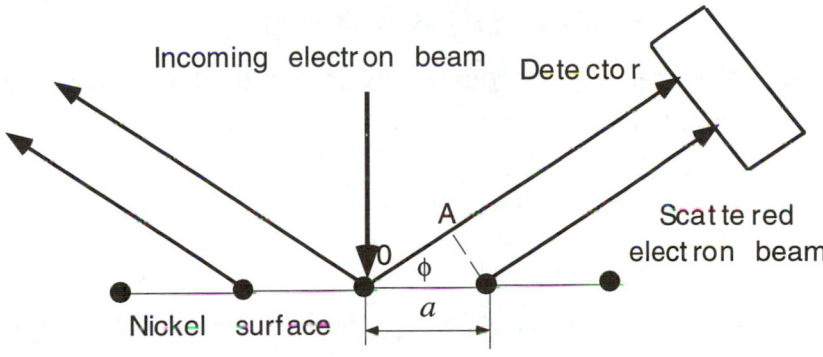

**Fig. 1.2.2.**  Electron beam scattered from the surface of nickel.

If electrons behaved like particles, then the intensity of the scattered electron beam would be a smooth continuous function of the scattering angle.  In fact, the intensity of the beam exhibited a maximum at a certain angle.  To understand why it happened and why this meant that electrons behave like waves, we have to consider how waves combine.  Let us consider two waves $\Phi_1$ and $\Phi_2$ propagating in the same direction $x$ with the same wavelength $\lambda = 2\pi/k_x$ and with the **phase shift** $\theta$:

$$\Phi_1(x,t) = A_1 \cos(k_x x - \omega t)$$
$$\Phi_2(x,t) = A_2 \cos(k_x x - \omega t + \theta) \qquad (1-2-6)$$

[Since we included the phase shift in the argument of the cosine function, we do not need to introduce complex numbers as in eq. (1-2-3).]  The intensity of the combined wave is equal to

$$I = |\Phi_1 + \Phi_2|^2 = |A_1 \cos(k_x x - \omega t) + A_2 \cos(k_x x - \omega t + \theta)|^2$$
$$= A_1^2 \cos^2(k_x x - \omega t) + A_2^2 \cos^2(k_x x - \omega t + \theta) \tag{1-2-7}$$
$$+ 2 A_1 A_2 \cos(k_x x - \omega t) \cos(k_x x - \omega t + \theta)$$

As can be seen from eq. (1-2-7), for

$$\theta = 2\pi n \tag{1-2-8}$$

where $n$ is any integer, the intensity is equal to

$$I = (A_1 + A_2)^2 \cos^2(k_x x - \omega t) \tag{1-2-9}$$

For $A_1 = A_2$, the intensity is four times larger than for each wave taken separately. However, for

$$\theta = (2n+1)\pi \tag{1-2-10}$$
$$I = (A_1 - A_2)^2 \cos^2(k_x x - \omega t) \tag{1-2-11}$$

the intensity is exactly zero when $A_1 = A_2$. Particles certainly do not behave that way!

As shown in Fig. 1.2.2, the difference in length traveled from the surface to the detector for the electron beams scattered from the adjacent atoms is equal to OA = $a \cos(\phi)$ where $a$ is the interatomic spacing and $\phi$ is the reflection angle. This difference corresponds to the phase shift

$$k_x a \cos(\phi) = \frac{2\pi a \cos(\phi)}{\lambda} \tag{1-2-12}$$

When this phase shift is equal to $2\pi n$, maxima of the reflected intensity are observed:

$$n\lambda = a \cos(\phi) \tag{1-2-13}$$

This is exactly what Davisson and Germer observed, and this is possible if and only if electrons have wavelike properties. May we then conclude that electrons always behave as waves? Not really. One can measure the electron mass, $m_e = 9.11 \times 10^{-31}$ kg, and the electronic charge, $q = 1.602 \times 10^{-19}$ C. These are typical particlelike properties. Hence, electrons behave both as particles and as waves.

**Example 1-2-1.**

Calculate and plot the combined intensity of the two waves, $\Phi_1(x,t)=A\cos(k_x x-\omega t)$ and $\Phi_2(x,t)=A\cos(k_x x-\omega t+\theta)$ as a function of $\theta$ for $0 < \theta < \pi$ at the moment of time when $k_x x - \omega t = 2\pi$ for $A = 1$ (relative units).

**Solution:**
For $A = 1$ and $k_x x - \omega t = 2\pi$, $\Phi_1 = 1$, $\Phi_2 = \cos(\theta)$, $I = [1 + \cos(\theta)]^2$. The resulting plot is shown below.

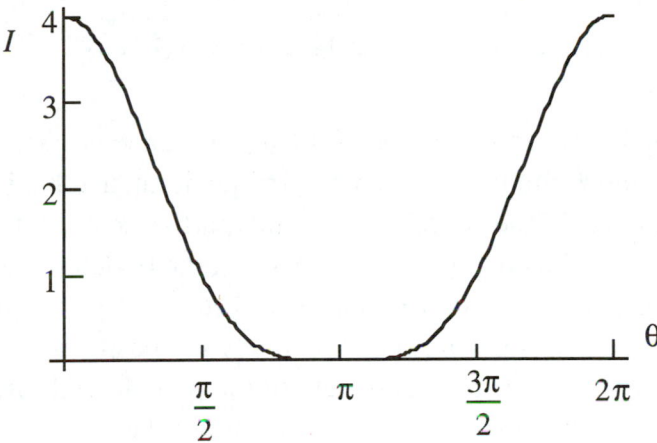

**Example 1-2-2.**
Consider an electron propagating in free space with velocity $v = 10^6$ m/s in direction $x$. The free electron mass is $9.11 \times 10^{-31}$ kg. Calculate the electron momentum, wave vector, de Broglie wavelength, and energy. Compare the electron energy with the energy band gap of Si (1.12 eV).

$$v = 10^6 \text{ m/s}$$

$$\circ\!\!\longrightarrow\ x$$

**Solution:**
The electron momentum and wave vector are directed along the $x$-axis. Hence, $p_x = mv = 9.11 \times 10^{-31} \times 10^6 = 9.11 \times 10^{-25}$ kgm/s, $p_y = p_z = 0$.
$k_x = p_x/\hbar = mv/\hbar = 9.11 \times 10^{-25}/1.054 \times 10^{-34} = 8.64 \times 10^9$ 1/m
$k_y = k_z = 0$
The de Broglie wavelength:

$\lambda = 2\pi/k = 7.27 \times 10^{-10}$ m $= 7.27$ Å

[1 meter (1 m) is equal to $10^{10}$ Angstroms ($10^{10}$ Å)]

The electron energy is kinetic energy:

$$E = \frac{mv^2}{2} = \frac{p^2}{2m} = \frac{\hbar^2 k^2}{2m} = \frac{9.11 \times 10^{-31} \times \left(10^6\right)^2}{2}$$

$$= 4.55 \times 10^{-19}\,(\text{J}) = \frac{4.55 \times 10^{-19}}{1.602 \times 10^{-19}} = 2.84\,(\text{eV})$$

This is 2.54 times larger than the energy band gap of Si (1.12 eV).

In classical mechanics, the motion of a particle is characterized by its coordinate and momentum or velocity. In quantum mechanics, the particle momentum is linked to the wavelength of the particle's wave function [see eq. (1-2-4)]. But a wave has no coordinate. To accurately determine a wavelength, we have to follow the wave in a region that is much larger than a wavelength. This means that we will not determine accurately the position of the particle. On the other hand, if we establish a coordinate of the particle with precision, we will not be able to accurately determine its wavelength related to its momentum. In 1927, Werner Heisenberg stated the famous **Uncertainty Principle**. According to this principle, the product of uncertainties, $\Delta p_x$ and $\Delta x$, of a particle's momentum and its coordinate must be larger than $\hbar/2$

$$\Delta p_x \Delta x \geq \hbar/2 \tag{1-2-14}$$

According to statistical physics, the average energy of an electron in a gas of free electrons at thermal equilibrium is $3k_B T/2$ where $T$ is temperature and $k_B = 1.38 \times 10^{-23}$ J/K is the Boltzmann constant. The velocity of random electronic thermal motion, $v_T$, can be found by equating the kinetic energy of this motion, $m_e v_T^2/2$, to $3k_B T/2$. The free electron mass is $m_e = 9.11 \times 10^{-31}$ kg, so that $v_T = (3k_B T/m_e)^{1/2} \approx 1.2 \times 10^5$ m/s, $p = m_e v_T \approx 1.1 \times 10^{-25}$ kgm/s, $k = p/\hbar \approx 10^9$ m$^{-1}$, and $\lambda = 2\pi/k \approx 6.3 \times 10^{-9}$ m $= 63$ Å. Hence, $\lambda$ is comparable to the dimensions of very small semiconductor devices (which may have certain dimensions as small as 50 Å), and quantum effects may play a very important role in such devices.

An incredibly small value of the de Broglie wavelength does not mean that quantum effects are noticeable only on a microscopic scale. There are many large–scale (macroscopic) phenomena that are manifestations of quantum phenomena.    The most famous effects of this kind include laser emission (considered in Chapter 8) and superconductivity (considered in Chapter 9).

**Example 1-2-3.**

Consider an electron propagating with velocity $v = 10^6$ m/s in direction $x$ in a 100 Å wide gap.   The free electron mass is $9.11\times10^{-31}$ kg.   Calculate the electron momentum and energy.

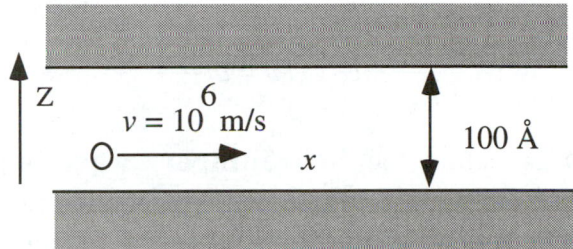

**Solution:**

The calculation of the $x$-components of the electron momentum and wave vector is identical to that in Example 1-2-2:   $p_x = mv = 9.11\times10^{-31}\times10^6 = 9.11\times10^{-25}$ kgm/s,  $k_x = p_x/\hbar = mv = 9.11\times10^{-25}/1.054\times10^{-34} = 8.64\times10^9$ 1/m. Also, $p_y = 0$ and $k_y = 0$, just like in Example 1-2-2.   However, since $\Delta z = 100$ Å and the Uncertainty Principle states $\Delta p_z \Delta z \geq \dfrac{\hbar}{2}$,

$$\Delta p_z \geq \frac{\hbar}{2\Delta z} = \frac{1.054\times10^{-34}}{2\times10^{-8}} = 5.27\times10^{-27}\,(\text{kgm/s})$$

$$\Delta E_z = \frac{\Delta p_z^2}{2m} = \frac{\left(5.27\times10^{-27}\right)^2}{2\times9.11\times10^{-31}} = 1.52\times10^{-23}\,(\text{J}) = 9.52\times10^{-5}\,(\text{eV})$$

## 1.2.2.   Schrödinger wave equation.

As was mentioned above, the dual particlelike and wavelike properties of matter are described by introducing a function, $\Phi(x, y, z, t)$, called a **wave function**. The probability of finding an object in the incremental volume $dxdydz$ at time $t$ is equal to $|\Phi|^2 dxdydz$.   Therefore, the wave function is also called a probability amplitude.   If the wave function of an object is concentrated in a certain region in space, this means that the object is likely to occupy this region.   The wave function, $\Phi$, for a free electron is given by eq. (1-2-3), which is a typical function describing a wave.   This function satisfies the following equation:

$$-\frac{\hbar^2 \nabla^2}{2m_e}\Phi = i\hbar\frac{\partial\Phi}{\partial t} \qquad (1\text{-}2\text{-}15)$$

[This can be easily checked substituting eq. (1-2-3) into eq. (1-2-15).] Equation (1-2-15) is very similar to the wave equation for electromagnetic waves. Schrödinger showed that in a more general case when a particle is moving in a certain potential $U(\mathbf{r})$, where $\mathbf{r}$ is the space vector, this equation becomes

$$\left[-\frac{\hbar^2 \nabla^2}{2m_e}+U(\mathbf{r})\right]\Phi = i\hbar\frac{\partial\Phi}{\partial t} \qquad (1\text{-}2\text{-}16)$$

Equation (1-2-16) is called the **Schrödinger wave equation**. From a mathematical point of view, this equation is a second-order linear differential equation with nonlinear coefficients for a complex function $\Phi(\mathbf{r}, t)$. We can search for the solution of this equation in the following form:

$$\Phi(\mathbf{r}, t) = \psi(\mathbf{r})\exp(-i\omega t) \qquad (1\text{-}2\text{-}17)$$

Substituting eq. (1-2-17) into eq. (1-2-16) and dividing both sides of the resulting equation by $\psi(\mathbf{r})\exp(-i\omega t)$, we obtain

$$-\frac{\hbar^2 \nabla^2 \psi(\mathbf{r})}{2m_e}+U(\mathbf{r})\psi(\mathbf{r}) = E\psi(\mathbf{r}) \qquad (1\text{-}2\text{-}18)$$

where

$$E = \hbar\omega \qquad (1\text{-}2\text{-}19)$$

Equation (1-2-18) is called the **time-independent Schrödinger equation**.

The time-independent **wave function** $\psi(\mathbf{r})$ and its derivatives with respect to space coordinates must be continuous. These requirements are used in order to establish boundary conditions for solutions of the time-independent Schrödinger equation. In this section, we will consider the solution of this equation for two important cases: for free space and for an infinitely deep one–dimensional potential well.

In free space, the potential energy is $U = 0$ and the Schrödinger equation becomes

$$-\frac{\hbar^2 \nabla^2 \psi(\mathbf{r})}{2m_e} = E\psi(\mathbf{r}) \qquad (1\text{-}2\text{-}20)$$

Let us consider the electron motion only in one direction, $x$, so that $\nabla^2 = d^2/dx^2$. Then eq. (1-2-20) becomes

$$-\frac{\hbar^2}{2m_e}\frac{d^2\psi(x)}{dx^2} = E\psi(x) \qquad (1\text{-}2\text{-}21)$$

The solution of this differential equation is given by

$$\psi = A_1 \exp\,(ikx) + B_1 \exp(-ikx) \qquad (1\text{-}2\text{-}22)$$

where the wave vector is

$$k = (2m_e E / \hbar^2)^{1/2} \qquad (1\text{-}2\text{-}23)$$

The de Broglie relation $\lambda = 2\pi\hbar / p$ [see eq. (1-2-4)] links the wavelength, $\lambda$, and the electron momentum, $p = m_e v$. Since $\lambda = 2\pi/k$, we obtain

$$p = \hbar k \qquad (1\text{-}2\text{-}24)$$

From this equation and eq. (1-2-23), we find that

$$E = \frac{p^2}{2m_e} = \frac{m_e v^2}{2} \qquad (1\text{-}2\text{-}25)$$

Hence, the constant $E$ in eq. (1-2-21) is the electron energy, and the solution given by eq. (1-2-22) is consistent with the de Broglie wave function for a free particle [see eq. (1-2-3)].

Substituting eq. (1-2-22) into eq. (1-2-17), we obtain

$$(1\text{-}2\text{-}26)$$
$$\Phi = A \exp\,[i(kx - \omega t)] + B \exp[i(-kx - \omega t)]$$

Hence, the free electron motion is described by a plane wave. The relationship between energy and frequency is the same as for photons: $E = \hbar\omega$. This is a reflection of the quantum mechanical duality that was so clearly illustrated for free electrons by the Davisson and Germer experiment.

### 1.2.3. Infinitely deep one-dimensional potential well.
In an infinitely deep one-dimensional potential well of width $a$ (see Fig. 1.2.3), the potential is given by

$$U = \infty \quad \text{for} \quad x < 0$$

$$U = 0 \quad \text{for} \quad 0 \leq x \leq a \tag{1-2-27}$$

$$U = \infty \quad \text{for} \quad x > a$$

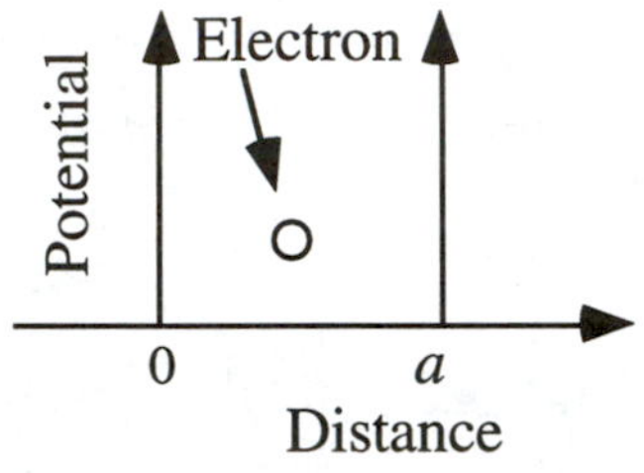

**Fig. 1.2.3.** Electron in an infinitely deep potential well.

Within the potential well, the solution for $\psi$ is given by eq. (1-2-22) since $U = 0$. In the regions where $U = \infty$, $\psi = 0$ (please explain why). Hence,

$$\psi = 0 \quad \text{for} \quad x = 0 \tag{1-2-28}$$

$$\psi = 0 \quad \text{for} \quad x = a \tag{1-2-29}$$

Using eq. (1-2-28), we find from eq. (1-2-22), $A_1 = -B_1$ and, hence,

$$\psi = A\sin(kx) \tag{1-2-30}$$

where $k = (2m_e E)^{1/2} / \hbar$ and $A = 2iA_1$. Equation (1-2-29) leads to the requirement $\sin(ka) = 0$ and, hence, $ka = \pi n$, that is,

$$E = E_n = \frac{\pi^2 \hbar^2 n^2}{2m_e a^2} \tag{1-2-31}$$

where $n = 1, 2, 3, 4, \ldots$ (any positive integer number) is called a quantum number. The **normalization constant**, $A$, in eq. (1-2-30) can be found from

$$\int_{-\infty}^{\infty} |\psi(x)|^2 dx = \int_{0}^{a} \left| A\sin\left(\frac{\pi n}{a} x\right) \right|^2 dx = 1 \tag{1-2-32}$$

This condition means that the particle is localized within the potential well, so that the probability of finding the particle in the potential well is equal to unity. Using eq. (1-2-32), we obtain $A = (2/a)^{1/2}$.  Hence,

$$\psi_n(x) = \left(\frac{2}{a}\right)^{1/2} \sin\left(\frac{\pi n}{a} x\right) \qquad (1\text{-}2\text{-}33)$$

Notice that the number of points where the wave function is zero, apart from $x = 0$ and $x = a$, is equal to $n - 1$ (these points are called **nodes**).  The energy of a particle in the potential well can have only discrete (**quantized**) values called **energy levels** (see Fig. 1.2.4).

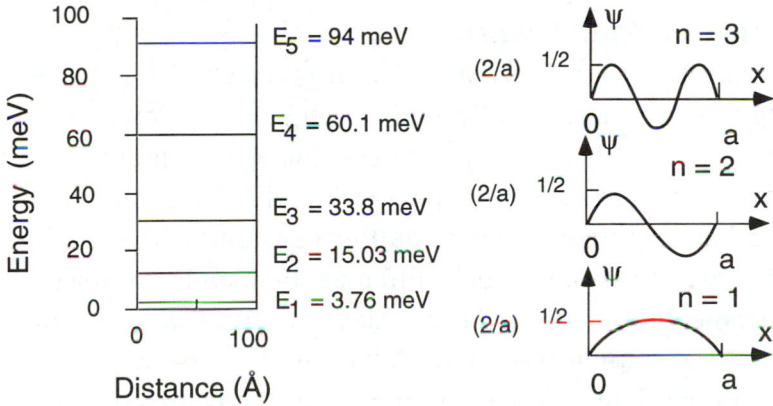

**Fig. 1.2.4.**  Energy levels and wave functions for an infinitely deep square potential well.  Energy levels and wave functions are calculated using eqs. (1-2-31) and (1-2-33) for $m = 9.11 \times 10^{-31}$ kg and $a = 100$ Å.

**Example 1-2-4.**

Consider an electron in the ground energy state of an infinitely deep potential well (see Fig. 1.2.3).  Find the dependence of the probability of finding an electron between 0 and $x$ as a function of $x$.

**Solution:**

The wave function in the ground state ($n = 1$) is given by

$$\psi_1(x) = \left(\frac{2}{a}\right)^{1/2} \sin\left(\frac{\pi}{a} x\right)$$

[see eq. (1-2-35)].  Hence, the probability density is

$$\frac{dP(x)}{dx}=|\psi(x)|^2=\frac{2}{a}\sin^2\left(\frac{\pi}{a}x\right)$$

and the probability, $P$, of finding an electron in an infinite potential well in the ground state between 0 and $x$ is given by

$$P(x)=\frac{2}{a}\int_0^x\sin^2\left(\frac{\pi}{a}x'\right)dx'=\frac{x}{a}-\frac{\sin\left(\frac{2\pi x}{a}\right)}{2\pi}$$

For $x = a$, $P(x) = 1$ as expected (the electron is somewhere in the potential well).

### *1.2.4.   Other potential wells.

While analyzing the wave function and energy levels for a potential well, we were solving a one–dimensional Schrödinger equation.  However, the real world is three–dimensional.  Most closely, the one–dimensional problem corresponds to a layered semiconductor structure shown in Fig. 1.2.5.  In this structure, a layer of a semiconductor material, such as **gallium arsenide (GaAs)**, for example, is sandwiched between two layers of a different semiconductor material for which the electron potential energy is larger, such as **aluminum gallium arsenide (AlGaAs)**.  As a consequence, the potential profile in the $y$ direction has the shape shown in the figure.  The electron motion in this direction is restricted.

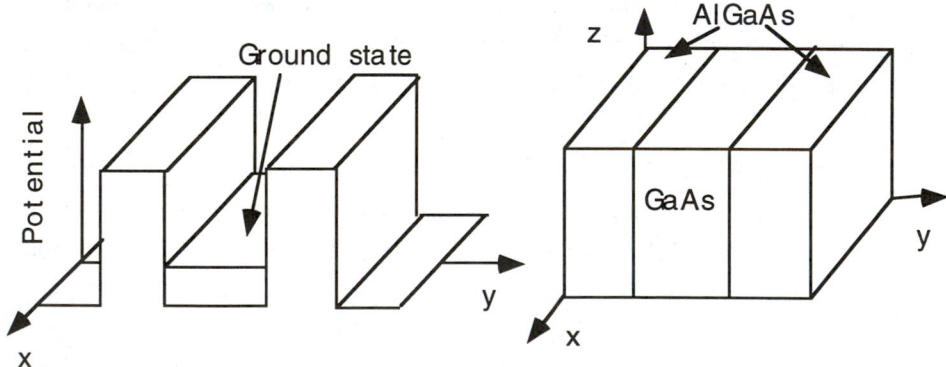

**Fig. 1.2.5.** Square potential in a layered semiconductor structure.

It can be described by a one–dimensional Schrödinger  equation, similar to how it was done above for  an infinitely deep potential well.  However, in the two other directions the electron motion is not restricted.   Hence, the dependence of the electron wave function on $x$ and $z$ is similar to that for free electrons. The kinetic energy of electron motion in these directions is a continuous function of the electron velocity and can have arbitrary values.

Two other important cases correspond to the situations when the electron motion is restricted in two directions (a quantum wire) or in all three directions (a quantum box); see Fig. 1.2.6. In a quantum wire, the dependence of the electron wave function on $x$, where $x$ is the direction of the wire, is similar to that for free electrons. In a quantum box, the electron energy levels are discrete.

Problems 1.2.7 and 1.2.8 address these two cases, which may find applications in modern solid–state devices.

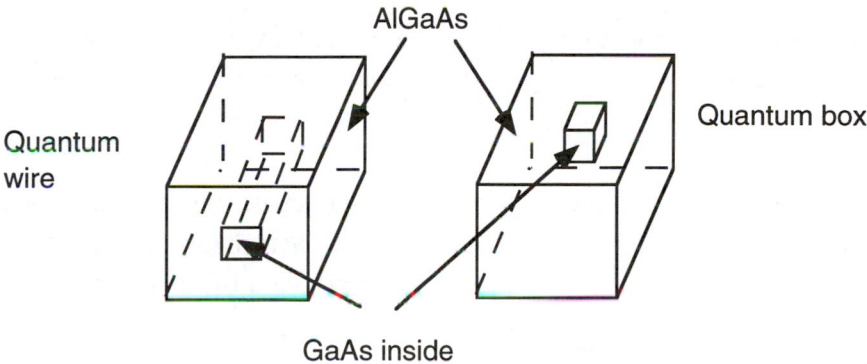

**Fig. 1.2.6.** Quantum wire and quantum box.

**Example 1-2-5.**

Electrons in the conduction band of GaAs behave as lighter particles than free electrons. Their effective mass, $m_n$, is 0.067 $m_e$, where $m_e$ is the free electron mass. Design a quantum well structure made from GaAs bound by AlGaAs layers such that the difference between the ground level in the subband ($n = 1$) and the state corresponding to the first excited level ($n = 2$) is equal to the energy of an infrared photon with the wavelength of $\lambda = 50$ μm in vacuum. The depth of this potential well is approximately 0.35 eV. Assume that, in this case, the potential well may be approximated by an infinitely deep potential well and comment on this assumption.

**Solution:**

Using eq. (1-2-25), where we substitute $m_e$ with $m_n$, we find that

$$E_2 - E_1 = \frac{3\pi^2\hbar^2}{2m_n a^2}$$

Equating $E_2 - E_1$ to the photon energy $\hbar\omega = \frac{2\pi\hbar c}{\lambda}$, where $c$ is the speed of light, we find the required width, $a$, of the quantum well:

$$a = \left(\frac{3\pi\hbar\lambda}{4mc}\right)^{1/2} = \left(\frac{3\pi\times1.054\times10^{-34}\times50\times10^{-6}}{4\times0.067\times9.11\times10^{-31}\times3\times10^8}\right)^{1/2}$$
$$= 2.59\times10^{-8}\,(\text{m}) = 259\text{Å}$$

Using eq.(1-2-25), we find $E_1 = 4.02\times10^{-21}$ J $= 0.025$ eV and $E_2 = 4\,E_1 = 0.1$ eV. Since $E_1$, $E_2$ are more than three times smaller than 0.35 eV, the approximation of an infinitely deep potential well may be acceptable.

We can also go beyond the approximation of an infinitely deep potential well. The solution of the Schrödinger equation for a finite potential well (see Fig. 1.2.7) is not that complicated, though it does involve a bit of algebra.

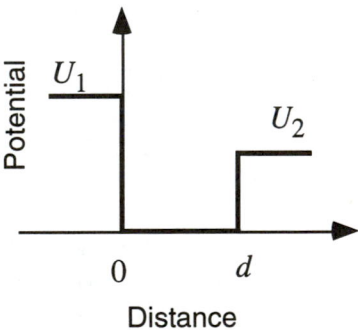

**Fig. 1.2.7.** Finite square potential well.

For the potential well shown in Fig. 1.2.7, the discrete energy levels must have energies $E < U_2 < U_1$. Hence, for $x \leq 0$

$$\psi = A_1 \exp(k_1 x) \tag{1-2-34}$$

For $x \geq d$

$$\psi = A_2 \exp(-k_2 x) \tag{1-2-35}$$

where $k_1 = \dfrac{\sqrt{2m(U_1 - E)}}{\hbar}$ and $k_2 = \dfrac{\sqrt{2m(U_2 - E)}}{\hbar}$.

Similar to what we obtained for an infinitely deep potential well, for $0 < x < d$,

$$\psi = A \sin (kx + \alpha) \tag{1-2-36}$$

where $k = (2mE/\hbar)^{1/2}$.

At the boundaries ($x = 0$ and $x = d$) we must have

$$\psi(+0) = \psi(-0), \quad d\psi/dx(+0) = d\psi/dx(-0) \tag{1-2-37}$$

$$\psi(d+0) = \psi(d-0), \quad d\psi/dx(d+0) = d\psi/dx(d-0) \tag{1-2-38}$$

or

$$\psi(+0)/d\psi/dx(+0) = \psi(-0)/d\psi/dx(-0)$$

$$\psi(d+0)/dy/dx(d+0) = \psi(d-0)/d\psi/dx(d-0)$$

These equations yield

$$k\cos(\alpha)/\sin(\alpha) = k_1 \tag{1-2-39}$$

$$k\cos(dk+\alpha)/\sin(dk+\alpha) = k_2$$

Hence,

$$\sin\alpha = k\hbar/(2mU_1)^{1/2} \tag{1-2-40}$$

$$\sin(dk+\alpha) = -k\hbar/(2mU_2)^{1/2}$$

Eliminating $\alpha$, we obtain the following equation for $k$:

$$kd = n\pi - \sin^{-1}\left(\frac{k\hbar}{\sqrt{2mU_1}}\right) - \sin^{-1}\left(\frac{k\hbar}{\sqrt{2mU_2}}\right) \tag{1-2-41}$$

where $n = 1, 2, 3, \ldots$ and the $\sin^{-1}$ functions vary between 0 and $\pi/2$.

For a symmetrical potential well ($U_2 = U_1 = U_o$), eq. (1-2-41) becomes

$$\frac{kd}{2} = \frac{n\pi}{2} - \sin^{-1}\left(\frac{k\hbar}{\sqrt{2mU_o}}\right) \tag{1-2-42}$$

For $n = 1$, this equation can be rewritten as

$$\sin\left(\frac{\pi}{2} - \frac{kd}{2}\right) = \frac{k}{k_o} \quad \text{or} \quad \cos\left(\frac{kd}{2}\right) = \frac{k}{k_o} \tag{1-2-43}$$

where $k_o = (2mU_o/\hbar)^{1/2}$.   For a very shallow potential well, $kd \ll 1$, and the cosine function in eq. (1-2-43) can be expanded into a Taylor series preserving only the constant and the squared terms.  This leads to

$$1 - \frac{1}{2}\left(\frac{kd}{2}\right)^2 = \frac{k}{k_o} \quad \text{or} \quad \left[1 - \frac{1}{2}\left(\frac{kd}{2}\right)^2\right]^2 = \left(\frac{k}{k_o}\right)^2 \tag{1-2-44}$$

Keeping only the squared terms $(kd)^2$ in eq. (1-2-44) we obtain

$$1 - (kd/2)^2 = (k/k_o)^2 \tag{1-2-45}$$

The substitution of $k=(2mE/\hbar)^{1/2}$ and $k_o = (2mU_o/\hbar)^{1/2}$ into eq. (1-2-45) yields

$$E = U_o - \frac{md^2 E U_o}{2\hbar^2} \approx U_o - \frac{md^2 U_o^2}{2\hbar^2} \tag{1-2-46}$$

This shows that even a shallow potential well ($kd \ll 1$) has at least one bound state.

An analysis similar to that for a square potential well can be also performed for other potential profiles.  In particular, triangular and parabolic potential wells are important for understanding modern solid–state devices (see Figs. 1.2.8 and 1.2.9).  Therefore, we give expressions for energy levels in such potential wells for reference purposes.

A triangular potential well (shown in Fig. 1.2.8) is often a good approximation for potential distributions near semiconductor interfaces.  For such a potential, the energy levels are given by

$$E_n = \left(\frac{\hbar^2}{2m}\right)^{1/3} \left(\frac{3\pi q F_s}{2}\right)^{2/3} \left(n + \frac{3}{4}\right)^{2/3} \tag{1-2-47}$$

where the quantum number $n = 0, 1, 2, 3, 4, \ldots$ (zero and any positive integer number), $qF_s$ is the potential slope, and $F_s$ is the surface field (see Fig. 1.2.8).

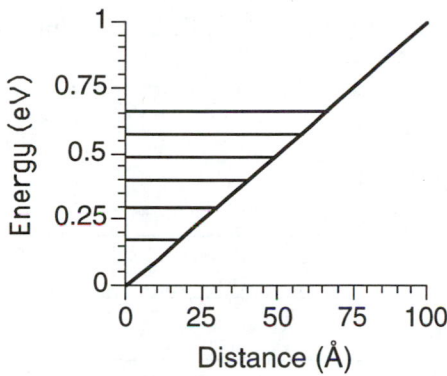

**Fig. 1.2.8.** Lowest energy levels for an infinitely deep triangular potential well. The energy levels are calculated using eq. (1-2-52) for $m = 9.11 \times 10^{-31}$ kg and for the field $F_s = (1/q)dU/dx = 10^8$ V/m. The energy levels are 0.168 eV, 0.297 eV, 0.401 eV, 0.493 eV, 0.577 eV, and 0.656 eV for $n = 0, 1, 2, 3, 4$, and 5, respectively.

A useful exercise is to check units in eq. (1-2-47):

$$\left(\frac{J^2 s^2}{kg}\right)^{1/3}\left(\frac{CV}{m}\right)^{2/3} = \left(\frac{J^2 s^2}{kg}\right)^{1/3}\left(\frac{J}{m}\right)^{2/3} = J \tag{1-2-48}$$

where we have taken into account that the unit of $\hbar$ is Js and that of J is $kg m^2/s^2$.

In classical mechanics, a parabolic potential, $U = f_s x^2/2$, describes a **harmonic oscillator** vibrating at a frequency

$$\omega = (f_s/m)^{1/2} \tag{1-2-49}$$

which is independent of the vibration amplitude (see Fig. 1.2.9). The constant $f_s$ is called a **force constant** and $x$ is the displacement of a particle from an equilibrium position. [Usually, such a parabolic dependence $U(x)$ is valid only if $x$ is sufficiently small]. The energy levels obtained for this potential are relevant for many kinds of vibrations and waves propagating in a semiconductor, such as vibrations of atoms in a crystal or electromagnetic waves. The solution of the Schrödinger equation for such a potential shows that the energy levels are given by

$$E_n = \hbar\omega\left(n + \frac{1}{2}\right)$$ (1-2-50)

where the quantum number $n = 0, 1, 2, 3, 4, 5 \ldots$ (zero and any positive integer number).

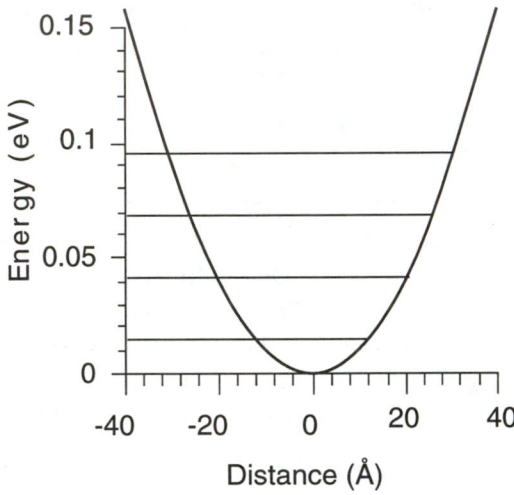

**Fig. 1.2.9.** Lowest energy levels for an infinitely deep parabolic potential well for the electron mass, $m = 9.11 \times 10^{-31}$ kg, and the force constant $f_s = 10^{-4}$ eV/Å$^2$. The energy difference between the levels is $\hbar\omega = 0.0276$ eV, the radian frequency, $\omega = 4.19 \times 10^{13}$ s.

We notice that for any shape of the potential well, particles cannot reach the bottom of the well. The lowest energy level (called the **ground state**) is higher for narrow potential wells. This result is a direct consequence of the Uncertainty Principle. If the ground state corresponded to zero energy, this would mean that the particle momentum would be equal to zero. However, the uncertainty in the particle position is determined by the size of the quantum well. This would have violated the uncertainty principle.

By restricting electron motion in space we can control the electron energies and the energies corresponding to electron transitions between different energy states. This idea is widely used in modern quantum semiconductor devices because modern semiconductor growth and device fabrication technologies allow us to restrict the electron motion within literally a few interatomic distances.

Table 1.2.1 lists a few important equations of quantum mechanics.

| | |
|---|---|
| Photon energy | $E = \hbar\omega$ |
| Photon momentum | $p = E/c$ |
| de Broglie wavelength | $\lambda = h/p = 2\pi\hbar/p$ |
| Uncertainty Principle | $\Delta p_x \Delta x \geq \hbar/2$ |
| | $\Delta p_y \Delta y \geq \hbar/2$ |
| | $\Delta p_z \Delta z \geq \hbar/2$ |
| Schrödinger wave equation | $\left[ -\dfrac{\hbar^2 \nabla^2}{2m} + U(\mathbf{r}) \right]\Phi = i\hbar\dfrac{\partial \Phi}{\partial t}$ |
| Wave function of stationary state | $\Phi(\mathbf{r}, t) = \psi(\mathbf{r})\exp(-i\omega t)$ |
| Time independent Schrödinger wave equation | $\left[ -\dfrac{\hbar^2 \nabla^2}{2m} + U(\mathbf{r}) \right]\psi = E\psi$ |
| Free electron wave function | $\psi = A\exp(i\mathbf{kr}) + B\exp(-i\mathbf{kr})$ |
| Relationship between momentum and wave vector | $\mathbf{p} = \hbar\mathbf{k}$ |
| Relationship between energy and wave vector for a free electron | $E = \dfrac{\hbar^2 k^2}{2m_e}$ |
| Relationship between energy and momentum for a free electron | $E = \dfrac{p^2}{2m_e}$ |
| Energy levels and wave functions for an infinitely deep square potential well | $E_n = \dfrac{\pi^2 \hbar^2 n^2}{2ma^2}, \; n = 1, 2, 3, 4 \ldots$ <br><br> $\psi_n(x) = \left(\dfrac{2}{a}\right)^{1/2} \sin\left(\dfrac{\pi n}{a}x\right)$ |
| Energy levels in an infinitely deep triangular potential well | $E_n = \left(\dfrac{\hbar^2}{2m}\right)^{1/3}\left(\dfrac{3\pi q F_s}{2}\right)^{2/3}\left(n + \dfrac{3}{4}\right)^{2/3}$ <br><br> $n = 0, 1, 2, 3, 4, \ldots$ |
| Energy levels of a harmonic oscillator | $E_n = \hbar\omega\left(n + \dfrac{1}{2}\right), \; n = 0, 1, 2, 3, 4, \ldots$ |

**Table 1.2.1.** Summary of basic equations of quantum mechanics.

## *1-3.   TUNNELING

Let us consider an electron propagating in a layer of the semiconductor material GaAs bound by two layers of AlGaAs (see Fig. 1.3.1).   As was discussed in Section 1.2,  the AlGaAs layers form potential barriers so that the electron is localized in the middle GaAs semiconductor layer by the bounding AlGaAs layers.  According to classical physics (and, perhaps, according to common sense, whatever it means), the electron with energy smaller than the potential barrier height must remain in the middle layer.  However, wavelike electronic properties mean that the electron wave function will penetrate into and beyond the barriers. Hence, there is a finite probability for an electron, initially placed inside this double barrier structure, to "tunnel" through the barriers.   The tunneling probability depends on the height and width of the potential barriers.   For infinitely high barriers, the wave function does not penetrate into the barriers at all (see Fig. 1.2.3), and the tunneling probability is zero.

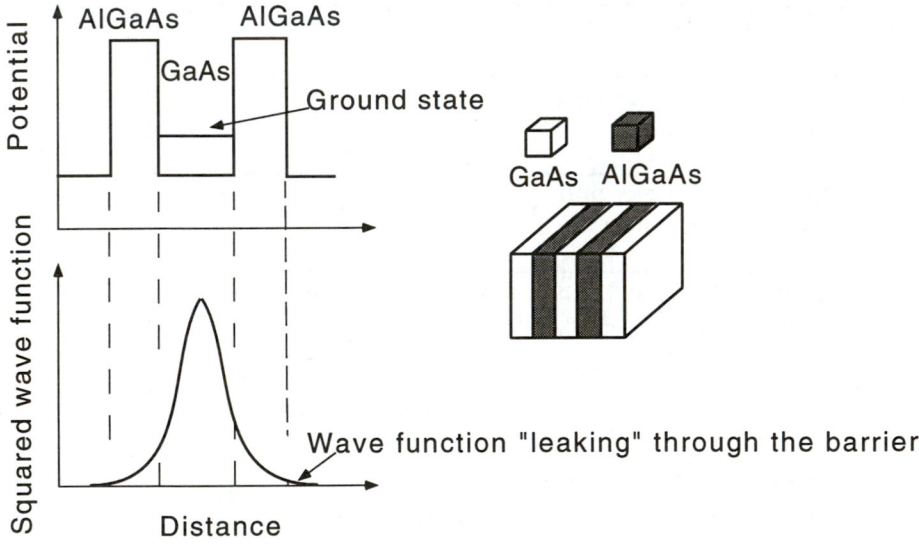

**Fig. 1.3.1**. Electron wave function in a double barrier structure.

In 1965, a famous artist, Magritte, painted a horse–back rider who seems to tunnel through trees in a forest (see Fig. 1.3.2).   In reality, for tunneling to occur, the de Broglie wavelength of the tunneling particle should be comparable with the width of the potential barrier (see Subsection 1.2.1 for a brief discussion

with the width of the potential barrier (see Subsection 1.2.1 for a brief discussion of the de Broglie wavelength).

**Fig. 1.3.2.**   Rene Magritte (1898-1967), The Blank Signature (1965), courtesy National Gallery of Art, Washington, DC.  Collection of Mr. and Mrs. Paul Mellon.

The quantitative analysis of tunneling should be based on the solution of the Schrödinger equation

$$-\frac{\hbar^2 \nabla^2 \psi(\mathbf{r})}{2m_e} + U(\mathbf{r})\psi(\mathbf{r}) = E\psi(\mathbf{r}) \tag{1-3-1}$$

[see eq. (1-2-7)].    Notice that we will be solving the time-independent Schrödinger equation.   Since $|\psi(\mathbf{r})|^2$ is the probability density of finding an electron at point $\mathbf{r}$, it is sufficient to determine $\psi(\mathbf{r})$ to find the probability of the electron penetration across the barrier.

We will first solve the Schrödinger equation for the simple potential barrier shown in Fig. 1.3.3 and then will use the results in order to describe tunneling through potential barriers of arbitrary shapes.

For the potential barrier shown in Fig. 1.3.3 we can rewrite the Schrödinger equation (1-3-1) as follows:

$$\frac{d^2\psi}{dx^2} + k^2\psi = 0 \qquad \text{for } x < 0 \tag{1-3-2}$$

$$\frac{d^2\psi}{dx^2} + k_b^2\psi = 0 \qquad \text{for } 0 \le x \le d_b \tag{1-3-3}$$

$$\frac{d^2\psi}{dx^2} + k^2\psi = 0 \quad \text{for } x > d_b \tag{1-3-4}$$

Here

$$k^2 = \frac{2m_e E}{\hbar^2} \tag{1-3-5}$$

$$k_b^2 = \frac{E - U_b}{E} k^2 = n_b^2 k^2 \tag{1-3-6}$$

where $n_b^2 = \dfrac{E - U_b}{E}$. Please notice that since $E < U_b$, $k_b^2$ is negative and $k_b$ is imaginary.

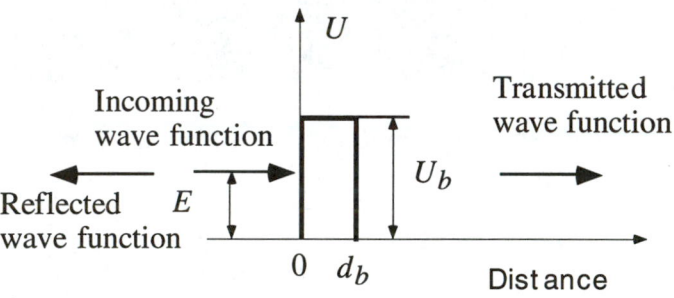

**Fig. 1.3.3.** Electron tunneling across a rectangular potential barrier.

The solutions of eqs. (1-3-2) to (1-3-4) in each region are given by

$$\psi = A_1 \exp(ikx) + B_1 \exp(-ikx) \quad \text{for } x < 0 \tag{1-3-7}$$

$$\psi = A_b \exp(ik_b x) + B_b \exp(-ik_b x) \quad \text{for } 0 \leq x \leq d_b \tag{1-3-8}$$

$$\psi = A_2 \exp(ikx) + B_2 \exp(-ikx) \quad \text{for } x > d_b \tag{1-3-9}$$

Indeed, for $x < 0$ and for $x > d_b$, these solutions are the same as for free space [see eq. (1-2-16) in Subsection 1.2.2]. Inside the barrier, the mathematical form of the solution is the same [since eq. (1-3-3) has the same form as eqs. (1-3-2) and (1-3-4)] but $k_b$ is imaginary.

We use the solutions given by eqs. (1-3-7) to (1-3-9) to find the transmission coefficient for the wave function of electrons incoming onto the barrier as shown in Fig. 1.3.2.  We consider an electron propagation from  left to right.  Then the term $A_1$ exp($ikx$) in eq. (1-3-7) describes the incoming wave function and the term $B_1$ exp($-ikx$) describes the reflected wave function.  The term $A_2$ exp($ikx$) in eq. (1-3-9) describes the transmitted wave function.  Since we are considering the situation when electrons propagate from left to right,  we must choose

$$B_2 = 0 \qquad\qquad (1\text{-}3\text{-}10)$$

(which means that there is no electron wave coming from the right to the left). The transmission coefficient is defined as

$$D = |A_2|^2/|A_1|^2 \qquad\qquad (1\text{-}3\text{-}11)$$

Since we are only interested in the ratio $A_2/A_1$ and $A_2$ is proportional to $A_1$, we can choose $A_1 = 1$.  The remaining four   constants, $B_1$, $A_b$, $B_b$, and $A_2$, have to be determined from the four requirements of the continuity of the wave function and its first derivative at the region boundaries:

$$\psi(0^-) = \psi(0^+) \qquad\qquad (1\text{-}3\text{-}12)$$

$$\frac{d\psi}{dx}\left(0^-\right) = \frac{d\psi}{dx}\left(0^+\right) \qquad\qquad (1\text{-}3\text{-}13)$$

$$\psi(d_b - 0) = \psi(d_b + 0) \qquad\qquad (1\text{-}3\text{-}14)$$

$$\frac{d\psi}{dx}(d_b - 0) = \frac{d\psi}{dx}(d_b + 0) \qquad\qquad (1\text{-}3\text{-}15)$$

We can also define the reflection coefficient

$$R = |B_1|^2/|A_1|^2 \qquad\qquad (1\text{-}3\text{-}16)$$

Since the incoming electrons are either reflected from  the barrier  or penetrate through the barrier, we must have

$$D + R = 1 \qquad\qquad (1\text{-}3\text{-}17)$$

Substituting eqs. (1-3-7) to (1-3-9) into eqs. (1-3-12) to (1-3-15), we obtain four algebraic equations for the constants $B_1$, $A_b$, $B_b$, and $A_2$. After quite a bit of algebra, we obtain [see Blokhintsev (1964)]

$$A_2 = \frac{4n_b \exp(-ikd_b)}{\exp(-ik_b d_b)(1+n_b)^2 - \exp(ik_b d_b)(1-n_b)^2} \qquad (1\text{-}3\text{-}18)$$

$$B_1 = \frac{[\exp(-ik_b d_b) - \exp(ik_b d_b)](1 - n_b^2)}{\exp(-ik_b d_b)(1+n_b)^2 - \exp(ik_b d_b)(1-n_b)^2} \qquad (1\text{-}3\text{-}19)$$

As was mentioned above, when $E < U_b$, $k_b^2$ is negative and $k_b$ and $n_b$ are imaginary [see eq. (1-3-6)]. When $\exp(k|n_b|d_b) << 1$, we find from eqs. (1-3-11) and (1-3-18)

$$D \approx D_o \exp\left\{ -\frac{2[2m_e(U_b-E)]^{1/2} d_b}{\hbar} \right\} \qquad (1\text{-}3\text{-}20)$$

where

$$D_o = \frac{16 |n_b|^2}{\left(1 + |n_b|^2\right)^2} \qquad (1\text{-}3\text{-}21)$$

Equation (1-3-20) can be rewritten as

$$D \approx D_o \exp\left\{ -\frac{p_{missing} d_b}{\hbar/2} \right\} \qquad (1\text{-}3\text{-}22)$$

where $p_{missing} = [2m_e(U_b-E)]^{1/2}$ is the momentum the particle is missing in order to have enough energy to make it over the barrier. According to the Uncertainty Principle $\frac{\Delta p \Delta x}{\hbar/2} \geq 1$. Comparing this expression to the exponent in eq. (1-3-22), we clearly see that the tunneling phenomenon is closely linked to the Uncertainty Principle.

Following Blokhintsev (1964), we now obtain an approximate expression for $D$ for a barrier of an arbitrary shape, considering such a barrier as a superposition of very thin rectangular barriers (see Fig. 1.3.4).

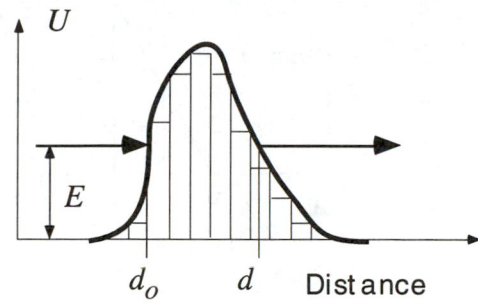

**Fig. 1.3.4.** Electron tunneling across a potential barrier of an arbitrary shape.

For each thin barrier we can use eq. (1-3-21), and the total transmission coefficient, $D$, will be approximately given by the product of transmission coefficients for all the rectangular barriers:

$$D = D_{b1}D_{b2}\ldots D_{bn} \tag{1-3-23}$$

Hence, we obtain

$$D \approx D_o \exp \int_{d_o}^{d} -2\frac{\sqrt{2m_e[U_b(x)-E]}}{\hbar}dx \tag{1-3-24}$$

where $d_o$ and $d$ are coordinates of the points where $U_b = E$ on both sides of the barrier. In this equation, we can crudely estimate $D_o$ in eq. (1-3-24) by using the same equation as we used for a rectangular barrier and substituting $U_b$ with the maximum value of the potential in the barrier, $U_{max}$.

**Example 1-3-1.**
Calculate the tunneling probability, $D_t$, for free electrons ($m_e = 9.11 \times 10^{-31}$ kg) incoming onto the triangular barrier shown in Fig. 1.3.5 and compare it with the tunneling probability of a rectangular barrier shown on the same figure.

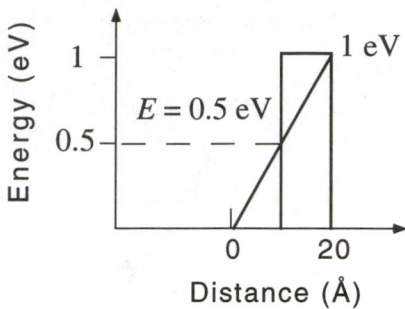

**Fig. 1.3.5.** Triangular and rectangular potential barriers.

**Solution:**

The triangular potential barrier is described by $U_b(x) = qF_s x$ where $F_s = 5 \times 10^8$ V/m, $q$ is the electronic charge, and $x$ is distance. Substituting $U_b = qF_s x$ into the integral in eq. (1-3-24) we obtain:

$$\exp\left[-\int_{d_o}^{d} 2\frac{\sqrt{2m_e[U_b(x)-E]}}{\hbar}dx\right] = \exp\left(-\frac{2\sqrt{2m_e qF_s}}{\hbar}J\right)$$

where $J = \int_{d_o}^{d}\sqrt{(x-d_o)}dx = \int_{0}^{d-d_o}\sqrt{t}dt = \frac{2t^{3/2}}{3}\Big|_{0}^{d-d_o} = \frac{2(d-d_o)^{3/2}}{3}$

and, hence,  $D_t = D_o \exp\left(-\frac{4\sqrt{2m_e qF_s(d-d_o)^3}}{3\hbar}\right)$

For the given values of parameters, $n_b^2 \approx (0.5 - 1)/0.5 = -1$, $|n_b| = 1$, $D_o \approx 4$, $d = 20$ Å, $d_o = 10$ Å, $F_s = 5 \times 10^8$ V/m, and we find $D_t \approx 0.032$. For the rectangular barrier shown in Fig. 1.3.5, we find using eq. (1-3-20), $D \approx 0.00284$.

It is also interesting to consider the case when $E > U_b$. According to classical mechanics, under such conditions, all particles should go over the barrier and none should be reflected. However, from eqs. (1-3-11) and (1-3-18), we find that in this case

$$D = \frac{16n_b^2}{\left(1+n_b\right)^4 + \left(1-n_b\right)^4 - 2\left(1-n_b^2\right)^2 \cos\left(2kn_bd_b\right)} \tag{1-3-25}$$

When the electron energy is large, $n_b^2 = \dfrac{E-U_b}{E} \approx 1$, and $D = 1$, just as expected from classical mechanics. However, for smaller values of $n_b$, the $\cos(2kn_bd_b)$ term in the denominator may be important. This term describes wave properties of an electron since it depends on how many half wavelengths of the electron are equal to the barrier width. If $\cos(2kn_bd_b) = 1$, $D = 1$. If $\cos(2kn_bd_b) = -1$, then

$$D = 4n_b^2 / \left(1 + n_b^2\right)^2 \tag{1-3-26}$$

(see Fig. 1.3.6).

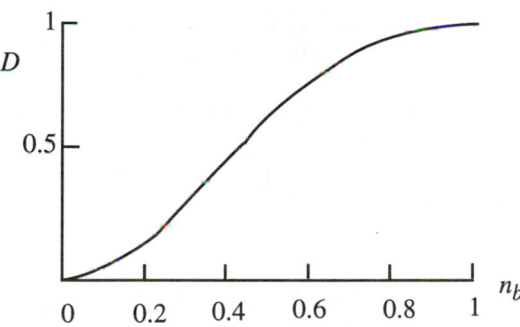

**Fig. 1.3.6.** $D$ versus $n_b$ for $\cos(2kn_bd_b) = -1$.

**Example 1-3-2.**
Calculate the reflection coefficient, $R$, for free electrons ($m_e = 9.11 \times 10^{-31}$ kg) incoming onto a barrier as shown in Fig. 1.3.7.

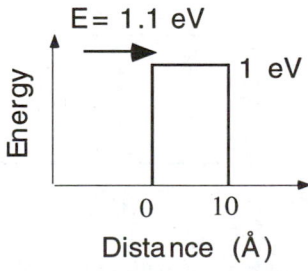

**Fig. 1.3.7.** Electron passage over potential barrier.

**Solution:**

For the given values of parameters, $n_b^2 \approx (1.1 - 1)/1.1 = 0.091$, $n_b = 0.301$, $d_b = 10$ Å,

$$2kn_bd_b=2\frac{\sqrt{2m_e(E-U_b)}}{\hbar}d_b=2\frac{\sqrt{2\times9.11\times10^{-31}\times1.602\times10^{-19}\times0.1}}{1.054\times10^{-34}}10\times10^{-10}=3.24,$$

$\cos(2kn_bd_b) \approx -0.995$, $D \approx 0.306$, $R \approx 0.694$.

The basic equations that describe tunneling are summarized in Table 1.3.1.

| Reflection and transmission coefficients for a potential barrier | $R + D = 1$ |
|---|---|
| Tunneling probability (transmission coefficient) for a rectangular barrier | $D \approx D_o \exp\left\{-\frac{2\left[2m_e\left(U_b-E\right)\right]^{1/2}d_b}{\hbar}\right\}$ <br><br> or <br><br> $D \approx D_o \exp\left\{-\frac{p_{missing}d_b}{\hbar/2}\right\}$ <br><br> $D_o = \frac{16\,\mid n_b\mid^2}{\left(1+\mid n_b\mid^2\right)^2}$  $\quad n_b^2 = \frac{E-U_b}{E}$ |
| Tunneling probability (transmission coefficient) for an arbitrary barrier | $D = D_o \exp \int_{d_o}^{d} -2\frac{\sqrt{2m_e\left(U_b-E\right)}}{\hbar}dx$ |
| Tunneling probability (transmission coefficient) for a triangular barrier | $D_t = D_o \exp\left(-\frac{4\sqrt{2m_eqF_s}\left(d-d_o\right)^3}{3\hbar}\right)$ |

**Table 1.3.1**. Summary of basic equations describing tunneling.

# 1-4.  ATOMIC STATES AND CHEMICAL BONDS

### 1.4.1.  Hydrogen atom.

To understand the properties of solid–state materials, we have to understand the properties of their constituent atoms.  We will first consider a hydrogen atom, the simplest atom of all, and then use the results to understand the chemical properties of more complex atoms, such as silicon.  A hydrogen atom has only one electron and a nucleus that has one positive elementary charge.  The nucleus has nearly all the mass of the hydrogen atom ($\approx 1.67 \times 10^{-27}$ kg, which is approximately 1,800 times larger than a free electron mass $m_e \approx 9.11 \times 10^{-31}$ kg; see Appendix A2).  However, the size of the nucleus ($\approx 10^{-13}$ cm) is much smaller than the size of the atom, which is on the order of 1 Å $= 10^{-8}$ cm.  The negatively charged electron in the hydrogen atom is attracted to the positive nucleus, and their interaction is described by the Coulomb law:

$$U(r) = -\frac{q^2}{4\pi\varepsilon_o r} \tag{1-4-1}$$

Here $q$ is the electronic charge, $\varepsilon_o$ is the permittivity of vacuum, and $r$ is the distance between the electron and the nucleus.  If electrons were described by classical mechanics, the electron would have fallen into the nucleus, and no hydrogen atom (or any other atoms for that matter) would exist.  Fortunately for us all, this would have violated the Uncertainty Principle.  In fact, the Coulomb potential forms a potential well in the three-dimensional space, and the energy levels of the electron in a hydrogen atom are quantized (see Fig. 1.4.1).  The solution of the time-independent Schrödinger equation [see eq. (1-2-22)] with the Coulomb potential energy given by eq. (1-4-1) shows that the energy levels for an electron in a hydrogen atom are given by

$$E_n = -\frac{E_B}{n^2} \tag{1-4-2}$$

where

$$n = 1, 2, 3, 4, \dots \tag{1-4-3}$$

is called the **principal quantum number,**

$$E_B = \frac{q^2}{8\pi\varepsilon_o a_B} \qquad (1\text{-}4\text{-}4)$$

is called the **Bohr energy** ($E_B$ = 13.6 eV = $2.18 \times 10^{-18}$ J), and

$$a_B = \frac{4\pi\varepsilon_o \hbar^2}{m_e q^2} \qquad (1\text{-}4\text{-}5)$$

is called the **Bohr radius** ($a_B$ = 0.52917 Å). (In the ground state, the wave function is primarily concentrated within the sphere with the radius that is approximately twice the Bohr radius, $a_B$.) The first three energy levels for a hydrogen atom are shown in Fig. 1.4.1.

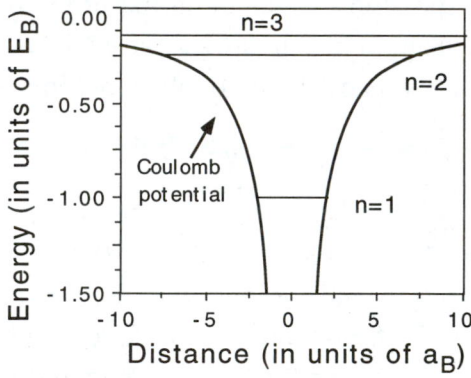

**Fig. 1.4.1.** Coulomb potential and the first three energy levels for a hydrogen atom (from Shur, 1990, © Prentice Hall, 1990, reproduced by permission of Prentice Hall, Inc., Englewood Cliffs, NJ).

**Example 1-4-1.**
Photons with the energy $\hbar\omega_{ik} = E_i - E_k$ are absorbed by hydrogen gas because they cause electron transitions between levels $E_k$ and $E_i$. Find the wavelength of the radiation absorbed due to the transitions between the second and third state.

**Solution:**

$$\hbar\omega_{32} = E_3 - E_2 = E_B\left(\frac{1}{2^2} - \frac{1}{3^2}\right) = 1.89 \ (\text{eV})$$

$$\omega_{32} = \frac{1.602\times10^{-19}\times1.89}{1.054\times10^{-34}} = 2.87\times10^{15}\ \left(\text{s}^{-1}\right)$$

$$\lambda=\frac{2\pi c}{\omega_{32}}=\frac{2\pi\times3\times10^{8}}{2.87\times10^{15}}=6.57\times10^{-7}(m)=0.657(\mu m)$$

The wave functions, $\psi(\mathbf{r})$, of the electron in a hydrogen atom are fairly complex. This is not surprising since they depend on three coordinates, that is, they are functions of three variables. Besides the principal quantum number, $n$, these functions depend on three more quantum numbers characterizing the electronic states – the **orbital quantum number**, $l$, the **magnetic quantum number**, $m$, and the **spin**, $S$. The orbital and magnetic quantum numbers determine the angular dependence of $\psi$. Very crudely, these numbers are linked to the electron rotation around the nucleus, and the electron spin could be associated with an internal electron rotation.

The analysis of the Schrödinger equation for a hydrogen atom shows that these three additional quantum numbers can accept the following values

$$l = 0, 1, 2, \ldots, n-1 \qquad (1\text{-}4\text{-}6)$$

$$m = -l, -l+1, \ldots, l-1, l \qquad (1\text{-}4\text{-}7)$$

$$S = \pm 1/2 \qquad (1\text{-}4\text{-}8)$$

Altogether, each electronic state in a hydrogen atom is characterized by the four quantum numbers. However, the electron energy of each state in a hydrogen atom depends only on $n$, [see eq. (1-4-2)].

### 1.4.2. Many electron atoms and the Periodic Table.

Atoms containing many electrons, such as a silicon atom, are somewhat similar to a hydrogen atom, with the modification that the nuclear charge is not $q$ but $Zq$, where $Z$ is the atomic number, equal to the number of electrons in the atom. We may still classify the electronic states by the same set of quantum numbers: $n$, $l$, $m$, and $S$. However, in a many electron atom the electron states with different orbital quantum numbers, $l$, have different energies, unlike those in a hydrogen atom. However, just like in a hydrogen atom, the states with the smallest values of the principal quantum number, $n$, have the lowest energies. Energy levels with the same value of $n$ and different values of $l$ tend to be close to each other. All electronic states having the same principal quantum number $n$ are referred to as a **shell**. Since electrons primarily are confined in the region of space where the potential energy is smaller than their energy, $E_n$, the average distance of an electron from the nucleus depends on $n$ (see Fig. 1.4.1). The innermost shell electrons have $n = 1$, the next shell corresponds to $n = 2$, and so on. (see Fig. 1.4.2). A shell is divided into "**subshells**" corresponding to different values of the orbital quantum number, $l$. The subshells are usually labeled like this:

$$2s^2$$

where the integer in front of the letter is equal to $n$, the superscript shows the total number of electrons in the subshell, and the lower case letter corresponds to the value of $l$ as follows

$$l = 0, 1, 2, 3, 4, 5 \ldots$$

$$\text{subshell} = s, p, d, f, g, h \ldots$$

[Only rarely will we refer to the states with the values of $l$ more than 3, and usually we will deal with s-electrons ($l = 0$) and p-electrons ($l = 1$) and, occasionally d-electrons ($l = 2$).] Each subshell has $2l + 1$ allowed states corresponding to different values of $m$, and for each $m$ there are two values of the spin quantum number, $S$ [see eqs. (1-4-6) to (1-4-8), Example 1-4-2, and Problem 1-4-1].

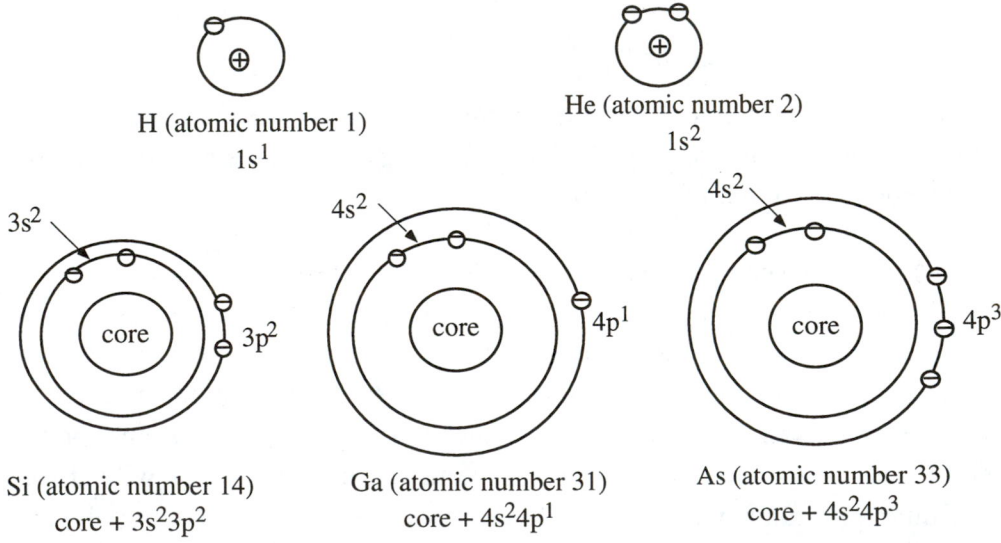

H (atomic number 1)
$1s^1$

He (atomic number 2)
$1s^2$

Si (atomic number 14)
core + $3s^2 3p^2$

Ga (atomic number 31)
core + $4s^2 4p^1$

As (atomic number 33)
core + $4s^2 4p^3$

**Fig. 1.4.2.** Electronic configurations of H, He, Si, Ga, and As. Here "core" means the combination of nucleus and all inner shells.

According to the Pauli exclusion principle, not more than two electrons can occupy an energy state with the same three quantum numbers $n$, $l$, and $m$.  (These two electrons have different values of spin $S = \pm 1/2$.)   Hence, an s-subshell can contain no more than two electrons and a p-subshell can have no more than six electrons.  Electrons occupy the lowest energy levels first.  Therefore, electronic configurations of many electron atoms look like those shown for He, Si, Ga, and As in Fig. 1.4.2.

Elements with a similar electronic structure for valence electrons usually have similar chemical properties.  For example, Si (the electronic structure core + $3s^2 3p^2$) and Ge (the electronic structure core + $4s^2 4p^2$) both have semiconducting properties.  The **Periodic Table of Elements** accounts for this similarity by listing the elements with a similar electronic structure for valence electrons in the same column (see Appendix A3).

Ga also has a valence electronic configuration (core + $4s^2 4p^1$) somewhat similar to that of Si (core + $3s^2 3p^2$), but it has one valence electron less than Si. As (valence electronic configuration core + $4s^2 4p^3$; see Fig. 1.4.2) has one valence electron more than Si.  Hence, in GaAs compound, each atom, on average, has the same number of the valence electrons as Si: 2 s electrons and 2 p electrons.   GaAs is a semiconductor material, like Si or Ge.   Many other semiconductor compounds can be "designed" in a similar way by combining other elements of column III of the Periodic Table (having 3 valence electrons: 2 s electrons and 1 p electron) with elements from column V (having 5 valence electrons: 2 s electrons and 3 p electrons).  These semiconductors (such as GaAs, InAs, InP, AlAs, GaP, AlP, and InSb) are called **III-V compound semiconductors**.

We can also combine elements from column II of the Periodic Table (having 2 valence electrons) with elements from column VI (having 6 valence electrons – 2 s electrons and 4 p electrons).   Many such compound semiconductors (such as CdS, ZnS, ZnSe, CdTe, and CdSe) have semiconducting properties.  These compounds are called **II–VI compound semiconductors**).

Another important characteristic of an atom is its size.  For example, Ga and Al atoms (which both belong to column III of the Periodic Table) have very similar dimensions (see Section 2.2).   Hence, Ga and Al can be mixed and matched, and GaAs and AlAs may form solid–state solutions, such as $Al_x Ga_{1-x} As$ where $x$ is a molar fraction of Al.   (Such materials are called **ternary compounds**.)   By varying $x$ from 0 to 1, one can change the properties of

$Al_xGa_{1-x}As$ from those of GaAs to those of AlAs. Other important examples of ternary compounds include $In_xGa_{1-x}As$, $GaIn_xP_{1-x}$, and $Al_xIn_{1-x}As$. **Quaternary compounds** consisting of four elements, such as $In_xGa_{1-x}As_yP_{1-y}$, can also be grown. This "material engineering" approach allows us to design semiconductor materials with desired properties, which may vary with distance, providing excellent opportunities for a creative device designer.

Compound semiconductors have found many important applications, especially in high-speed, high-performance devices and optoelectronic devices (see Chapters 7 and 8) but, thus far, they have not been able to challenge silicon supremacy as a semiconductor material of choice for nearly all electronic devices.

**Example 1-4-2.**

Write the electron configuration for the atom with 10 electrons.

**Solution:**

For $n = 1$, the only allowed value of $l$ is 0 (s state) and the only allowed value of $m$ is 0. Hence, we can place 2 electrons (with two opposite values of spin) into the shell with $n = 1$. For $n = 2$, the allowed values of $l$ are 0 and 1. For $l = 0$, the allowed value of $m$ is 0. We can place 2 electrons (with two opposite values of spin) into the subshell with $n = 2$, $l = 0$ (s state). For $l = 1$ (p state), the allowed values of $m$ are –1, 0, and 1. Into each state with given values of $n$, $l$, and $m$, we can place two electrons (with two opposite values of spin) for a total of 6 placed into the subshell with $n = 2$, $l = 1$; see the Fig. 1.4.3.

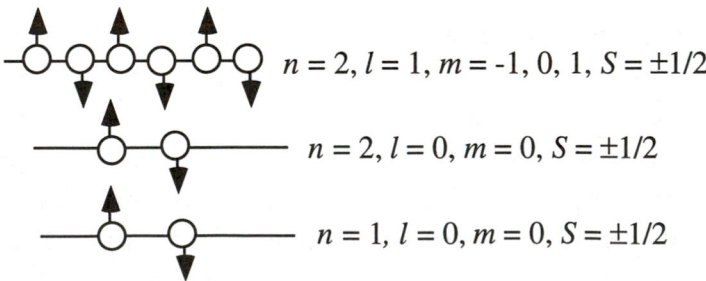

$$n = 2, l = 1, m = -1, 0, 1, S = \pm 1/2$$

$$n = 2, l = 0, m = 0, S = \pm 1/2$$

$$n = 1, l = 0, m = 0, S = \pm 1/2$$

**Fig. 1.4.3.** Electronic configuration of an atom with 10 electrons (Ne).

The total number of electrons in the shell with $n = 2$ is 8. This element is an inert gas. The electron configuration is $1s^2 2s^2 2p^6$.

## 1-4-3.    Chemical bonds.

Elements, such as Si, Ga, As, Ge, and N, have only s and p electrons in their valence shells (shells with the largest value of the principal quantum number, $n$, for a given atom).  As was discussed above, it takes 8 valence electrons to fill up all the states in these 2 valence subshells (2 s electrons and 6 p electrons).  If all the electronic states in the valence s and p subshells are filled, the element will be chemically inert (such as Ne; see Example 1-4-2 and Fig. 1.4.3).  In the Periodic Table, the elements with completely filled s and p valence subshells correspond to inert gases.  For example,

$$\text{Ar}\quad 1s^2 2s^2 2p^6 3s^2 3p^6$$
$$\text{Kr}\quad \text{core} + 4s^2 4p^6$$
$$\text{Xe}\quad \text{core} + 5s^2 5p^6$$

When atoms are combined together in a solid, they may share valence electrons, forming **chemical bonds**.  In silicon, germanium, and related compound semiconductors, these bonds are formed in such a way that neighboring atoms share their valence electrons having (on average)  completed s and p subshells of the valence shell.  In these semiconductors, each atom forms four bonds with four other atoms (four nearest neighbors) and shares 2 valence electrons with each of them (i. e., it shares 8 valence electrons with all four nearest neighbors).  Since each atom has four nearest neighbors, it is **tetrahedrally coordinated** (see Fig. 1.4.4).

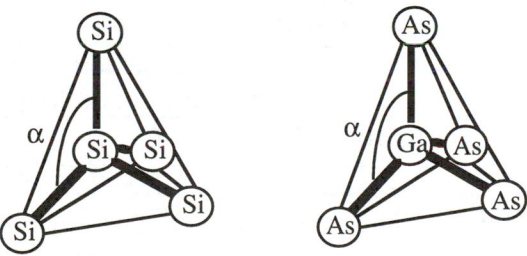

**Fig. 1.4.4.**  Tetrahedral bond configuration in Si and GaAs.  $\alpha = 108^\circ 29'$.

If all atoms in a crystal are identical (as in Si or Ge, for example) the electrons shared in a bond must spend, on average, the same time on each atom.  This corresponds to a purely homopolar (covalent) bond.  In a compound semiconductor, such as GaAs, bonding electrons spend a greater fraction of  time on the anions (i. e., negatively charged atoms).  This situation corresponds to a partially heteropolar (partially ionic) bond.

The quantitative measure of the type of a chemical bond is ionicity, $f_i$. The ionicity is zero for a purely covalent bond and unity for a purely heteropolar bond. The ionicity values $0 \leq f_i \leq 1$ can be assigned to any binary chemical compound. Elemental semiconductors, such as Si, Ge, or diamond, have zero ionicity. According to the ionicity scale proposed by Phillips (1973), $f_i = 0.177$ for silicon carbide (SiC) and 0.31 for GaAs.

Many compound semiconductors (such as GaAs, InP, AlGaAs, and HgTe) also have the tetrahedral bond configuration (see Fig. 1.4.4).

Table 1.4.1 gives a summary of basic equations that describe atomic states and chemical bonds.

| | |
|---|---|
| Energy levels in a hydrogen atom | $E_n = -\dfrac{E_B}{n^2}$ where $n = 1, 2, 3, 4, \ldots$ is the principal quantum number |
| Bohr energy | $E_B = \dfrac{q^2}{8\pi\varepsilon_o a_B} = 13.6 \text{ eV} = 2.18 \times 10^{-18} \text{ J}$ |
| Bohr radius | $a_B = \dfrac{4\pi\varepsilon_o \hbar^2}{m_e q^2} = 0.52917 \text{ Å}$ |
| Orbital quantum number, $l$<br>Magnetic quantum number, $m$<br>Spin quantum number, $S$ | $l = 0, 1, 2, \ldots, n-1$<br>$m = -l, -l+1, \ldots, l-1, l$<br>$S = \pm 1/2$ |
| Relationship between energy levels and the frequency of absorbed light | $\hbar\omega_{ik} = E_i - E_k$ |
| Pauli exclusion principle | No two electrons can have the same set of quantum numbers (including the spin) |
| Notation for atomic subshells | $l = 0, 1, 2, 3, 4, 5 \ldots$<br>subshell $= $ s, p, d, f, g, h $\ldots$ |
| Electron configuration for Si | $1s^2 2s^2 2p^6 3s^2 3p^2$ which means<br>Two s-electrons in the first shell ($n = 1$)<br>Two s-electrons in the second shell ($n = 2$)<br>Six p-electrons in the second shell ($n = 2$)<br>Two s-electrons in the third shell ($n = 3$)<br>Two p-electrons in the third shell ($n = 3$) |
| Bond ionicity | $f_i = 0$ for covalent (homopolar) bonds<br>$f_i = 1$ for ionic (heteropolar) bonds |

**Table 1.4.1.** Basic equations and definitions for atomic states.

# REFERENCES

D. I. Blokhintsev, *Principles of Quantum Mechanics*, Allyn and Bacon, Boston (1964)

J. C. Phillips, *Bonds and Bands in Semiconductors*, Academic Press, New York (1973)

# BIBLIOGRAPHY

H. Kroemer, *Quantum Mechanics*, Prentice Hall, Englewood Cliffs, NJ (1994)

M. E. LEVINSHTEIN AND G. S. SIMIN, *Getting to Know Semiconductors*, World Scientific, Singapore (1992)

*A popular introduction into the physics of semiconductor devices.*

R. F. PIERRET, *Semiconductor Fundamentals*, Second Edition, Modular Series on Solid State Devices, Vol. 1, Addison-Wesley, Reading, MA (1988)

*A good undergraduate text on semiconductor physics.*

C. M. WOLFE, N. J. HOLONYAK, AND G. E. STILLMAN, *Physical Properties of Semiconductors*, Prentice Hall, Englewood Cliffs, NJ (1989)

*A graduate level text on semiconductor physics.*

R. A. SMITH, *Semiconductors*, Second Edition, Cambridge University Press, Cambridge, UK (1978)

*Basic introduction into semiconductor physics.*

K. SEEGER, *Semiconductor Physics, An Introduction,* Third Edition, Series on Solid-State Sciences, Vol. 40, Springer Verlag, New York (1985)

*A more advanced text on semiconductor physics.*

L. SOLIMAR AND D. WALSH, *Lectures on Electrical Properties of Materials*, Fifth Edition, Oxford University Press, Oxford (1993)

*An excellent introduction to material science.*

B. G. STREETMAN, *Solid State Electronics Devices*, Fourth Edition, Prentice Hall, Englewood Cliffs, NJ (1995)

*One of the most popular textbooks on semiconductor devices.*

# REVIEW QUESTIONS

**1.** What is the de Broglie wavelength of a particle with energy 0.1 eV in a semiconductor conduction band with effective mass of $10^{-31}$ kg?

☐ 1 point

**2.** What is a wave function?

☐ 1 point

**3.** What is the relationship between a photon's momentum and wave vector?

☐ 1 point

**4.** What is the relationship between a photon's momentum and wavelength?

☐ 1 point

**5.** How do you reconcile the expression for the wave function of a free electron in space

$$\Phi(x,y,z,t) = A\exp[i(k_x x + k_y y + k_z z)]\exp(-i\omega t)$$

with the Uncertainty Principle?

☐ 1 point

**6.** Find the magnitude of the wave vector of a free electron that has the same energy as a photon with the wavelength of 0.55 μm. Compare it with the magnitude of the wave vector for this photon.

☐ 2 points

**7.** How will the energy of the lowest energy state in an infinitely deep one–dimensional potential well change if the width of the potential well is doubled?

☐ 1 point

**8.** A layered AlGaAs/GaAs quantum well structure is a part of a device that can detect infrared radiation (called an infrared detector). Choose the thickness of the GaAs layer in such a way that the separation between the ground state and the bottom of the first subband is equal to 100 meV. You can estimate the GaAs layer thickness by using equations for an infinitely deep potential well and using $m_n = 6.1 \times 10^{-32}$ kg instead of the electron mass, $m_e$.

☐ 2 points

**9.** a. The tunneling transmission coefficient for a certain electron energy and a certain rectangular potential barrier is 0.01. How much will this transmission coefficient change if the barrier width is doubled? (Assume that the coefficient, $D_o$, in front of the tunneling transmission exponent is unity.)

☐ 2 points

b. How much will the reflection coefficient change?

☐ 1 point

c. How much will the transmission coefficient change if the barrier width is kept the same but the difference between the top energy in the barrier and the energy of an incoming electron is doubled?

☐ 2 points

**10.** How will the Bohr energy and radius change if the electron is made 10 times lighter?

☐ 1 point

**11.** Explain the following notation $1s^2 2s^2 2p^6 3s^2 3p^6$. What element has this electronic structure?

☐ 1 point

**12.** The electronic structure of the valence electrons for element A in an AB semiconductor is $4s^2 4p^1$. What is a possible electronic structure of the valence electrons for element B?

☐ 1 point

# PROBLEMS

**1-2-1.**  Find the momentum and velocity of a particle with mass of $10^{-31}$ kg and de Broglie wavelength of 200 Å.

**1-2-2.**  Find the radian frequency and period of the infrared radiation with wavelength of 1 μm. Find the energy of the infrared photon. Express its energy both in joules and electronvolts.

**1-2-3.**  The uncertainty in a particle position is 10 Å. Find the uncertainty in the particle momentum and the corresponding uncertainty in the kinetic energy. The particle mass is $10^{-31}$ kg. Express the uncertainty in energy both in joules and eV.

**1-2-4.**  The velocity of a free particle is $3\times10^5$ m/s. The mass of the particle is $10^{-31}$ kg. Find the particle energy, de Broglie wavelength, wave vector, and momentum.

**1-2-5.**  Consider a free electron bound in a cubic cavity with a volume of 1000 $\text{Å}^3$. Use the uncertainty principle to estimate the minimum energy (ground state) of this electron. Give the answer in both joules and electron-volts.

**1-2-6.**  In certain cases, the equation describing energy levels in an infinitely deep potential well may be applied to a quantum well formed in a layer of one semiconductor material bound by layers of a different semiconductor (heterostructure quantum well). In these cases, the free electron mass, $m_e$, in the equation for energy levels in a potential well should be substituted with the so-called effective mass, $m_n$. What should be the thickness of the semiconductor layer to ensure that the difference between the ground (i.e., the lowest) energy level and the first excited level is equal to the thermal energy ($k_B T$) at room temperature ($T = 300$ K)? Assume $m_n = 0.1\ m_e$.

**1-2-7.**  A free electron is located in an infinitely deep two-dimensional potential well such that
$U = 0$   for $0 < x < 50$ Å and  $0 < y < 100$ Å

$U = \infty$  outside of this region

Find the allowed electron energy levels and the electron wave functions.  Compare the results with the energy levels and wave functions shown in Fig. 1.2.3 and comment on the similarities and the differences between these two problems.

**1-2-8.**  A free electron is located in an infinitely deep three-dimensional potential well such that

U = 0   for $0 < x < 50$ Å, $0 < y < 100$ Å, and $0 < z < 150$ Å

$U = \infty$  outside of this region

Find the allowed electron energy levels and the electron wave function.  Compare the results with the energy levels and wave functions shown in Fig. 1.2.3 and comment on the similarities and on the differences between these two problems.

**1-3-1.**  Estimate the tunneling probability for an electron in gallium arsenide ($m_n \approx 0.067\ m_e = 6.1 \times 10^{-32}$ kg), tunneling through a rectangular barrier with a barrier height $U_b = 1$ eV and a barrier width of 20 Å. The electron energy is 0.25 eV.

**1-3-2.**  Estimate tunneling probability, $D_t$, for free electrons with energy of 0.5 eV  incoming onto a parabolic barrier shown in the figure ($m_e = 9.11 \times 10^{-31}$ kg).

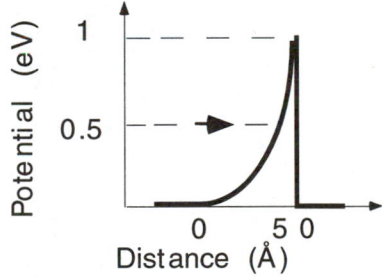

**1-3-3.**  The potential energy of electrons in a metal with a surface electric field is shown in the figure.  The electron concentration is $10^{23}$ cm$^{-3}$; the electronic mass is $9.11 \times 10^{-31}$ kg.  The velocity of electrons impinging on the metal surface is $\dfrac{1}{4}\sqrt{\dfrac{2E}{m_e}}$ where $E = 4$ eV.  (a) Find the strength of the electric field at the surface.  (b) Calculate the electric current density of electrons escaping the metal.

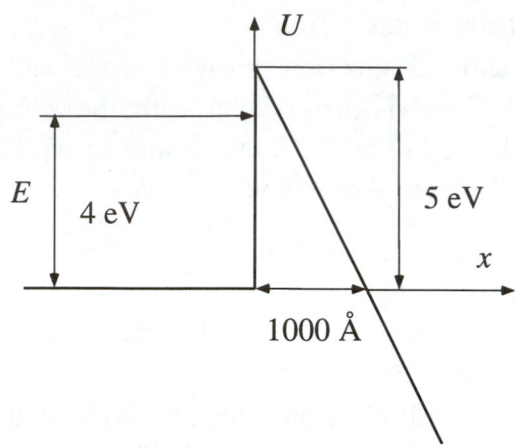

**1-4-1.** Compare electronic configurations for Si, Ga, Al, P, As, N, Ne. Example: Si $(1s^2 2s^2 2p^6 3s^2 3p^2)$. Which is the most stable electron configuration and why? **Hint**: $1s^2$ means that the principal quantum number $n = 1$, orbital quantum number $l = 0$, and this state is occupied by 2 electrons; $2s^2$ corresponds to 2 s electrons ($l = 0$) with $n = 2$; $2p^2$ means that for $n = 2$ there are 2 p electrons ($l = 1$). Use the Periodic Table to find the electron configurations for the other elements (see Appendix A3).

# 2

# Basics of Solid–State Physics

## 2-1.  INTRODUCTION

Materials used in electronics include semiconductors, dielectrics, metals, and superconductors.  Solid-state physics describes properties of all these materials.  The basics of solid-state physics have become especially important since so many new materials are now competing for applications in electronics and since new technologies of material growth hold promise of custom design material properties.  This chapter will explain why material properties are so different and how they can be changed.

## 2-2.  CRYSTAL STRUCTURE

### 2.2.1.  Crystal symmetry.

Most semiconductor devices are made of **crystalline** silicon.  In a crystal, atoms form a regular, periodic structure.  Figure 2.2.1a shows the simplest possible crystal structure – a **simple cubic lattice**.  Since the distances between atoms are different in different directions in a crystal, the crystal properties in these different directions may be different as well.  (Crystals may have **anisotropic properties**.)

In a **polycrystalline** sample (see Fig. 2.2.1b), many small crystals (grains) of irregular sizes are separated by grain boundaries.  Over large distances, anisotropic crystal properties are averaged out in a polycrystalline material.

In an **amorphous** material, the interatomic distances between the nearest atoms are more or less the same but the **long–range order** is absent. For electronic applications, the most important amorphous material is **amorphous silicon,** which may be inexpensively deposited as a thin film over very large areas (as large as several square feet). Amorphous silicon transistors are used in liquid crystal displays and in solar cells (see Chapter 8).

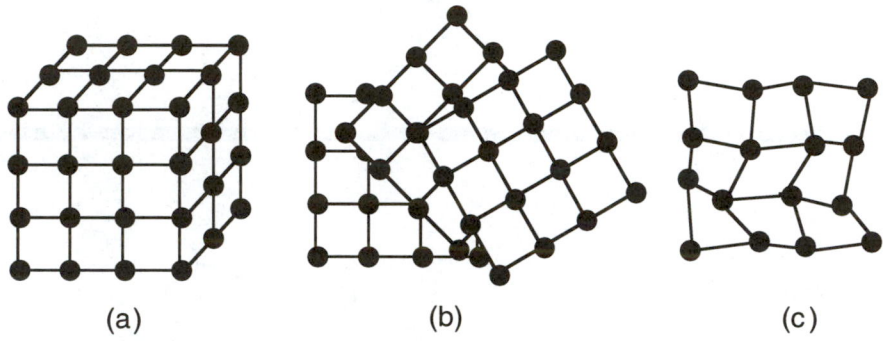

(a)                          (b)                          (c)

**Fig. 2.2.1.** Atomic structure of (a) single crystal, (b) polycrystalline material, and (c) amorphous material. Solid dots represent atoms. For clarity, the polycrystalline and amorphous materials are shown in two dimensions.

The most striking feature of a crystal structure is a periodic repetition of the same pattern of atoms in space. An infinitely large crystal has **translational symmetry**, which means that nothing changes if all atoms are shifted by one or more periods. If we do not pay attention to the boundaries of finite objects, we find numerous examples of translational symmetry in nature or art. For example, the famous artist Andy Warhol placed nearly identical images of Marilyn Monroe in a periodic array; see Fig. 2.2.2 (this work is exhibited in the National Gallery of Art in Washington, DC). (Warhol's other famous work, also on display in the National Gallery of Art in Washington, DC, shows a periodic array of Campbell Soup cans.)

To describe different crystals we introduce the concept of a **crystal lattice**, which is a three-dimensional array of periodically located points in space. This periodic array of points can be reproduced by using **primitive basis vectors**, $a_1$, $a_2$, and $a_3$, which are three independent shortest vectors connecting lattice points. (Here "independent" means that neither of the three vectors can be obtained by adding or subtracting any combination of the other two vectors.)  In other words, all points belonging to a crystal lattice are defined by vectors

$$\mathbf{R}_{k,l,m} = k\mathbf{a}_1 + l\mathbf{a}_2 + m\mathbf{a}_3 \qquad\qquad (2\text{-}2\text{-}1)$$

where $k$, $l$, and $m$ are integers (see Fig. 2.2.3).

**Fig. 2.2.2.**  Andy Warhol's Silk-screen photograph of Marilyn Monroe.  The Tate Gallery, London/Art Resource.

Primitive vectors form a parallelepiped called a **primitive cell**.  A primitive cell does not contain any lattice points inside.  Replicating primitive cells can reproduce the entire crystal lattice.  However, a primitive cell may not have the full rotational symmetry of the crystal lattice.  The smallest cell of a crystal lattice that still retains its rotational symmetry is called a **unit cell**.  Fig. 2.2.3 shows unit cells for three cubic lattices: **simple cubic, body–centered cubic (bcc), and face-centered cubic (fcc)** along with their corresponding primitive vectors and the primitive cell for the fcc lattice.  As can be seen from the figure, the bcc and fcc unit cells have a cubic symmetry but the bcc and fcc primitive cells do not.

A **crystal structure** is formed by placing an identical group of atoms (called a **basis**) into the same position with respect to each point in a crystal lattice as shown in Fig. 2.2.3.  Actually a more realistic model should represent the atoms as larger balls touching each other.  However, it is difficult to illustrate crystal symmetry using such a model.

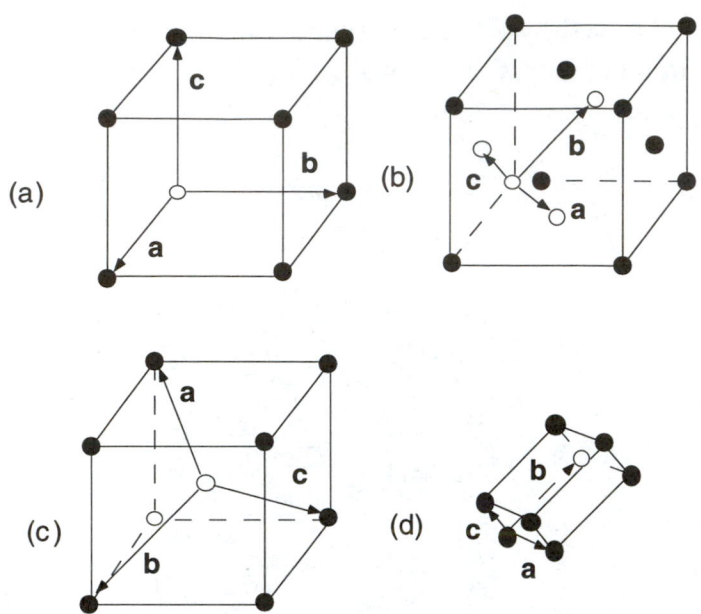

**Fig. 2.2.3.** Unit cells and primitive vectors for (a) simple cubic, (b) face–centered cubic (fcc), (c) body–centered cubic (bcc) lattices and, (d) primitive cell for fcc lattice. Circles represent atoms.

As discussed in Section 1.4 and shown in Fig 2.2.4, each silicon atom forms four bonds with its nearest neighbors (this bonding arrangement is called the **tetrahedral configuration** since each silicon atom is located at the center of the tetrahedron formed by four other silicon atoms.) In Si, such an arrangement is formed by two interpenetrating face-centered cubic sublattices of silicon atoms, shifted with respect to each other by one fourth of the body diagonal. A diamond crystal (one of the crystalline modifications of carbon) has the same crystal structure, which is traditionally called the **diamond crystal structure**. (Another important semiconductor – germanium (Ge) – also has the diamond crystal structure.)

As we discussed in Section 1.4, many compound semiconductors (such as GaAs, InP, AlGaAs, and HgTe) also have the tetrahedral bond configurations. As a consequence, most III–V compounds have the **zinc blende** crystal structure (see Fig. 2.2.5) that is very similar to the diamond structure. This structure contains two kinds of atoms, A and B, with each species forming a face-centered cubic lattice. These mutually penetrating face-centered cubic (fcc) lattices of element A and element B are shifted relative to each other by a quarter of the body diagonal of the unit cell cube.

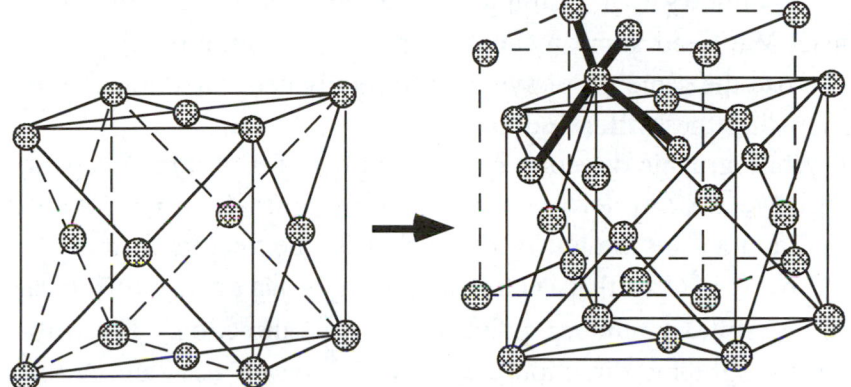

**Fig. 2.2.4.** The fcc and diamond structure (formed by two interpenetrating fcc lattices shifted by one quarter body diagonal). Thick lines show the bonds formed between an atom and its four nearest neighbors (tetrahedral bond configuration). Silicon and germanium have diamond crystal structure.

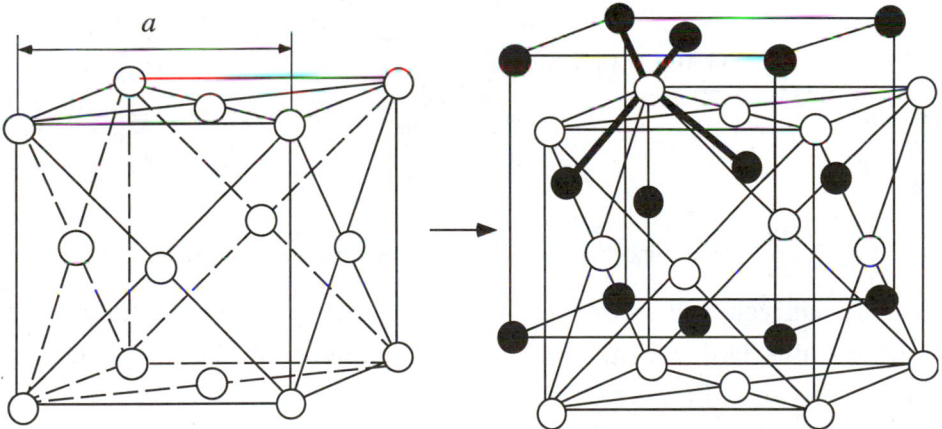

**Fig. 2.2.5.** The fcc and zinc blende structure [formed by two interpenetrating fcc lattices shifted by one quarter of the body diagonal (similar to diamond structure)]. Thick lines show the bonds between an atom and its four the nearest neighbors (tetrahedral bond configuration). $a$ is the lattice constant.

## *2.2.2.   Miller indices.

As was mentioned above, crystals have anisotropic properties, that is, properties that may depend on directions in the crystal (referred to as **crystallographic directions**). However, in certain directions, the crystal properties are the same because of symmetry (for example, the directions $z$ and $y$ in Fig. 2.2.1a are equivalent). We need some means to specify crystallographic directions that allows us to use the same set of symbols for equivalent directions. Traditionally, this is done using the **Miller indices**.

A crystallographic direction is specified by a set of three integers, $u$, $v$, and $w$, defining a vector $u\mathbf{a}_1 + v\mathbf{a}_2 + w\mathbf{a}_3$, which points along the given direction. Here $\mathbf{a}_1$, $\mathbf{a}_2$, and $\mathbf{a}_3$ are the unit vectors, that is, the vectors forming a unit cell; see Fig. 2.2.6. It is clear that vectors $u\mathbf{a}_1 + v\mathbf{a}_2 + w\mathbf{a}_3$ and $k(u\mathbf{a}_1 + v\mathbf{a}_2 + w\mathbf{a}_3)$, where $k$ is an arbitrary number, point in the same direction. The Miller indices for this crystallographic direction are a set of integers $u$, $v$, and $w$ that have no common integral divisor and are enclosed in square brackets. Examples are [100], [111], [110]. A negative integer is represented by placing a bar above the integer (for example $[\bar{1}\,\bar{1}\,\bar{1}]$).

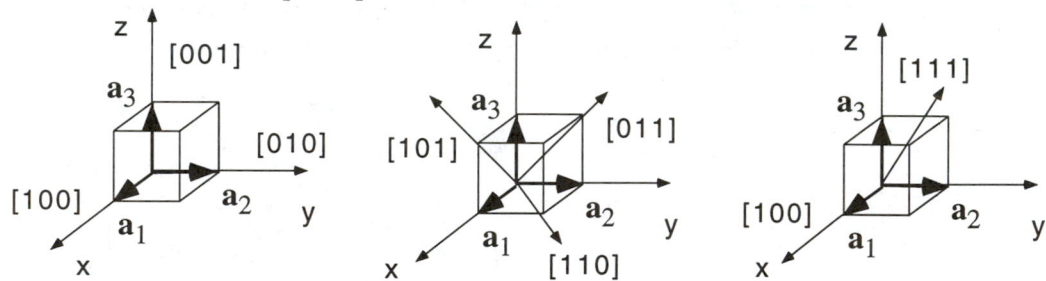

**Fig. 2.2.6.** Some important crystal directions for a cubic crystal.

A set of all equivalent directions is denoted as <$uvw$>. For instance, symbol <111> for a cubic lattice stands for all 8 directions along the body diagonals of the unit cell shown in Fig. 2.2.6.

The Miller indices are also used to specify **crystal planes**. For a crystal plane, the Miller indices are defined using intercepts of the crystal plane with the crystallographic axes pointing in directions of the unit vectors $\mathbf{a}_1, \mathbf{a}_2, \mathbf{a}_3$. These intercepts are vectors $n_1\mathbf{a}_1$, $n_2\mathbf{a}_2$, and $n_3\mathbf{a}_3$ where $n_1$, $n_2$, and $n_3$ are integers. The Miller indices for a crystal plane are integers $u$, $v$, and $w$ that are proportional to $1/n_1$, $1/n_2$, and $1/n_3$ and have no common integral divisor. These indices are enclosed in parentheses (see Fig. 2.2.7). For a cubic crystal (such as silicon or gallium arsenide), the plane $(uvw)$ is perpendicular to the direction $[uvw]$. A set of equivalent planes is denoted as $\{uvw\}$. Using a simple cubic lattice as an example, you can check that crystal planes with the smallest Miller indices, such as {100}, {110}, {111} have the largest density of atoms. Usually crystals are cleaved along these planes and are grown in the directions perpendicular to these planes.

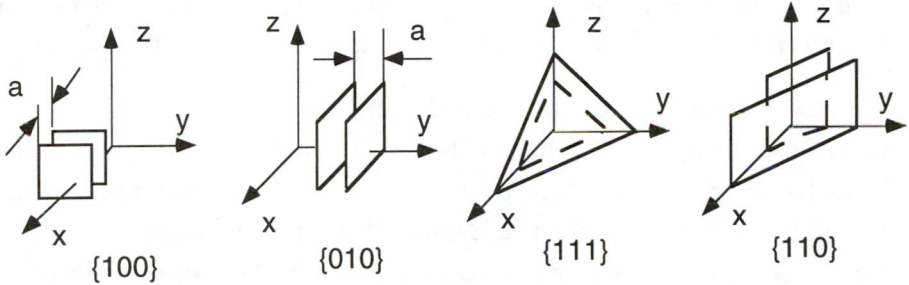

{100}          {010}          {111}          {110}

**Fig. 2.2.7.** Miller indexes for important crystal planes for a cubic crystal. The distances between the plane intercepts with the coordinate axes are equal to the period of the crystal lattice, $a$.

**Example 2-2-1.**

Show the (132) plane and the [131] direction in a cubic crystal.

**Solution:**

In order to determine the [131] direction, we simply find the point with coordinates $1a$, $3a$, $1a$ (where $a$ is the lattice constants) and connect this point with the origin (see Fig. 2.2.8a). In order to determine the (132) plane, we first find the numbers reciprocal to the Miller indices: 1, 1/3, 1/2 and multiply these numbers by 6 to obtain the plane intercepts $6a$, $2a$, and $3a$ (see Fig. 2.2.8b).

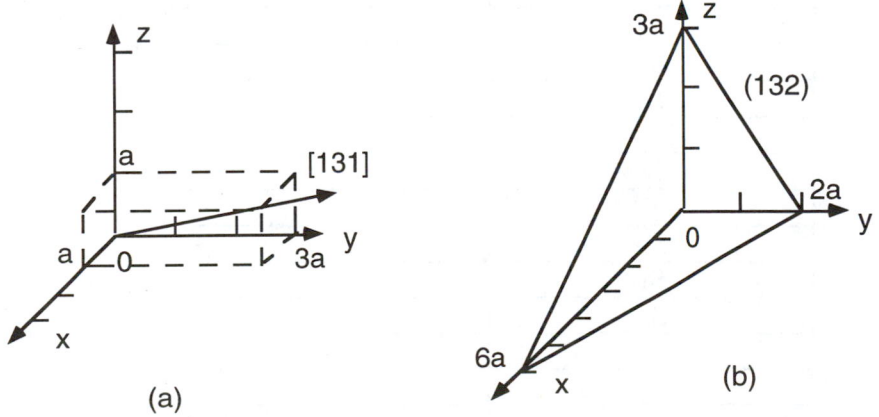

(a)

(b)

**Fig. 2.2.8.** (a) Direction [131] and (b) plane (132) in a cubic crystal.

As can be seen from Fig. 2.2.4 and Fig. 2.2.5, the bonds between the nearest Si atoms in silicon and between the nearest Ga and As atoms in gallium arsenide are in the <111> directions. The closest equivalent atoms in gallium arsenide (Ga and Ga or As and As) are in the <110> directions. Silicon and germanium crystals break (cleave) easily along {111} planes. GaAs films are grown in direction [100] and GaAs wafers are cleaved along {011} planes.

### 2.2.3.  Ternary and quaternary compounds.

The same atoms occupy approximately the same volume in different compounds. This allows us to represent atoms in a crystal as spheres touching the nearest neighbors and introduce an **atomic radius** for each element. The distances between the nearest atoms may be found as the sum of the atomic radii (see Table 2.2.1). Table 2.2.2 lists the lattice constants of several important semiconductors with diamond or zinc blende structure. (For a cubic lattice, the lattice constant is equal to the side of the unit cell cube.) As we can see from Table 2.2.2, the distances between the nearest neighbors are fairly close to the sums of the atomic radii. We also notice that the atomic radii of Ga and Al are equal (1.26 Å).

| Element | Covalent radius (Å) |
|---|---|
| Al | 1.26 |
| As | 1.18 |
| B | 0.88 |
| Bi | 1.46 |
| C | 0.77 |
| Ga | 1.26 |
| Cd | 1.48 |
| Ge | 1.22 |
| Hg | 1.48 |
| In | 1.44 |
| Mn | 1.27 |
| N | 0.70 |
| P | 1.10 |
| Sb | 1.36 |
| Si | 1.17 |
| Sn | 1.40 |
| Te | 1.47 |
| Zn | 1.31 |

**Table 2.2.1.**  Atomic (covalent) radii.

| Material | Lattice constant $a$ (Å) at 25°C | Nearest neighbor distance, $a\sqrt{3}/4$(Å) | Sum of covalent radii (Å) |
|---|---|---|---|
| Si | 5.434 | 2.353 | 2.34 |
| Ge | 5.657 | 2.450 | 2.44 |
| **$A_3B_5$** | | | |
| AlAs | 5.661 | 2.451 | 2.440 |
| AlP | 5.451 | 2.360 | 2.360 |
| AlSb | 6.136 | 2.657 | 2.620 |
| BAs | 4.776 | 2.068 | 2.060 |
| BN | 3.615 | 1.565 | 1.580 |
| BP | 4.538 | 1.965 | 1.980 |
| BSb | 5.170 | 2.239 | 2.240 |
| GaAs | 5.653 | 2.448 | 2.440 |
| GaP | 5.451 | 2.360 | 2.360 |
| GaSb | 6.095 | 2.639 | 2.620 |
| InAs | 6.058 | 2.623 | 2.620 |
| InP | 5.867 | 2.540 | 2.540 |
| InSb | 6.479 | 2.805 | 2.800 |
| **$A_2B_6$** | | | |
| CdTe | 6.482 | 2.807 | 2.950 |
| HgS | 5.841 | | |
| HgSe | 6.084 | | |
| HgTe | 6.462 | 2.798 | 2.950 |
| ZnS | 5.415 | | |
| ZnSe | 5.653 | | |
| ZnTe | 6.101 | 2.642 | 2.780 |

**Table 2.2.2.** Lattice constants (from Shur, 1990, copyright © Prentice Hall, 1990, reproduced by permission of Prentice Hall, Inc., Englewood Cliffs, NJ).

Since Al and Ga have very close atomic radii, the lattice constants of GaAs and AlAs are very close (at 300 K, the lattice constants of GaAs and AlAs are 5.6533 Å and 5.6605 Å, respectively).   For the same reason, these compound semiconductors may form solid-state solutions, such as $Al_xGa_{1-x}As$, where $x$ is a molar fraction of Al.  (Such materials are called **ternary compounds**.)  By varying $x$ from 0 to 1 one can change the properties of $Al_xGa_{1-x}As$ from those of GaAs to those of AlAs.  AlGaAs can be easily grown on GaAs forming a **heterostructure**, that is, a structure including two different semiconductor materials in intimate contact (see, for example, Fig. 1.3.1).  Other examples of ternary compounds are $In_xGa_{1-x}As$, $Al_xIn_{1-x}As$, and $Hg_xCd_{1-x}Te$. By changing the composition of a ternary compound, its lattice constant can be matched to the lattice constant of an appropriate binary compound. The lattice constant, $a_{ter}$, of a ternary compound $A_xC_{1-x}B$ varies roughly linearly with composition

$$a_{ter} \approx a_{bin1}\, x + a_{bin2}\, (1 - x) \tag{2-2-2}$$

where $a_{bin1}$ is the lattice constant of the binary compound AB and $a_{bin2}$ is the lattice constant of the binary compound CB. For example, from eq. (2-2-2) and Table 2.2.2 we find that for $In_{0.47}Ga_{0.53}As$ $a_{ter} \approx 5.84$ Å, which matches quite well with the lattice constant of InP ($a = 5.86$ Å). As a consequence, $In_{0.47}Ga_{0.53}As$ can be grown on InP substrates forming a high-quality heterostructure.

We can also grow **quaternary compounds** consisting of four elements, such as $In_xGa_{1-x}As_yP_{1-y}$. In principle, such a "material engineering" approach allows us to design semiconductors with desired properties. However, the material quality of many of these new materials is not yet good enough for large-scale applications.

Since we model atoms as spheres touching each other, we are naturally interested to know how many atoms we can fit in a given space. The arrangement with the largest density of spheres is called a **close-packed structure**. Let us first consider spheres in one plane. We will achieve the largest density of spheres when we have six spheres touching each ball in the center (see Fig. 2.2.9a). (Bees use a very similar structure in honeycombs; see Fig. 2.2.9b.)

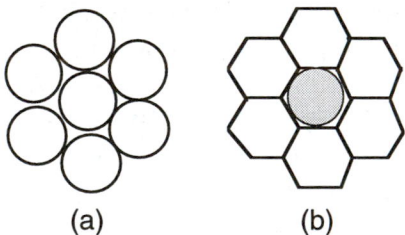

(a)                          (b)

**Fig. 2.2.9.** (a) Closed-packed balls on a plane and (b) honeycomb structure. Shaded circle in Fig. 2.2.9b represents one of the spheres that form a closed-packed structure when put into a honeycomb.

If we use the same arrangement for perpendicular crystal planes, each ball will have 12 nearest neighbors. Hence, atoms in three-dimensional close-packed structures have 12 nearest neighbors.  Two crystal lattices that correspond to close-packed structures are the face–centered cubic (fcc) lattice (see Problem 2-2-10) and a **hexagonal structure**.  Fig. 2.2.10 shows the bonding arrangement in a **wurtzite** hexagonal crystal structure (compare to Fig. 2.2.4).  (Wide band semiconductors such as SiC or GaN often have **wurtzite** hexagonal crystal structure.)  The majority of useful semiconductor crystals correspond to the close-packed structures.

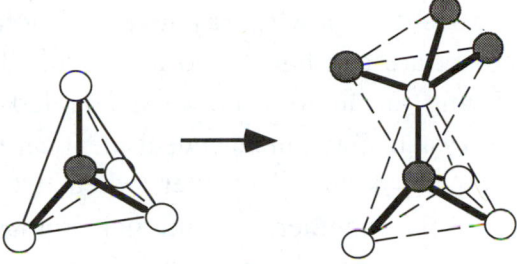

**Fig. 2.2.10.**  Bonding tetrahedron and two interpenetrating tetrahedra in wurtzite crystal structure.  Notice the same bonding arrangement (with four nearest neighbors) as for diamond and zinc blende structures.

The summary of basic equations and concepts related to crystal structure is given in Table 2.2.3.

| | |
|---|---|
| Coordinates of points belonging to a crystal lattice | $\mathbf{R}_{k,l,m} = k\mathbf{a}_1 + l\mathbf{a}_2 + m\mathbf{a}_3$ where $k$, $l$, and $m$ are integers |
| Primitive basis vectors, $\mathbf{a}_1$, $\mathbf{a}_2$, and $\mathbf{a}_3$ | Three independent shortest vectors connecting lattice sites |
| Primitive cell | Cell formed by primitive vectors |
| Unit cell | The smallest cell that still retains lattice symmetry |
| Miller indices for a crystal direction | $[uvw]$.  A set of all equivalent directions is denoted as $<uvw>$ |
| Miller indices for a crystal plane | $(uvw)$.  A set of all equivalent planes is denoted as $\{uvw\}$ |
| Crystal structures of Si and GaAs | Diamond and zinc blende |
| Lattice constant, $a_{ter}$, of a ternary compound $A_xC_{1-x}B$ | $a_{ter} \approx a_{bin1}\, x + a_{bin2}\,(1-x)$ |
| Closed-packed structures | Touching spheres placed into fcc or hexagonal crystal lattices |

**Table 2.2.3**.  Basic equations and concepts related to crystal structure.

## 2-3. ENERGY BANDS. DIELECTRICS, SEMICONDUCTORS, AND METALS

### 2.3.1. Energy bands.

As was discussed in Section 1-4, electrons in an atom can have only certain discrete values of energy (called **energy states** or **energy levels**). When atoms form solids, each atomic level splits into many close energy levels forming **energy bands**. Fig. 2.3.1 shows how energy levels of isolated atoms split into energy bands when atoms are combined into a crystal. This splitting can be understood by considering an analogy between coupled identical resonance circuits and atoms in a crystal. Each independent identical resonance circuit has the same resonance frequency, $\omega_o$. However, when two identical resonance circuits are brought closely together, the mutual conductance leads to the appearance of two close resonance frequencies. This is analogous to splitting atomic energy levels into a band of energy states when atoms are brought together in a crystal.

According to the Pauli exclusion principle only two electrons (with different spins) may occupy an atomic energy level with a given set of quantum numbers $n$, $l$, and $m$ (see Section 1.4). Accordingly, $2N$ electrons may occupy an energy band containing $N$ energy levels. Just like in an atom, where the lowest energy levels are filled and higher energy levels are empty, the lowest energy bands in a crystal are filled and higher energy bands are empty. The bands of allowed energy states are separated by a forbidden range of energy states. The empty or partially filled bands are called **conduction bands**. The bands completely filled by valence electrons are called **valence bands**.

In semiconductors and dielectrics, the lowest conduction band is separated from the highest filled valence band by an energy gap, $E_g$. In a dielectric, $E_g$ is very large compared to the thermal energy, $k_B T$, so that the valence bands are completely filled and conduction bands are totally devoid of electrons.

Typically, for a dielectric, $E_g$ is on the order of or larger than 5 to 6 eV. Normally, almost no current flows in a dielectric when an electric field is applied. In semiconductors, energy gaps vary from 0.1 eV or less (in narrow gap semiconductors) to 3.5 eV (in wide band gap semiconductors such as GaN). [There are even some special semiconductors (such as HgTe) that have a zero energy gap! These materials are called gapless semiconductors.] The values of energy gaps in semiconductors are more comparable to a thermal energy.

Therefore, thermal excitations and/or **donor** and **acceptor** impurities (which we will discuss in Section 3.4) may supply electrons to the conduction band and/or electron vacancies (called **holes**) to the valence band (see the Introduction). These mobile charge carriers allow the electric current to flow when an electric field is applied.

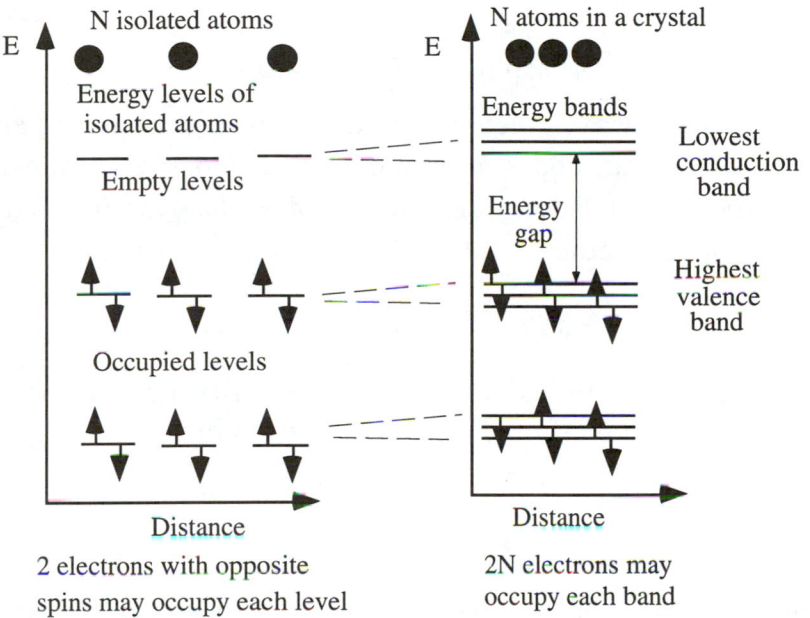

**Fig. 2.3.1.**  Splitting of energy levels of isolated atoms into energy bands in a crystal.  Arrows pointing up and down represent electrons with different values of spin (S = ±1/2) (after Shur 1990, copyright © Prentice Hall, 1990, reproduced by permission of Prentice Hall, Inc., Englewood Cliffs, NJ).

In a metal, the lowest conduction band is only partially filled.  Hence, many electrons (typically one electron per atom) can carry an electric current in an electric field.  This explains why metals have a small resistance.

Figure I.1 schematically shows the band structures of a metal, a semiconductor, a gapless semiconductor (with zero energy gap, such as HgTe), and a dielectric.  The dependencies of the electronic energies (such as $E_c$, $E_v$, etc.) in a crystal on position, such as shown in Fig. I.1, are called **band diagrams**.  In Fig. I.1, these band diagrams are very simple – just lines parallel to the $x$-axis since here we are considering uniform crystals in a zero electric field.  However, in a semiconductor device, the energy bands, generally speaking, vary with distance, and the band diagrams become useful for illustrating properties and understanding behavior of semiconductor materials and devices.

If an electron is placed into a conduction band, it can move freely in a crystal since the electron energy can change just like in free space. As was shown in Section 1.2, the electron wave function $\psi$ of a free electron propagated in direction $x$ is given by

$$\psi = A_1 \exp(ikx) + B_1 \exp(-ikx) \tag{2-3-1}$$

where $k = p/\hbar$ is the **wave vector**, $p$ is the momentum, and $A_1$ and $B_1$ are constants. This wave function is the sum of two functions: $\psi_k = A_1 \exp(ikx)$ and $\psi_k = \quad B_1 \exp(-ikx)$. In a three–dimensional space, the wave functions of a free electron have a similar form:

$$\psi_{\mathbf{k}}(\mathbf{r}) = A \exp(i\mathbf{k} \cdot \mathbf{r}) \tag{2-3-2}$$

where $\mathbf{r}$ is the **space vector** and $\mathbf{k}$ is the wave vector (which may be positive or negative). The energy of a free electron, $E_n$, is given by

$$E_n = \frac{m_e v^2}{2} = \frac{p^2}{2m_e} = \frac{\hbar^2 k^2}{2m_e} \tag{2-3-3}$$

Figure 2.2.2 compares the energy spectrum of a free electron with the energy spectrum of Si and GaAs. Silicon is an **indirect gap semiconductor** where the minimum of the conduction band and the maximum of the valence band correspond to different values of the wave vector, $\mathbf{k}$. GaAs is a **direct gap semiconductor** where the minimum of the conduction band and the maximum of the valence band both occur at $\mathbf{k} = 0$.

### 2.3.2.    Effective mass.

Near the minimum of the conduction band, the dependence of the electron energy on the wave vector can be approximated by a parabolic function similar to that for an electron in free space [see eq. (2-3-3) and Fig. 2.3.2]. However, the curvature of this dependence is usually quite different from that for an electron in free space. Also, different crystallographic directions are not equivalent, and this curvature may depend on direction. These features can be accounted for by introducing an inverse effective mass tensor with components defined as follows:

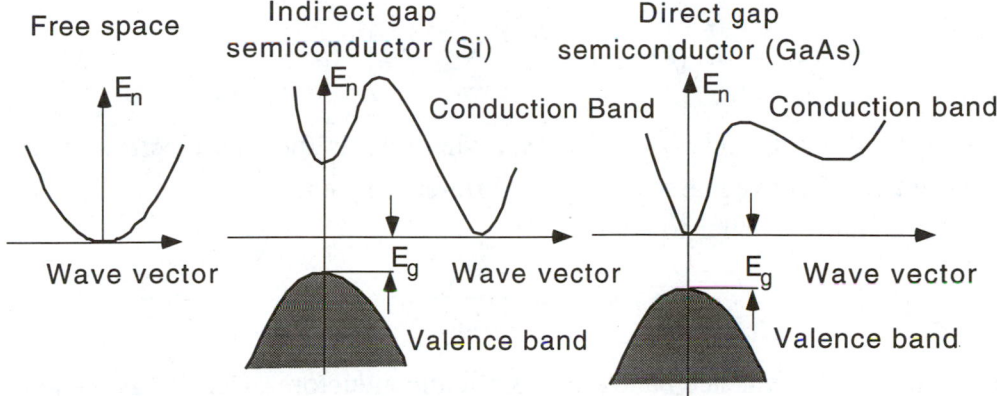

**Fig. 2.3.2.** Qualitative energy spectra for electrons in free space, Si, and GaAs. Shaded states in valence bands are filled with valence electrons.

$$\frac{1}{m_{i,j}} = \frac{1}{\hbar^2}\frac{\partial^2 E_n}{\partial k_i \partial k_j} \tag{2-3-4}$$

where $k_i$ and $k_j$ are the projections of the wave vector $\mathbf{k}$. When $E(\mathbf{k})$ depends only on the magnitude of $\mathbf{k}$, and not on the $\mathbf{k}$ direction, this tensor reduces to a scalar inverse effective mass, $1/m_n$. This is the case for GaAs, where dependence of the electron energy on the wave vector near the lowest minimum of the conduction band, can be approximated by the following function:

$$E_n(k) = E_c + \frac{\hbar^2 k^2}{2m_n} \tag{2-3-5}$$

However, in Si, the lowest minimum of the conduction band is quite anisotropic. In this case, the simplest equation for $E(\mathbf{k})$ is obtained by expressing $\mathbf{k}$ in terms of two components: $k_l$ and $k_t$, which is perpendicular to $k_l$. These components are called **longitudinal and transverse components** of $\mathbf{k}$, respectively. (This takes into account the crystal symmetry, which makes all possible directions of $k_t$ equivalent.) Now the inverse effective mass tensor reduces to two components: $1/m_l$ and $1/m_t$, where $m_l$ and $m_t$ are called the **longitudinal effective mass** and **transverse effective mass**, respectively. In this case, the equation for $E(\mathbf{k})$ is given by

$$E_n(k) = E_c + \frac{\hbar^2}{2}\left(\frac{k_l^2}{m_l} + \frac{k_t^2}{m_t}\right)$$

(2-3-6)

For Si, $m_l \approx 0.98\ m_e$ and $m_t \approx 0.189\ m_e$, where $m_e$ is the free electron mass, so that these two effective masses are quite different indeed!

An equation similar to eq. (2-3-5)

$$E_p(k) = E_v - \frac{\hbar^2 k^2}{2m_p}$$

(2-3-7)

can be used for the valence bands in cubic semiconductors. Here $E_v$ is the energy corresponding to the top of the valence band, and $m_p$ is called a **hole** effective mass. (The concept of a hole will be discussed in Section 3.2.)

### *2.3.3.   Electric field distributions in a metal, a dielectric, and a semiconductor.

In a metal, the concentration of free carriers (and, hence, the conductivity, $\sigma$) is so large that the electric field inside a metal sample is very small. This is easy to understand by referring to **Ohm's law** for the electric current density, **j**:

$$\mathbf{j} = \sigma\mathbf{F}$$

(2-3-8)

Here, **F** is the electric field. Since $\sigma$ is very high, **F** must be very small.[1]

In a dielectric, the electric field must be constant since there are no charges (charge density, $\rho = 0$). This means that according to Poisson's equation, the divergence of the electric field must be zero:

$$\nabla \cdot \mathbf{F} = \frac{\rho}{\varepsilon_s} = 0$$

(2-3-9)

where $\varepsilon_s$ is the dielectric permittivity.

The magnitude of the electric field, $F_{dielectric}$, in a dielectric can be related to the outside electric field, $F_{outside}$, using the boundary condition of the continuity of the electric induction, $\varepsilon F$:

---

[1] Equation (2-3-8) implies that $\sigma$ is the same for all crystallographic directions. This may not be valid for noncubic crystals. For example, in certain modifications of SiC, the conductivity in different crystallographic directions may differ by a factor of four.

$$F_{dielectric} = F_{outside} \frac{\varepsilon_{outside}}{\varepsilon_{dielectric}} \qquad (2\text{-}3\text{-}10)$$

Here $\varepsilon_{dielectric}$ and $\varepsilon_{outside}$ are dielectric permittivities of the dielectric and the surrounding medium, respectively.

The interface electric field in a semiconductor, $F_{si}$, is related to the outside field by the same boundary condition of the continuity of the electric field:

$$F_{si} = F_{outside} \frac{\varepsilon_{outside}}{\varepsilon_{s}} \qquad (2\text{-}3\text{-}11)$$

where $\varepsilon_s$ is the dielectric permittivity of the semiconductor. Inside a semiconductor, the conductivity can vary dramatically. Under certain conditions, a semiconductor could behave similar to a metal and under other conditions, it may behave more like a dielectric. Typically, however, a semiconductor behaves in its own unique way, and the distribution of the electric field inside a semiconductor should be found from the solution of **Poisson's equation**

$$\nabla \cdot \mathbf{F} = \frac{\rho}{\varepsilon_s} \qquad (2\text{-}3\text{-}12)$$

where $\rho$ is the space charge density. This leaves us with the question of how to find $\rho$? The answer to this question will be given in the end of Chapter 3 and in Chapter 4. It is clear, however, that electrons in the conduction band of a semiconductor will move in an electric field in the direction opposite to the electric field direction. As a consequence, electrons will shift from one side of the sample to the other, creating an electron depletion (positive charge) at one of the semiconductor interfaces and an electron accumulation (negative charge) at the other interface. (Since the electric field does not create or destroy electrons but just moves them around, the semiconductor sample as a whole will remain neutral.) Equilibrium is reached when the electric field created by these shifted electron charges will exactly compensate the electric field inside the semiconductor. Hence, the electric field in the bulk will be equal to zero, as in a metal. But the electric field at the semiconductor interfaces obeys the same law as for a dielectric. A semiconductor truly lives up to its name! Figure 2.3.3 shows electric field profiles for a metal, in a semiconductor, and in a dielectric.

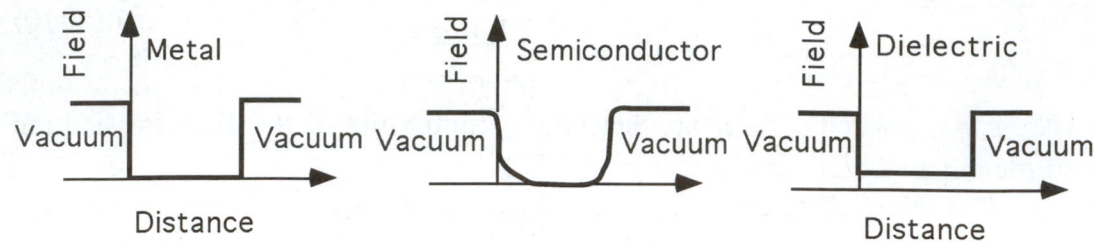

**Fig. 2.3.3.** Schematic field distributions in a metal, a semiconductor, and a dielectric at zero current density. In a semiconductor and in a dielectric, the field at the interface is equal to $\varepsilon_o F_{vacuum}/\varepsilon_s$. The field in a metal and deep inside of a semiconductor equals zero.

### *2.3.4.   Brillouin zone.

In order to understand the $E_n$ versus **k** dependencies in semiconductors better, we will consider a case when the periodic crystal potential is relatively small so that, to the first order, electrons are expected to behave as free electrons whose motion is perturbed by the crystal potential. Let us first consider the reflection of an electron wave [described by eq. (2-3-2)] incoming on a set of parallel crystal planes (see Fig. 2.3.4).

If the electron waves reflected from each crystal plane are in phase, then the reflected waves add up in each point of space (this is called **constructive interference**). The constructive interference of the reflected waves leads to a total reflection of an incoming electron wave. For this to happen, the phase difference between the waves reflected from adjacent planes (which is equal to $2a\cos\varphi$, see Fig. 2.3.4) must be equal to an integer number of the electron wavelengths, $n\lambda$, where $\lambda = 2\pi/|\mathbf{k}|$, **k** is the electron wave vector, and $n = \pm1, \pm2, \pm3 \ldots$. Thus we obtain the reflection condition

$$2a\cos\varphi = n\lambda \qquad (2\text{-}3\text{-}13)$$

Such a reflection is called a **Bragg reflection**. (All of this is very similar to the Davisson and Germer experiment discussed in Section 1.2.) For the normal incidence of the electron wave ($\varphi = 0$), the Bragg reflection occurs when

$$\lambda = 2a/n \qquad (2\text{-}3\text{-}14)$$

or

$$k = \pi n/a = nK/2 \quad \text{where } K = 2\pi/a \qquad (2\text{-}3\text{-}15)$$

and $n = \pm 1, \pm 2, \pm 3, \ldots$ Electrons with such (or close) values of $k$ are reflected from the parallel crystal planes and cannot propagate in a crystal lattice. For these values of $k$, the **energy spectrum** (i. e., the dependence of energy, $E_n$, on $k$) will change drastically compared to that for a free electron. The most important change is the appearance of an energy gap that separates the energy bands almost completely filled with electrons (**valence bands**) from the energy bands, which are nearly empty (conduction bands).

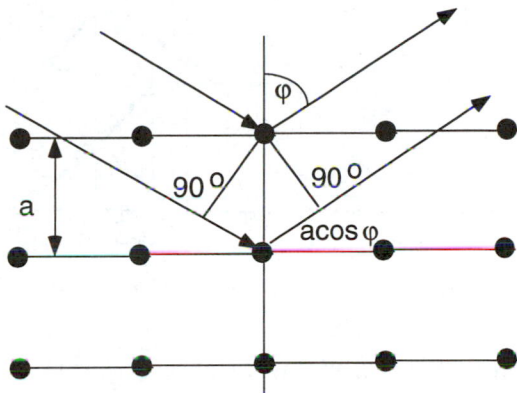

**Fig. 2.3.4.** Reflection of electron waves from crystal planes.

In free space, the dependence of the electron energy, $E_n$, on **k** is **parabolic** (see eq. 2-3-5). This dependence is shown in Fig. 2.3.5 a. In a crystal, the Bragg reflection of the electron waves from the crystal planes makes the propagation of electrons with certain energies impossible, leading to the gaps in the energy spectrum.

Let us now compare the electron wave functions with wave vectors $k$ and $k + K$ for the one-dimensional case for points $x = na$ where $n$ is an integer [see eq. (2-3-15)]:

$$\psi_{k+K}(na) = A\exp\left[i(k+K)na\right] = A\exp\left[i(ka) + i(Ka)\right] =$$
$$A\exp(ika + i2\pi n) = A\exp i(ka) = \psi_k(a) \qquad (2\text{-}3\text{-}16)$$

At these points, the two wave functions are identical! It is also possible to show that $\psi_k$ and $\psi_{k+K}$ have to coincide for any value of $x$. This means that the energy spectrum shown in Fig. 2.3.5 a can be "folded" into the region of the values of $k$ from $-K/2$ to $K/2$. This "folding" is achieved by deducting from the wave vector, $k$, as many values of $K$ as needed to bring $k$ into the interval from $-K/2$ to $K/2$. For example, if $k = 4.75\pi/a$, deducting $5\pi/a$ yields $k = -0.25\,\pi/a$ (greater than $-K/2 = -\pi/a$ and smaller than $K/2 = \pi/a$). As follows from eq. (2-3-16), the resulting energy spectrum will still correspond to all possible wave functions.

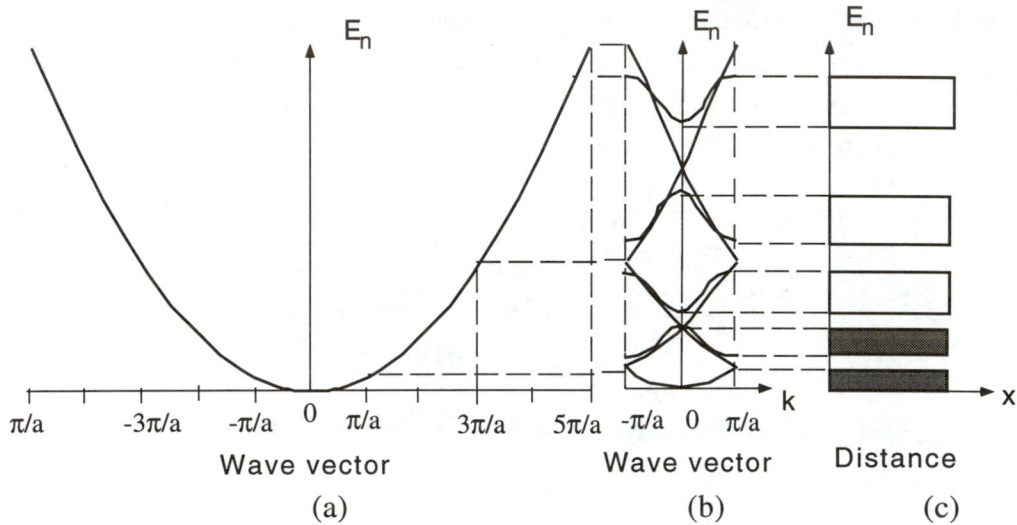

**Fig. 2.3.5.** $E_n$ versus $k$ and $E_n$ versus $x$. (a) $E_n(k)$ for free electrons (see eq. (2-3-2)). (b) $E_n(k)$ for a free electron "folded" into region $-K/2 < k < K/2$ where $K = 2\pi/a$ (thin lines). Bold lines represent a nearly free electron model for a crystal. (c) Allowed and forbidden energy bands versus $x$. (The filled bands are shaded.)

The most important feature of the energy spectrum of an electron in a crystal is the presence of the allowed energy bands separated by **energy gaps** as shown in Figs. 2.3.1, 2.3.2, and 2.3.4.

### Example 2-3-1.

What is the electron wavelength corresponding to the Bragg reflection from (110) planes for the [110] direction of propagation of electron waves in a simple cubic lattice with the lattice constant $a$?

### Solution:

The distance between adjacent (110) planes is equal to $a\sqrt{2}$. Therefore, the value of $a$ in eqs. (2-3-13) and (2-3-14) should be replaced with $a\sqrt{2}$ (compare with Fig. 2.3.4). Thus we obtain the reflection condition

$$2\,a\sqrt{2}\,\cos\varphi = n\lambda$$

For the normal incidence of the electron wave ($\varphi = 0$), the Bragg reflection occurs when

$$\lambda = 2a\sqrt{2}/n \quad \text{or} \quad k = \pi n/(a\sqrt{2})$$

For a three-dimensional crystal, all directions of **k** are possible.  We use wave vectors, **k**, to describe the electronic motion in a crystal, since in an ideal crystal, electrons behave almost like plane waves, with their wave functions changing with time and in space as sin and cos functions.  As any other vectors, vectors **k** can be defined as vectors emanating from the origin of the coordinate system in a mathematical space.  For space vectors **r**, with components $r_x$, $r_y$, $r_z$, such a mathematical space is a real three-dimensional space.  For wave vectors **k**, with components $k_x$, $k_y$, $k_z$, such a space is called the **k-space** or the **reciprocal space** (since distances in $k$-space are measured in 1/meters).  The relationship between real space (where distances are measured in meters) and reciprocal space (where distances are measured in 1/m) is very similar to that between time (measured in seconds) and frequency (measured in 1/s).  In this analogy, a crystal lattice (discrete points in space) corresponds to time-discrete sampling, and regions in $k$-space correspond to frequency bands.

In each direction in the reciprocal space, the energy spectrum can be defined within the region from $-\mathbf{K}/2$ to $\mathbf{K}/2$ where vector **K** corresponds to the Bragg reflection from a set of the crystal planes perpendicular to this particular direction (just  like that shown in Fig. 2.3.4).  This defines a certain region in the $k$-space, which contains the energy spectrum.  This region is called the **first Brillouin zone**.

The first Brillouin zone for the face-centered cubic, diamond and the zinc blende structures is shown in Fig. 2.3.6.  It is the same for all three structures.  Also shown are symmetry points and directions of the first Brillouin zone.  These points and directions play a special role in determining the electronic properties of crystals as discussed below.

The approach to the description of the electron energy spectrum considered above is called a **nearly free electron model**, since we viewed this spectrum as essentially a free electron energy spectrum given by eq. (2-3-3) and modified by Bragg reflections.  This approach  shows  that the energy spectrum, $E(\mathbf{k})$, of a

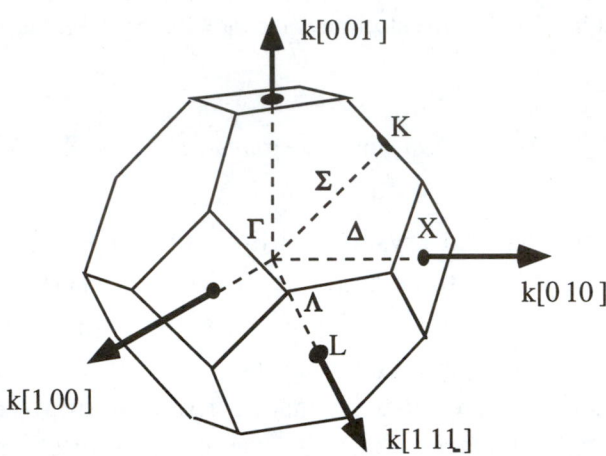

**Fig. 2.3.6.** The first Brillouin zone for the face-centered cubic, diamond
and the zinc blende structures.

crystal can be contained in the region of the **k**-space called the first Brillouin
zone. The wave vector magnitude is given by

$$k = 2\pi / \lambda$$

(2-3-17)

where $\lambda$ is the wavelength. As we discussed in Section 1.2, the value of $k$ is
proportional to the electron momentum, $p$:

$$\hbar k = p$$

(2-3-18)

At certain values of $\lambda$, electrons experience the Bragg reflection from crystal
planes. The corresponding values of $k$ bound a region in the $k$-space called the
first Brillouin zone.

The electron energy is a function of $k_x$, $k_y$, and $k_z$, which will vary within
the first Brillouin zone. Semiconductor properties are primarily determined by
regions in the first Brillouin zone corresponding to lines and points of symmetry.
These symmetry points and lines are marked in Fig. 2.3.6. Points $\Gamma$ (in the
center of the first Brillouin zone), X, and L, and directions $\Delta$ and $\Lambda$ are
particularly important. The dependencies of the electronic energy on the wave
vector (energy bands) are usually calculated for these lines and points of high
symmetry in the first Brillouin zone.

## *2.3.5.  $E$ (k)  dependence  for  Si,  Ge,  III–V,  and  most  II–VI  semiconductors.

The most important features of the band structures are represented by the lowest minima of the conduction band, $E_c$, and by the highest maxima of the valence band, $E_v$, since the states with energies much higher than the minimum of the conduction band are empty and the states with energies well below the maximum of the valence band are completely filled.  (These completely filled and totally empty states do not contribute to the semiconductor conductivity.)  Bands shown in Fig. 2.3.7 are typical for Si, Ge, III–V, and most II–VI semiconductors. (When you compare points X and L shown in Fig. 2.3.6 and in Fig. 2.3.7, please keep in mind that directions [001] and [010], directions [111] and [11$\bar{1}$] are equivalent because of the cubic crystal symmetry.)  The top valence band and the lowest conduction band are separated by an energy gap.  In all important cubic semiconductors, the top of the highest filled (valence) band is located at the Γ point of the first Brillouin zone (see Fig. 2.3.6), that is, at the point **k** = (0,0,0). Two of the valence bands, called the **light** and the **heavy hole bands**, have the same energy at this point (i.e., are degenerate), while the third valence band is separated from the other two and is called the **split-off band**.  The lowest minimum of the conduction band is located at different points in the first Brillouin zone in different semiconductors.  For example, in germanium, the L minimum (corresponding to the L point in the first Brillouin zone) is the lowest. The lowest minimum of the conduction band in silicon is found along the Δ axis [corresponding to the direction $(k,0,0)$], close to the X point of the **first Brillouin zone**.  In GaAs, the lowest minimum of the conduction band is at the Γ point, that is, at the same value of the wave vector **k** as the top of the valence band.  GaAs is therefore called a **direct gap semiconductor**.  Silicon and germanium, however, are **indirect gap semiconductors**.

When we shine light on a semiconductor material, an electron in a valence band can gain energy from a photon and experience a transition into a conduction band.  (Such a transition, called an **interband transition**, leads to **photon absorption**.)  Both energy and momentum must be conserved during this transition.  Hence, the photon energy should be larger than the energy gap ($\hbar\omega \geq E_g$), and the energy gap determines what minimum frequency of electromagnetic radiation is absorbed in a semiconductor material.  We recall that a photon momentum, $p = \hbar\omega/c$ where $c$ is the velocity of light (close to $3\times10^8$ m/s); see Section 1.2.  Therefore, photon momenta involved in transitions from valence to

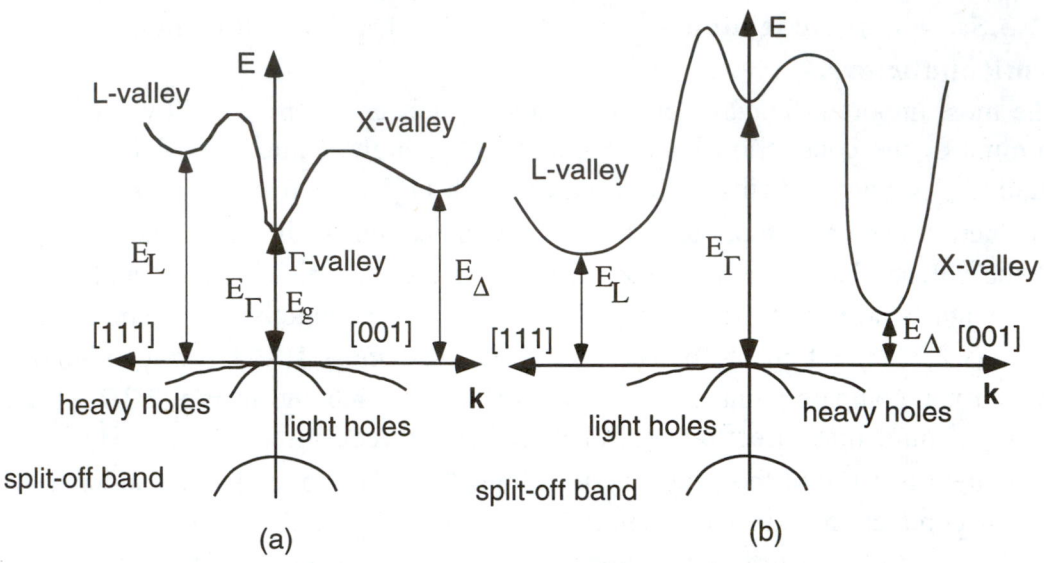

**Fig. 2.3.7.** Schematic band diagrams for cubic semiconductors for (a) direct gap semiconductor, such as GaAs, and (b) for indirect gap semiconductor, such as Si.

conduction bands are on the order of $E_g/c$. Maximum electron momenta are on the order of $\hbar\pi/a$ (this value is related to the Bragg reflection condition as was discussed above). Thus the ratio of the wave vector of the photons involved in interband transitions ($k_{light} = \omega/c$) to the maximum electron wave vector, $k_{max}$, is on order of

$$\frac{k_{light}}{k_{max}} = \frac{E_g a}{\hbar c \pi} \qquad (2\text{-}3\text{-}19)$$

Using typical values of $E_g \approx 1$ eV and of a lattice constant $a = 5$ Å, we find that this ratio is on the order of $10^{-3}$. On the scale of Fig. 2.3.7 where the maximum relevant values of $k$ are on the order of $\pi/a$, the electron transition from the valence into the conduction band looks practically vertical, since the conservation of momentum requires that the electron momentum should only change due to the photon momentum. That is why direct gap semiconductors, which have the top of the valence band and the bottom of the conduction band aligned in the same **k** point, are better suited for electron–hole pair generation and generally better suited for applications in optoelectronic and light emitting devices.

　　Gray tin and HgTe are examples of gapless semiconductors. The absence of the energy gap leads to very interesting and unusual properties of these materials.  Using, for example, $Hg_xCd_{1-x}Te$ alloys, one may create materials with narrow energy gaps, since the energy gap of this ternary compound decreases from the energy gap of GdTe to zero as $x$ changes from 0 to 1.  Such alloys are of interest as sensitive infrared detectors, since they can be used to register interband transitions caused by infrared photons.  Such detectors are used in night vision devices, since objects emit infrared radiation at night.

**Example  2-3-2.**

The energy gap of GaAs is 1.42 eV.  What is the minimum frequency of light causing transitions from the valence into the conduction band of GaAs?  What is the wavelength of this radiation?

**Solution:**

$$\hbar\omega = E_g = 1.42(eV)$$

$$\omega = \frac{1.602\times10^{-19}\times1.42}{1.054\times10^{-34}} = 2.14\times10^{15}\left(s^{-1}\right)$$

$$f = \frac{\omega}{2\pi} = 3.40 \times 10^{14}\,(Hz)$$

$$\lambda = \frac{2\pi c}{\omega} = \frac{2\pi\times3\times10^8}{2.14\times10^{15}} = 8.81\times10^{-7}(m) = 0.881(\mu m)$$

Since it is difficult to indicate the angular dependence of $E(\mathbf{k})$ using conventional $E(k)$ diagrams of the kind shown in Fig. 2.3.7, this dependence is usually illustrated by drawing the surfaces in the $\mathbf{k}$-space corresponding to the same electron energy (see Fig. 2.3.8).  These surfaces are called **surfaces of equal energy**.  Equation (2-3-5) describes **spherical surfaces of equal energy**.  Equation (2-3-6) describes **ellipsoidal surfaces of equal energy**. In this graphical representation, the value of $E$ is not important since our goal is to describe the angular dependence, which is represented by the shape of the surfaces of equal energy and not by their dimensions.  As shown in Fig. 2.3.8, there may be several equivalent minima of equal energy for equivalent directions in the $\mathbf{k}$-space (six  lowest minima in Si located close to the X points of the first Brillouin zone but only one lowest minimum in GaAs located in point $\Gamma$ of the first Brillouin zone).

**Example 2-3-3.**

The $E_n$ versus **k** in Si is given by eq. (2-3-6) where $m_l \approx 0.98\ m_e$ and $m_t \approx 0.189\ m_e$. What is the ratio of axes in the ellipsoidal surfaces of equal energy?

**Solution:**

In the $k_l$ and $k_t$ directions, the dependence $E_n(\mathbf{k})$ given by eq. (2-3-6) reduces to

$$E_n(k) = E_c + \frac{\hbar^2 k_l^2}{2m_l} \quad \text{and} \quad E_n(k) = E_c + \frac{\hbar^2 k_t^2}{2m_l}$$

respectively. For the equal energy surface, these two energies must be equal, and $k_l$ and $k_t$ coincide with the half axes of the ellipsoid, $k_{le}$ and $k_{te}$. Hence,

$$\frac{\hbar^2 k_{te}^2}{2m_t} = \frac{\hbar^2 k_{le}^2}{2m_l} \quad \text{and} \quad \frac{k_{te}}{k_{le}} = \left(\frac{m_t}{m_l}\right)^{1/2} = \left(\frac{0.19}{0.98}\right)^{1/2} = 0.44$$

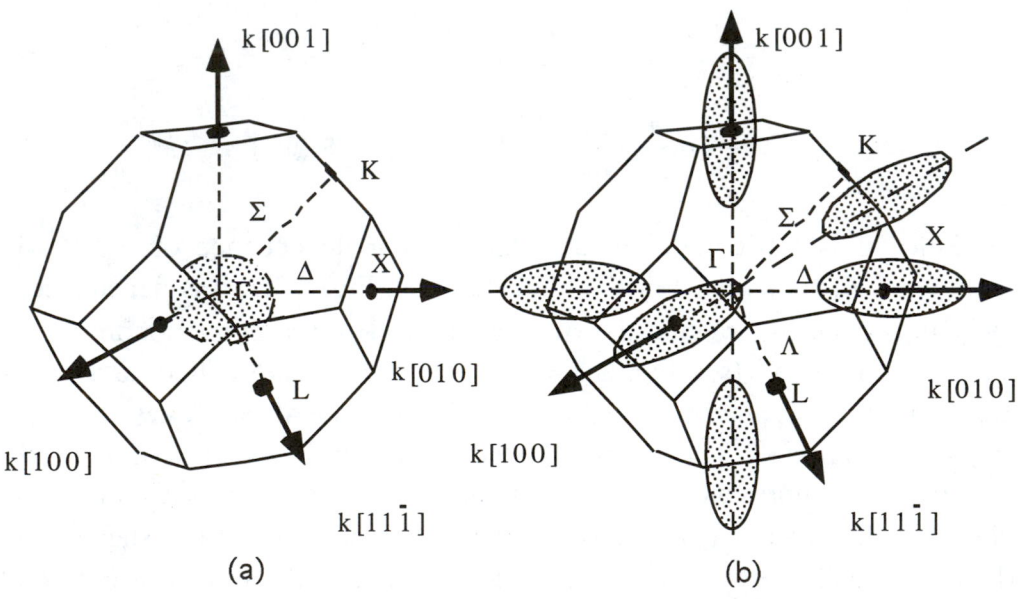

**Fig. 2.3.8.** Surfaces of equal energy for $\Gamma$ and X minima of the conduction band.

The wave vector **k** and the kinetic energy $E_n - E_c$ in eqs. (2-3-5) and (2-3-6) are measured from the minimum of the conduction band. As mentioned above, eq. (2-3-5) represents a parabolic dependence of energy on the wave vector, similar to the dependence for an electron in free space, $E_n = \hbar^2 k^2/(2m_e)$. However, the value of $m_n$ (which is called the effective mass of the electron) can be very much different from the free electron mass, $m_e$.  (In gallium arsenide $m_n \approx 0.067\ m_e$.)  This large difference is caused by the periodic crystal potential.

The energy versus wave vector dependence for both light and heavy holes can be approximated by eq. (2-3-7).  The effective mass $m_p$ is quite different for light and heavy holes (in Si, $m_{pl} \approx 0.16\ m_e$ and $m_{ph} \approx 0.53\ m_e$ for light and heavy holes, respectively).

Table 2.3.1 provides the band structure parameters for important semiconductors.

| | $E_\Gamma$ (eV) | $E_L$ (eV) | $E_\Delta$ (eV) | $E_{so}$ (eV) | Electron effective masses | | | Hole effective masses | |
|---|---|---|---|---|---|---|---|---|---|
| | | | | | $m_l$ | $m_n$ | $m_t$ | $m_{ph}$ | $m_{pl}$ |
| Si | 4.08 | 1.87 | 1.13 | 0.04 | 0.98 | -- | 0.19 | 0.53 | 0.16 |
| Ge | 0.89 | 0.67 | 0.96 | 0.29 | 1.64 | -- | 0.082 | 0.35 | 0.043 |
| AlP | 3.3 | 3.0 | 2.1 | 0.05 | -- | -- | -- | 0.63 | 0.2 |
| AlAs | 2.95 | 2.67 | 2.16 | 0.28 | 2.0 | -- | -- | 0.76 | 0.15 |
| AlSb | 2.5 | 2.39 | 1.6 | 0.75 | 1.64 | -- | 0.23 | 0.94 | 0.14 |
| GaP | 2.24 | 2.75 | 2.38 | 0.08 | 1.12 | -- | 0.22 | 0.79 | 0.14 |
| GaAs | 1.42 | 1.71 | 1.90 | 0.34 | -- | 0.067 | -- | 0.62 | 0.074 |
| GaSb | 0.715 | 1.07 | 1.30 | 0.77 | -- | 0.045 | -- | 0.49 | 0.046 |
| InP | 1.35 | 2.0 | 2.3 | 0.13 | -- | 0.080 | -- | 0.85 | 0.089 |
| InAs | 0.35 | 1.45 | 2.14 | 0.38 | -- | 0.023 | -- | 0.6 | 0.027 |
| InSb | 0.17 | 1.5 | 2.0 | 0.81 | -- | 0.014 | -- | 0.47 | 0.015 |
| ZnS | 3.8 | 5.3 | 5.2 | 0.07 | -- | 0.28 | -- | -- | -- |
| ZnSe | 2.9 | 4.5 | 4.5 | 0.43 | -- | 0.14 | -- | -- | -- |
| ZnTe | 2.56 | 3.64 | 4.26 | 0.92 | -- | 0.18 | -- | -- | -- |
| CdTe | 1.80 | 3.40 | 4.32 | 0.91 | -- | 0.096 | -- | -- | -- |

**Table 2.3.1**.  Energy band gaps and effective masses (in units of $m_e$) of some cubic semiconductors [after C. Jacoboni and L. Reggiani, *Advances in Physics*, **28**, no. 4, pp. 493–553 (1979)].

Equation (2-3-5) is also the simplest possible model of the band structure and is frequently used for crude estimates of transport properties.  The similarity with the free electron motion is very useful for providing an insight into the physics of electronic motion.  However, a more realistic $E_n$ versus **k** relationship is given by

$$E_n\left(1 + \alpha E_n\right) = \frac{\hbar^2 k^2}{2m_n} \tag{2-3-20}$$

Equation (2-3-20) accounts for **nonparabolicity**, that is, for an increase of the electron effective mass at higher energies, which can be noticed as a decrease in the curvature of the $E_n$ versus $k$ relationship with increasing $E_n$ (see Fig. 2.3.7). A detailed analysis shows that the value of the nonparabolicity constant, $\alpha$, depends on the energy gap, $E_g$, and effective mass, $m_n$:

$$\alpha \approx \frac{1}{E_g}\left(1 - \frac{m_n}{m_e}\right)^2 \tag{2-3-21}$$

When a high electric field is applied to a semiconductor sample, charge carriers may gain energy from the electric field and occupy higher energy states. Figure 2.3.9 shows the electron distribution in the **Brillouin** zone of GaAs in a low electric field (Fig. 2.3.9a), when electrons primarily occupy energy states in the lowest minimum of the conduction band ($\Gamma$ minimum) and in a high electric field (Fig. 2.3.9b), when electrons primarily occupy energy states in the upper (L) minima of the conduction band (compare with Figs. 2.3.6, 2.3.7, and 2.3.8).

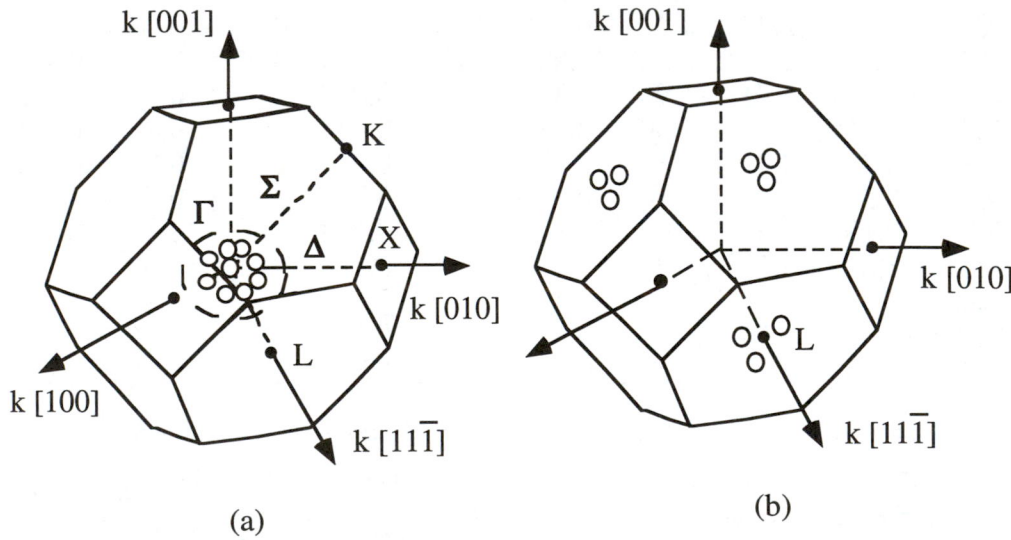

(a) (b)

**Fig. 2.3.9.** Electrons (represented by open dots) populating (a) central ($\Gamma$) and (b) higher (L) valleys of conduction band in GaAs. In still higher electric fields, the electrons will also populate the X minima.

Higher minima of the conduction band are very important in modern small semiconductor devices where electric fields are typically quite large.

A summary of properties of energy bands is given in Table 2.3.2.

| | |
|---|---|
| Direct gap | The top maximum of the valence band and the bottom minimum of the conduction band are in the same point of $\mathbf{k}$-space ($k = 0$) |
| Indirect gap | The top maximum of the valence band and the bottom minimum of the conduction band are in different points of $\mathbf{k}$-space |
| Inverse effective mass tensor | $$\frac{1}{m_{i,j}} = \frac{1}{\hbar^2}\frac{\partial^2 E_n}{\partial k_i \partial k_j}$$ |
| Energy versus wave vector for a simple conduction band | $$E_n(k) = E_c + \frac{\hbar^2 k^2}{2m_n}$$ |
| Energy versus wave vector for conduction band in silicon | $$E_n(k) = E_c + \frac{\hbar^2}{2}\left(\frac{k_l^2}{m_l} + \frac{k_t^2}{m_t}\right)$$ |
| Energy versus wave vector for nonparabolic bands | $$E_n\left(1 + \alpha E_n\right) = \frac{\hbar^2 k^2}{2m_n}$$ |
| Nonparabolicity constant | $$\alpha \approx \frac{1}{E_g}\left(1 - \frac{m_n}{m_e}\right)^2$$ |
| Bragg reflection condition | $2a \cos \varphi = n\lambda$ |
| Reciprocal space | Mathematical space of vectors $\mathbf{k}$ ($\mathbf{k}$-space) |
| First Brillouin zone | Region in the $\mathbf{k}$-space that contains the energy spectrum of a crystal and bounded by values of $\mathbf{k}$ at which electrons experience the Bragg reflection from crystal planes |

**Table 2.3.3.** Basic equations and concepts related to energy bands.

## 2-4.  DISTRIBUTION FUNCTIONS AND DENSITIES OF STATES

### 2.4.1.  Distribution function and electron concentration.

By now we have a basic idea of how electron energy states look in atoms and crystals. We understand that the electron energy states consist of allowed bands separated by forbidden bands. In a semiconductor, an energy gap separates the highest nearly completely filled band (valence band) from the lowest nearly totally empty conduction band. The number of filled levels in the conduction band and unoccupied levels in the valence band depends on temperature. At higher temperatures, more electrons acquire enough thermal energy to be promoted into higher energy states in the conduction band. As was explained in the introduction to the book, if the number of electrons in the conduction band and the number of unoccupied energy levels in the valence bands are zero, electrons are not able to change their energy in the electric field, and the material behaves like a dielectric.

In this section, we will calculate the electron concentration in the conduction band. Let us first consider a system with a very large number of particles, $N$, such as gas molecules, for example. The system is such that each particle can have only certain values of energy $E_1, E_2, \ldots E_i \ldots$. Let us consider the situation when the number of states is much greater than the number of particles and the probability of finding a particle in a given state is much smaller than unity. In this case, the Pauli exclusion principle is not important (since it is very unlikely that two particles will occupy the same energy level), and the probability of finding a particle in the state with energy $E_i$ is given by

$$P(E_i) = \frac{N_i}{N} \tag{2-4-1}$$

where $N_i$ is the total number of particles in this state. The average particle energy can be found as

$$<E> = \sum_i \frac{N_i E_i}{N} \tag{2-4-2}$$

In equilibrium, the probabilities of having particles in two energy states, $E_k$ and $E_i$, are related via the **Boltzmann factors:**

$$\frac{P(E_i)}{P(E_k)} = \exp\left(\frac{E_k - E_i}{k_B T}\right) \tag{2-4-3}$$

This equation means that the probability of finding a particle in a given energy state, $E_i$, decreases exponentially with $E_i$. For a continuous energy spectrum, the probability of finding a particle with the energy between $E$ and $E+dE$ is given by

$$f dE = A \exp\left(-\frac{E}{k_B T}\right) dE \tag{2-4-4}$$

The function $f$ is called the **Boltzmann distribution function**.

For electrons, the Pauli exclusion principle states that no more than two electrons (with opposite spins) can occupy a given energy level. Electrons tend to occupy states with low energies first. Hence, all the states with low energies are filled exactly in the same way – one electron in each energy state (counting the two states with the same energy available for electrons with opposite spins as two separate states). At such low energies, the electron probability function, $f$, must be equal to unity since all these states are occupied. However, at high values of energy, when the probability of occupying an energy state is much smaller than unity, the Pauli principle presents no limitation, and the distribution function should reduce to the Boltzmann distribution function given by eq. (2-4-4).

A more detailed analysis shows that the electron distribution function is given by

$$f_n(E) = \frac{1}{1 + \exp\left(\dfrac{E - E_F}{k_B T}\right)} \tag{2-4-5}$$

This function is called the **Fermi-Dirac distribution function**. The energy $E_F$ is called the **Fermi energy** (or the **Fermi level**). It can be defined as the energy at which the Fermi-Dirac distribution function is equal to 1/2. When $E$ is greater than $E_F$ by a few $k_B T$, $f_n(E)$ is much smaller than 1 and the Fermi-Dirac distribution function reduces to the Boltzmann distribution function. When $E$ is smaller than $E_F$ by a few $k_B T$, $f_n(E)$ is equal to 1, in agreement with the Pauli exclusion principle (see Fig. 2.4.1).

As can be seen from Fig. 2.4.1 and from eq. (2-4-5), when $T \to 0$, the Fermi-Dirac distribution function approaches the step function, which is equal to

1 for $E < E_F$ and equal to zero for $E > E_F$.

The distribution function, $f(E)$, allows us to find the probability that a given state is filled, provided that we know how to figure out the Fermi energy, $E_F$. We will address the important issue of how to find the Fermi level, $E_F$, later.

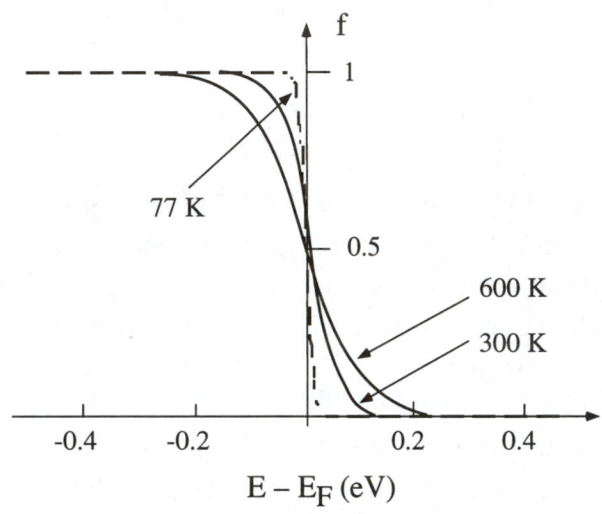

**Fig. 2.4.1.** Fermi-Dirac distribution function.

For now, we assume that the Fermi level is given. In this case, we can calculate the total electron concentration in the conduction band by multiplying the occupation function, $f(E)$, by the number of states per unit volume and per unit energy, $g(E)$, in the energy interval between $E$ and $E + dE$ and integrating over all energies in the conduction band:

$$n = \int_{E_c}^{E_{ct}} f_n(E)g_n(E)dE \tag{2-4-6}$$

Here $E_c$ and $E_{ct}$ are the energies corresponding to the bottom and top energies in the conduction band, $n$ is measured in m$^{-3}$, and $g_n(E)$ is measured in m$^{-3}$J$^{-1}$ if the electron energy is measured in joules or in m$^{-3}$eV$^{-1}$ if the electron energy is measured in electron-volts. [The function $g_n(E)$ is called the **density of states**.] If the Fermi level is many thermal energies, $k_BT$, below $E_{ct}$, then the upper limit in the integral in eq. (2-4-6) can be taken as infinity since, at energies well above the Fermi level, the function $f(E)$ is very small and the contribution to this integral from this energy range is insignificant:

$$n = \int_{E_c}^{\infty} f_n(E) g_n(E) dE \qquad (2\text{-}4\text{-}7)$$

## *2.4.2.   Density of states.

To calculate the density of states, $g_n(E)$, we have to calculate the total density of states in a crystal per unit energy, $G_n(E)$, and divide by the crystal volume, $V$:

$$g_n(E) = \frac{G_n(E)}{V} \qquad (2\text{-}4\text{-}8)$$

Since a larger crystal contains more atoms and more electrons, the total density of states, $G_n(E)$, should be proportional to the crystal volume. In an infinite crystal ($V \Rightarrow \infty$), $G_n(E)$ is infinite. Hence, we must consider a finite crystal to determine $g_n(E)$. This poses a real problem since, strictly speaking, the energy states and wave functions in a crystal depend on the boundary conditions used for the solution of the Schrödinger equation. However, it can be proven that, for a large enough crystal, the boundary conditions should not affect the energy states too much, even though such a proof is beyond the scope of this book. Perhaps it is also clear intuitively. If the boundary conditions do not matter, we may as well impose artificial boundary conditions that make the solution of this problem easier. We consider a large crystal cube with side $L$ and demand that the electron wave functions be periodical in space with the periodicity determined by the size of the cube:

$$\psi(x+L, y+L, z+L) = \psi(x,y,z) \qquad (2\text{-}4\text{-}9)$$

Even though these conditions (called Born-von Karman boundary conditions) are very convenient from a mathematical point of view, they don't make sense from a physics point of view. Why, indeed, should all electronic wave functions in a crystal repeat themselves within each cube of size $L$ where $L$ is only vaguely defined as being large? However, as stated above, these boundary conditions do not affect the actual density of states too much, provided that the crystal is sufficiently large so that electrons deep inside do not feel what is happening at the crystal boundaries. However, when the crystal dimensions become comparable with the electron de Broglie wavelength, these boundary conditions become totally inappropriate.

Now we recall that, to the first order, electron wave functions in a crystal are described by eq. (2-3-2):

$$\psi(\mathbf{r}) = A \exp(i\mathbf{k} \cdot \mathbf{r}) \tag{2-4-10}$$

Hence, eq. (2-4-9) demands that

$$e^{ik_x x + ik_y y + ik_z z} = e^{ik_x(x+L) + ik_y(y+L) + ik_z(z+L)} \tag{2-4-11}$$

Equation (2-4-11) requires that

$$e^{ik_x L} e^{ik_y L} e^{ik_z L} = 1 \tag{2-4-12}$$

leading to

$$k_x L = 2\pi n_1 \tag{2-4-13}$$

$$k_y L = 2\pi n_2 \tag{2-4-14}$$

$$k_z L = 2\pi n_3 \tag{2-4-15}$$

where $n_1$, $n_2$, and $n_3$ are integers. Hence, the difference between two closest allowed $k_x$ values (or the difference between two closest allowed $k_y$ or $k_z$ values) is $2\pi/L$, and each allowed value of $\mathbf{k}$ with coordinates $k_x$, $k_y$, and $k_z$ occupies the volume of $(2\pi/L)^3 = (2\pi)^3/V$ in the $\mathbf{k}$-space where $V = L^3$ is the crystal cube volume.

Let us now calculate the total number of electronic states, $dN$, with values of $k$ between $k$ and $k + dk$. We first calculate the incremental spherical volume in the $\mathbf{k}$-space contained between $k$ and $dk$, which is equal to $4\pi k^2 dk$. We then divide this incremental spherical volume by the volume $(2\pi)^3/V$ occupied in the $\mathbf{k}$-space by one state with a given value of $k$. Finally, we multiply the result by a factor of two, since we can place two electrons with opposite values of spin into each state with a given value of $k$:

$$dN = \frac{2 \times 4\pi k^2 dk}{\left[(2\pi)^3 / V\right]} = \frac{V k^2 dk}{\pi^2} \qquad (2\text{-}4\text{-}16)$$

Taking into account that for parabolic bands

$$k = \frac{\sqrt{2m_n \left(E - E_c\right)}}{\hbar} \qquad (2\text{-}4\text{-}17)$$

and, hence,

$$kdk = m_n dE / \hbar^2 \qquad (2\text{-}4\text{-}18)$$

we express $k$ and $dk$ through $E$ in eq. (2-4-16) and obtain the density of allowed energy states (including spin) per unit volume:

$$g_n(E) = \frac{1}{V}\frac{dN}{dE} = 4\pi \left(\frac{2m_n}{h^2}\right)^{3/2} \left(E - E_c\right)^{1/2} \qquad (2\text{-}4\text{-}19)$$

Here $h = 2\pi\hbar$. As was mentioned above, the dimension of $g_n$ is in m$^{-3}$/eV if $E$ is in eV (see Problem 2-4-2).

Substituting eq. (2-4-19) into eq. (2-4-7), we express the electron concentration in the conduction band in terms of the position of the Fermi level (see also Fig 2.4.2):

$$n = \int_{E_c}^{\infty} g_n(E)f_n(E)dE = N_c F_{1/2}(\eta_n) \qquad (2\text{-}4\text{-}20)$$

where

$$N_c = 2\left(\frac{m_n k_B T}{2\pi\hbar^2}\right)^{3/2} \qquad (2\text{-}4\text{-}21)$$

is called the **effective density of states** for the conduction band,

$$\eta_n = \left(E_F - E_c\right)/k_B T \qquad (2\text{-}4\text{-}22)$$

and

$$F_{1/2}(\eta_n) = \frac{2}{\sqrt{\pi}} \int_0^\infty \frac{x^{1/2}dx}{1+\exp(x-\eta_n)} \tag{2-4-23}$$

is the **Fermi integral**. For $\eta_n \leq -3$,

$$F_{1/2} \approx \exp(\eta_n) \tag{2-4-24}$$

For $\eta_n \geq 3$,

$$F_{1/2} \approx \frac{4\eta_n^{3/2}}{3\sqrt{\pi}} \tag{2-4-25}$$

and the **Fermi wave vector**, $k_F = \sqrt{2m_n(E_F - E_c)}/\hbar$, can be expressed as

$$k_F = \left(3\pi^2 n\right)^{1/3} \tag{2-4-26}$$

For $-10 < \eta_n < 10$, the Fermi integral, $F_{1/2}(\eta_n)$, can be interpolated by the following expression [see Shur (1990)]

$$\ln\left[F_{1/2}(\eta_n)\right] = -0.32881 + 0.74041\eta_n$$
$$-0.045417\eta_n^2 - 8.797\times10^{-4}\eta_n^3 + 1.5117\times10^{-4}\eta_n^4 \tag{2-4-27}$$

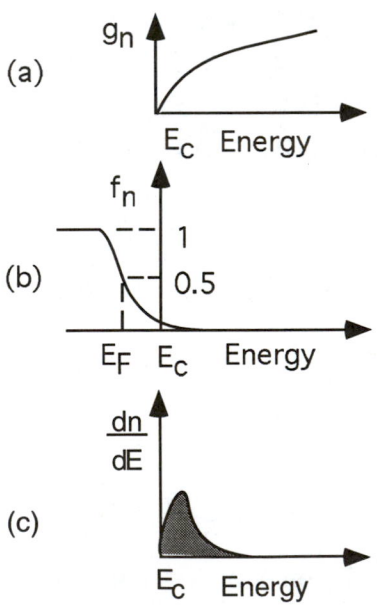

**Fig. 2.4.2.** (a) Density of states, (b) distribution function, and (c) electron density per unit energy. The dashed area is equal to the electron concentration in the conduction band.

**Example 2-4-1.**

Express the value of energy corresponding to the peak of the $dn/dE$ distribution in Fig. 2.4.2c in terms of temperature, $T$, assuming that $E_c - E_F \gg k_B T$.

**Solution:**

$$\frac{dn}{dE} = g_n(E) f_n(E)$$

For $E_c - E_F \gg k_B T$

$$f_n = \exp\left(\frac{E_F - E}{k_B T}\right)$$

Hence, using eq. (2-4-19), we obtain

$$\frac{dn}{dE} = \exp\left(\frac{E_F - E}{k_B T}\right) 4\pi \left(\frac{2m_n}{h^2}\right)^{3/2} (E - E_c)^{1/2}$$

The peak energy, $E_p$, is determined from the following equation:

$$\frac{d}{dE}\left(\frac{dn}{dE}\right)\Bigg|_{E=E_p} = 0$$

which yields

$$-\frac{1}{k_B T}\left(E_p - E_c\right)^{1/2} + \frac{1}{2\left(E_p - E_c\right)^{1/2}} = 0$$

Hence,

$$E_p - E_c = \frac{k_B T}{2}$$

In semiconductors, such as Si or Ge, there are several equivalent minima in the conduction band with anisotropic effective masses (see Fig. 1.6.8). In this case, the effective electron mass, $m_n$, in eq. (2-4-21) should be replaced by

$$m_{dn} = Z^{2/3}\left(m_l m_t^2\right)^{1/3} \tag{2-4-28}$$

where $m_l$ and $m_t$ are the longitudinal and transverse effective masses, $Z$ is the number of equivalent valleys, and $m_{dn}$ is the **density of states effective mass**.

Equation (2-4-21) can be rewritten as

$$N_c = 2.512\times10^{19}\left(\frac{m_{dn}}{m_e}\right)^{3/2}\left(\frac{T}{300}\right)^{3/2}\left(cm^{-3}\right) \tag{2-4-21a}$$

where $m_e$ is the free electron mass. For GaAs, $m_{dn}/m_e \approx 0.067$ and, hence, $N_c \approx 4.36\times10^{17}$ cm$^{-3}$ for $T = 300$ K. [This value is slightly inaccurate since eqs. (2-4-21) and (1-17-21a) do not account for nonparabolicity.]

A summary of basic equations related to distribution functions and densities of states is given in Table 2.4.1.

| | |
|---|---|
| Boltzmann distribution function | $f = A\exp(-E/k_B T)$ |
| Fermi-Dirac distribution function | $f_n(E) = \dfrac{1}{1+\exp\left[(E-E_F)/k_B T\right]}$ |
| Electron concentration in the conduction band | $n = \displaystyle\int_{E_c}^{\infty} g_n(E)f_n(E)dE = N_c F_{1/2}(\eta_n)$ |
| Density of states for a parabolic isotropic band | $g_n(E) = 4\pi\left(2m_n/h^2\right)^{3/2}(E-E_c)^{1/2}$ |
| Effective density of states | $N_c = 2\left(\dfrac{m_n k_B T}{2\pi\hbar^2}\right)^{3/2}$ $N_c = 2.512\times10^{19}\left(\dfrac{m_{dn}}{m_e}\right)^{3/2}\left(\dfrac{T}{300}\right)^{3/2}\left(cm^{-3}\right)$ |
| Fermi integral | $F_{1/2}(\eta_n) = \dfrac{2}{\sqrt{\pi}}\displaystyle\int_0^{\infty}\dfrac{x^{1/2}dx}{\left[1+\exp(x-\eta_n)\right]}$ where $\eta_n = (E_F-E_c)/k_B T$ |
| Density of states effective mass for anisotropic conduction band | $m_{dn} = Z^{2/3}\left(m_l m_t^2\right)^{1/3}$ |

**Table 2.4.1.** Distribution functions and densities of states.

# *2-5.    DENSITIES OF STATES FOR TWO-DIMENSIONAL AND ONE-DIMENSIONAL ELECTRON GASES

### *2-5-1.    Two–dimensional electron gas.

As was mentioned in Section 1.2, quantum well structures have found important applications in novel semiconductor devices.  In such structures, a thin region of a narrow gap semiconductor is sandwiched between layers of a wide band gap semiconductor or surrounded by a wide band gap semiconductor (see Fig. 2.5.1).

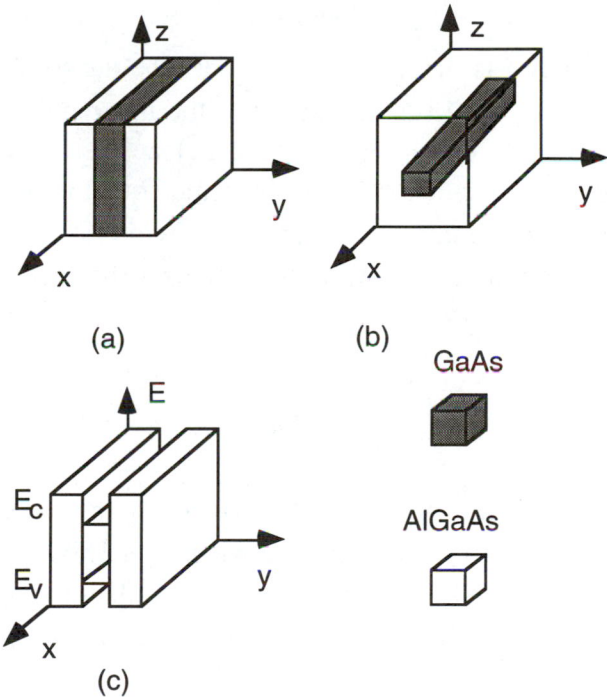

**Fig. 2.5.1.**  (a) GaAs layer sandwiched between AlGaAs layers, (b) GaAs region surrounded by AlGaAs, and (c) corresponding band diagram. (AlGaAs has a wider energy gap than GaAs.)

Let us first consider electrons in a narrow gap semiconductor layer, such as is shown in Fig. 2.5.1a.  If this layer is thin enough, the motion of carriers in the direction perpendicular to the heterointerfaces is **quantized**, meaning that this motion involves discrete (quantum) energy levels.  In this case, electrons propagating in the narrow gap semiconductor are often referred to as a **two–dimensional electron gas**. Electrons in an unrestricted semiconductor are sometimes called a **three-dimensional electron gas**. Electrons propagating in the structure shown in Fig. 2.5.1b (often called a **quantum wire**) are called a **one-dimensional electron gas**.

The lowest energy levels for a square potential well can be estimated as follows:

$$E_j - E_c = \frac{\pi^2 \hbar^2}{2m_n d^2} j^2 \qquad (2\text{-}5\text{-}1)$$

[see eq. (1-2-31)]. Here $j$ is the quantum number labeling the levels, and $d$ is the thickness of the quantum well (the thickness of the layer shown in Fig. 2.5.1a in the $y$ direction). [Strictly speaking, eq. (1-2-31) applies to an infinitely deep potential well. However, we can still use the same equation as long as $E_j$ is well below the bottom of the conduction band in the wide band material]. For the quantization to be important, the difference between the levels should be much larger then the thermal energy $k_B T$, that is,

$$\frac{\pi^2 \hbar^2}{2m_n d^2} >> k_B T$$
$$(2\text{-}5\text{-}2)$$

Using this condition, we find, for example, that in GaAs where $m_n/m_e \approx 0.067$, the levels are quantized at room temperature when $d \le 150$ Å.

In the direction parallel to the heterointerfaces, the electronic motion is not restricted. Hence, the wave function for a two–dimensional electron gas can be presented as

$$\psi = f(y)\exp(ik_x x + ik_z z) \qquad (2\text{-}5\text{-}3)$$

where $f(y)$ may be approximated by eq. (1-2-33):

$$f(y) = \left(\frac{2}{d}\right)^{1/2} \sin\left(\frac{\pi j}{d} y\right) \qquad (2\text{-}5\text{-}4)$$

The term $\exp(ik_x x + ik_z z)$ in the wave function describing the electronic motion in directions $x$ and $z$ is similar to that of free electrons (see Section 1.2). This is understandable since electrons move freely in these directions. The dependence of the electron energy on the wave vector for a two–dimensional electron gas is given by

$$E - E_j = \frac{\hbar^2 \left( k_x^2 + k_z^2 \right)}{2m_n} \tag{2-5-5}$$

The $k_y$-component is absent in eq. (2-5-5) since the motion in the $y$-direction is quantized. Each quantum level, $E_j$, given by eq. (2-5-1) corresponds to an **energy subband** described by eq. (2-5-5); see Fig. 2.5.2.

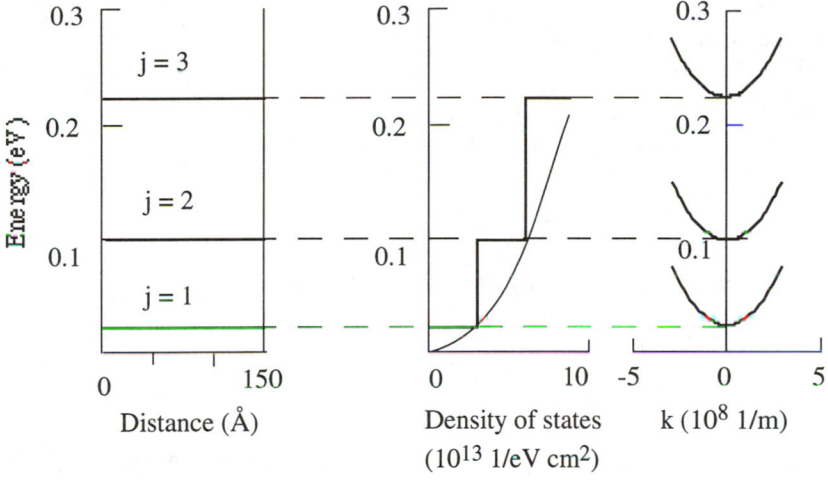

**Fig. 2.5.2.** Energy levels (bottoms of subbands), density of states for quantum well structure, and energy versus $k = (k_x^2 + k_z^2)^{1/2}$ for two–dimensional electron gas in GaAs quantum well.

The density of states for each subband can be found using an approach similar to that used above for a three-dimensional density of states, that is, by counting the number of states with wave vectors $k$ between $k$ and $dk$. The corresponding area in **k**-space is equal to $2\pi k dk$. The density of allowed states is equal to the number of allowed values of $k$ in this area in $k$-space times two (the factor of 2 takes into account the two possible values of spin). The density of allowed points in $k$-space for the unit size sample is $1/(2\pi)^2$ (see Section 2.4). Hence, the total number of states with values of $k$ between $k$ and $k + dk$ is

$$dN = \frac{2 \times 2\pi k dk}{(2\pi)^2}$$

$$(2\text{-}5\text{-}6)$$

Taking into account that

$$k = \frac{\sqrt{2m_n\left(E - E_j\right)}}{\hbar}$$

$$(2\text{-}5\text{-}7)$$

where $E_j$ is the bottom of the $j$th subband and

$$kdk = m_n dE / \hbar^2$$

$$(2\text{-}5\text{-}8)$$

[see eq. (2-3-16)] we obtain

$$dN = \frac{4\pi\left[2m_n\left(E-E_j\right)/\hbar^2\right]^{1/2} 2m_n dE / \hbar^2}{2\left[2m_n\left(E-E_j\right)/\hbar^2\right]^{1/2}(2\pi)^2} = DdE$$

$$(2\text{-}5\text{-}9)$$

where the density of states, $D$, for one subband is given by

$$D = \frac{m_n}{\pi\hbar^2}$$

$$(2\text{-}5\text{-}10)$$

The states of the first (bottom) subband overlap with the states of the second (from the bottom) subband for energies larger than the second energy level, and so on. As a consequence, the overall density of states has a "staircase" shape, shown in Fig. 2.5.2. With an increase in the well thickness, $d$, the steps in Fig. 2.5.2 gradually decrease and merge into an envelope parabolic function, which is equal to the three–dimensional density of states function multiplied by $d$ (as shown by the thin line in Fig. 2.5.2).

Using eq. (2-5-10), we can relate the sheet electron density, $n_{sj}$, in the $j$th subband of the two-dimensional electron gas to the Fermi level, $E_F$, subband bottom energy, $E_j$, and temperature, $T$:

$$n_{sj} = \int_{E_j}^{\infty} Df(E)dE = D \int_{E_j}^{\infty} \frac{1}{1 + \exp\left(\dfrac{E - E_F}{k_B T}\right)} dE$$

$$(2\text{-}5\text{-}11)$$

The integral in eq. (2-5-11) can be easily evaluated by introducing variable $t = \exp\left(\dfrac{E - E_F}{k_B T}\right)$. Then we obtain $dE = \dfrac{k_B T}{t} dt$ and

$$n_{sj} = D k_B T \int_{t_j}^{\infty} \frac{dt}{(1+t)t} \tag{2-5-12}$$

where $t_j = \exp\left(\dfrac{E_j - E_F}{k_B T}\right)$. Furthermore,

$$\int \frac{dt}{(1+t)t} = \int \frac{dt}{t} - \int \frac{dt}{1+t} = -\ln\left(1 + \frac{1}{t}\right)$$

and, hence,

$$n_{sj} = D k_B T \ln\left[1 + \exp\left(\frac{E_F - E_j}{k_B T}\right)\right] \tag{2-5-13}$$

The total density of the two-dimensional gas is

$$n_s = D k_B T \sum_{j=1}^{l} \ln\left[1 + \exp\left(\frac{E_F - E_j}{k_B T}\right)\right] \tag{2-5-14}$$

where $l$ is the total number of subbands.

### *2-5-2.   One–dimensional electron gas.

Let us now consider a one-dimensional quantum wire where electron motion in two directions ($y$ and $z$) is quantized and in one direction ($x$) electrons are free to move (see Fig. 2.5.1b). The wave function $\psi(x, y, z)$ and dispersion relation $E_{n1,n2,k}$ are now given by

$$\psi = f(y)f(z)\exp(ik_x x) \tag{2-5-15}$$

where $f(x)$ and $f(y)$ are functions localized within the cross section of the quantum wire,

$$E - E_c = E_{i1,i2} + \frac{\hbar^2 k_x^2}{2m_n} = \frac{\pi^2 \hbar^2}{2m_n d_y^2} i_1^2 + \frac{\pi^2 \hbar^2}{2m_n d_z^2} i_2^2 + \frac{\hbar^2 k_x^2}{2m_n} \qquad (2\text{-}5\text{-}16)$$

where $i_1$ and $i_2$ are quantum numbers related to quantization in the $y$ and $z$ directions, $d_y$ and $d_z$ are dimensions of the quantum wire in the $y$ and $z$ directions, and where we assume that the dispersion relation for the electron energy in each subband is parabolic.

The density of states for a one–dimensional subband can be found by using an approach similar to that used above for a two-dimensional density of states and in Section 2.4 for a three-dimensional density of states, that is, by counting the number of states with wave vectors $k$ between $k$ and $dk$. The density of allowed states is equal to the number of allowed values of $k$ in this area in $k$-space times two (the factor of 2 takes into account two possible values of spin). For a one–dimensional system, the density of allowed points in $k$-space for a unit size sample is $1/(2\pi)$. Hence, the total number of states with $k$ between $k$ and $k + dk$ is

$$dN = 2 \times 2 \frac{dk}{2\pi} = \frac{2dk}{\pi} \qquad (2\text{-}5\text{-}17)$$

[An additional factor of 2 appears in eq. (2-5-17) because there are two directions of $k$: positive and negative.] Using eq. (2-5-7), we obtain

$$dk = \frac{m_n^{1/2} dE}{\hbar \sqrt{2(E - E_{i1,i2})}} \qquad (2\text{-}5\text{-}18)$$

where $E_{i1,i2}$ is to the bottom of the subband corresponding to the quantum numbers $i_1$ and $i_2$. Hence,

$$dN = \frac{\sqrt{2} m_n^{1/2} dE}{\pi \hbar \sqrt{E - E_{i1,i2}}} = \Omega dE \qquad (2\text{-}5\text{-}19)$$

where the density of states, $\Omega$, is given by

$$\Omega(E) = \frac{\sqrt{2} m_n^{1/2}}{\pi \hbar \sqrt{E - E_{i1,i2}}} \qquad (2\text{-}5\text{-}20)$$

Using eq. (2-5-20), we can relate the linear electron density, $n_{li1,i2}$, in one subband to the Fermi level, $E_F$, and temperature, $T$:

$$n_{li1,i2} = \int_0^\infty \Omega(E)f(E)dE = \frac{\sqrt{2m_n}}{\hbar\pi} \int_{E_{i1,i2}}^\infty \frac{1}{\sqrt{E - E_{i1,i2}}\left[1 + \exp\left(\dfrac{E - E_F}{k_B T}\right)\right]} dE \quad (2\text{-}5\text{-}21)$$

This expression can be rewritten as

$$n_{li1,i2} = \frac{\sqrt{2m_n k_B T}}{\hbar\pi} J(u_f) \quad (2\text{-}5\text{-}22)$$

where $u_f = \dfrac{E_F - E_{i1,i2}}{k_B T}$ and

$$J(u_f) = \int_0^\infty \frac{du}{\sqrt{u}\left[1 + \exp(u - u_f)\right]} \quad (2\text{-}5\text{-}23)$$

The integral in this equation (2-5-23) can be evaluated analytically in the limiting cases of $-u_f \gg 1$ and $u_f \gg 1$. For the intermediate values of $u_f$, the integral has to be evaluated numerically (see Problem 2-5-2). The total linear density is given by

$$n_s = \sum_{i_1}^{l_1} \sum_{i_2}^{l_2} n_{si1,i2} \quad (2\text{-}5\text{-}24)$$

It may be a useful exercise to check units in eq. (2-5-22): $\dfrac{\text{kg}^{1/2} \times \text{joule}^{1/2}}{\text{joule} \times \text{s}} = \dfrac{1}{\text{m}}$ (as expected for an electron density per unit length).

**Example 2-5-1.**

Choose reasonable parameter values and compare the densities of states for three–dimensional, two–dimensional and one–dimensional electron gases in GaAs. Consider only the two lowest subbands for the two–dimensional and one–dimensional electron gases.

**Solution:**

The electron effective mass in GaAs, $m_n \approx 0.067\ m_e$ where $m_e$ is the free electron mass. Using eq. (2-4-19), we find the density of states for the three-dimensional electron gas in GaAs:

$$g(E) = 1.183 \times 10^{20} \left[ (E - E_c)(\text{eV}) \right]^{1/2} \left( \frac{1}{\text{cm}^3 \text{eV}} \right)$$

Using eq. (2-5-10), we find the density of states for each subband for the two dimensional electron gas in GaAs:

$$D = \frac{m_n}{\pi \hbar^2} = 2.80 \times 10^{13}\ (1/\text{cm}^2 \text{eV})$$

Using eq. (2-5-20), we find the density of states for each subband for the one dimensional electron gas in GaAs:

$$\Omega(E) = \frac{\sqrt{2} m_n^{1/2}}{\pi \hbar \sqrt{E - E_{i1,i2}}} = \frac{4.22 \times 10^6}{\sqrt{E - E_{i1,i2}(\text{eV})}} \left( \frac{1}{\text{cm eV}} \right)$$

We choose the thickness of the GaAs layer containing the two-dimensional gas to be equal to 100 Å, (which is comparable to the de Broglie wavelength in GaAs at room temperature, see Section 1.2). The lowest energies in the two lowest subbands are equal to 0.056 eV and 0.224 eV [determined from eq. (2-5-1) by substituting $i = 1$ and $i = 2$, respectively].

We also choose the cross section of the GaAs quantum wire containing the one–dimensional gas to be equal to 100 Å $\times$ 100 Å. The lowest energies in the two lowest subbands are equal to 0.112 eV and 0.280 eV [determined from eq. (2-5-16) by substituting $k_x = 0$ and $i_1 = 1$, $i_2 = 1$ and $i_1 = 1$, $i_2 = 2$, respectively]. Figure 2.5.3 shows the dependencies of these densities of states on energy. Of course, the densities of states for one-, two-, and three–dimensional electron gases have different dimensions. If we make, for example, $10^5$ parallel identical quantum wires per cm then the two-dimensional density of states in all these wires will be $10^5$ greater than $\Omega$ and will be more comparable to $D$. In similar way, we can consider many identical parallel layers containing two–dimensional gas and then multiplying $D$ by the number of layers per cm, we can obtain the three–dimensional density of states, which will be more comparable to $g$.

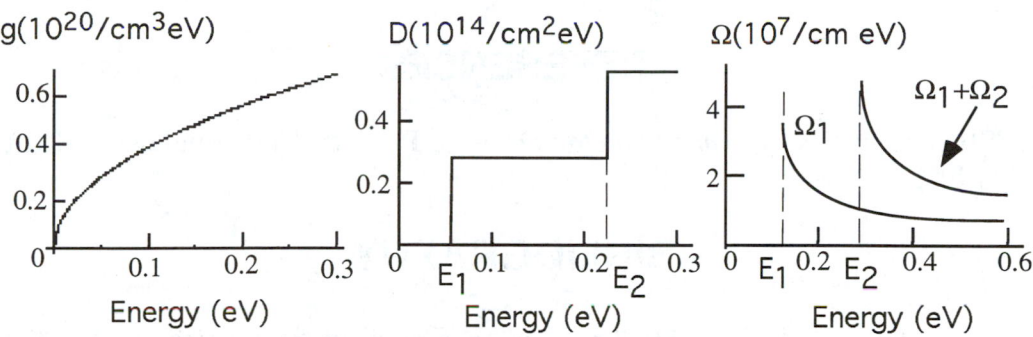

**Fig. 2.5.3.** Densities of states versus energy for three–dimensional ($g$), two–dimensional ($D$), and one–dimensional ($\Omega$) electron gases in GaAs conduction band. Only the two lowest subbands are accounted for two–dimensional and one–dimensional electron gases.

As was shown in this section, once we find the densities of states we can calculate the electron concentration in the conduction band. However, energy states in the valence band may play an equally important role. Fortunately, as will be discussed in the next chapter, all we learned about the conduction band states applies to valence bands as well.

Table 2.5.1 gives a summary of the most important equations related to two–dimensional and one–dimensional electron gases.

| | |
|---|---|
| Electron energies for two–dimensional electron gas in a square potential well | $E - E_j = \hbar^2 \left( k_x^2 + k_z^2 \right) / 2m_n$ |
| Density of states for one subband for two-dimensional electron gas | $D = \dfrac{m_n}{\pi \hbar^2}$ |
| Density of two-dimensional electron gas | $n_s = \sum\limits_{j=1}^{l} Dk_B T \ln\left[ 1 + \exp\left( \dfrac{E_F - E_j}{k_B T} \right) \right]$ |
| Energies for one-dimensional electron gas for a rectangular quantum wire with cross section $d_y \times d_z$ | $E - E_c = \dfrac{\pi^2 \hbar^2}{2m_n d_y^2} i_1^2 + \dfrac{\pi^2 \hbar^2}{2m_n d_z^2} i_2^2 + \dfrac{\hbar^2 k_x^2}{2m_n}$ |
| Density of states for one subband for one-dimensional electron gas | $\Omega(E) = \dfrac{\sqrt{2} m_n^{1/2}}{\pi \hbar \sqrt{\left( E - E_{i1,i2} \right)}}$ |

**Table 2.5.1.** Densities of states for two–dimensional and one–dimensional electron gases.

# REFERENCES

M. SHUR, *Physics of Semiconductor Devices*, Prentice Hall, Englewood Cliffs, NJ (1990)

# BIBLIOGRAPHY

L. SOLIMAR AND D. WALSH, *Lectures on Electrical Properties of Materials*, Fifth Edition, Oxford University Press, Oxford (1993)

*An excellent introduction to material science.*

M. E. LEVINSHTEIN AND G. S. SIMIN, *Getting to Know Semiconductors*, World Scientific, Singapore (1992)

*Popular introduction into the physics of semiconductors.*

C. KITTEL, *Introduction to Solid State Physics*, Sixth Edition, Wiley (1986)

*A popular text on solid-state physics.*

N. W. ASHCROFT AND N. D. MERMIN, *Solid State Physics*, Holt, Rinehart, and Winston, Philadelphia (1986)

*A more advanced text on solid-state physics.*

C. M. WOLFE, N. J. HOLONYAK, AND G. E. STILLMAN, *Physical Properties of Semiconductors*, Prentice Hall, Englewood Cliffs, NJ (1989)

*A graduate level text on semiconductor physics.*

R. A. SMITH, *Semiconductors*, Second Edition, Cambridge University Press, New York (1978)

*Basic introduction into semiconductor physics.*

K. SEEGER, *Semiconductor Physics, An Introduction,* Third Edition, Series on Solid-State Sciences, Vol. 40, Springer Verlag, New York (1985)

*A more advanced text on semiconductor physics.*

# REVIEW QUESTIONS

**1.** Give the definitions of crystalline, polycrystalline, and amorphous materials.

☐ 1 point

**2.** Give the definitions of primitive basis vectors, primitive cell, and unit cell.

☐ 1 point

**3.** The crystal structure of a compound semiconductor AB shown in Fig. R3.1 consists of atoms A and B with atomic radii of 1.26 Å and 1.18 Å.

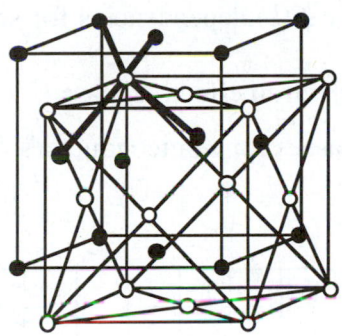

**Fig. R3.1.**

a. What is the name of this crystal structure? ☐ 1 point

b. What would it be called if all the atoms were identical? ☐ 1 point

c. What is the smallest interatomic distance? ☐ 1 point

d. What is the lattice constant? ☐ 1 point

e. What are the Miller indices for the crystal plane with the highest density of atoms for the zinc blende structure?

☐ 1 point

f. How many atoms are in the unit cell of the zinc blende structure?

☐ 1 point

g. Consider the structure in Fig. R3.1 but with identical atoms with atomic radii of 1 Å. What is the electron wavelength corresponding to the Bragg reflection from (111) planes for the [111] direction of propagation of electron waves?

□ 3 points

**4. a.** What is the relation between electron energy and electron momentum in a parabolic conduction band?

□ 1 point

**b.** What is the relation between electron energy and electron wave vector in a parabolic conduction band?

□ 1 point

**5.** Fig. R5.1 shows parabolic $E(k)$ dependencies for semiconductors 1 and 2.

**a.** Which one has a larger electron effective mass (1 or 2)?

□ 1 point

**b.** What are the effective masses for semiconductors 1 and 2 (in units of the free electron mass)?

□ 2 points

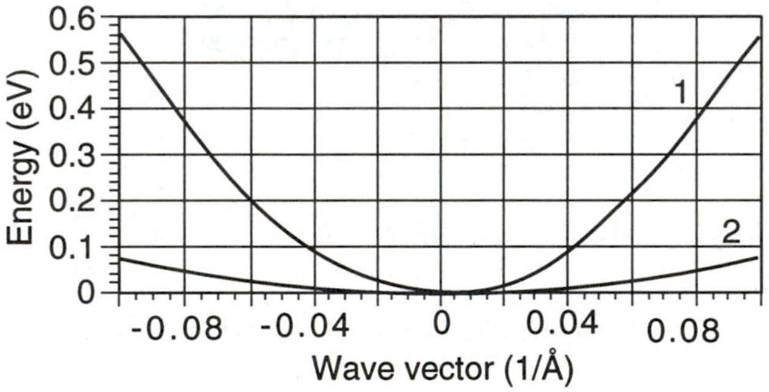

**Fig. R5.1.**

**c.** What is the de Broglie wavelength of an electron with energy 0.2 eV in semiconductor 1?

□ 1 point

**d.** What is the meaning of energy $E = 0$ in this figure?

□ 1 point

**e.** In reality, the $E(k)$ dependence is never exactly parabolic. Write an equation that relates $E$ and $k$ and accounts for a deviation from the parabolicity.

□ 1 point

f.  Which semiconductor (1 or 2) is expected to have the larger nonparabolicity constant (assuming that they have the same energy band gap)?

☐ 1 point

g.  What is the ratio of the effective densities of states for electrons in the conduction band for semiconductors 1 and 2?

☐ 1 point

**6.** What composition, $x$, of the ternary compound $Al_xIn_{1-x}As$ will you use to obtain a material with nearly the same lattice constant as InP?
**Hint:** Use Table 2.2.2.

☐ 2 points

**7.** a. How many lowest equivalent minima of the conduction band does Si have?

☐ 1 point

b.  What is the shape of the surfaces of equal energy for these minima and how are these surfaces oriented in the $k$-space (specify Miller indices)?

☐ 2 points

c.  Which effective mass is larger for these minima (longitudinal or transverse)?

☐ 1 point

**8.** The photons with energies smaller than the energy gap of a semiconductor practically are not absorbed in a semiconductor, and the photons with energies larger than the energy gap experience strong absorption.   Therefore, semiconductor light sensors are usually made from semiconductors with energy gaps smaller than the photon energy.  Which semiconductors (from those listed in Table 2.3.1) can be used for "solar-blind" radiation detectors (assuming that the solar radiation has wavelengths larger than 0.39 µm).

☐ 2 points

**9.** In the two potential quantum wells shown in Fig. R9.1, the layers between the barriers are made of semiconductors 1 and 2, respectively.  Also shown are energies of the ground states for the electronic subbands.   Electrons in semiconductor 1 have a smaller effective mass.

a.  Which band diagram corresponds to semiconductor 1 (A or B) and why?

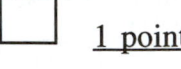 1 point

b. Show the energy levels of an electron with a kinetic energy of 0.1 eV in diagrams A and B.

2 points

c. What is the ratio of the densities of states for electrons in the conduction subbands in Fig. R9.1 for semiconductors 1 and 2?

1 point

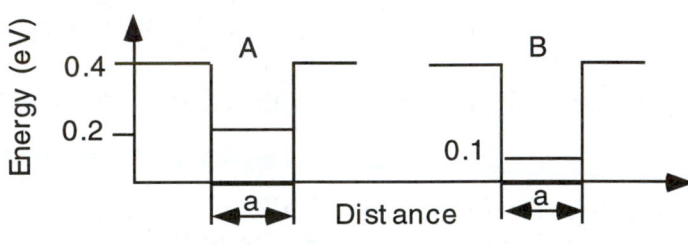

**Fig. R9.1.**

**10.** Fig. R10.1 shows the density of states as a function of energy for a two-dimensional electron gas (which looks like a staircase).

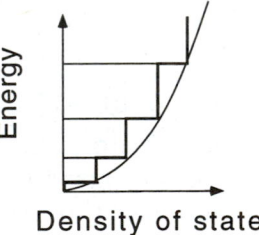

**Fig. R10.1.**     Density of states

a. What is the cause for these steps in the density of states?

2 points

b. What distances on the graph are supposed to be equal and why?

2 points

c. What does the thin parabolic line shown in the graph represent?

1 point

# PROBLEMS

**2-2-1.** It is natural to assume that the volume occupied by an atom is roughly proportional to the number of electrons in the atoms. To check this assumption, use the data in Table 2.2.1 and the Periodic Table to plot the covalent radii versus $N_{atom}^{1/3}$ where $N_{atom}$ is the atomic number. Using the linear regression (i. e., a straight line approximation) for the plot, deduce the value of the covalent radius extrapolated to $N_{atom} = 1$ and compare with the Bohr radius.

**2-2-2.** Which crystal lattice has the largest number of nearest neighbors – simple cubic, fcc, bcc, or diamond? Which one has the smallest number of nearest neighbors?

**2-2-3.** Which crystal lattice has the largest number of the second nearest neighbors – simple cubic, fcc, bcc, or diamond? Which one has the smallest number of the second nearest neighbors?

**2-2-4.** For the same lattice constant, which crystal lattice had the largest volume of the primitive cell – simple cubic, fcc, or bcc? Which one has the smallest volume? How many atoms does each of these primitive cells contain?

**2-2-5.** Sketch and compare primitive lattice vectors and identify the basis atoms for the diamond crystal structure and for fcc.

**2-2-6.** What are the Miller indices for the plane shown in the figure?

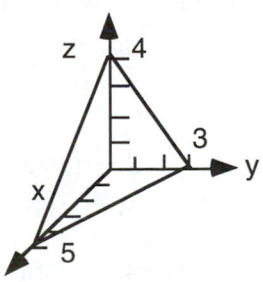

**2-2-7.** Show the (311) plane and the [311] direction in a cubic crystal lattice.

**2-2-8.** Prove that the angle $\alpha$ between two nearest bonds in the silicon crystal (see the figure below) is approximately equal to 109.5°.

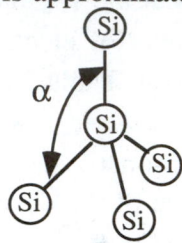

**2-2-9.** Covalent radii of Ga and As are 1.26 Å and 1.18 Å, respectively. Find the lattice constant of GaAs and the volume of the primitive cell.

**2-2-10.** Assume that atoms in an fcc lattice can be modeled as hard touching balls and calculate the percentage of the crystal volume occupied by the atoms.

**2-2-11.** Which plane (100), (110), or (111) in diamond structure has the highest density of atoms?

**2-3-1.** Consider a semiconductor crystal with a separation between the adjacent crystal planes of 3 Å. What is the energy and momentum of a free electron at which it will experience Bragg reflection from the crystal planes, assuming that the direction of the electronic motion is perpendicular to the crystal planes? Give your answer for the electron energy in electron-volts.

**2-3-2.** Consider an insulating epoxy that incorporates many randomly distributed metal balls. Does this medium behave as a dielectric or as a metal? (This problem is related to electron transport in amorphous semiconductors.)

**2-3-3.** In many semiconductors, the dependence of the electron energy, $E$, on the wave vector, $k$, for the lowest minimum of the conduction band differs from the simple parabolic relation $E = \dfrac{\hbar^2 k^2}{2m_n}$. A more accurate equation linking $E$ and $k$ is given by

$$E(1+\alpha E) = \frac{\hbar^2 k^2}{2m_n}$$

where $m_n$ is the effective mass for $E = 0$, $k$ is the wave vector, and $\alpha$ is a constant (called a nonparabolicity constant). In this case, the electron effective mass, $m^*$, defined as $m^* = \dfrac{\hbar^2}{\partial^2 E / \partial k^2}$, is a function of energy. Calculate the dependence of the effective mass, $m^*$, on energy.

**2-3-4.** Estimate a GaAs mole fraction, $x$, in a ternary compound $GaAs_xSb_{1-x}$, such that this compound will be lattice matched to InP.
**Hint:** Assume a linear variation of the lattice constant with the mole fraction, $x$, and assume that the lattice constants of InP, GaAs, and GaSb are 5.867 Å, 5.653 Å, and 6.095 Å, respectively.

**2-4-1.** Calculate the derivative of the Fermi-Dirac distribution function with respect to energy. Plot the dependence of the Fermi-Dirac distribution function and its derivative with respect to energy for $T = 77$ K, 300 K, and 600 K.

**2-4-2.** Express the density of states, $g_n$, (in cm$^{-3}$/eV) as a function of $m_n/m_e$, and $E - E_c$ (in eV) where $m_e$ is the free electron mass and $E$ is the electron energy. Plot $g_n$ versus $E - E_c$ for GaAs ($m_n/m_e = 0.067$).

**2-4-3** Calculate and plot the density states, $g_n$, (in cm$^{-3}$/eV) versus $E - E_c$ (in eV) and the electron density per unit energy, $dn/dE$, (in cm$^{-3}$/eV) for Si at $T = 300$ K for $E_F - E_c = -0.2$ eV and $E_F - E_c = 0.2$ eV. The effective mass of density states in Si is 1.18 $m_e$. Compare with Fig. 2.4.2.

**2-5-1.** Prove that the sheet concentration of the two-dimensional electron gas, $n_s$, in one subband depends on $u_f = (E_F - E_o)/k_B T$ (where $E_F$ is the Fermi level and $E_o$ is the bottom of the subband) as follows:

$$n_s = D(E_F - E_o) \text{ for } u_f >> 1 \text{ and } n_s = DkT \exp\left(\frac{E_F - E_o}{k_B T}\right) \text{ for } -u_f >> 1$$

**2-5-2.** Prove that the linear concentration of the one–dimensional electron gas, $n_l$, in one subband depends on $u_f = (E_F - E_o)/k_B T$ (where $E_F$ is the Fermi level and $E_o$ is the bottom of the subband) as follows:

$$n_l = \frac{\sqrt{2}m_n^{1/2}\sqrt{k_B T}}{\hbar\pi}\int_0^\infty \frac{du}{\sqrt{u}\left[1 + \exp\left(u - u_f\right)\right]}$$

Calculate the dependence of $n_l$ on $u_f$ for $-5 \le u_f \le 5$ by evaluating numerically the integral in this equation. (For extra credit, derive analytical expressions linking $n_l$ and $u_f$ for $u_f < -5$ and $u_f > 5$.) Assume $m_n = 0.067m_e$, $T = 300$ K.

**2-5-3.** Consider a periodic structure consisting of quantum wells of widths $d_1$ separated by the barriers of width $d_2$ (such a structure is called a **superlattice**); see the figure below. The electron effective mass is $m_n$. The lattice temperature is $T$. Assume that there is only one subband, $E_o$, in each quantum well. Find the average volume electron density, $n$, in this system as a function of the Fermi level assuming that the Fermi level is below the top of the barriers.

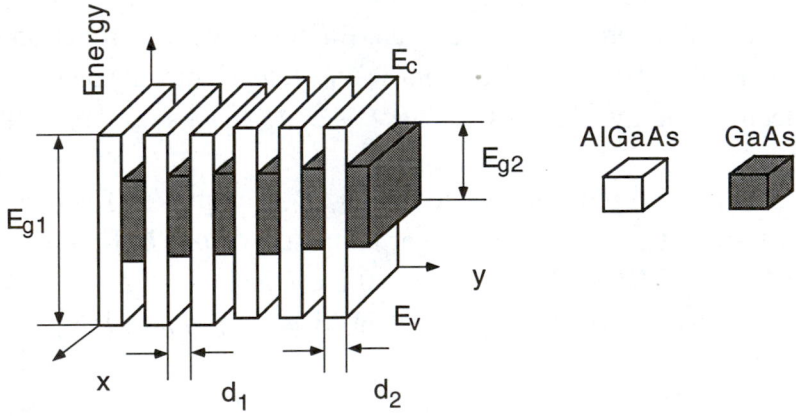

# 3

# Electrons and Holes in Semiconductors[1]

## 3-1. INTRODUCTION

To understand and describe the operation of semiconductor devices, we need to know how the carriers of the electric current – electrons and holes – are distributed in a semiconductor material in energy and space, how their concentrations can be changed, how electrons and holes move in the applied electric field, and what basic equations describe such a motion. All these issues are addressed in this chapter.

## 3-2. ELECTRONS AND HOLES

To describe the behavior of electrons in the valence band, we have to consider the important concept of a hole. (This concept was briefly introduced in the Introduction to the book). In contrast to the conduction band where most of the states are usually empty, the valence band states are nearly all filled. Electrons in the valence band of a crystal come from outer shells of atoms constituting the crystal. These valence electrons form bonds between the atoms. Due to thermal excitation or for some other reason, an electron may leave a bond and get promoted into the conduction band, becoming free to roam around in the crystal. This electron becomes a free electron in the conduction band. The opening created in the bond can be occupied by another electron from an adjacent bond, and so on. Rather than following the motion of electrons jumping from a bond to a vacancy (which creates a new vacancy on the bond), then a motion of another

---

[1]The title of this chapter coincides with the title of one of the most famous books in the semiconductor field: William Shockley, *Electrons and Holes in Semiconductors*, D. van Nostrand Co., Inc., New York (1950). Students who are seriously interested in semiconductor devices may greatly benefit from reading that book.

electron jumping into this new vacancy, and so on, we can follow the motion of the bond vacancy for an electron. This can be visualized as the motion of a vacancy on a bond in crystal lattice that propagates when it is filled by electrons from neighboring bonds (see Fig. 3.2.1). In terms of energy band diagrams, this bond vacancy corresponds to an unoccupied electron state in the valence band.

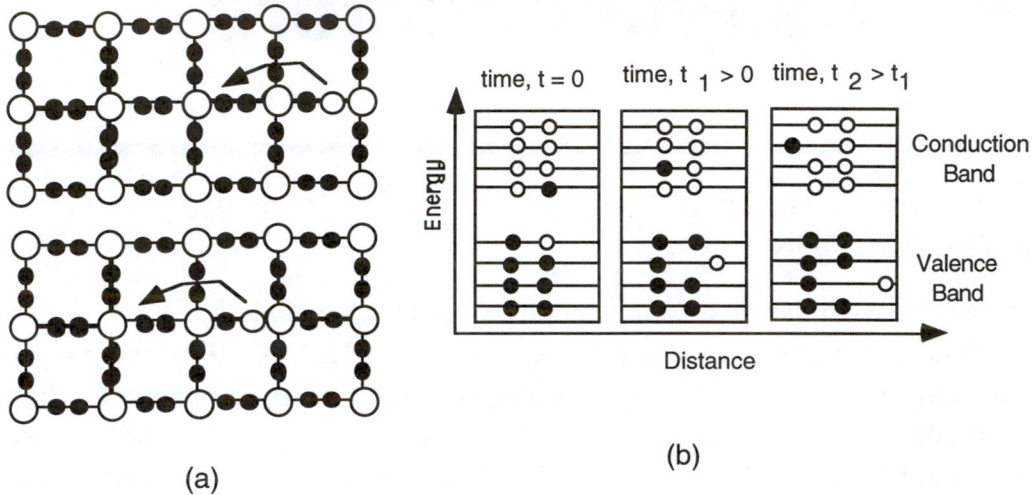

(a)   (b)

**Fig. 3.2.1.**   (a) Propagation of a hole in a crystal lattice and (b) the corresponding band diagram. Large empty circles – atoms, small dark circles – electrons, small open circles – holes.

As can be seen from Fig. 3.2.1, the changes in the electron level occupancy in the valence band are easier to follow by identifying the motion of the vacant energy level. Such a motion can be represented as a fictitious particle (called a **hole**) representing the absence of an electron. The absence of the negative electronic charge, $-q$, corresponds to the positive charge $q = -(-q)$. Hence, the hole has a positive charge with the magnitude equal to the electronic charge. The Fermi-Dirac occupation function for a hole can be calculated as

$$f_p(E) = 1 - f_n(E) = \frac{1}{1 + \exp\left[(E_F - E)/k_B T\right]} \qquad (3\text{-}2\text{-}1)$$

Figure 3.2.2 shows the simplest possible dependence of energy on the wave vector for the conduction and valence bands. In this case, the dependence of the energy in the valence band on the wave vector is given by

$$E_p(k) = E_v + \frac{\hbar^2 k^2}{2m_p} \tag{3-2-2}$$

where $m_p$ is the hole effective mass and $E_v$ is the top of the valence band [see eq. (2-3-18)]. (As we recall from Chapter 2, such a dependence on the wave vector is called a parabolic dependence.) Generally speaking, the hole effective mass is proportional to the curvature of the dependence of the valence band energy, $E_p$, on the wave vector, $k$ [compare with eq. (2-3-15)].

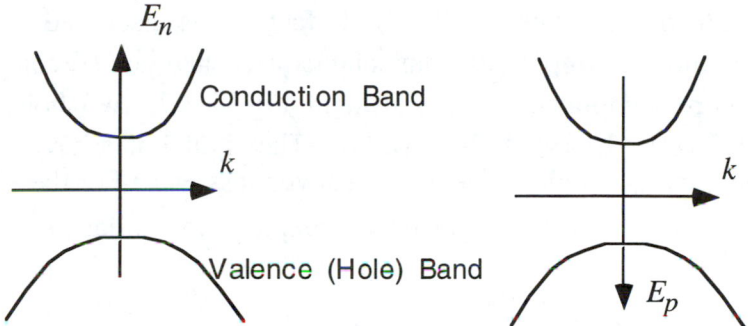

**Fig. 3.2.2.** Simplistic model of band structure. The curvatures (second derivatives of the energy with respect to the wave vector) of the conduction and valence band are proportional to the inverse effective masses for electrons and holes, respectively.

In full analogy with the conduction band electrons (see Section 2.4), we find that the density of allowed energy states per unit volume in a parabolic valence band (the hole density of states) is given by

$$g_p(E) = 4\pi \left(\frac{2m_p}{h^2}\right)^{3/2} (E_v - E)^{1/2} \tag{3-2-3}$$

The hole density (hole concentration) is given by

$$p = \int_{-\infty}^{E_v} g_p(E) f_p(E)\, dE = N_v F_{1/2}(\eta_p) \tag{3-2-4}$$

where

$$N_v = 2 \left( \frac{m_p k_B T}{2\pi\hbar^2} \right)^{3/2} \tag{3-2-5}$$

is the effective density of states for the valence band and

$$\eta_p = \frac{E_v - E_F}{k_B T} \tag{3-2-6}$$

This analogy between electrons and holes relies on the simplest possible model of a band structure (see Fig. 3.2.2). In fact, as we discussed in Section 2.3, the band structure in important semiconductors, such as silicon or gallium arsenide, is more complicated. As shown in Fig. 2.3.7, two hole bands exist (called the **light** and **heavy hole bands**). The light holes have the effective mass $m_{pl}$, and the heavy holes have the effective mass $m_{ph}$. For these hole bands, the effective densities of states is given by an expression similar to eq. (3-2-5):

$$N_{vl} = 2 \left( \frac{m_{pl} k_B T}{2\pi\hbar^2} \right)^{3/2} \qquad N_{vh} = 2 \left( \frac{m_{ph} k_B T}{2\pi\hbar^2} \right)^{3/2} \tag{3-2-7}$$

As a consequence, the total effective hole density of states is given by

$$N_v = N_{vl} + N_{vh} = 2 \left( \frac{m_{pl} k_B T}{2\pi\hbar^2} \right)^{3/2} + 2 \left( \frac{m_{ph} k_B T}{2\pi\hbar^2} \right)^{3/2}$$

$$= 2 \left( \frac{m_{pd} k_B T}{2\pi\hbar^2} \right)^{3/2} \tag{3-2-8}$$

which coincides with eq. (3-2-5) if the hole effective mass, $m_p$, in the equation for the density of states is substituted by the density of states effective mass

$$m_{dp} = \left( m_{ph}^{3/2} + m_{pl}^{3/2} \right)^{2/3} \tag{3-2-9}$$

The concept of a hole is very convenient, since it allows us to apply the results obtained for the conduction band electrons to a $p$-type semiconductor with very few modifications. In particular, all the results obtained in Section 2.5 for two-dimensional and one-dimensional electron gases also apply to two-dimensional and one-dimensional hole gases, with apparent changes in notation.

**Example 3-2-1**

Calculate the hole density of states effective mass, $m_{dp}$, in Si ($m_{ph} = 0.53\ m_e$ and $m_{pl} = 0.16\ m_e$) and compare with the electron density of states effective mass, $m_{dn}$, in Si (the longitudinal and transverse electron effective masses in Si, $m_l = 0.98\ m_e$ and $m_t = 0.189\ m_e$, and there are 6 equivalent minima in the conduction band; see Section 2.3).

**Solution:**

$m_{dp} = (0.53^{3/2} + 0.16^{3/2})^{2/3} m_e = 0.587 m_e$

Using eq. (1-7-28), we find:

$m_{dn} = Z^{2/3}(m_l m_t^2)^{1/3} = 6^{2/3}(0.98 \times 0.189^2)^{1/3}\ m_e = 1.08\ m_e$

Table 3.2.1 summarizes important equations describing electron and hole properties.

| | |
|---|---|
| Electron and hole distribution functions | $f_n(E) = \dfrac{1}{1 + \exp[(E - E_F)/k_B T]}$ <br><br> $f_p(E) = 1 - f_n(E) = \dfrac{1}{1 + \exp[(E_F - E)/k_B T]}$ |
| Electron and hole concentrations | $n = \displaystyle\int_{E_c}^{\infty} g_n(E) f_n(E) dE = N_c F_{1/2}(\eta_n)$ <br><br> $p = \displaystyle\int_{-\infty}^{E_v} g_p(E) f_p(E) dE = N_v F_{1/2}(\eta_p)$ <br><br> where $\eta_n = (E_F - E_c)/k_B T$, $\eta_p = (E_v - E_F)/k_B T$ |
| Electron and hole densities of states | $g_n(E) = 4\pi(2m_n/h^2)^{3/2}(E - E_c)^{1/2}$ <br><br> $g_p(E) = 4\pi\left(\dfrac{2m_p}{h^2}\right)^{3/2}(E_v - E)^{1/2}$ |
| Electron and hole effective densities of states | $N_c = 2\left(\dfrac{m_n k_B T}{2\pi\hbar^2}\right)^{3/2}$ $\quad N_v = 2\left(\dfrac{m_p k_B T}{2\pi\hbar^2}\right)^{3/2}$ |
| Density of states effective hole mass | $m_{dp} = \left(m_{ph}^{3/2} + m_{pl}^{3/2}\right)^{2/3}$ |

**Table 3.2.1.** Important equations describing properties of holes.

## 3-3.  ELECTRON AND HOLE CONCENTRATIONS

When the Fermi level, $E_F$, is in the energy gap and $E_c - E_F$, $E_F - E_v$ are larger than several thermal energies, $k_B T$, the semiconductor is said to be **nondegenerate**. When $E_c - E_F > 3k_B T$, the Fermi integral, $F_{1/2}\left(\dfrac{E_F - E_c}{k_B T}\right)$, reduces to an exponential function

$$F_{1/2} \approx \exp\left(\frac{E_F - E_c}{k_B T}\right) \tag{3-3-1}$$

[see eq. (2-4-24)] and we obtain from eq. (2-4-20)

$$n = N_c \; \exp\left(\frac{E_F - E_c}{k_B T}\right) \tag{3-3-2}$$

A similar expression describes a hole concentration when $E_F - E_v > 3k_B T$:

$$p = N_v \; \exp\left(\frac{E_v - E_F}{k_B T}\right) \tag{3-3-3}$$

As can be seen from eqs. (3-3-2) and (3-3-3), the electron and hole concentrations, $n$ and $p$, in a nondegenerate semiconductor are much smaller than the effective densities of states, $N_c$ and $N_v$, in the conduction and valence bands, respectively. Hence, the number of electrons in the conduction band, for example, is much smaller than the number of available energy states, and the Pauli exclusion principle [which requires that not more than two electrons (with different spins) occupy the same energy state] is not important.

When the Fermi level is in the conduction band and $E_F - E_c > 3k_B T$ or when the Fermi level is in the valence band and $E_v - E_F > 3k_B T$, the semiconductor is called **degenerate.** For an $n$-type degenerate semiconductor

$$F_{1/2} \approx \frac{4}{3\sqrt{\pi}}\left(\frac{E_F - E_c}{k_B T}\right)^{3/2} \tag{3-3-4}$$

[see eq. (2-4-25)] and, using eq. (2-4-20), we find

$$n \approx \frac{1}{3\pi^2}\left[\frac{2m_{dn}\left(E_F - E_c\right)}{\hbar^2}\right]^{3/2} \tag{3-3-5}$$

Similarly, for a $p$-type degenerate semiconductor

$$p \approx \frac{1}{3\pi^2}\left[\frac{2m_{dp}\left(E_v - E_F\right)}{\hbar^2}\right]^{3/2} \tag{3-3-6}$$

Using eqs. (3-3-5) and (3-3-6), we can express the position of the Fermi level in degenerate $n$-type and $p$-type semiconductors through the electron and hole concentrations, respectively

$$E_F - E_c = \frac{\hbar^2}{2m_{dn}}\left(3\pi^2 n\right)^{2/3} \tag{3-3-7}$$

$$E_v - E_F = \frac{\hbar^2}{2m_{dp}}\left(3\pi^2 p\right)^{2/3} \tag{3-3-8}$$

Figure 3.3.1 shows occupied states (dashed regions) in degenerate $n$-type and $p$-type semiconductors.

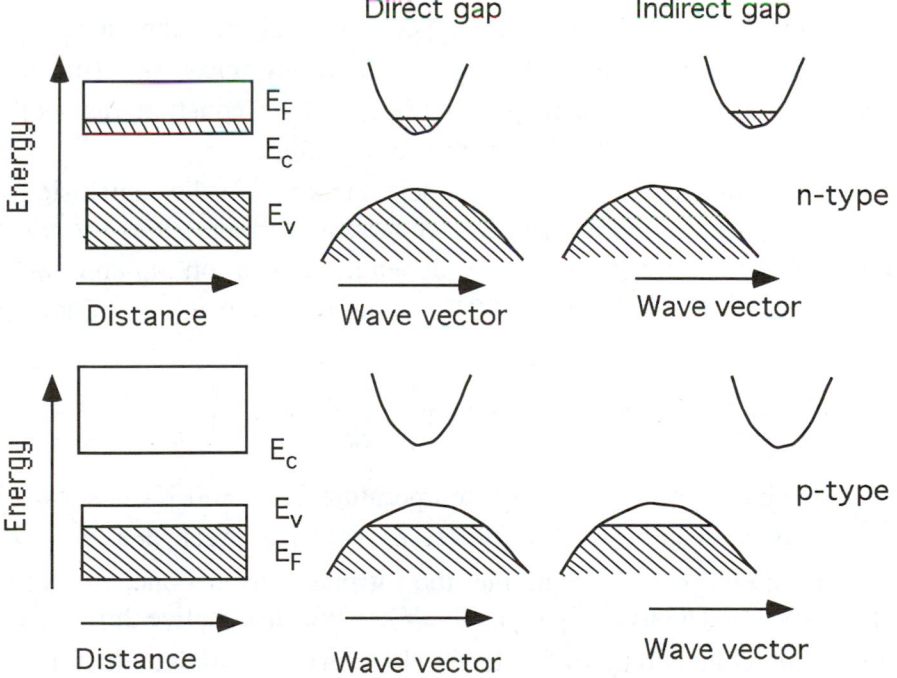

**Fig. 3.3.1.** Occupied states in degenerate semiconductors.

Figure 3.3.2 shows the density of states, the distribution function, and the carrier density per unit energy for nondegenerate and degenerate $n$-type and $p$-type semiconductors.

Multiplying eqs. (3-3-2) and (3-3-3), we obtain for a nondegenerate semiconductor:

$$np = N_c N_v \exp\left(-E_g / k_B T\right) \qquad (3\text{-}3\text{-}9)$$

where $E_g = E_c - E_v$ is the energy gap. This equation can be rewritten as

$$np = n_i^2 \qquad (3\text{-}3\text{-}10)$$

where

$$n_i = \left(N_c N_v\right)^{1/2} \exp\left(-\frac{E_g}{2k_B T}\right) \qquad (3\text{-}3\text{-}11)$$

is called the **intrinsic carrier concentration.**

As can be seen from eqs. (3-3-9), the $np$ product for a nondegenerate semiconductor is independent of the position of the Fermi level and is determined by the densities of states in the valence and conduction bands, the energy gap, and the temperature. Equation (3-3-10) is called the **mass-action law** (the same law as the mass-action law for chemical reactions). This equation (as well as the concept of the Fermi level) is valid at thermal equilibrium.

For a crude estimate of the dependence of the intrinsic carrier concentration, $n_i$, on the energy gap, $E_g$, we assume certain values for $m_n$ and $m_p$, independent of the energy gap, since the dependence of $n_i$ on $m_n$ and $m_p$ is much weaker than that on $E_g$. Taking $m_n = 0.3\ m_e$ and $m_p = 0.6\ m_e$, to be specific, we find

$$n_i = 1.34 \times 10^{21} \times T^{3/2} (\text{K}) \exp\left(-\frac{E_g}{2k_B T}\right) (\text{m}^{-3}) \qquad (3\text{-}3\text{-}12)$$

This dependence of $n_i$ on $E_g$ for room temperature ($T = 300$ K) is shown in Fig. 3.3.3. As can be seen from the figure, $n_i$ varies a great deal for different semiconductors: compare, for example, the intrinsic carrier concentrations for Si ($E_g = 1.11$ eV) and GaAs ($E_g = 1.42$ eV). We also notice how small $n_i$ is compared to the concentration of valence electrons in a semiconductor (which is on the order of $10^{23}$ cm$^{-3}$).

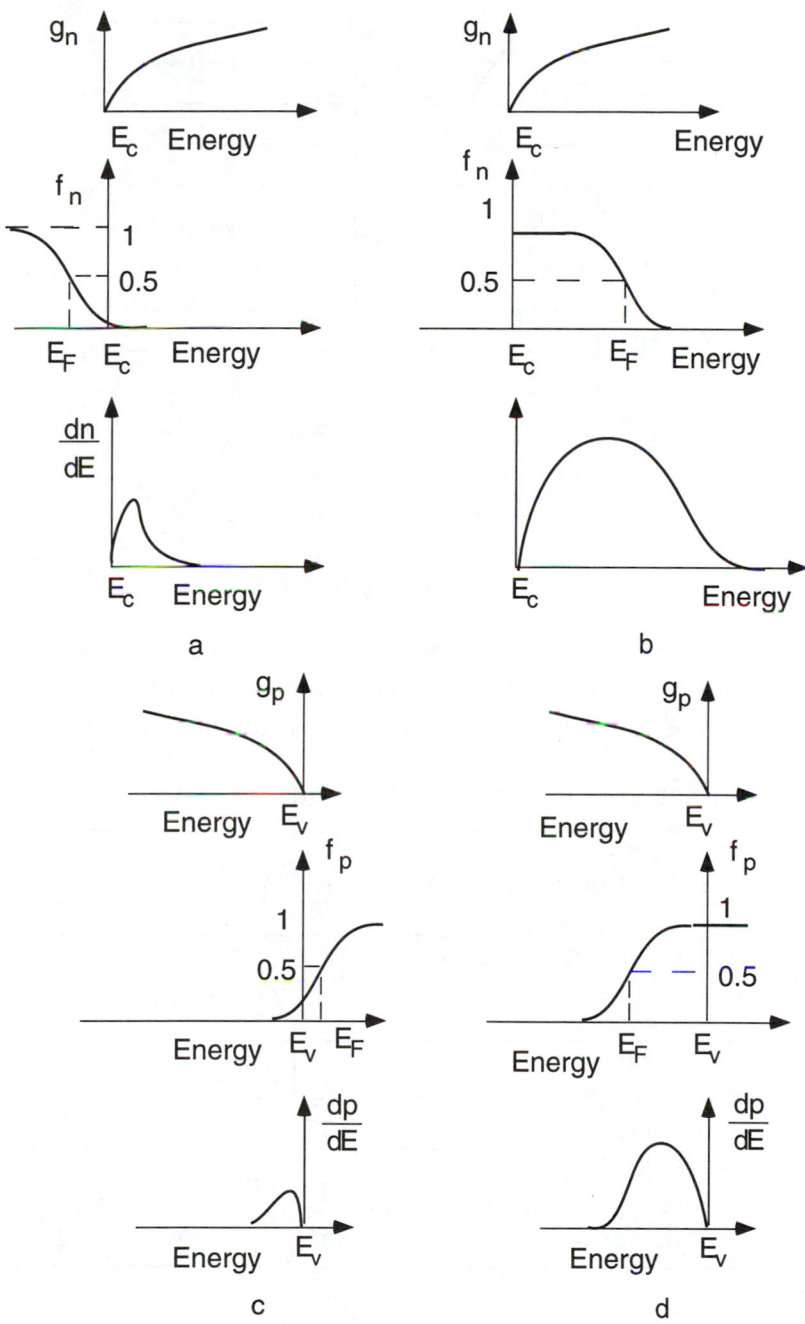

**Fig. 3.3.2.** Density of states, distribution function, and electron density per unit energy for *n*-type (a) nondegenerate and (b) degenerate semiconductor, and for *p*-type (c) non-degenerate and (d) degenerate semiconductor.

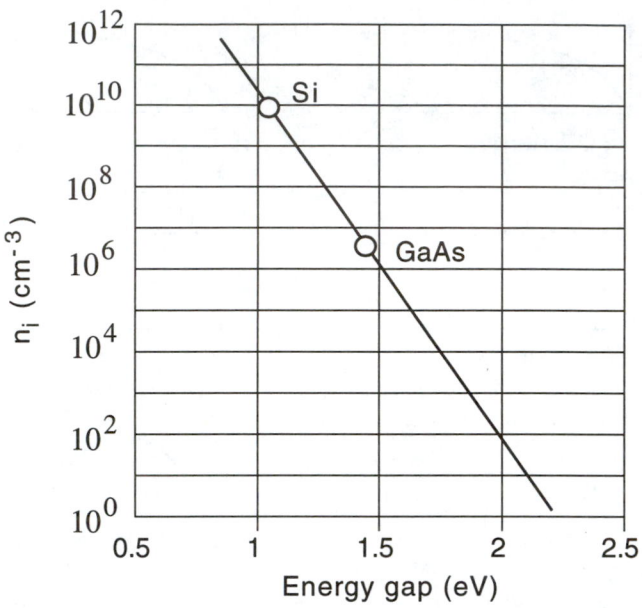

**Fig. 3.3.3.** Intrinsic carrier concentration versus energy gap at room temperature. (The values of $n_s$ for Si and GaAs are approximate since we used the same values of $m_n$ and $m_p$ in the calculation of this dependence.)

Table 3.3.1 summarizes important equations for carrier concentrations.

| | |
|---|---|
| Relation between electron concentration and the Fermi level position for a nondegenerate semiconductor | $n = N_c \exp\left(\dfrac{E_F - E_c}{k_B T}\right)$ |
| Relation between hole concentrations and the Fermi level position for a nondegenerate semiconductor | $p = N_v \exp\left(\dfrac{E_v - E_F}{k_B T}\right)$ |
| Relation between electron concentration and the Fermi level position for a degenerate semiconductor | $n \approx \dfrac{1}{3\pi^2}\left[\dfrac{2m_{dn}(E_F - E_c)}{\hbar^2}\right]^{3/2}$ |
| Relation between hole concentrations and the Fermi level position for a degenerate semiconductor | $p \approx \dfrac{1}{3\pi^2}\left[\dfrac{2m_{dp}(E_v - E_F)}{\hbar^2}\right]^{3/2}$ |
| Mass-action law | $np = n_i^2$ <br> where $n_i = (N_c N_v)^{1/2} \exp\left(-\dfrac{E_g}{2k_B T}\right)$ |

**Table 3.3.1.** Equations relating carrier concentrations to Fermi level.

# 3-4. INTRINSIC, DOPED, AND COMPENSATED SEMICONDUCTORS

### 3.4.1.  Donors and acceptors.

In a pure neutral semiconductor, the concentration of negatively charged electrons must be equal to the concentration of positively charged holes. In this case, the semiconductor is said to be **intrinsic**. Using the condition $n = p$ and eqs. (3-3-10) and (3-3-11), we find the electron and hole concentrations for an intrinsic nondegenerate semiconductor:

$$n = p = n_i = \left(N_c N_v\right)^{1/2} \exp\left(-\frac{E_g}{2k_B T}\right) \tag{3-4-1}$$

The corresponding Fermi level is called the **intrinsic Fermi level** (which is usually denoted as $E_i$). Equating eq. (3-3-2) and eq. (3-3-3), solving the resulting equation for $E_F$, and using eqs. (2-4-21) and (3-2-5) to express $N_c$ and $N_v$ through the effective masses of densities of states, we find:

$$E_i = \frac{E_c + E_v}{2} + \frac{3}{4}k_B T \ln\left(\frac{m_{dp}}{m_{dn}}\right) = E_c - \frac{E_g}{2} + \frac{3}{4}k_B T \ln\left(\frac{m_{dp}}{m_{dn}}\right) \tag{3-4-2}$$

$E_g$ is typically on the order of an electron-volt (1.11 eV for Si), and $k_B T$ expressed in eV is approximately 26 meV at room temperature. Hence, the last term in the right-hand side of eq. (3-4-2) is much smaller than $E_g/2$, and the $E_i$ is located approximately in the middle of the energy gap (see Fig. 3.4.1).

As was discussed in the Introduction to the book, electron and hole concentrations and the position of the Fermi level may all be changed by doping, that is, by introducing into the semiconductor impurities that supply electrons or impurities that supply holes. (These impurities are called **donors** or *n*-type **dopants** and **acceptors** or *p*-type **dopants**, respectively.) Doped semiconductors are called **extrinsic semiconductors**.

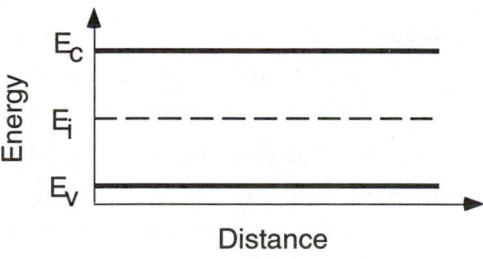

**Fig. 3.4.1.** Intrinsic Fermi level.

As we discussed in Section 1.4, silicon belongs to the fourth column of the Periodic Table (see Appendix A3), and silicon atoms have four valence electrons. An atom belonging to the fifth column of the Periodic Table has five valence electrons. When such an impurity atom belonging to the fifth column of the Periodic Table substitutes for a silicon atom in a silicon crystal, it uses only four out of five valence electrons to form four bonds with its nearest neighbors. The fifth electron is weakly attached to its host atom. Hence, it can easily gain enough energy from a thermal excitation to transfer into the conduction band. In other words, such an impurity is a donor, and **donors introduce energy levels into the energy gap that are close to the bottom of the conduction band** (see the band diagram in Fig. 3.4.2a). Antimony (Sb), phosphorus (P), and arsenic (As) are examples of donors in silicon. **A donor is neutral when occupied by an electron and becomes positively charged after it "donates" its excess electron to the conduction band.**

Impurity atoms belonging to the third column of the Periodic Table [for example, boron (B), aluminum (Al), gallium (Ga), and indium (In)] act in silicon as acceptors, since they lack one valence electron in order to form four bonds with the nearest Si neighbors. Such an impurity atom provides an empty available energy level for an electron, that is, creates a hole. **The energy levels of shallow acceptors are above and close to the top of a valence band** (see Fig. 3.4.2b). **An acceptor is negatively charged when occupied by an electron and becomes neutral after it "accepts" an electron from the valence band.**

Energies $E_c - E_d$ for donors and $E_a - E_v$ for acceptors (where $E_d$ and $E_a$ are **donor** and **acceptor energy levels**) are called donor and acceptor **ionization energies**. Donors and acceptors are called **shallow** when their ionization energies are smaller than or close to a thermal energy $k_B T$. Shallow donors and acceptors are nearly **fully ionized**, that is, nearly all of them supply electrons to the conduction band and holes into the valence band, respectively.

Figure 3.4.3 shows the model that can be used to estimate the **donor ionization energy**, $E_c - E_d$. This model compares a donor in a semiconductor to a hydrogen atom in vacuum. The electron in a hydrogen atom is in vacuum (with the dielectric permittivity $\varepsilon_o$) and has a free electron mass, $m_e$. The semiconductor surrounding a donor atom has the dielectric permittivity $\varepsilon_s$, and electrons in the semiconductor conduction band have an effective electron mass, $m_n$ (if we ignore the anisotropy in the effective mass). Otherwise these two situations are similar, thus allowing us to use the results obtained for the energy levels in a hydrogen atom (see Subsection 1.4.1) to the problem of determining the donor energy levels in the energy gap of a semiconductor.

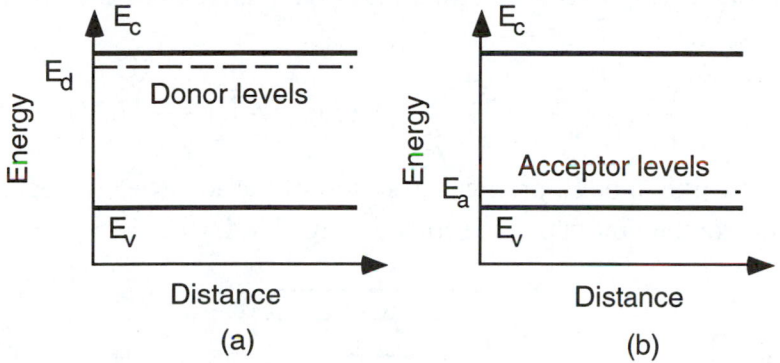

**Fig. 3.4.2.** (a) Donor and (b) acceptor levels.

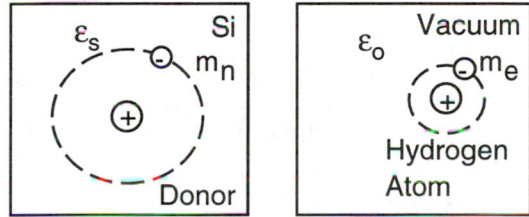

**Fig. 3.4.3.** Comparison between donor and hydrogen atom.

The solution of the Schrödinger equation for a hydrogen atom (see Subsection 1.4.1) shows that the energy levels are given by

$$E_n = -E_B / n^2 \qquad (3\text{-}4\text{-}3)$$

where $n = 1, 2, 3, 4 \ldots$ is called the principal quantum number,

$$E_B = \frac{q^2}{8\pi\varepsilon_o a_B} = \frac{m_e q^4}{32\pi^2 \varepsilon_o^2 \hbar^2}$$

is called the **Bohr energy** ($E_B = 13.6$ eV $= 2.18\times10^{-18}$ J), and

$$a_B = \frac{4\pi\varepsilon_o \hbar^2}{m_e q^2}$$

is called the **Bohr radius** ($a_B = 0.52917$ Å).    (The Bohr radius could be interpreted as a characteristic radius of the electron orbit in a hydrogen atom.)

The energy of the ground state is determined by the Bohr energy where the permittivity of free space is replaced with the dielectric permittivity of the semiconductor, and the free electron mass, $m_e$, is replaced with the electron effective mass, $m_n$:

$$E_c - E_d \approx \left(\frac{\varepsilon_o}{\varepsilon_s}\right)^2 \left(\frac{m_n}{m_e}\right)\frac{q^4 m_e}{32\pi^2\varepsilon_o^2\hbar^2} \approx 13.6\left(\frac{\varepsilon_o}{\varepsilon_s}\right)^2 \left(\frac{m_n}{m_e}\right) (eV) \qquad (3\text{-}4\text{-}4)$$

This hydrogen-like model predicts a series of donor energy levels near the bottom of the conduction band, as shown in Fig. 3.4.4.

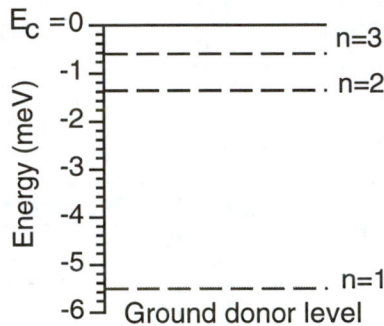

**Fig. 3.4.4.** Hydrogenlike donor levels in GaAs ($m_n = 0.067\ m_e$), $\varepsilon_s =$ 12.9 $\varepsilon_o$. Zero energy corresponds to the bottom of the conduction band.

Compare with Fig. 3.4.2a, where only ground donor level is shown.

The characteristic radius of the electron orbit is given by the Bohr radius where, once again, the permittivity of free space is replaced with the dielectric permittivity of the semiconductor, and the free electron mass, $m_e$, is replaced with the effective mass, $m_n$:

$$a_{Beff} = \frac{4\pi\varepsilon_o\hbar^2}{q^2 m_e}\left(\frac{m_e}{m_n}\right)\left(\frac{\varepsilon_s}{\varepsilon_o}\right) \qquad (3\text{-}4\text{-}5)$$

or

$$a_{Beff}(\text{Å}) = 0.52917\left(\frac{m_e}{m_n}\right)\left(\frac{\varepsilon_s}{\varepsilon_o}\right) \qquad (3\text{-}4\text{-}6)$$

This model is meaningful only if the effective Bohr radius of this electron, $a_{Beff}$, is large compared to the lattice constant and $E_c - E_d << E_g$. (In GaAs, $a_{Beff} \approx$ 102 Å, which is approximately 18 lattice constants.)

For a concentration $N_d$ of donor atoms, the concentration of **ionized** (or **empty**) **donors**, $N_d^+$, is given by

$$N_d^+ = \frac{N_d}{1 + g_d \exp\left[(E_F - E_d)/(k_B T)\right]} \tag{3-4-7}$$

where $E_d$ is the donor energy level and $g_d$ is called the donor **degeneracy factor**. The ratio of the concentrations of the empty ($N_d^+$) and **filled** ($N_d^o = N_d - N_d^+$) donors is found from

$$\frac{N_d^o}{N_d^+} = g_d \exp\left(\frac{E_F - E_d}{k_B T}\right) \tag{3-4-8}$$

The theory that allows us to estimate the degeneracy factor is beyond the scope of this book. In the simplest possible case, this theory predicts $g_d = 2$.

Similarly, we have for acceptor levels

$$\frac{N_a^-}{N_a^o} = \frac{1}{g_a} \exp\left(\frac{E_F - E_a}{k_B T}\right) \tag{3-4-9}$$

Here $E_a$ is the acceptor energy level and (in the simplest case) the acceptor degeneracy factor $g_a = 4$. For a shallow hydrogenlike acceptor level, we have

$$E_a - E_v \approx \left(\frac{\varepsilon_o}{\varepsilon_s}\right)^2 \left(\frac{m_p}{m_e}\right) \frac{q^4 m_e}{32\pi^2 \varepsilon_o^2 \hbar^2} \tag{3-4-10}$$

(More reliable values of $E_a$ and $g_a$ should be found from the experimental data.)

### 3.4.2.    Electron and hole concentrations.

A uniform semiconductor sample in zero electric field must be neutral (have a zero electric charge density) since electric charges create an electric field. Hence, the position of the Fermi level, $E_F$, can be found from the condition of neutrality, that is,

$$p + N_d^+ - N_a^- - n = 0 \qquad (3\text{-}4\text{-}11)$$

where $N_d^+$ and $N_a^-$ are the concentrations of ionized donors and acceptors, respectively. When a semiconductor is doped only by donors so that $N_a^- = 0$, we find from eq. (3-4-11)

$$n = p + N_d^+ \qquad (3\text{-}4\text{-}12)$$

Solving this equation together with eq. (3-3-10) (which is called the mass-action law), we find the concentration of electrons, $n_n$, in an $n$-type semiconductor (here we added the subscript $n$ to indicate the conductivity type):

$$n_n = \frac{1}{2}\left( \sqrt{N_d^{+2} + 4n_i^2} + N_d^+ \right) \qquad (3\text{-}4\text{-}13)$$

For shallow ionized donors $N_d^+ \approx N_d$ and eq. (3-4-13) may be simplified:

$$n_n \approx \frac{1}{2}\left( \sqrt{N_d^2 + 4n_i^2} + N_d \right) \qquad (3\text{-}4\text{-}14)$$

The concentration of holes in the $n$-type material is found from the mass-action law [see eq.(3-3-10)]:

$$p_n = n_i^2/n_n \qquad (3\text{-}4\text{-}15)$$

In most cases, in an $n$-type semiconductor, $N_d^+ \approx N_d \gg n_i$, and these equations are reduced to:

$$n_n \approx N_d \qquad (3\text{-}4\text{-}16)$$

$$p_n \approx n_i^2/N_d \qquad (3\text{-}4\text{-}17)$$

$$E_F \approx E_c - k_B T \ln\left( \frac{N_c}{N_d} \right) \qquad (3\text{-}4\text{-}18)$$

For a $p$-type semiconductor, the charge neutrality condition is given by

$$n + N_a^- = p \qquad (3\text{-}4\text{-}19)$$

and equations for the hole concentration are similar to those derived above for the electron concentration in the $n$-type material:

$$p_p = \frac{1}{2}\left(\sqrt{N_a^{-2} + 4n_i^2} + N_a^-\right) \tag{3-4-20}$$

where we added the subscript $p$ to indicate the conductivity type. For shallow acceptors, $N_a^- \approx N_a$, and

$$p_p = \frac{1}{2}\left(\sqrt{N_a^2 + 4n_i^2} + N_a\right) \tag{3-4-21}$$

When $N_a^- \approx N_a \gg n_i$ and

$$p_p \approx N_a \tag{3-4-22}$$

$$n_p \approx n_i^2/N_a \tag{3-4-23}$$

$$E_F \approx E_v + k_B T \ln\left(\frac{N_v}{N_a}\right) \tag{3-4-24}$$

[Equations (3-4-17) to (3-4-18) and (3-4-23) to (3-4-24) are valid only for non-degenerate semiconductors.]

Fig. 3.4.5 shows the position of the Fermi level for different concentrations of shallow impurities as a function of temperature. At high temperatures, the Fermi level approaches the intrinsic Fermi level. At low temperatures, the Fermi level in an $n$-type semiconductor is closer to the bottom of the conduction band, and the Fermi level in a $p$-type semiconductor is closer to the top of the valence band. (Please also notice the dependence of the energy gap $E_g = E_c - E_v$ on temperature shown in Fig. 3.4.5.)

The temperature dependence of the electron concentration in Si is shown in Fig. 3.4.6. In an intrinsic material, $n = n_i$, which is proportional to $\exp(-E_g/2k_BT)$; see eq. (3-4-1). At low temperatures, not all donors are fully ionized, and the electron concentration is proportional to $\exp\left(-\dfrac{E_c - E_d}{k_B T}\right)$. At intermediate temperatures (in particular at room temperature, $T = 300$ K), the electron concentration is nearly independent of temperature.

Practically all semiconductors, even those of extreme purity, contain many different impurities. It is virtually impossible to reach impurity concentrations smaller than $10^{12}$ cm$^{-3}$. (This, of course, corresponds to a tremendously pure material, since there is only one impurity atom per $10^{11}$ or so host atoms!)

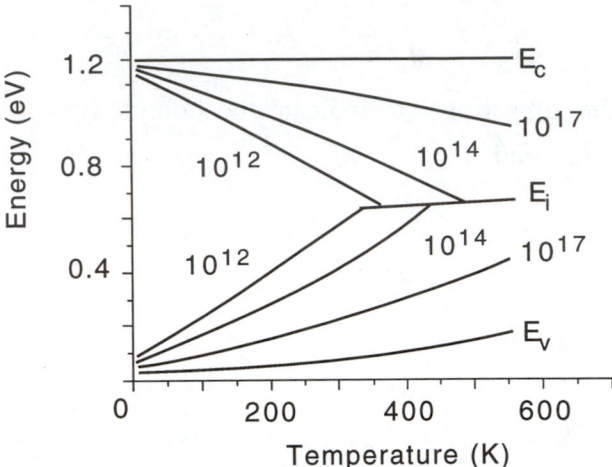

**Fig. 3.4.5.** Fermi level versus temperature for silicon doped with shallow donors (phosphorus) and shallow acceptors (boron). Donor and acceptor ionization energies and $g$-factors are 4.5 meV, $g_d = 2$ and 4.5 meV, $g_a = 4$. Numbers near the curves in the upper and lower half of the energy gap are the values of $N_d$ and $N_a$ in cm$^{-3}$, respectively. (From Shur, 1990, copyright © Prentice Hall, 1990, reproduced by permission of Prentice Hall, Inc., Englewood Cliffs, NJ.)

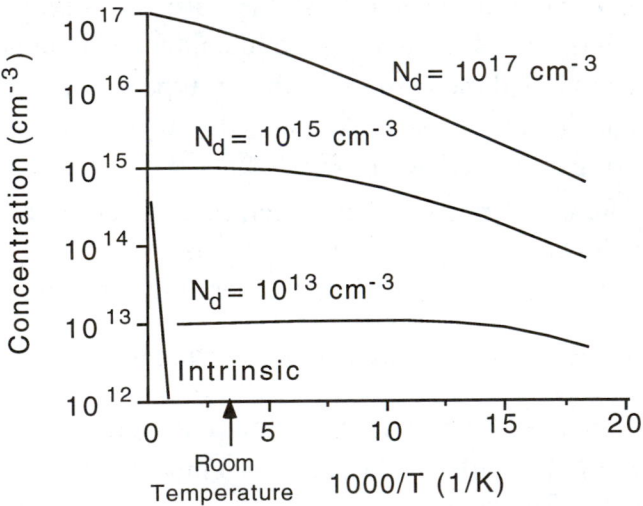

**Fig. 3.4.6.** Electron concentration in $n$-type silicon versus temperature for different doping levels. Donor activation energy 4.5 meV, $g_d = 2$. (From Shur, 1990, copyright © Prentice Hall, 1990, reproduced by permission of Prentice Hall, Inc., Englewood Cliffs, NJ.)

Often both donors and acceptors are present in a semiconductor; then the semiconductor is said to be **compensated**. In a compensated semiconductor, the **conductivity type** ($n$ or $p$) is determined by the larger concentration of ionized impurities. If $N_d > N_a$, the effective donor density becomes

$$N_{deff} = N_d - N_a \qquad (3\text{-}4\text{-}25)$$

In the opposite case, when $N_a > N_d$, we have

$$N_{aeff} = N_a - N_d \qquad (3\text{-}4\text{-}26)$$

(see Fig. 3.4.7). If the impurities are fully ionized, we can still use eqs. (3-4-14) and (3-4-21) for the carrier concentrations, $n$ and $p$, if we substitute $N_d$ with $N_{deff}$ and $N_a$ with $N_{aeff}$.

The value of

$$K = \begin{cases} N_a / N_d & \text{for } N_a < N_d \\ N_d / N_a & \text{for } N_d < N_a \end{cases} \qquad (3\text{-}4\text{-}27)$$

is called the compensation ratio. As was mentioned above, practically all semiconductor samples are compensated to some extent, but in most cases the term "a compensated semiconductor" is used when the value of $K$ is not negligibly small (let us say $K \approx 0.05$ or larger).

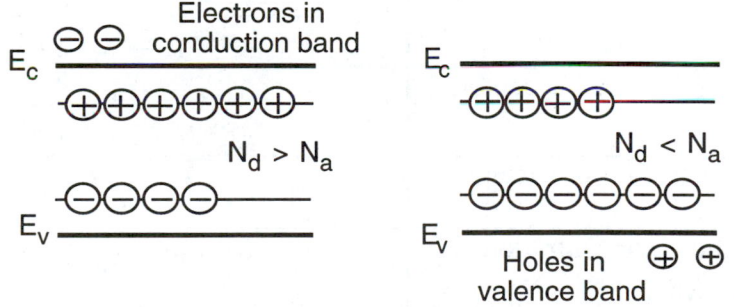

**Fig. 3.4.7.** Schematic band diagrams for compensated semiconductor for $N_d > N_a$, and $N_a > N_d$.

**Example 3-4-1.**

Silicon is doped with phosphorus atoms (concentration $10^{16}$ cm$^{-3}$) and boron atoms (concentration $5 \times 10^{16}$ cm$^{-3}$). Estimate the concentration of electrons and holes in this material at 600 K.

**Solution:**

In Si, phosphorus and boron atoms act as donors and acceptors, respectively. Hence, the material is compensated (*p*-type). The effective acceptor concentration is $5 \times 10^{16} - 10^{16} = 4 \times 10^{16}$ cm$^{-3}$. Since at 600 K all acceptors are ionized, $p \approx 4 \times 10^{16}$ cm$^{-3}$. The electron concentration is found from the mass action law: $n = n_i^2 / p$. At $T = 300$ K ($k_B T \approx 0.02584$ eV), $n_i = (N_c N_v)^{1/2} \exp(-E_g / 2k_B T) \approx 10^{10}$ cm$^{-3}$. At 600 K,

$$n_i \approx \left(\frac{600}{300}\right)^{3/2} 10^{10} \exp\left(\frac{1.11}{2 \times 0.02584}\right) \exp\left(-\frac{1.11}{2 \times 0.05168}\right) \approx 1.3 \times 10^{15} \left(\text{cm}^{-3}\right). \text{ Hence,}$$

$n = (1.3 \times 10^{15})^2 / 4 \times 10^{16} \approx 4.23 \times 10^{13}$ (cm$^{-3}$).

Table 3.4.1 lists the important equations related to doping.

| | |
|---|---|
| Intrinsic Fermi level | $E_i = (E_c + E_v)/2 + 3k_B T \ln\left(m_{dp}/m_{dn}\right)/4$ |
| Donor levels predicted by hydrogen-like model | $E_c - E_d \approx \left(\dfrac{\varepsilon_o}{\varepsilon_s}\right)^2 \left(\dfrac{m_n}{m_e}\right) \dfrac{q^4 m_e}{32\pi^2 \varepsilon_o^2 \hbar^2}$ <br><br> $E_c - E_d \text{ (eV)} \approx 13.6 \left(\varepsilon_o/\varepsilon_s\right)^2 (m_n/m_e)$ |
| Donor Bohr radius | $a_{Beff} = \dfrac{4\pi\varepsilon_o \hbar^2}{q^2 m_e}\left(\dfrac{m_e}{m_n}\right)\left(\dfrac{\varepsilon_s}{\varepsilon_o}\right)$ <br><br> $a_{Beff}\left(\text{Å}\right) = 0.52917(m_e/m_n)(\varepsilon_s/\varepsilon_o)$ |
| Concentration of ionized donors | EMBED "Equation" \* mergeformat <br><br> $N_d^+ = \dfrac{N_d}{1 + g_d \exp\left[(E_F - E_d)/(k_B T)\right]}$ |
| Concentration of ionized acceptors | $N_a^- = \dfrac{N_a}{1 + g_a \exp\left[(E_a - E_F)/(k_B T)\right]}$ |
| Electron concentration in *n*–type sample | $n_n = \dfrac{1}{2}\left(\sqrt{N_d^{+2} + 4n_i^2} + N_d^+\right)$ |
| Electron and hole concentrations in *n*-type sample when $N_d^+ \approx N_d$ | $n_n \approx N_d, \quad p_n \approx n_i^2/N_d \text{ for } N_d \gg n_i$ |
| Hole concentration in *p*-type sample | $p_p = \dfrac{1}{2}\left(\sqrt{\left(N_a^-\right)^2 + 4n_i^2} + N_a^-\right)$ |
| Hole and electron concentrations in *p*-type sample when $N_a^- \approx N_a$, | $p_p \approx N_a, \quad n_p \approx n_i^2/N_a \text{ for } N_a \gg n_i$ |

**Table 3.4.1.** Equations related to intrinsic, doped, and compensated semiconductors.

## 3-5. ELECTRON AND HOLE MOBILITIES AND DRIFT VELOCITIES

Electrons in a conduction band and holes in a valence band are able to move in a semiconductor. Since they carry a unit of elementary charge ($q = 1.602 \times 10^{-19}$ C) each, they are often called **charge carriers** or **free carriers**. Even in the absence of an electric field, charge carriers experience a chaotic random thermal motion.

The average kinetic energy of thermal motion per one electron is $3k_BT/2$, where $T$ is temperature in degrees Kelvin and $k_B$ is the Boltzmann constant. The electron **thermal velocity**, $v_{thn}$, is found by equating the electron kinetic energy to $3k_BT/2$:

$$\frac{m_n v_{thn}^2}{2} = \frac{3k_BT}{2} \tag{3-5-1}$$

where $m_n$ is the electron effective mass. (A similar equation, where $v_{thn}$ is replaced with the hole thermal velocity, $v_{thp}$, and $m_n$ is replaced with the hole effective mass, $m_p$, applies to holes.) From eq. (3-5-1), we obtain

$$v_{thn} = \left(\frac{3k_BT}{m_n}\right)^{1/2} \tag{3-5-2}$$

Thermal velocities $v_{thn}$ and $v_{thp}$ represent average magnitudes of carrier velocities caused by the thermal motion. Since the directions of this thermal motion are random, the same number of carriers crosses any device cross section in any direction so that the electric current is equal to zero. This situation changes when we apply an electric field.

The electron **drift velocity**, $v_n$, caused by an applied electric field, is superimposed on this chaotic thermal motion. At room temperature, the electron velocity due to the thermal motion is usually greater than or at least comparable to the drift velocity. Therefore, an exact description of the electronic motion in a semiconductor has to account for the randomness of the electron velocity. However, an approximate description of the drift velocity can be obtained from Newton's second law of motion for an electron moving in an electric field **F**. A free electron in space is accelerated by the electric field as follows:

$$m_e \frac{d\mathbf{v}_n}{dt} = -q\mathbf{F} \qquad (3\text{-}5\text{-}3)$$

In a semiconductor, the free electron mass has to be replaced by the effective mass, $m_n$:

$$m_n \frac{d\mathbf{v}_n}{dt} = -q\mathbf{F} \qquad (3\text{-}5\text{-}4)$$

The velocity, $\mathbf{v}_n$, in this equation represents the velocity caused by the electric field, that is, the drift velocity. Equation (3-5-4) represents a highly idealized case. In fact, such accelerated electron motion in a semiconductor is impeded by electron collisions caused by the vibrations of atoms near their equilibrium positions, impurities (especially by charged impurities such as ionized donors or acceptors), and crystal imperfections. To the first order, these collisions can be described by adding to the equation of motion a term that is proportional to the electron drift velocity:

$$m_n \frac{d\mathbf{v}_n}{dt} = -q\mathbf{F} - m_n \frac{\mathbf{v}_n}{\tau_{np}} \qquad (3\text{-}5\text{-}5)$$

The second term in the right-hand side of eq. (3-5-5) limits the electron drift velocity (and, hence, the electron momentum, $m_n\mathbf{v}_n$). Therefore, the time constant, $\tau_{np}$, is called the momentum relaxation time. In low electric fields, $\tau_{np}$ is independent of the electric field. Generally speaking, this time becomes a function of the electron energy when the electric field is sufficiently high and electrons acquire energy from the field.

A more accurate representation of the electron motion would be to consider electrons accelerating in an electric field with this acceleration interrupted by a random scattering process caused by impurities or lattice vibrations. In this picture, $\tau_{np}$ can be related to an average time between random collisions interrupting these free electron flights. The average distance that an electron travels between two collisions is called the **mean free path**. In relatively weak electric fields when the electron drift velocity is much smaller than the thermal velocity, the mean free path is given by

$$\lambda_n = v_{thn}\tau_{np} \qquad (3\text{-}5\text{-}6)$$

Usually, the momentum relaxation time, $\tau_{np}$, is between $10^{-12}$ to $10^{-14}$ s. The electron mean free path in GaAs at room temperature is around 1,500 Å.

At low frequencies, $\omega \ll 1/\tau_{np}$, the left-hand side of eq. (3-5-5) is small compared to either term on the right-hand side of this equation so that

$$q\mathbf{F} \approx -m_n \frac{\mathbf{v}_n}{\tau_{np}} \tag{3-5-7}$$

and, hence,

$$\mathbf{v}_n = -\frac{q\tau_{np}}{m_n}\mathbf{F} = -\mu_n\mathbf{F} \tag{3-5-8}$$

where

$$\mu_n = \frac{q\tau_{np}}{m_n} \tag{3-5-9}$$

is called the electron **low field mobility** or just **mobility**. [The negative sign in eq. (3-5-8) corresponds electronic motion in the direction opposite to the direction of the electric field.]

### Example 3-5-1.

The electron mobility in GaAs at temperatures $T = 77$ K and $T = 300$ K is equal to 300,000 cm$^2$/Vs and 9,000 cm$^2$/Vs, respectively. The electron effective mass is equal to 0.067 $m_e$ where $m_e$ is the free electron mass. Find the electron mean free path at these temperatures.

### Solution:

The momentum relaxation time is given by

$$\tau_{np} = \frac{m_n\mu_n}{q} \tag{3-5-10}$$

Hence, the mean free path

$$\lambda_n = v_{thn}\tau_{np} = \left(\frac{3k_BT}{m_n}\right)^{1/2}\frac{m_n\mu_n}{q} = (3k_BTm_n)^{1/2}\frac{\mu_n}{q} \tag{3-5-11}$$

Substituting the parameter values, we find: $\lambda_n(77$ K$) = 2.67$ μm, $\lambda_n(300$ K$) = 1,570$ Å.

The electron drift current density is given by

$$\mathbf{j} = -qn\mathbf{v}_n = q\mu_n n\mathbf{F} \tag{3-5-12}$$

As mentioned above, the electron low field mobility, $\mu_n$, is determined by electron collisions with phonons and impurities. The momentum relaxation time, $\tau_{np}$, can be approximately expressed as

$$\frac{1}{\tau_{np}} = \frac{1}{\tau_{ii}} + \frac{1}{\tau_{ni}} + \frac{1}{\tau_{lattice}} + \dots \tag{3-5-13}$$

where the terms on the right-hand side represent momentum relaxation times due to different scattering processes such as **ionized impurity scattering** ($\tau_{ii}$), **neutral impurity scattering** ($\tau_{ni}$), and **lattice vibration scattering** ($\tau_{lattice}$). (Similar considerations also apply to holes.)

Since in a low electric field, $\tau_{np}$ and $m_n$ are independent of the electric field, the electron drift velocity, $\mathbf{v}_n$, is proportional to the electric field. However, in high electric fields when electrons may gain a considerable energy from the electric field, $\tau_{np}$ and, in certain cases, $m_n$ become strong functions of the electron energy and, hence, of the electric field. In this case, the energy balance equation has to be added to eq. (3-5-5):

$$\frac{dE_n}{dt} = -q\mathbf{F} \cdot \mathbf{v}_n - \frac{E_n - E_o}{\tau_{nE}} \tag{3-5-14}$$

and in eq. (3-5-5) $\tau_{np}$ and $m_n$ become functions of $E_n$. Here, $E_n$ is the electron energy, $E_o = 3k_BT_o/2$ is the electron energy under thermal equilibrium conditions, $T_o$ is the semiconductor temperature, and $\tau_{nE}$ is the effective energy relaxation time. The typical values of $\tau_{nE}$ are of the order of $10^{-11}$ to $10^{-13}$ s. We can also introduce an **electron temperature**, such that $E_n = 3k_BT_n/2$. In a steady state, $dE_n/dt = 0$ and we have

$$E_n = E_o + q\tau_{nE} \, \mathbf{F} \cdot \mathbf{v}_n \text{ or } T_n = T_o + \frac{2}{3}\frac{q}{k_B}\tau_{nE} \, \mathbf{F} \cdot \mathbf{v}_n \tag{3-5-15}$$

In high electric fields, $T_n$ can greatly exceed $T_o$. In this case, the electrons are called **hot electrons**. In modern devices, dimensions are often small, electric fields are high, and electrons are hot.

Hot electrons transfer energy into thermal vibrations of the crystal lattice.

Such vibrations can be modeled as harmonic oscillations with a certain frequency, $\omega_l$. As was discussed in Section 1.2, the energy levels of a harmonic oscillator are equidistant with the energy difference between the levels equal to $E_l = \hbar\omega_l$. Hence, often, the dominant scattering process for a hot electron can be represented as follows. The electron accelerates in the electric field until it gains enough energy to excite lattice vibrations:

$$\frac{m_n v_{n\,\max}^2}{2} = E_n - E_o \approx \hbar\omega_l \qquad (3\text{-}5\text{-}16)$$

where $v_{n\max}$ is the maximum electron drift velocity. Then the scattering process occurs, and the electron loses all the excess energy and all the drift velocity. Hence, the electron drift velocity varies between zero and $v_{n\max}$, and average electron drift velocity ($v_n = v_{n\max}/2$) becomes nearly independent of the electric field:

$$v_n \approx \sqrt{\frac{\hbar\omega_l}{2m_n}} = v_{sn} \qquad (3\text{-}5\text{-}17)$$

($v_{sn}$ is called the **electron saturation velocity**.) Typically, $v_{sn} \approx 10^5$ m/s. Indeed, the measured drift velocity becomes nearly constant in high electric fields (see Fig. 3.5.1).

As can be seen from Fig. 3.5.1, in many semiconductors, such as GaAs, InP, and InGaAs, the electron velocity decreases with an electric field in a certain range of high electric fields. (In this range of the electric fields, the electron differential mobility, $\mu_{dn} = dv_n/dF$, is negative.) In these semiconductors, the central valley of the conduction band ($\Gamma$) is the lowest (see Section 2.3). In high electric fields, hot electrons acquire enough energy to transfer from the central ($\Gamma$) valley of the conduction band (where the effective mass and, hence, the density of states is relatively small) into the satellite valleys (X and L), where electrons have a higher effective mass and, hence, a larger density of states but a smaller drift velocity. This process is schematically illustrated in Fig. 3.5.2.

As can be seen from Fig. 3.5.1, compound semiconductors have a potential for a higher speed operation than silicon because electrons in these materials may move faster. The negative differential mobility observed in many compound semiconductors can be used for generating microwave oscillations at very high frequencies (see Ch. 4).

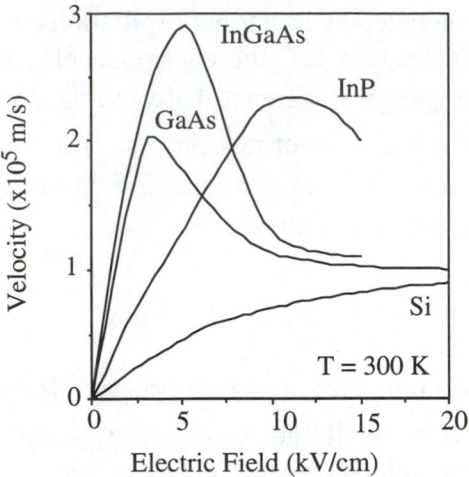

**Fig. 3.5.1.** Velocity versus electric field for several semiconductors.

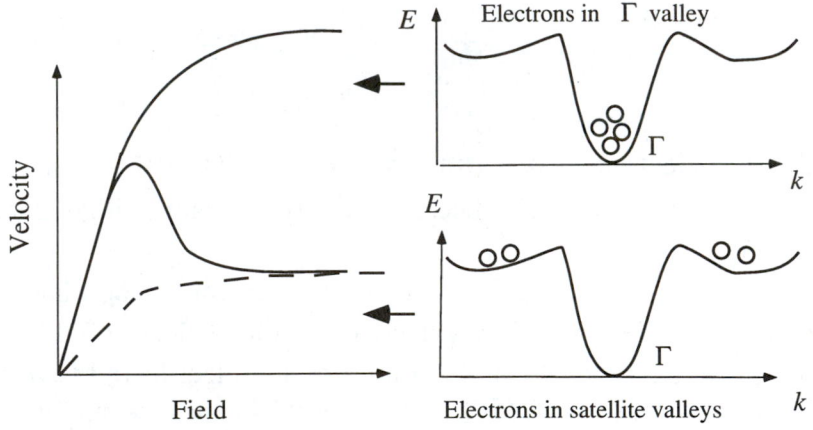

**Fig. 3.5.2.** Explanation of negative differential mobility region. Dashed lines represent the electron velocity in the central ($\Gamma$) valley (the top curve) and in upper ("heavy") valleys (the bottom curve). In a low electric field when electrons are in the central valley, the top curve applies. In high electric fields, electrons are in the upper valleys, and the bottom curve applies. The solid line represents $v(F)$ with negative differential mobility.

Figure 3.5.1 implies that the electron drift velocity depends on the electric field alone. However, this is true only for relatively long samples and small electric fields when the transit times $t = L/v$ and $t_{th} = L/v_{th}$ (where $L$ is the sample length) are much greater than the momentum and energy relaxation times. Modern semiconductor devices are often so small that their dimensions are comparable to or even smaller than the electron mean free path. In this case, electrons may acquire higher velocities than those shown in Fig. 3.5.1, since transient effects associated with acceleration of carriers become important. In the limiting case of very short devices, the electron transit time may become so small that most electrons will not experience any collisions during the transit. Such a mode of electron transport is called **ballistic transport**.

Figure 3.5.3 compares the dependencies of the electron and hole drift velocities in Si on the electric field at room temperature. (Since the valence bands are more similar in different semiconductors than conduction bands, the hole drift velocities are not as dramatically different for different semiconductors as electron velocities.) As can be seen from Fig. 3.5.3, the hole mobility is considerably smaller than the electron mobility in Si. (This is generally true for other semiconductor materials as well.)

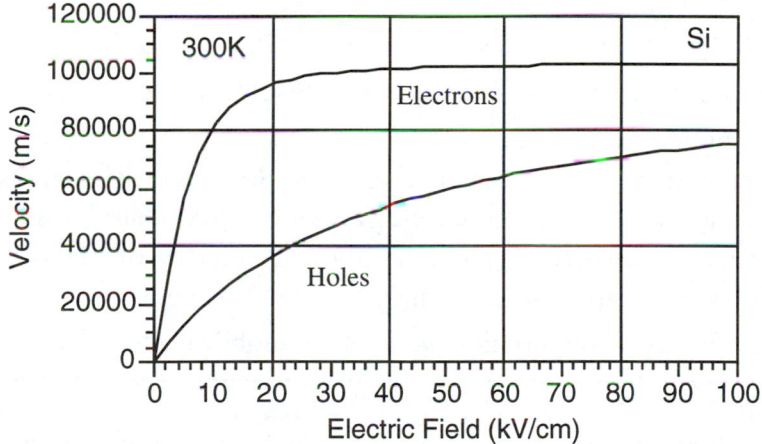

**Fig. 3.5.3.** Electron and hole drift velocities in Si at room temperature.

The electron and hole drift velocities in high electric fields depend on temperature and on the total concentration of charged impurities, $N_T = N_a + N_d$. These dependencies are often conveniently described by empirical analytical expressions. For reference purposes, we give here approximate equations describing the electron and hole velocities, $v_n(F)$ and $v_p(F)$, in bulk silicon:

$$v_n = \frac{\mu_n F}{\sqrt{1+\left(\mu_n F / v_s\right)^2}} \qquad v_p = \frac{\mu_p F}{1+\mu_p F / v_s} \qquad (3\text{-}5\text{-}18)$$

where the electron and hole mobilities

$$\mu_n = \mu_{mn} + \frac{\mu_{on}}{1+(N_T/N_{cn})^\nu} \qquad \mu_p = \mu_{mp} + \frac{\mu_{op}}{1+(N_T/N_{cp})^\nu} \qquad (3\text{-}5\text{-}19)$$

$$\mu_{mn}=88\left(\frac{T}{300}\right)^{-0.57}\left(\frac{cm^2}{Vs}\right) \qquad \mu_{mp}=54\left(\frac{T}{300}\right)^{-0.57}\left(\frac{cm^2}{Vs}\right) \qquad (3\text{-}5\text{-}20)$$

$$\mu_{on}=1250\left(\frac{T}{300}\right)^{-2.33}\left(\frac{cm^2}{Vs}\right) \qquad \mu_{op}=407\left(\frac{T}{300}\right)^{-2.33}\left(\frac{cm^2}{Vs}\right) \qquad (3\text{-}5\text{-}21)$$

$$N_{cn}=1.26\times10^{17}\left(\frac{T}{300}\right)^{2.4}\left(cm^{-3}\right) \qquad N_{cp}=2.35\times10^{17}\left(\frac{T}{300}\right)^{2.4}\left(cm^{-3}\right) \qquad (3\text{-}5\text{-}22)$$

Here $\nu = 0.88\,(T/300)^{-0.146}$ and the saturation velocity, $v_s$, is given by

$$v_s=\frac{2.4\times10^7}{1+0.8\exp(T/600)} \quad (cm/s) \qquad (3\text{-}5\text{-}23)$$

that is, the exponent $\nu$ and the saturation velocity, $v_s$, are the same for holes and electrons in silicon.

The expressions for the hole and electron drift velocities given above are obtained for majority carriers (i. e., electrons in $n$-type material and holes in $p$-type material) in bulk silicon. The mobility and velocity of electrons in $p$-type material and holes in $n$-type material may be quite different.

Table 3.5.1 gives important equations for mobilities and drift velocities.

| | |
|---|---|
| Thermal velocities | $v_{thn}=(3k_BT/m_n)^{1/2} \qquad v_{thp}=(3k_BT/m_p)^{1/2}$ |
| Electron and hole mobilities | $\mu_n=q\tau_{np}/m_n \qquad \mu_p=q\tau_{pp}/m_p$ |
| Total electron momentum relaxation time | $\dfrac{1}{\tau_{np}}=\dfrac{1}{\tau_{ii}}+\dfrac{1}{\tau_{ni}}+\dfrac{1}{\tau_{lattice}}+\ldots$ |
| Electron temperature | $T_n = T_o + \dfrac{2}{3}\dfrac{q}{k_B}\tau_{nE}\,\mathbf{F}\cdot\mathbf{v}_n$ |
| Electron saturation velocity | $v_{sn}\approx\sqrt{\hbar\omega_l/(2m_n)}$ |
| Electron and hole velocities in Si versus electric field | $v_n = \dfrac{\mu_n F}{\sqrt{1+(\mu_n F/v_s)^2}} \qquad v_p = \dfrac{\mu_p F}{1+\mu_p F/v_s}$ |

**Table 3.5.1.** Equations for mobilities and drift velocities.

## 3-6.  DIFFUSION

The diffusion process is related to random thermal motion.  It may be illustrated using a simple one-dimensional model that assumes that carriers may move in random to the left or right, with equal probability.  Let us consider two adjacent regions of the length $\lambda_x$ each where $\lambda_x = v_{thx}\tau_{np}$ is a mean free path in the direction $x$, that is, the average length traveled by an electron in this direction between two scattering events, and $v_{Tx}$ is the $x$-component of the thermal velocity, $v_T$.  We assume that the left region has more electrons than the right region (see Fig. 3.6.1).  On average, half of the electrons in each region will move to the left and half to the right.  Hence, more electrons move from the left region to the right than vice versa, and a net **diffusion flux** will tend to equalize the electron concentrations in these two regions.

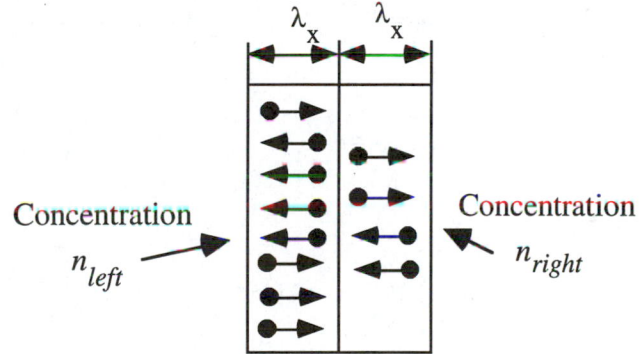

**Fig. 3.6.1**.  Illustration of diffusion process.

The **diffusion current density** (flowing from the left to the right) can be calculated as

$$j_{diff} = -q\frac{N_{left} - N_{right}}{\tau_{np}} = q\frac{\lambda_x\left(n_{right} - n_{left}\right)}{2\tau_{np}} = q\frac{\Delta n\lambda_x}{2\tau_{np}} \qquad (3\text{-}6\text{-}1)$$

where  $\Delta n = n_{right} - n_{left}$, $n_{right}$, $n_{left}$, are the electron concentrations in the left and right region, and  $N_{left} = \lambda_x n_{left}/2$,  $N_{right} = \lambda_x n_{right}/2$  are the numbers of electrons crossing the boundary between the left and right regions in Fig. 3.6.1 in opposite directions (here a factor of 2 accounts for the fact that, on average, half of the electrons move to the right and half of the electrons move to the left.)  If $\lambda_x$ is small compared with a characteristic distance of the variation of the electron concentration, $n(x)$, then

$$\Delta n = 2\frac{dn}{dx}\lambda_x \tag{3-6-2}$$

Hence, we obtain

$$j_{diff} = q\frac{dn}{dx}\frac{\lambda_x^2}{\tau_{np}} = qD_n\frac{dn}{dx} \tag{3-6-3}$$

where the **diffusion coefficient** is

$$D_n = \lambda_x^2 / \tau_{np} = v_{Tx}^2\tau_{np}^2 / \tau_{np} = v_{Tx}^2\tau_{np} \tag{3-6-4}$$

Since

$$\frac{m_n v_{thn}^2}{2} = \frac{m_n v_{thnx}^2}{2} + \frac{m_n v_{thny}^2}{2} + \frac{m_n v_{thnz}^2}{2} = \frac{3k_B T}{2} \tag{3-6-5}$$

$$v_{thnx}^2 = \frac{k_B T}{m_n} \tag{3-6-6}$$

and we finally obtain

$$D_n = v_{thnx}^2\tau_{np} = \frac{k_B T\tau_{np}}{m_n} \tag{3-6-7}$$

The low field mobility, $\mu$, is given by

$$\mu_n = q\tau_{np} / m_n \tag{3-6-8}$$

(see Section 3.5), and eq. (3-6-7) can be rewritten as

$$D_n = \frac{k_B T\mu_n}{q} \tag{3-6-9}$$

This important formula is called the **Einstein relation**. The Einstein relation may be also derived as follows. Let us consider an $n$-type sample with nonuniform carrier concentration in direction $x$ (which may be related to a nonuniform doping) and no electric current [$j = j_{drift} + j_{diff} = 0$, see eqs. (3-5-12) and (3-6-3)] so that

$$0 = n\mu_n F + D_n \partial n / \partial x \tag{3-6-10}$$

The electric field is

$$F = -\frac{\partial V}{\partial x} = \frac{1}{q}\frac{\partial E_c}{\partial x} \tag{3-6-11}$$

where $V = -E_c/q$ is the electric potential, and $E_c$ is the bottom of the conduction band (i. e., the electron potential energy). The carrier concentration is

$$n = N_c \exp\left(\frac{E_F - E_c}{k_B T}\right) \tag{3-6-12}$$

where $N_c$ is the effective density of states and $E_F$ is the Fermi level. Differentiating $n$ with respect to $x$, we obtain

$$\frac{\partial n}{\partial x} = -N_c \exp\left(\frac{E_F - E_c}{k_B T}\right)\frac{1}{k_B T}\frac{\partial E_c}{\partial x} = -\frac{n}{k_B T}\frac{\partial E_c}{\partial x} \tag{3-6-13}$$

Substituting eqs. (3-6-11) to (3-6-13) into eq. (3-6-10), we obtain $D_n = \mu_n k_B T/q$.

**Example 3-6-1.**

The electron and hole mobilities in Si at room temperature $(T = 300$ K$)$ are 1,000 cm²/Vs and 300 cm²/Vs, respectively. Calculate the electron and hole diffusion coefficients.

**Solution:**

Since for $T = 300$ K, $k_B T/q = 0.02584$ eV, $D_n = \mu_n k_B T/q = 0.02584 \times 1000 \approx 25.8$ cm²/s and $D_p = \mu_p k_B T/q = 0.02584 \times 300 \approx 7.75$ cm²/s.

Equations (3-6-11) and (3-6-13) show that under equilibrium conditions (that is, when $j = 0$) a non-uniform doping leads to the electric field

$$F_{bi} = -\frac{k_B T}{qn}\frac{\partial n}{\partial x} \tag{3-6-14}$$

so that the conduction current caused by this electric field exactly compensates for the diffusion current. This field is called a **built-in electric field**. The idea of creating a built-in electric field using a non-uniform doping finds important applications in bipolar junction transistors (see Chap. 5).

**Example 3-6-2.**

The electron concentration in a 1 μm long device varies linearly with distance from $10^{16}$ cm$^{-3}$ to $1.1 \times 10^{17}$ cm$^{-3}$. Plot the built-in electric field as a function of distance at $T = 300$ K.

**Solution:**

For a linear profile, $n = n_o + \Delta n \, x/L$, the built-in electric field is given by

$$F_{bi} = -\frac{k_B T}{qn}\frac{\partial n}{\partial x} = -\frac{1}{\frac{n_o}{\Delta n}+\frac{x}{L}}\frac{k_B T}{qL} = -0.258\frac{1}{0.1+x(\mu m)}\left(\frac{kV}{cm}\right)$$

The negative sign of the electric field means that the direction of the field is in the direction of negative $x$. The plot of the electric field and the doping profile are shown in Fig. 3.6.2.

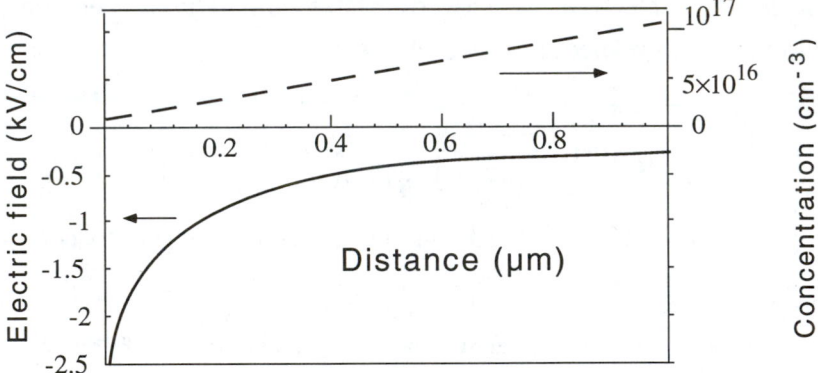

**Fig. 3.6.2.** Linear carrier profile and corresponding built-in electric field distribution.

Table 3.6.1 gives a summary of important equations related to diffusion.

| Relationship between diffusion, mean free path and thermal velocity | $D_n = \lambda_x^2 / \tau_{np} = v_{thnx}^2 \tau_{np} = v_{thn}^2 \tau_{np}/3$ |
| --- | --- |
| Einstein relation | $D_n = k_B T \mu_n / q$ |
| Built-in electric field | $F_{bi} = -(k_B T / qn)\partial n / \partial x$ |

**Table 3.6.1.** Summary of important equations related to diffusion.

## 3-7.  BASIC SEMICONDUCTOR EQUATIONS

### 3.7.1.  Drift-diffusion model.

The bottom of the conduction band and the top of the valence band correspond to potential energies of electrons and holes, respectively.  They are related to the electric potential, $\phi$, as follows:

$$E_c = -q\phi + const$$
$$E_v = -q\phi - E_g + const \qquad\qquad (3\text{-}7\text{-}1)$$

When an electric field is applied to a semiconductor, the bands are "tilted" as shown in Fig. 3.7.1, and electrons and holes move in opposite directions leading to an electric current.  As was discussed in the Introduction to the book, Fig. 3.7.1 represents an example of a band diagram, which shows the dependence of electron and hole potential energies on distance.

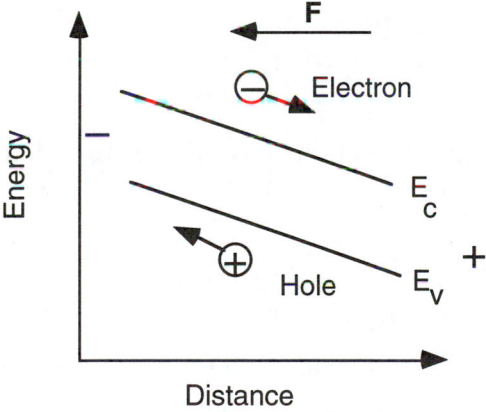

**Fig. 3.7.1.**  Band diagram of a semiconductor in an applied electric field. "+" and "-" signs represent external electric charges creating the electric field.

In a low electric field, the **conductivity current densities** for electrons and holes are given by the sum of the drift current density (proportional to the electric field) and the diffusion current density (proportional to a carrier concentration gradient, $\nabla n$):

$$\mathbf{j}_n = q\left(n\mu_n \mathbf{F} + D_n \nabla n\right) \qquad\qquad (3\text{-}7\text{-}2)$$

$$\mathbf{j}_p = q\left(p\mu_p\mathbf{F} - D_p\nabla p\right) \tag{3-7-3}$$

In low electric fields, the diffusion coefficients are related to the mobilities via the Einstein relations (see Section 3.6).

For a uniform semiconductor ($\nabla n = 0$, $\nabla p = 0$), these equations reduce to

$$\mathbf{j}_n = qn\mu_n\mathbf{F} = \sigma_n\mathbf{F} \tag{3-7-4}$$

$$\mathbf{j}_p = qp\mu_p\mathbf{F} = \sigma_p\mathbf{F} \tag{3-7-5}$$

so that the total conductivity current density is

$$\mathbf{j} = \sigma\mathbf{F} \tag{3-7-6}$$

where $\sigma = \sigma_n + \sigma_p$. Equation (3-7-6) represents **Ohm's law**. The electric current, $I$, is given by

$$I = \frac{\sigma S}{L}FL = \frac{U}{R} \tag{3-7-7}$$

where $S$ is the sample cross section, $L$ is the sample length, $U$ is the applied voltage, and

$$R = \frac{L}{\sigma S} \tag{3-7-8}$$

is the sample resistance (see Problem 3-7-1).

In a high electric field, carriers acquire energy from the electric field and the carrier energy becomes larger than the average thermal energy, $3k_BT/2$, where $T$ is the sample temperature. As a consequence, the electron and hole velocities are no longer proportional to the electric field when the electric field is high (see Figs. 3.5.2 and 3.5.3). The diffusion coefficients also become dependent on the electric field.

Since electron and hole drift velocities and diffusion coefficients depend on the electric field, the following phenomenological equations are frequently used in semiconductor device modeling:

$$\mathbf{j}_n = q\left[-n\mathbf{v}_n(\mathbf{F}) + D_n(\mathbf{F})\nabla n\right] \tag{3-7-9}$$

$$\mathbf{j}_p = q[p\mathbf{v}_p(\mathbf{F}) - D_p(\mathbf{F})\nabla p] \qquad (3\text{-}7\text{-}10)$$

Here, $\mathbf{v}_n(\mathbf{F})$, $\mathbf{v}_p(\mathbf{F})$, $D_n(\mathbf{F})$, and $D_p(\mathbf{F})$ are assumed to be the same functions of electric field as computed or measured for a uniform sample under steady-state conditions. These equations are referred to as **drift-diffusion equations** or the **drift-diffusion model**. In a low electric field, these equations reduce to eqs. (3-7-2) and (3-7-3), but they also allow us to describe electric currents in relatively short devices, where electric fields are high.

The advantage of using eqs. (3-7-9) and (3-7-10), or even similar equations with field-independent diffusion coefficients, is their relative simplicity, which allows one to achieve some insight into the device physics. In fact, these equations are often used in commercial device simulators. However, the drift-diffusion model may lead to considerable errors in describing hot electron behavior or even near equilibrium transport in simple systems with large built-in electric fields such as $p\text{-}n$ junctions. Also, at high frequencies, the velocity and diffusion do not follow instantaneously the variations of the electric field. Therefore, the mobility and diffusion become frequency-dependent. This is not accounted for by the drift-diffusion equations. Therefore, a realistic simulation of modern semiconductor devices often requires more sophisticated approaches, which we briefly discuss in the end of this section.

Equations (3-7-9) and (3-7-10) [or eqs. (3-7-2) and (3-7-3) in a low electric field] allow us to calculate electric current densities provided electric fields and carrier concentrations are known. However, the electric field itself depends on the charges distributed in the semiconductor (called **space charges**). The dependence of the electric field on the space charge density, $\rho$, is described by Poisson's equation

$$\nabla \cdot \mathbf{F} = \frac{\rho}{\varepsilon} \qquad (3\text{-}7\text{-}11)$$

where

$$\rho = q\left(N_d^+ - N_a^- - n + p\right) + q\left(p_t - n_t\right) \qquad (3\text{-}7\text{-}12)$$

$N_d^+$ and $N_a^-$ are the concentrations of ionized (i. e., positively charged) donors and ionized (i. e., negatively charged) acceptors, respectively, and $qp_t$ and $qn_t$ are the charge densities of any other positively and negatively charged impurities that may be present in the semiconductor. Often (but not always, unfortunately) these impurity charges may neglected so that

$$\rho \approx q\left(N_d^+ - N_a^- - n + p\right) \tag{3-7-13}$$

What is now left is to determine $n$ and $p$. This is simple for a uniform semiconductor under thermal equilibrium conditions when $n \approx N_d^+ - N_a^-$ and $p \approx n_i^2 / n$ in an $n$-type sample (i. e., when $N_d > N_a$); see Section 3.4. However, semiconductor devices are rarely uniform and usually operate under nonequilibrium conditions. For example, a photon absorbed in the semiconductor may promote an electron from the valence band into the conduction band, thus creating both an electron in the conduction band and a hole in the valence band (called an **electron-hole pair**). Quantitatively, this process is described by introducing the **generation rate**, $G$, which is equal to the concentration of electron-hole pairs produced in one second. On the other hand, a conduction band electron may fall down into a vacant space in the valence band. In this process, called **recombination**, electron-hole pairs are destroyed. The quantitative measure of recombination is the **recombination rate**, $R$, which is equal to the concentration of electron-hole pairs recombining in one second. In a uniform semiconductor,

$$\frac{dp}{dt} = \frac{dn}{dt} = G - R \tag{3-7-14}$$

so that, in steady-state

$$G = R \tag{3-7-15}$$

As discussed below, the recombination rate may be often represented as

$$R = \frac{p_n - p_{no}}{\tau_{pl}} \tag{3-7-16}$$

where $p_n$ is the concentration of minority carriers (holes) in an $n$-type semiconductor, $p_{no}$ is the equilibrium hole concentration, and $\tau_{pl}$ is the hole **recombination time** or the hole **lifetime**. In this case, the solution of eq. (3-7-14) is given by

$$p_n = p_{no} + G\tau_{pl} \tag{3-7-17}$$

For a nonuniform semiconductor, eq. (3-7-14) has to be modified in order to account for changes in the current density. [It is similar to counting the number of cars in Detroit. In order to establish the rate of change for the number of cars in Detroit, it is not enough to find out how many cars are produced in Detroit (which is a great many) and how many cars are scrapped. We also must account for cars shipped out from Detroit and for cars brought to Detroit from outside.] For an incremental volume, such a change in the current is proportional to $\nabla \cdot \mathbf{j}$, and eq. (3-7-14) should be replaced by

$$\frac{\partial n}{\partial t} = \frac{1}{q}\nabla \cdot \mathbf{j}_n + G - R \tag{3-7-18}$$

$$\frac{\partial p}{\partial t} = -\frac{1}{q}\nabla \cdot \mathbf{j}_p + G - R \tag{3-7-19}$$

These equations are called the **continuity equations**. Once again, we notice that $G$ and $R$ are the same for electrons and holes, since generation and recombination processes produce and destroy electron-hole pairs. The continuity equation (3-7-18) is illustrated by Fig. 3.7.2.

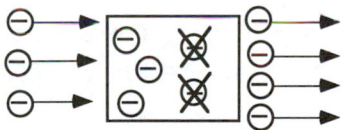

**Fig. 3.7.2.**   Illustration of the continuity equation. If, in a steady state, three electrons enter an incremental volume where three electrons are generated and two electrons recombine, four electrons will leave this incremental volume. (The continuity equation is the equation of balance for the number of particles.)

We also notice that in a nonuniform semiconductor, excess concentrations of electrons or holes may exist even if the generation rate is zero: the additional carriers may come from outside, for example, from contacts made to a semiconductor sample.

Equations (3-7-9) to (3-7-11), (3-7-13), (3-7-18) and (3-7-19) form a complete system of equations of the drift-diffusion model. To understand how they work we have to consider a few examples.

**Example 3-7-1.**

Find the electric field profile in a $p$-type semiconductor with a hole mobility $\mu = 100$ cm$^2$/Vs and the hole concentration $p_o = 10^{15}$ cm$^{-3}$, carrying a current density $j = 1$ A/cm$^2$. The $p$-type region is next to a very highly doped region (with $p_b = 10^{17}$ cm$^{-3}$); see Fig. 3.7.3a. (A highly doped region is called a $p^+$ **region**.) Neglect the diffusion current and assume that the hole drift velocity $v_p = \mu_p F$, where $F$ is the electric field. The semiconductor dielectric permittivity, $\varepsilon_s = 10^{-10}$ F/m.

**Solution:**

The hole concentration must be a continuous function of distance. It varies from $p = 10^{17}$ cm$^{-3}$ at the boundary with the $p^+$ region to $10^{15}$ cm$^{-3}$ far from the boundary. Its variation is described by the continuity equation (3-7-19) where we put $G = 0$, $R = 0$, $\partial p / \partial t = 0$, and $\nabla \cdot \mathbf{j} = dj / dx$. Hence, we obtain

$$d\left(qp\mu_p F\right)/dx=0$$

This is a first-order ordinary differential equation with the solution

$$qp\mu_p F=const=j$$

which is the condition of the continuity of the current density along the sample. This equation has two unknowns: $F$ and $p$. The second equation linking $F$ and $p$ is Poisson's equation. Since in this case, $\rho = q(p - p_o)$, Poisson's equation becomes:

$$\frac{dF}{dx}=\frac{q\left(p-p_o\right)}{\varepsilon_s}$$

Substituting $p=j/\left(q\mu_p F\right)$ into Poisson's equation, we obtain

$$\frac{dF}{dx}=\frac{q}{\varepsilon_s}\left(\frac{j}{q\mu_p F}-p_o\right)$$

Now we have an ordinary differential equation with the boundary condition $F(0) = j/(q\mu_p p_b)$. It is relatively easy to solve this equation numerically using a personal computer. However, we can also obtain an analytical solution. Introducing the dimensionless field $f = F/F_o$ and the dimensionless coordinate $t = x/x_o$ where $F_o=\dfrac{j}{q\mu_p p_o}$ and $x_o=\dfrac{\varepsilon_s F_o}{qp_o}$, we can rewrite this equation as

$$\frac{df}{dt} = \frac{1-f}{f}$$

or as

$$\frac{f df}{1-f} = dt$$

Integrating the left-hand side with respect to $f$ and the right-hand side with respect to $t$, we obtain

$$f_b - f - \ln\frac{1-f}{1-f_b} = t$$

where $f_b = F(0)/F_o$. (Please check by differentiation and by substitution of $t = 0$, $f = f_b$ that this solution satisfies both the differential equation and the boundary condition.) For the given values of parameters, $F_o = 6242$ V/m, $x_o = 3.9 \times 10^{-9}$ m. The calculated field profile is shown in Fig. 3.7.3b.

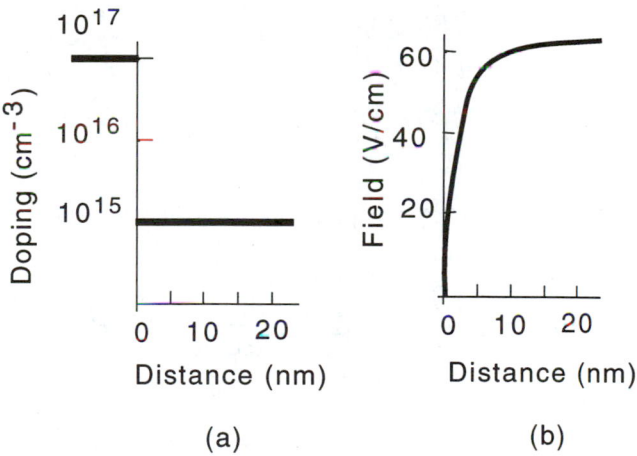

(a)                         (b)

**Fig. 3.7.3.** (a) Doping and (b) field profiles.

### Example 3-7-2.

The generation rate, $G$, in an $n$-type semiconductor is uniform and equal to $10^{20}$ cm$^{-3}$s$^{-1}$. The semiconductor is doped with shallow donors with concentration $N_d = 10^{15}$ cm$^{-3}$. The hole lifetime is 1 μs. What is the steady-state electron concentration?

**Solution:**

Since electrons and holes are generated in pairs, the excess electron and hole concentrations, $\Delta n$ and $\Delta p$, are equal. From eq. (3-7-17), $\Delta p = G\tau_{pl} = 10^{20} \times 10^{-6} = 10^{14}$ cm$^{-3}$. Assuming that all shallow donors are ionized, we find $n = N_d + \Delta n = 1.1 \times 10^{15}$ cm$^{-3}$.

### 3.7.2. Quasineutral semiconductor.

When electrons and holes are generated in pairs in a uniform semiconductor, the semiconductor remains neutral or nearly neutral (**quasineutral**). An example would be a piece of an $n$-type semiconductor where extra carriers are generated by light. This situation also occurs in a semiconductor diode, in a Bipolar Junction Transistor, in a solar cell, and in many other devices. In this subsection, we apply basic semiconductor equations to a quasineutral semiconductor.

For simplicity, we consider a one-dimensional steady-state situation. In this case, eqs. (3-7-18) and (3-7-19) become

$$\frac{1}{q}\frac{\partial j_n}{\partial x} + G - R = 0 \tag{3-7-20}$$

$$-\frac{1}{q}\frac{\partial j_p}{\partial x} + G - R = 0 \tag{3-7-21}$$

Substituting eqs. (3-7-2) and (3-7-3) into eqs. (3-7-20) and (3-7-21), we obtain

$$D_n \frac{\partial^2 n_n}{\partial x^2} + \mu_n F \frac{\partial n_n}{\partial x} + \mu_n n_n \frac{\partial F}{\partial x} + G - R = 0 \tag{3-7-22}$$

$$D_p \frac{\partial^2 p_n}{\partial x^2} - \mu_p F \frac{\partial p_n}{\partial x} - \mu_p p_n \frac{\partial F}{\partial x} + G - R = 0 \tag{3-7-23}$$

The assumption that the semiconductor is quasineutral means that

$$n_n - n_{no} \approx p_n - p_{no} \tag{3-7-24}$$

Hence, $\partial n_n/\partial x \approx \partial p_n/\partial x$ and $\partial^2 n_n / \partial x^2 \approx \partial^2 p_n / \partial x^2$, which allows us to simplify eqs. (3-7-22) and (3-7-23) and eliminate the $\partial F/\partial x$ term to obtain

$$D_a \frac{\partial^2 p_n}{\partial x^2} - \mu_a F \frac{\partial p_n}{\partial x} + G - R = 0 \qquad (3\text{-}7\text{-}25)$$

where

$$D_a = \frac{\mu_p \, p_n D_n + \mu_n \, n_n D_p}{\mu_n \, n_n + \mu_p \, p_n} \qquad (3\text{-}7\text{-}26)$$

is called the **ambipolar diffusion coefficient** and

$$\mu_a = \frac{\mu_n \, \mu_p (n_n - p_n)}{\mu_n \, n_n + \mu_p \, p_n} \qquad (3\text{-}7\text{-}27)$$

is called the **ambipolar mobility**. When $n_n \gg p_n$, $D_a \approx D_p$, $\mu_a \approx \mu_p$ and eq. (3-7-25) reduces to

$$D_p \frac{\partial^2 p_n}{\partial x^2} - \mu_p F \frac{\partial p_n}{\partial x} + G - R = 0 \qquad (3\text{-}7\text{-}28)$$

Equation (3-7-28) is the continuity equation for minority carriers (holes) in an $n$-type sample under steady-state and quasineutral conditions.

### 3.7.3.  Displacement and total current densities.

The continuity equations lead to another useful semiconductor equation. Subtracting eq. (3-7-18) from eq. (3-7-19), we obtain

$$q \frac{\partial}{\partial t}(p - n) + \nabla \cdot \left( \mathbf{j}_n + \mathbf{j}_p \right) = 0 \qquad (3\text{-}7\text{-}29)$$

From eq. (3-7-13), $\rho \approx q \left( N_d^+ - N_a^- - n + p \right)$. Hence,

$$q \frac{\partial}{\partial t}(p - n) = \frac{\partial \rho}{\partial t} \qquad (3\text{-}7\text{-}30)$$

Differentiating Poisson's equation [eq. (3-7-12)] with respect to time, we find $\partial \rho / \partial t = \varepsilon_s \partial \nabla \cdot \mathbf{F} / \partial t$. Substituting $\partial \rho / \partial t$ from this equation into eq. (3-7-30) and the result into (3-7-29), we obtain

$$\varepsilon \frac{\partial}{\partial t} \nabla \cdot \mathbf{F} + \nabla \cdot \left( \mathbf{j}_n + \mathbf{j}_p \right) = 0 \qquad (3\text{-}7\text{-}31)$$

Integrating eq. (3-7-31) over the space coordinates, we obtain

$$\mathbf{j}(t) = \mathbf{j}_n + \mathbf{j}_p + \varepsilon \frac{\partial}{\partial t} \mathbf{F} \qquad (3\text{-}7\text{-}32)$$

where $\mathbf{j}(t)$ is the **total current density**, and the $\varepsilon \partial \mathbf{F}/\partial t$ term is called the **displacement current density**. The total current is the integral of the current density over the sample cross section:

$$I = \int_S \mathbf{j}\, d\mathbf{s} \qquad (3\text{-}7\text{-}33)$$

For a sample with a constant cross section $S$ and a uniform current density, we obtain

$$I = jS \qquad (3\text{-}7\text{-}34)$$

### *3.7.4.   More advanced models.

As was mentioned above, the drift-diffusion model discussed above (see Subsection 3.7.1) may lead to large errors in the simulation of modern semiconductor devices with very short feature sizes (as small as $0.1\ \mu\text{m}$ or less), which may operate at very high frequencies (up to 300 GHz or so). For an accurate analysis of such devices, it is not enough to calculate the average concentrations of electrons and holes as functions of distance and time. Rather, we must follow the variations of the electron and hole **distribution functions** in time and space. The equation describing such a variation is called the **Boltzmann transport equation**. Thus, the analysis of very small semiconductor devices (with sizes smaller than a micron or so) may require the solution of the Boltzmann transport equation together with Poisson's equation. (The description of even smaller devices may require a direct solution of the Schrödinger equation.) Often, the solution of the Boltzmann transport equation is obtained using the **Monte Carlo technique**. This technique, which borrowed its name from a famous gambling resort, simulates the electron transport by generating large sequences of random numbers, which represent random events of electron scattering by impurities and lattice vibrations [see, for example, Shur (1990) for further details]. The probabilities of these events are chosen to represent the probabilities of electron scattering. In the past, such a method usually required a supercomputer. Monte Carlo programs are now running even on workstations as well because of a rapid improvement in computer hardware.

Semiconductor equations called **balance equations** represent a meaningful compromise between the drift-diffusion equations and full-blown Monte Carlo simulation. These equations account for the finite momentum and energy relaxation times and are suitable for the simulation of submicron semiconductor devices. Just a few years ago, all commercial device simulators used the drift-diffusion equations. More modern device simulators, such as PISCES from Stanford or ATLAS-II from Silvaco, Inc. (see Section 4.13), solve the energy transport equations as well, and thus are better suited for the simulation of small semiconductor devices.

Table 3.7.1 summarizes the most important semiconductor equations.

| | |
|---|---|
| Electron and hole current densities in low electric field | $\mathbf{j}_n = q\left(n\mu_n\mathbf{F} + D_n\nabla n\right)$ <br> $\mathbf{j}_p = q\left(p\mu_p\mathbf{F} - D_p\nabla p\right)$ |
| Ohm's law | $\mathbf{j} = \sigma\mathbf{F} \quad I = \dfrac{\sigma S}{L}FL = \dfrac{U}{R}$ |
| Sample resistance | $R = L / \sigma S$ |
| Electron and hole current densities in arbitrary electric field | $\mathbf{j}_n = q\left[-n\mathbf{v}_n(\mathbf{F}) + D_n(\mathbf{F})\nabla n\right]$ <br> $\mathbf{j}_p = q\left[p\mathbf{v}_p(\mathbf{F}) - D_p(\mathbf{F})\nabla p\right]$ |
| Poisson's equation | $\nabla\cdot\mathbf{F} = \rho / \varepsilon$ |
| Continuity equations for electrons and holes | $\dfrac{\partial n}{\partial t} = \dfrac{1}{q}\nabla\cdot\mathbf{j}_n + G - R$ <br> $\dfrac{\partial p}{\partial t} = -\dfrac{1}{q}\nabla\cdot\mathbf{j}_p + G - R$ |
| Ambipolar diffusion coefficient and ambipolar mobility | $D_a = \dfrac{\mu_p\, p_n D_n + \mu_n\, n_n D_p}{\mu_n\, n_n + \mu_p\, p_n}$ <br> $\mu_a = \dfrac{\mu_n\, \mu_p\left(n_n - p_n\right)}{\mu_n\, n_n + \mu_p\, p_n}$ |
| Total current density (including displacement current) | $\mathbf{j}(t) = \mathbf{j}_n + \mathbf{j}_p + \varepsilon\dfrac{\partial}{\partial t}\mathbf{F}$ |

**Table 3.7.1.** Summary of the most important semiconductor equations.

## 3-8. QUASI-FERMI LEVELS.  GENERATION AND RECOMBINATION

### 3.8.1. Quasi-Fermi levels.

Under thermal equilibrium conditions, in a nondegenerate semiconductor, the electron and hole concentrations can be expressed via the Fermi level

$$n = N_c \exp\left(\frac{E_F - E_c}{k_B T}\right) \tag{3-8-1}$$

$$p = N_v \exp\left(\frac{E_v - E_F}{k_B T}\right) \tag{3-8-2}$$

(see Section 3.3).  These concentrations are related by the mass-action law

$$pn = n_i^2$$

[which is obtained by multiplying the carrier concentrations given by eqs. (3-8-1)]. Under nonequilibrium conditions, which occur when an electric current flows or when extra electron-hole pairs are generated by light, these equations are no longer valid.  For example, when we generate electron-hole pairs, both the electron and hole concentrations increase and, hence, the $pn$ product exceeds $n_i^2$.

Under nonequilibrium conditions, it may be useful to represent the electron and hole concentrations as

$$n = N_c \exp\left(\frac{E_F^{(n)} - E_c}{k_B T}\right) \qquad p = N_v \exp\left(\frac{E_v - E_F^{(p)}}{k_B T}\right) \tag{3-8-3}$$

Equations (3-8-3) may be considered as definitions of $E_F^{(n)}$ and $E_F^{(p)}$, which are called the **electron** and **hole quasi-Fermi levels**, respectively (sometimes they are also called **Imrefs**, i. e., Fermi spelled backwards).  Under equilibrium conditions, $E_F^{(p)} = E_F^{(n)} = E_F$.  Under nonequilibrium conditions, $E_F^{(p)}$ is not equal to $E_F^{(n)}$ and both may be functions of position and/or time.  From eqs. (3-8-3), we obtain

$$np = n_i^2 \exp\left(\frac{E_F^{(n)} - E_F^{(p)}}{k_B T}\right) \tag{3-8-4}$$

Since in equilibrium $np = n_i^2$, the difference $E_F^{(n)} - E_F^{(p)}$ can be used as the measure of the deviation from equilibrium.

The introduction of quasi-Fermi levels is very useful because carrier concentrations in a semiconductor device may vary as functions of position and/or bias by many orders of magnitude, whereas the quasi-Fermi levels change only within the energy gap or just inside the bands near the band edges. This variation is much easier to visualize.

**Example 3-8-1.**

We shine light on an $n$-type GaAs sample with doping density $N_d$. The light is uniformly absorbed and electron-hole pairs with density $P$ are produced. Find the dependence of the electron and hole quasi-Fermi levels on $P$ and sketch this dependence for $10^{11}$ cm$^{-3}$ $< P < 10^{17}$ cm$^{-3}$, $N_d = 10^{15}$ cm$^{-3}$, $T = 300$ K.

**Solution:**

Since electrons and holes are generated in pairs, the additional electron and hole concentrations are equal to the concentration of the electron-hole pairs, $P$. Hence, the total electron concentration in the sample equals

$$n \approx P + N_d \qquad (3\text{-}8\text{-}5)$$

and the hole concentration is

$$p \approx P + n_i^2 / N_d \qquad (3\text{-}8\text{-}6)$$

The electron and hole quasi-Fermi levels calculated from eqs. (3-8-5) and (3-8-6) are shown in Fig. 3.8.1 versus the electron-hole pair density, $P$. As can be seen from Fig. 3.8.1, when $P$ varies from $10^{11}$ to $10^{17}$ cm$^{-3}$, $E_c - E_F^{(n)}$ varies from 1.31 to 1.35 eV and $E_F^{(p)} - E_v$ varies from 0.52 eV to approximately 0.15 eV. $E_F^{(p)}$ changes that much because of the large relative change in the hole concentration, from $n_i^2 / N_d$ to $P + n_i^2 / N_d$.

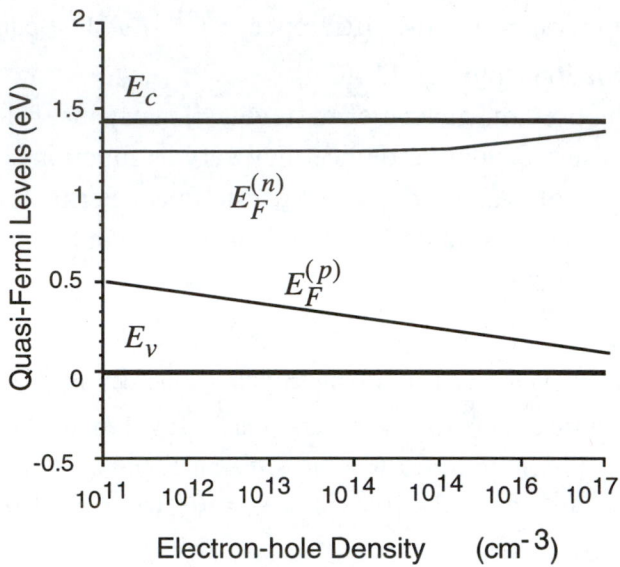

**Fig. 3.8.1.** Electron and hole quasi-Fermi levels versus concentration of light-generated electron–hole pairs in $n$-type GaAs. Parameters used: $T = 300$ K, $N_d = 10^{15}$ cm$^{-3}$, $N_c = 4.7 \times 10^{17}$ cm$^{-3}$, $N_v = 7 \times 10^{18}$ cm$^{-3}$, $n_i = 1.79 \times 10^6$ cm$^{-3}$ (from Shur, 1990, copyright © Prentice Hall, 1990, reproduced by permission of Prentice Hall, Inc., Englewood Cliffs, NJ).

In low electric fields, the electron current density may be expressed through the electron quasi-Fermi level, $E_F^{(n)}$, as follows:

$$
j_n = q\mu_n n F + q D_n \frac{\partial n}{\partial x}
$$

$$
= q\mu_n n \frac{1}{q}\frac{\partial E_c}{\partial x} + q D_n \frac{\partial}{\partial x}\left[ N_c \exp\left( \frac{E_F^{(n)} - E_c}{k_B T} \right) \right] \tag{3-8-7}
$$

$$
= \mu_n n \frac{\partial E_F^{(n)}}{\partial x}
$$

(Here we used the Einstein relation: $D_n = \mu_n k_B T/q$; see Section 3.6.) In a similar fashion, we find that

$$
j_p = \mu_p p \frac{\partial E_F^{(p)}}{\partial x} \tag{3-8-8}
$$

Equations (3-8-7) and (3-8-8) clearly show that the Fermi level concept does not apply when a current flows in a semiconductor since the Fermi level must be constant throughout the device under thermal equilibrium!

### 3.8.2. Generation.

As was mentioned before, generation of electron-hole pairs in a semiconductor can be achieved by illuminating a semiconductor sample with light with photon energies larger than the energy gap of the semiconductor. The light is partially reflected from the semiconductor surface and partially absorbed in the semiconductor (see Fig. 3.8.2).

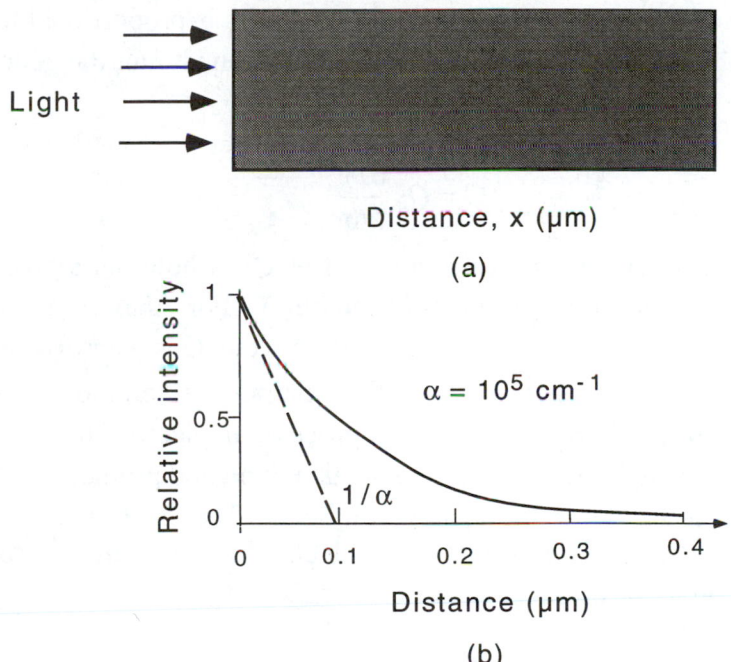

**Fig. 3.8.2.**   (a) Light beam shining onto a semiconductor and (b) the relative light intensity versus distance.

The number of photons absorbed in one second in the region between $x$ and $x + dx$ is proportional to the total number of photons in this region. Hence, we can write the following equation for the light intensity $P_l$ (measured in W/m$^2$):

$$dP_l / dx = -\alpha P_l \tag{3-8-9}$$

where $\alpha$ is an **absorption coefficient** (measured in 1/m). The solution of this equation is given by

$$P_l = P_o \exp(-\alpha x) \tag{3-8-10}$$

where $P_o$ is the light intensity at the surface. The distance $1/\alpha$ is called the **light penetration depth** (see Fig. 3.8.2b). If $1/\alpha \gg L$ where $L$ is the sample dimension, the generation rate of electron-hole pairs is nearly uniform within the sample. Absorption coefficient, $\alpha$, is a strong function of the wavelength. (Typically, $\alpha$ is measured for the wavelengths corresponding to the values of $\alpha$ from $10^2$ to $10^6$ cm$^{-1}$.) The number of generated electron-hole pairs is proportional to the number of absorbed photons. Since the energy of each photon is $\hbar\omega$, the generation rate is given by

$$G = Q_e \frac{\alpha P_l}{\hbar\omega} \tag{3-8-11}$$

where $Q_e$ is equal to the average number of electron-hole pairs produced by one photon. ($Q_e$ is called the **quantum efficiency**.) For $\hbar\omega > E_g$, the quantum efficiency $Q_e$ is often fairly close to unity. (Show that $G$ is measured in m$^{-3}$s$^{-1}$.)

In high electric fields, electron-hole pairs in a semiconductor are often generated by **impact ionization**. In this process, an electron (or a hole) acquires enough energy from the electric field to break a bond and promote another electron from the valence band into the conduction band. The electron impact ionization generation rate, $G_{ni}$, is proportional to the electron concentration, $n$, to the electron velocity, $v_n$, and to the **impact ionization coefficient**, $\alpha_{ni}$:

$$G_{ni} = \alpha_{ni} n v_n \tag{3-8-12}$$

where $\alpha_{ni}$, is given by the following empirical expression:

$$\alpha_{ni} = \alpha_{no} \exp\left[-(F_{in}/F)^{m_{in}}\right] \tag{3-8-13}$$

In a similar fashion, when the impact ionization is caused by holes, we have

$$G_{pi} = \alpha_{pi} p v_p \tag{3-8-14}$$

$$\alpha_{pi} = \alpha_{po} \exp\left[-\left(F_{ip} / F\right)^{m_{ip}}\right] \tag{3-8-15}$$

Here $F$ is an electric field and $\alpha_{no}$, $\alpha_{po}$, $F_{in}$, $F_{ip}$, $m_{in}$, and $m_{ip}$ are constants that depend on semiconductor material and, in certain cases, even on the direction of the electric field. The impact ionization coefficients for Si calculated by using eqs. (3-8-14) and (3-8-16) are shown in Fig. 3.8.3.

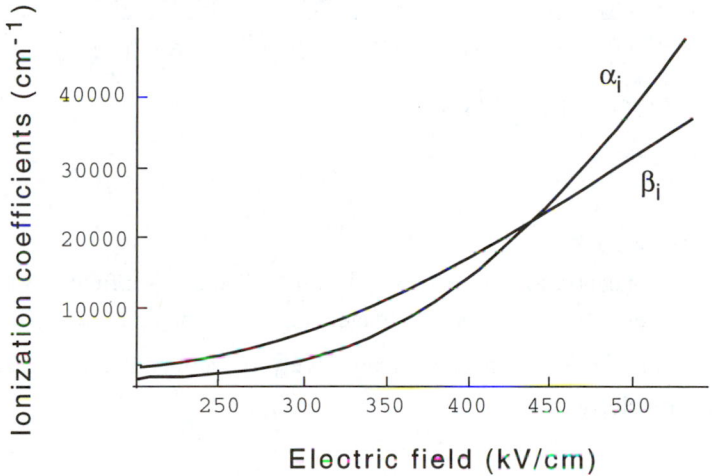

Electric field (kV/cm)

**Fig. 3.8.3.** Impact ionization coefficients for Si calculated using eqs. (3-8-14) and (3-8-16). $\alpha_o = 3{,}318$ cm$^{-1}$, $F_{no} = 1{,}174$ kV/cm, $m_{in} = 1$ [data from W. Maes, K. de Meyer, R. van Overstraeten, *Solid State Electronics*, Vol. 33, No. 6, pp. 705–718 (1990)], $\beta_o = 2000$ cm$^{-1}$, $F_{po} = 1{,}970$ kV/cm, $m_{ip} = 1$ [data from W. N. Grant, *Solid State Electronics*, Vol. 16, No. 10, pp. 1189–1203 (1990)].

**Example 3-8-2.**

A silicon sample is illuminated with a monochromatic light with intensity of 100 mW/cm$^2$. (This approximately equals the intensity of solar radiation at sea level when the sun is at zenith.) The wavelength of light is $\lambda = 0.6$ μm. The quantum efficiency is 0.9. The absorption coefficient $\alpha = 6 \times 10^3$ cm$^{-1}$. The sample thickness, $L = 0.1$ μm. Find the generation rate of electron-hole pairs (neglecting reflections).

**Solution:**

Since $\alpha L = 6 \times 10^3 \times 10^{-5} = 0.06 \ll 1$, the generation rate is nearly uniform within the sample and can be estimated as

$$G=Q_e\frac{\alpha P_l(0)}{\hbar\omega}$$

The frequency $\omega = 2\pi c/\lambda$, where $c$ is the velocity of light. Substituting the parameter values into this equation, we obtain

$$G=\frac{Q_e\alpha P_l(0)}{2\pi\hbar c/\lambda}=\frac{0.9\times6\times10^5\left(\frac{1}{m}\right)\times10^3\left(\frac{W}{m^2}\right)}{2\pi\times1.055\times10^{-34}(Js)\times3\times10^8\left(\frac{m}{s}\right)/0.6\times10^{-6}(m)}$$

$$= 1.63\times10^{27}\left(m^{-3}s^{-1}\right)$$

**Example 3-8-3.**

The electron concentration in GaAs is $10^{13}$ cm$^{-3}$. The electric field is 300 kV/cm. The electron velocity is $10^5$ m/s. Parameters $\alpha_{no}$, $F_{in}$, and $m_{in}$ for GaAs are equal to $2.2\times10^6$ cm$^{-1}$, 300 kV/cm, and 1, respectively. Calculate the impact ionization rate caused by electrons.

**Solution:**

$$G_{ni}=\alpha_{ni}nv_n=2.2\times10^8\left(m^{-1}\right)\exp\left(-\frac{300}{300}\right)\times10^{19}\left(m^{-3}\right)\times10^5(m/s)=$$
$$7.36\times10^{31}\left(m^{-3}s^{-1}\right)$$

### 3.8.3. Recombination and diffusion equation.

Under equilibrium conditions, $G = R$ where R is the recombination rate, that is, the generation of electron-hole pairs is balanced by different recombination processes which include **direct (band-to-band) radiative recombination, radiative band-to-impurity recombination, nonradiative recombination via impurity (trap) levels, Auger recombination**, and **surface recombination**. These processes (except for the Auger recombination discussed below) are schematically shown in Fig. 3.8.4.

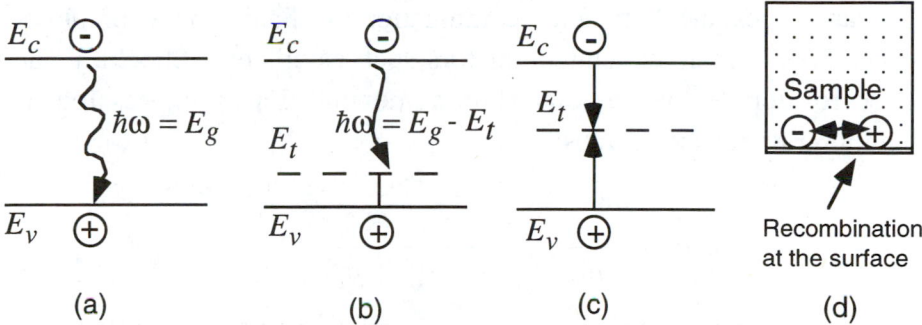

(a)    (b)    (c)    (d)

**Fig. 3.8.4.** (a) Direct (band-to-band) radiative recombination, (b) radiative band-to-impurity recombination, (c) nonradiative recombination via impurity (trap) levels, and (d) surface recombination.

In the radiative recombination processes, recombining electron-hole pairs emit light, and the energy released during this recombination process is transformed into the energy of photons. Since during band-to-band recombination the energy released by each recombining electron-hole pair is either equal to the energy gap or is slightly higher, the frequency of the emitted radiation can be estimated as

$$\hbar\omega = E_g \tag{3-8-16}$$

The radiative band-to-band recombination rate is proportional to the $np$ product:

$$R = A\left(np - n_i^2\right) \tag{3-8-17}$$

since both electrons and holes are required for the band-to-band recombination (just like the number of marriages in a city should be proportional to both the number of eligible men and to the number of eligible women). In practical light-emitting semiconductor devices, radiative band-to-impurity recombination may be more important than radiative band-to-band recombination. The radiative band-to-impurity recombination rate is given by

$$R = p/\tau_r \tag{3-8-18}$$

where $p$ is the concentration of the electron-hole pairs, $\tau_r = 1/(B_r N_t)$, $N_t$ is the concentration of impurities involved in this radiative recombination process, and $B_r$ is a constant.

In many cases, the dominant recombination mechanism is recombination via traps, especially in indirect semiconductors such as silicon. Shockley and Reed were first to derive the famous (but somewhat long) expression for the recombination rate related to traps:

$$R = \frac{pn - n_i^2}{\tau_{pl}(n + n_l) + \tau_{nl}(p + p_l)} \tag{3-8-19}$$

Here, $n_i$ is the intrinsic carrier concentration, and $\tau_{nl}$ and $\tau_{pl}$ are electron and hole lifetimes given by

$$\tau_{nl} = \frac{1}{v_{thn}\sigma_n N_t} \qquad \tau_{pl} = \frac{1}{v_{thp}\sigma_p N_t} \tag{3-8-20}$$

where $v_{thn}$ and $v_{thp}$ are electron and hole thermal velocities, $\sigma_n$ and $\sigma_p$ are effective capture cross sections for electrons and holes, $N_t$ is the trap concentration, and

$$n_l = N_c \exp\left(\frac{E_t - E_c}{k_B T}\right) \qquad p_l = N_v \exp\left(\frac{E_v - E_t}{k_B T}\right) \tag{3-8-21}$$

$N_c$ and $N_v$ are the effective densities of states in the conduction and valence band, respectively, and $E_t$ is the energy of the trap level. When electrons are minority carriers ($n = n_p \ll p \approx N_a = p_p$ where $N_a$ is the concentration of shallow ionized acceptors, $p \gg p_l$, $p \gg n_l$), eq. (3-8-19) reduces to

$$R = \frac{n_p - n_{po}}{\tau_{nl}} \tag{3-8-22}$$

where $n_{po} = n_i^2/N_a$. When holes are minority carriers ($p = p_n \ll n \approx N_d = n_n$ where $N_d$ is the concentration of shallow ionized donors, $n \gg n_l$, $n \gg p_l$)

$$R = \frac{p_n - p_{no}}{\tau_{pl}} \tag{3-8-23}$$

where $p_{no} = n_i^2/N_d$. The electron lifetime, $\tau_{nl}$, in $p$-type silicon and the hole lifetime, $\tau_{pl}$, in $n$-type silicon decrease with increasing doping. At low doping levels, this decrease may be explained by higher trap concentrations in the doped semiconductor. We can crudely estimate electron and hole lifetimes in Si as follows:

$$\tau_{nl}(s) = \frac{10^{12}}{N_a(cm^{-3})} \qquad \tau_{pl}(s) = \frac{3 \times 10^{12}}{N_d(cm^{-3})} \tag{3-8-24}$$

However, at high doping levels, the lifetimes decrease faster than the inverse doping concentration. The reason is that a different recombination mechanism, called **Auger recombination**, becomes important. In an Auger process, an electron and a hole recombine without involving trap levels, and the released energy (of the order of the energy gap) is transferred to another carrier (a hole in $p$-type material and an electron in $n$-type material). Auger recombination is the inverse of impact ionization, where energetic carriers cause the generation of electron-hole pairs. Since two electrons (in $n$-type material) or two holes (in $p$-type material) are involved in the Auger recombination process, the recombination lifetime associated with this process is inversely proportional to the square of the majority carrier concentration, that is,

$$\tau_{nl} = \frac{1}{G_p N_a^2} \quad \tau_{pl} = \frac{1}{G_n N_d^2} \tag{3-8-25}$$

for $p$-type material and for $n$-type material, respectively, where $G_p = 9.9 \times 10^{-32}$ cm$^6$/s and $G_n = 2.28 \times 10^{-32}$ cm$^6$/s for silicon.

As stated above, in many cases the recombination rate can be presented as $R = (p_n - p_{no})/\tau_{pl}$ (in $n$-type material where holes are minority carriers). In this case, eq. (3-7-28) describing the diffusion and drift of the minority carriers can be simplified. When the electric field, $F$, and generation rate, $G$, are equal to zero, this equation becomes

$$D_p \frac{\partial^2 p_n}{\partial x^2} - \frac{p_n - p_{no}}{\tau_{pl}} = 0 \tag{3-8-26}$$

Equation (3-8-26) is called the **diffusion equation**. The solution of this second-order linear differential equation is given by

$$\Delta p = A \exp(x/L_p) + B \exp(-x/L_p) \tag{3-8-27}$$

where $\Delta p = p_n - p_{no}$, $A$ and $B$ are constants to be determined from the boundary conditions, and

$$L_p = \sqrt{D_p \tau_{pl}} \tag{3-8-28}$$

is called the **hole diffusion length**. A similar diffusion equation applies to electrons in $p$-type material, where electrons are minority carriers. The **electron diffusion length** is then given by

$$L_n = \sqrt{D_n \tau_{nl}} \tag{3-8-29}$$

In many semiconductor devices, the recombination rate is very high at the semiconductor surface, where extra defects and traps are present. This **surface recombination** can be accounted for by using the following boundary condition for the diffusion equation (3-8-26) at this surface:

$$D_p \frac{\partial p_n}{\partial x}\bigg|_{x=x_s} = -S_p\big[p_n(x=x_s) - p_{no}\big] \qquad (3\text{-}8\text{-}30)$$

where $D_p$ is the hole diffusion coefficient, $p_n$ is the hole concentration, $p_{no} = n_i^2/N_d$ is the equilibrium hole concentration, and $x = x_s$ corresponds to the surface of the sample. $S_p$ is called the **surface recombination rate** and is given by

$$S_p = \sigma_p v_{thp} N_{st} \qquad (3\text{-}8\text{-}31)$$

where $N_{st}$ is the density of the surface traps and $\sigma_p$ is the trap cross section.

**Example 3-8-4.**

Find the solution of eq. (3-7-28) for zero electric field, constant generation rate, and recombination rate given by eq. (3-8-23). Assuming $G = 10^{27}$ m$^{-3}$s$^{-1}$, $D_p = 7$ cm$^2$/s, $\tau_{pl} = 10^{-7}$ s, calculate the hole concentration in the middle of a 10 $\mu$m $n$-type Si film doped at $10^{16}$ cm$^{-3}$ ($n_i = 10^{10}$ cm$^{-3}$) for the surface recombination rate $S_p = 10^5$ m/s.

**Solution:**

This solution is given by

$$p = A\exp\left(\frac{x}{L_p}\right) + B\exp\left(-\frac{x}{L_p}\right) + p_{no} + G\tau_{pl}$$

(Please check by substitution.) We choose the origin at $x = 0$. Then $A = B$ because of symmetry. From

$$D_p \frac{\partial p_n}{\partial x}\bigg|_{x=d/2} = -S_p\big[p_n(x=d/2) - p_{no}\big]$$

where $d/2 = 5$ $\mu$m, we find $A$ and obtain

$$p(0) \approx G\tau_{pl} - \frac{2G\tau_{pl}^2 S_p \exp\big[d/(2L_p)\big]}{L_p\big[\exp(d/L_p)-1\big] + S_p\tau_{pl}\big[1+\exp(d/L_p)\big]} \approx 1.55\times10^{13}\,\big(\text{cm}^{-3}\big)$$

The summary of equations related to quasi-Fermi levels and generation and recombination of carriers is given in Table 3.8.1.

| | |
|---|---|
| Definition of electron and hole quasi-Fermi levels | $n = N_c \exp\left(\dfrac{E_F^{(n)} - E_c}{k_B T}\right) \qquad p = N_v \exp\left(\dfrac{E_v - E_F^{(p)}}{k_B T}\right)$ |
| $np$ product | $np = n_i^2 \exp\left(\dfrac{E_F^{(n)} - E_F^{(p)}}{k_B T}\right)$ |
| Generation rate of electron-hole pairs by light | $G = Q_e \dfrac{\alpha P_l}{\hbar \omega}$ |
| Generation rate of electron-hole pairs by impact ionization by electrons | $G_{ni} = \alpha_{ni} n v_n$ where $\alpha_{ni} = \alpha_{no} \exp\left[-\left(\dfrac{F_{in}}{F}\right)^{m_{in}}\right]$ |
| Generation rate of electron-hole pairs by impact ionization by holes | $G_{pi} = \alpha_{pi} p v_p$ where $\alpha_{pi} = \alpha_{po} \exp\left[-\left(\dfrac{F_{ip}}{F}\right)^{m_{ip}}\right]$ |
| Radiative recombination rate | $R = A\left(np - n_i^2\right)$ |
| Recombination rates related to traps for minority electrons and minority holes | $R = \dfrac{n_p - n_{po}}{\tau_{nl}} \qquad R = \dfrac{p_n - p_{no}}{\tau_{pl}}$ |
| Auger recombination times for $p$-type and $n$-type materials | $\tau_{nl} = \dfrac{1}{G_p N_a^2} \qquad \tau_{pl} = \dfrac{1}{G_n N_d^2}$ |
| Diffusion equation for minority carriers (holes) in $n$-type material | $D_p \dfrac{\partial^2 p_n}{\partial x^2} - \dfrac{p_n - p_{no}}{\tau_{pl}} = 0$ |
| Diffusion equation for minority carriers (electrons) in $p$-type material | $D_n \dfrac{\partial^2 n_p}{\partial x^2} - \dfrac{n_p - n_{po}}{\tau_{nl}} = 0$ |
| Electron and hole diffusion lengths | $L_n = \sqrt{D_n \tau_{nl}} \qquad L_p = \sqrt{D_p \tau_{pl}}$ |
| Surface recombination boundary condition for diffusion equation | $D_p \dfrac{\partial p_n}{\partial x}\bigg|_{x=0} = -S_p\left[p_n(x=0) - p_{no}\right]$ |
| Surface recombination rate for minority carriers (holes) | $S_p = \sigma_p v_{thp} N_{st}$ |

**Table 3.8.1.** Equations related to quasi-Fermi levels and generation and recombination of carriers.

# *3-9. HALL EFFECT AND MAGNETORESISTANCE

### *3.9.1. Hall effect.

So far we have considered the electron and hole motion in electric fields. A magnetic field, $\mathbf{B}$, may also cause electron and hole motion. It exerts a force $\mathbf{f}_L = q\mathbf{v}_p \times \mathbf{B}$ (called the **Lorentz force**) on a charge $q$ moving with velocity $\mathbf{v}_p$. From the definition of the vector product, we recall that the direction of the Lorentz force is perpendicular to the plane containing vectors $\mathbf{v}_p$ and $\mathbf{B}$ and that the magnitude of the Lorentz force, $f_L = qv_p B\sin(\theta)$. The direction of $\mathbf{f}_L$ is given by the left-hand rule illustrated in Fig. 3.9.1: when the streamlines of the magnetic field strike the open palm of the left hand and the fingers show the direction of motion of a positive charge, the thumb points in the direction of the Lorentz force.

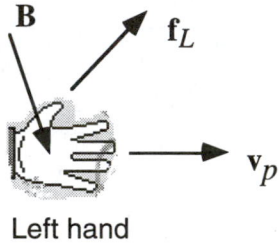

**Fig. 3.9.1.** Left-hand rule.

For a negative charge, $\mathbf{f}_L$ has the opposite direction. Hence, by finding out which way charge carriers in a semiconductor are steered by a magnetic field (i. e., the direction of the Lorentz force), we can establish whether these carriers are electrons or holes. Measuring the magnitude of the Lorentz force allows us to determine the magnitude of the electron or hole velocity. If we combine such a measurement with the measurement of the electric current density, $j$, which is proportional to both carrier concentration and carrier velocity, we can figure out the carrier concentration (for example, for a uniform $n$-type semiconductor in a uniform electric field, $j = qnv_n$ where $v_n$ is the electron drift velocity where $n$ is the electron concentration). This explains why standard experimental techniques of measuring mobilities include measurements in a magnetic field.

Let us consider a $p$-type semiconductor sample placed into a magnetic field (see Fig. 3.9.2). When a positive voltage $V$ is applied to contact 2 with respect to contact 1, holes move from contact 2 to contact 1 with the drift velocity

$$v_p = \mu_p F = \mu_p V / L \qquad (3\text{-}9\text{-}1)$$

where $F$ is the electric field, $L$ is the sample length, and $\mu_p$ is the hole mobility. The magnetic field, $\mathbf{B}$, acts on the holes with a Lorentz force $\mathbf{f}_L = q\mathbf{v}_p \times \mathbf{B}$ (deflecting them toward side 3 as shown in the figure).

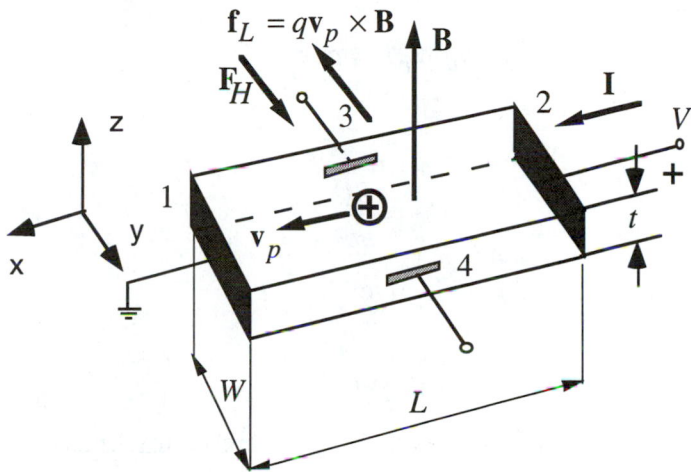

**Fig. 3.9.2.** Hall effect in $p$-type semiconductor sample in a magnetic field (from Shur, 1990, copyright © Prentice Hall, 1990, reproduced by permission of Prentice Hall, Inc., Englewood Cliffs, NJ). (As an exercise, prove that in $n$-type semiconductors, the Lorentz force will have the same direction but the Hall electric field will have the opposite direction.)

As a consequence, holes accumulate on this side and create a net positive charge there. Side 4 becomes depleted with holes. At this side, the negative charge of the ionized acceptors is not fully compensated by positively charged holes. These charges at sides 3 and 4 create an electric field, $\mathbf{F}_H$, (similar to that in a parallel plate capacitor). This field acts on holes with a force in the direction opposite to the Lorentz force. The process of the charge separation stops when this field (called a **Hall electric field**) compensates the Lorentz force so that the total force acting on the holes in the direction perpendicular to their drift velocity becomes equal to zero. This happens when

$$F_H = v_p B \qquad (3\text{-}9\text{-}2)$$

and the voltage difference (called **Hall voltage**)

$$V_H = F_H W = v_p B W \qquad (3\text{-}9\text{-}3)$$

develops between contacts 3 and 4.

This effect is called the **Hall effect** (after an American physicist, Hall, who discovered this effect more than a century ago). Contacts 3 and 4 are called the **Hall contacts**.

In a uniform sample, the electric current density is related to the drift velocity as follows:

$$j = qp\mu_p F = qpv_p \tag{3-9-4}$$

and, hence,

$$F_H = \frac{Bj}{qp} \tag{3-9-5}$$

and the Hall voltage, $V_H = F_H W$, is given by

$$V_H = RBI / t \tag{3-9-6}$$

where $I = jtW$ is the electric current, $t$ is the sample thickness, $W$ is the sample width, and

$$R = \frac{1}{qp} \tag{3-9-7}$$

is called the **Hall constant**. The hole mobility can be determined from the measured sample conductivity $\sigma_p = jL/V = qp\mu_p$ and the Hall constant

$$\mu_p = \sigma_p R \tag{3-9-8}$$

However, eq. (3-9-8) is not totally accurate since we have assumed that all holes in the sample move with the same velocity $v_p$. In fact, this velocity is the average drift velocity superimposed on the random thermal motion. Hence, some holes move more slowly and some move faster. The Lorentz forces acting on holes moving with different velocities are different, but the Hall electric field acts with the same force on all holes, independently of their velocities. A more accurate analysis of this problem shows that eq. (3-9-7) has to be modified as follows:

$$R = \frac{r_H}{qp} \tag{3-9-9}$$

where $r_H$ is called the **Hall factor**. Hence, the measured mobility

$$\mu_{Hp} = \sigma_p R = r_H \mu_p \qquad (3\text{-}9\text{-}10)$$

(called the **Hall mobility** of holes) is $r_H$ times greater than the usual drift mobility. The value of $r_H$, which depends on temperature, doping, magnetic field, and other factors, typically varies between 1 and 2. When impurity scattering is dominant, $r_H \approx 1.93$. When phonon scattering is dominant, $r_H$ is closer to unity (typically, 1.2 to 1.4).

For an $n$-type sample, the Hall constant is negative (due to the opposite direction of the Lorentz force)

$$R = -\frac{r_H}{qn} \qquad (3\text{-}9\text{-}11)$$

$$\mu_{Hn} = \sigma_n R = r_H \mu_n \qquad (3\text{-}9\text{-}12)$$

where $\mu_{Hn}$ is the Hall electron mobility.

**Example 3-9-1.**

The Hall measurements at 77 K on a $p$-type sample yielded the following values: current, $I = 100$ μA, Hall voltage, $V_H = 10$ mV, applied voltage $V = 10$ mV. The sample length, width, and thickness are $L = 100$ μm, $W = 10$ μm, and $t = 1$ μm, respectively. The magnetic field $B = 1{,}000$ Gs (0.1 Tesla). What is the carrier concentration and Hall mobility?

**Solution:**

First, we calculate the Hall constant

$$R = \frac{V_H t}{BI} = \frac{0.01 \times 10^{-6}}{0.1 \times 10^{-4}} = 10^{-3} \left(\frac{m^3}{C}\right)$$

Then we find the carrier concentration

$$p = \frac{1}{qR} = \frac{1}{1.602 \times 10^{-19} \times 10^{-3}} = 6.24 \times 10^{21} \left(m^{-3}\right) = 6.24 \times 10^{15} \left(cm^{-3}\right)$$

The sample conductivity

$$\sigma = \frac{IL}{VWt} = \frac{10^{-4} \times 10^{-4}}{10^{-2} \times 10^{-5} \times 10^{-6}} = 1{,}000 \left(\frac{1}{\Omega m}\right)$$

The carrier Hall mobility $\mu_{Hp} = \sigma_p R = 10^{-3} \times 1{,}000 = 1 (m^2/Vs) = 10{,}000$ cm²/Vs.

### *3.9.2.  Hall angle and geometric magnetoresistance.

Let us now consider the hole velocity in the Hall sample shown in Fig. 3.9.2 at the moment of time $t = 0$ when the magnetic field is turned on.  Since the Lorentz force, $\mathbf{f}_L = q\mathbf{v}_p \times \mathbf{B}$, the $y$-component of the velocity $v_y = -\mu v_x B$ where $v_x$ is the $x$-component of the hole velocity.  Hence, at first, the holes move under the angle

$$\theta = \left|\tan^{-1}\left(\frac{v_y}{v_x}\right)\right| = \left|\tan^{-1}\left(-\frac{\mu v_x B}{v_x}\right)\right| = \tan^{-1}(\mu B) \qquad (3\text{-}9\text{-}13)$$

called the **Hall angle**.  After holes accumulate at one of the sample side surfaces, depleting the other side, the Hall electric field compensates the Lorentz force and holes stop moving in the direction perpendicular to the current flow.

However, the Hall electric field fully develops only if  $L \gg W$ (see Fig. 3.9.2).  In the opposite limiting case,  $W \gg L$,  most of the holes moving under the Hall angle arrive at the contact, as illustrated by Fig. 3.9.3 and, hence, will not create the Hall electric field.  As a result, carriers travel the distance

$$L_H = L\left(1 + \tan^2\theta\right)^{1/2} \qquad (3\text{-}9\text{-}14)$$

which exceeds the sample length $L$.  For a zero Hall electric field

$$\mathbf{v} = \mu(\mathbf{F} + \mathbf{v} \times \mathbf{B}) \qquad (3\text{-}9\text{-}15)$$

where $\mu$ is the low field mobility where $F = F_x$ is the applied electric field and $F_y = 0$.  Hence,

$$v_x = \mu\left(F_x + v_y B\right) \qquad (3\text{-}9\text{-}16)$$

$$v_y = -\mu v_x B \qquad (3\text{-}9\text{-}17)$$

Substituting eq. (3-9-17) into eq. (3-9-16), we find that

$$v_x = \frac{\mu F_x}{1 + \mu^2 B^2} \qquad (3\text{-}9\text{-}18)$$

Since we define the low field mobility as the coefficient of proportionality between a drift velocity and an electric field, the effective low field mobility, $\mu_B$, for a sample with $W/L \gg 1$ in the magnetic field $B$, is given by

$$\mu_B = \frac{\mu}{1 + \mu^2 B^2} \qquad (3\text{-}9\text{-}19)$$

Hence, the sample resistivity, $\rho = 1/(q\mu_B p)$, measured in the magnetic field, $B$, is higher than the resistivity, $\rho_o = 1/(q\mu p)$ in zero magnetic field:

$$\frac{\rho - \rho_o}{\rho_o} = (\mu B)^2 \qquad (3\text{-}9\text{-}20)$$

where $p$ is the carrier concentration. ($\Delta\rho = \rho - \rho_o$ is called the **geometric magnetoresistance**.)

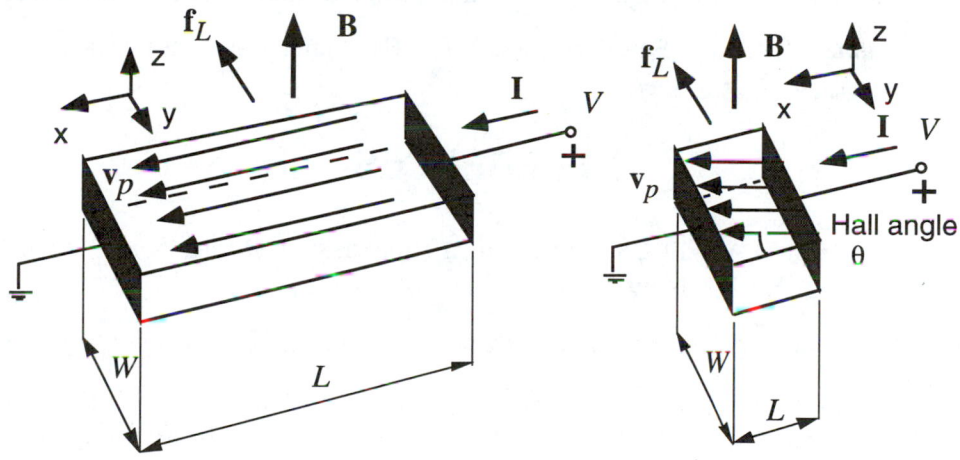

**Fig. 3.9.3.** Streamlines of hole velocity in $p$-type semiconductor samples with $W \ll L$ and $W \gg L$.

**Example 3-9-2.**

The electron mobility in a GaAs sample is 8,000 cm$^2$/Vs. Find the Hall angle and the geometric magnetoresistance $\Delta\rho/\rho_o$ in a magnetic field $B = 1$ Tesla.

**Solution:**

$\tan(\theta) = \mu B = 0.8 \times 1 = 0.8$. $\theta \approx 0.68 \approx 39°$. $\Delta\rho/\rho_o = (\mu B)^2 = 0.64$.

In addition to the geometric magnetoresistance, a so-called physical magnetoresistance is also observed in semiconductors of different geometry. Often this magnetoresistance is smaller than the geometric magnetoresistance.

Table 3.9.1 summarizes equations describing the Hall effect and geometric magnetoresistance.

| Hall electric field and Hall voltage | $F_H = v_p B$ <br> $V_H = v_p BW$ |
|---|---|
| Hall voltage and Hall constant | $V_H = RBI/t$ <br> $R = r_H/qp$ |
| Hall mobility | $\mu_{Hp} = \sigma_p R$ |
| Drift velocity | $\mathbf{v} = \mu(\mathbf{F} + \mathbf{v} \times \mathbf{B})$ |
| Geometric magnetoresistance | $\Delta\rho/\rho_o = (\mu B)^2$ |

**Table 3.9.1.** Equations describing Hall effect and geometric magnetoresistance.

# REFERENCES

B. SHKLOVSKI AND A. EFROS, *Electronic Properties of Doped Semiconductors*, Springer-Verlag, Berlin, New York (1984)

M. SHUR, *Physics of Semiconductor Devices*, Prentice Hall, Englewood Cliffs, NJ (1990)

S. M. SZE, *Semiconductor Devices. Physics and Technology*, John Wiley & Sons, New York (1985)

# BIBLIOGRAPHY

M. E. LEVINSHTEIN AND G. S. SIMIN, *Getting to Know Semiconductors*, World Scientific, Singapore (1992)

*Popular introduction to the physics of semiconductor devices.*

R. F. PIERRET, *Semiconductor Fundamentals*, Second Edition, Modular Series on Solid State Devices, Vol. 1, Addison-Wesley, Reading, MA, 1988

*A good undergraduate text on semiconductor physics.*

C. M. WOLFE, N. J. HOLONYAK, AND G. E. STILLMAN, *Physical Properties of Semiconductors*, Prentice Hall, Englewood Cliffs, NJ (1989)

*A graduate level text on semiconductor physics.*

R. A. SMITH, *Semiconductors*, Second Edition, Cambridge University Press, Cambridge, London, New York, Melbourne, Sydney (1978)

*Basic introduction to semiconductor physics.*

K. SEEGER, *Semiconductor Physics, An Introduction,* Third Edition, Series on Solid-State Sciences, Vol. 40, Springer Verlag, New York (1985)

*A more advanced text on semiconductor physics.*

B. G. STREETMAN, *Solid State Electronics Devices*, Fifth Edition, Prentice Hall, Englewood Cliffs, NJ (1995)

*One of the most popular textbooks on semiconductor devices.*

# REVIEW QUESTIONS

**1.** What is the relationship between Fermi-Dirac distribution functions for electrons and holes?

☐ 1 point

**2.** a. The semiconductor temperature is 300 K. At what value of energy (counted from the Fermi level) is the hole distribution function equal to 0.8 under thermal equilibrium conditions?

☐ 3 points

b. How do you describe a hole in your own words? ☐ 2 points

**3.** Derive eq. (3-2-1). ☐ 3 points

**4.** a. Which number could correspond to a shallow donor Bohr radius, 7 Å or

100 Å? Why? ☐ 2 points

b. The electronic configuration of the valence electrons in a certain element is $4s^2 4p^3$. Will this element be a donor, an acceptor, or neither in Si? Why?

☐ 2 points

c. What is the value of the energy gap in Si in eV at room temperature?

☐ 1 point

d. A silicon sample is doped with $10^{16}$ cm$^{-3}$ donors and $10^{16}$ cm$^{-3}$ acceptors. What is the position of the Fermi level with respect to the bottom of the conduction band at room temperature?

☐ 1 point

e. The electron concentration in a GaAs sample at room temperature is $10^{19}$ cm$^{-3}$. What is the position of the Fermi level with respect to the bottom of the conduction band? Higher? Lower? Why? (The effective density of states in the conduction band of GaAs at room temperature is $4.7 \times 10^{17}$ cm$^{-3}$.)

☐ 1 point

f. The effective densities of states in the conduction and valence bands of GaAs at room temperature are $4.7 \times 10^{17}$ cm$^{-3}$ and $7 \times 10^{18}$ cm$^{-3}$, respectively. How many electrons do you expect to find in a one millimeter cube of undoped GaAs at room temperature? How many holes? The energy gap of GaAs is 1.42 eV.

☐ 2 points

g. What if the same GaAs cube is doped with shallow donor with concentration $10^{13}$ cm$^{-3}$. How many electrons do you expect to find now? How many holes?

☐ 2 points

h. If the donor ionization energy is increased and all other parameters are kept the same, do you expect the electron concentration to increase or decrease?

☐ 1 point

i. What energy ratio will this depend on?

☐ 1 point

**5.** a What is the relation between the electron and hole concentrations in a nondegenerate semiconductor under equilibrium conditions?

☐ 1 point

b. Explain in your own words what the Fermi level is.

☐ 1 point

c  The figure below shows the dependence of the intrinsic carrier concentration on inverse temperature for a semiconductor material.  What is the energy gap of this semiconductor?  **Hint:**  You may neglect the temperature dependence of the densities of states for an estimate of the energy gap based on this graph.

☐ <u>2 points</u>

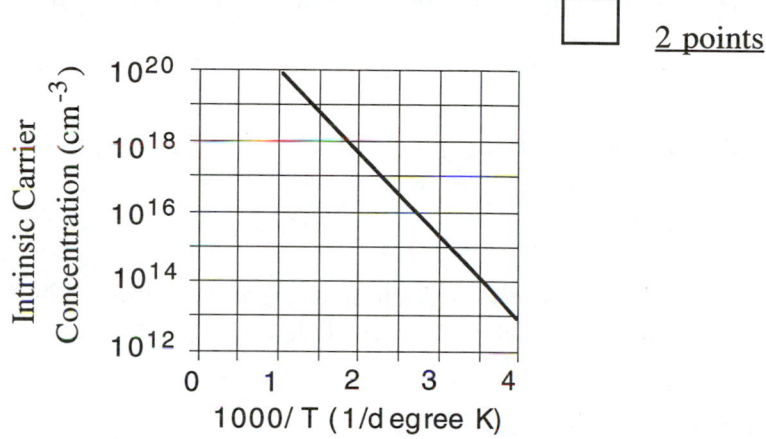

**6.**  Two degenerate GaAs samples 1 and 2 have electron concentrations $n_1$ and $n_2$ such that $n_2/n_1 = 8$.  The Fermi level in sample 1 is 40 meV above the bottom of the conduction band.  (The sample temperature is low so that $k_B T/q \ll 40$ meV.)  What is the position of the Fermi level in sample 2 with respect to the bottom of the conduction band?

☐ <u>1 point</u>

**7.**  The concentration of shallow donors in a semiconductor is $10^{15}$ cm$^{-3}$.  The concentration of shallow acceptors in the same semiconductor is $2\times10^{15}$ cm$^{-3}$. The intrinsic carrier concentration is  $10^{10}$ cm$^{-3}$.   What is the electron concentration?

☐ <u>1 point</u>

What is the hole concentration?

☐ <u>1 point</u>

**8.**  The concentration of ionized donors in a semiconductor is $10^{13}$ cm$^{-3}$.  The intrinsic carrier concentration is also $10^{13}$ cm$^{-3}$ (no acceptors).  What is the electron concentration?

☐ <u>1 point</u>

What is the hole concentration?

☐ <u>1 point</u>

**9.** Which material has a higher electron mobility – Si or GaAs?  Why?

☐ 2 points

**10.** What is the mechanism of the negative differential conductivity in GaAs?

☐ 1 point

**11.** The mobility at 150 K is 10,000 $cm^2$/Vs.  What is the diffusion coefficient?

☐ 1 point

**12.** The hole diffusion coefficient in a certain semiconductor is 10 $cm^2$/s.  The hole momentum relaxation time is $10^{-13}$ s.

a. What is the hole thermal velocity?

☐ 1 point

b. If the temperature is 300 K, what is the hole effective mass?

☐ 1 point

c. If the hole lifetime is 20 μs, what is the hole diffusion length?

☐ 1 point

**13.** Electrons in a semiconductor are affected by two scattering mechanisms: impurity scattering with the momentum relaxation time of $10^{-13}$ s, and lattice scattering with the momentum relaxation time of $2 \times 10^{-13}$ s.

a. What is the total effective momentum relaxation time?     ☐ 1 point

b. If  the effective mass is 0.1 $m_e$ where $m_e$ is the free electron mass, what is the low field mobility for this semiconductor?

☐ 1 point

c. What is the electron diffusion coefficient at 77 K?     ☐ 2 points

d. What is the electron thermal velocity at 77 K?

☐ 1 point

**14.** A light source is turned on at $t = 0$ creating a *uniform* generation rate $G = 10^{14}$ cm$^{-3}$s$^{-1}$ in a piece of an $n$-type semiconductor material doped at $10^{13}$ cm$^{-3}$. The hole life time is $5 \times 10^{-6}$ s. The intrinsic carrier density is $10^{10}$ cm$^{-3}$. Sketch the time dependence of the hole concentration and explain.

☐ 5 points

# PROBLEMS

**3-3-1.** Calculate and plot the intrinsic carrier concentration in Si for temperatures between 300 K and 400 K. Assume the effective masses of density of states for electrons and holes to be $m_n = 1.18\ m_e$ and $m_p = 0.81\ m_e$, respectively. (a) Assume that the energy gap $E_g = 1.12$ eV. (b) account for the temperature dependence of the energy gap: $E_g = E_{go} - \alpha T^2/(T + T_\beta)$ where $E_{go} = 1.17$ eV, $\alpha = 4.73 \times 10^{-4}$ eV/K, and $T_\beta = 636$ K. **Hint:** If you are using a personal computer, you may find it necessary to calculate $\ln(n_i)$ first and then $n_i$ to avoid numerical problems related to very small numbers.

**3-3-2.** Calculate and plot the intrinsic carrier concentration, $n_i$, in GaAs for 300 K $< T <$ 400 K. The effective masses of density of states for electrons and holes are $m_n = 0.067\ m_e$ and $m_p = 0.48\ m_e$, respectively. (a) Assume that the energy gap $E_g = 1.42$ eV. (b) Assume that $E_g = E_{go} - \alpha T^2/(T + T_\beta)$ where $E_{go} = 1.519$ eV, $\alpha = 5.405 \times 10^{-4}$ eV/K, and $T_\beta = 204$ K. **Hint:** If you are using a personal computer, you may find necessary to calculate $\ln(n_i)$ first and then $n_i$ to avoid numerical problems related to very small numbers.

**3-4-1.** Using the condition $n = p$, derive the equation linking the intrinsic Fermi level to the energy gap and to the effective masses of densities of states for electrons and holes.

**3-4-2.** A GaAs sample is doped with donors ($N_d = 10^{16}$ cm$^{-3}$) and acceptors ($N_a = 5 \times 10^{15}$ cm$^{-3}$). Calculate and plot the position of the Fermi level on temperature for temperatures between 100 K and 450 K. Assume that both donors and acceptors are fully ionized. Assume the effective mass of density of states for electrons $m_n = 0.067\ m_e$.

**3-5-1.** Plot the electron drift velocity in silicon versus electric field, $F$, for temperature $T = 77$ K and for $0 < F < 10$ kV/cm. Show on the same graph the electron thermal velocity given by $v_{th} = (3k_B T / m_n)^{1/2}$ where $m_n = 0.32\ m_e$. Assume the total impurity concentration to be equal to $10^{14}$ cm$^{-3}$. The electron drift velocity, $v_n$, in silicon can be approximated as

$$v_n = \frac{\mu_n F}{\sqrt{1 + (\mu_n F / v_s)^2}}$$

where

$$\mu_n(N_T, T) = \mu_{mn} + \frac{\mu_{on}}{1 + (N_T / N_{cn})^\nu}$$

$$\mu_{mn} = 88(T/300)^{-0.57} \quad (\text{cm}^2/\text{Vs})$$

$$\mu_{on} = 1.25 \times 10^3 (T/300)^{-2.33} \quad (\text{cm}^2/\text{Vs})$$

$$\nu = 0.88 (T/300)^{-0.146} \qquad N_{cn} = 1.26 \times 10^{17} (T/300)^{2.4} \quad (\text{cm}^{-3})$$

$$v_s = \frac{2.4 \times 10^7}{1 + 0.8 \exp(T/600)} \quad (\text{cm/s})$$

**3-5-2.** Calculate the electron temperature as a function of the electric field, $F$, for $0 \le F \le 5$ kV/cm if the device temperature is 77 K, the electron energy relaxation time is $10^{-12}$ s, the low field mobility, $\mu_n$, is 10,000 cm$^2$/Vs, the electron velocity, $v_n$, is given by

$$v_n = \frac{\mu_n F}{\sqrt{1+(\mu_n F / v_s)^2}}$$

and the electron saturation velocity $v_s = 1.2 \times 10^5$ m/s.

**3-6-1.** Define the field dependent mobility as $\mu_{field} = v_n/F$ where $v_n = \dfrac{\mu_n F}{\sqrt{1+(\mu_n F / v_s)^2}}$ is the electron velocity, and $F$ is the electric field. Assume $\mu_n = 1{,}000$ cm$^2$/Vs and $v_s = 10^5$ m/s. Assume the modified Einstein relationship $D_{eff} = \mu_{field} k_B T_e/q$ and calculate the dependence of $D_{eff}$ on $F$ for $0 \le F \le 100$ kV/cm assuming that $T_e = T_o + \dfrac{2}{3}\dfrac{q}{k_B}\tau_E F v_n$, $T_o = 300$ K, and $\tau_e = 10^{-12}$ s.

**3-6-2.** Choose a doping profile that will create a built-in field of 1 kV/cm in a 0.2 μm long sample. The sample temperature is 300 K.

**3-6-3.** In a GaAs field effect transistor, the doping concentration varies as $N_{do}\exp(x/l)$ for $0 < x < 0.1$ μm where $l = 0.1$ μm and $N_{do} = 10^{17}$ cm$^{-3}$. Plot the diffusion current density as a function of position. Calculate the built-in electric field that leads to the conduction current density compensating the diffusion current density. The low field mobility is 0.4 m$^2$/Vs. The device temperature is 300 K.

**3-7-1.** Calculate the resistance of the semiconductor sample shown in the figure.

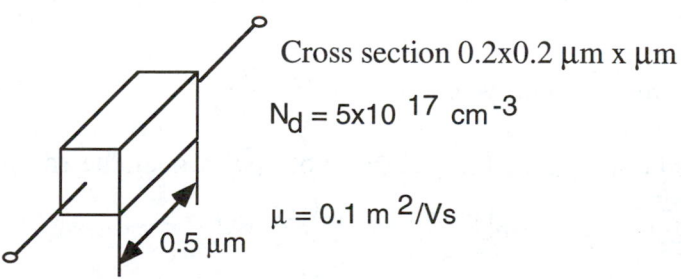

Cross section 0.2x0.2 μm x μm

$N_d = 5\times10^{17}$ cm$^{-3}$

$\mu = 0.1$ m$^2$/Vs

0.5 μm

**3-7-2.** Using the velocity-field characteristic for electrons in GaAs shown in Fig. 3.5.1, estimate (a) the low field mobility, (b) the maximum negative differential mobility, and (c) the doping level to obtain the resistivity of 1 $\Omega$cm.

**3-8-1.** The generation rate, $G$, for an $n$-type GaAs sample doped at $10^{15}$ cm$^{-3}$ is $10^{21}$ cm$^{-3}$s$^{-1}$. The recombination rate is $\Delta p/\tau$ where $\Delta p$ is the density of electron-hole pairs and $\tau = 10^{-8}$ s is the lifetime. Find the position of the electron and hole quasi-Fermi levels at room temperature. Assume the effective masses of density of states for electrons and holes to be $m_n = 0.067\ m_e$ and $m_p = 0.48\ m_e$, respectively.

**3-8-2.** The illumination of a semiconductor sample doped with shallow donors ($N_d = 10^{16}$ cm$^{-3}$) produces electron-hole pairs with the generation rate of $10^{23}$ cm$^{-3}$/s. The recombination rate is $R = b(np - n_i^2)$ where $n_i = 10^{10}$ cm$^{-3}$ and $b = 10^{-8}$ cm$^3$/s. What are the steady-state concentrations of electrons and holes?

**3-8-3.** Ultraviolet light with wavelength of 0.3 $\mu$m in vacuum and intensity $P_l = 50$ mW/cm$^2$ is shining at the surface of an $n$-type semiconductor sample (see the figure). The absorption coefficient, $\alpha = 10^6$ cm$^{-1}$. The quantum efficiency, $Q_e \approx 1$. The backside contact has a surface recombination rate $S_p$. Calculate and plot the steady-state hole distribution in the sample for the values of $S_p = 10^3$ cm/s, $10^5$ cm/s, $10^7$ cm/s. Neglect the intrinsic concentration of holes in the sample. Hole diffusion coefficient $D_p = 30$ cm$^2$/Vs, hole lifetime $\tau_p = 10^{-7}$ s, device length $L = 10$ $\mu$m.

**Hint:** Solve the continuity equation for holes $D_p \dfrac{\partial^2 p}{\partial x^2} - \dfrac{p - p_{no}}{\tau_p} = 0$

with the boundary conditions $D_p \dfrac{\partial^2 p}{\partial x^2}\bigg|_{x=0} = G$ where $G$ is the generation rate of the electron-hole pairs near the sample surface and

$D_p \dfrac{\partial p}{\partial x}\bigg|_{x=L} = -S_p[p(L) - p_{no}] \approx -S_p p(L).$

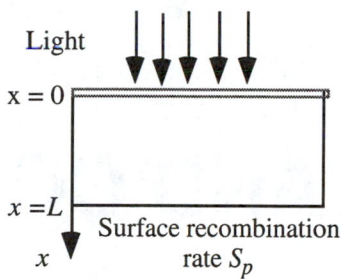

Light

$x = 0$

$x = L$

Surface recombination rate $S_p$

$x$

**3-8-4.** What is a possible explanation of the dependence of the lifetime, $\tau$, on the doping density in an $n$-type semiconductor shown in the figure?

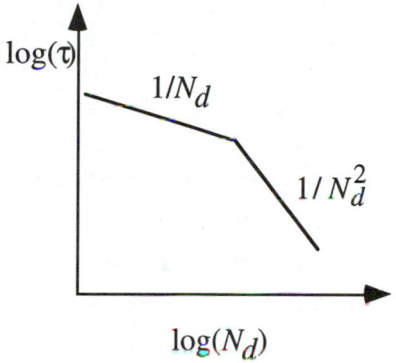

$\log(\tau)$

$1/N_d$

$1/N_d^2$

$\log(N_d)$

**3-8-5.** For the parameters given in Example 3-8-4, sketch the hole concentration profile. Repeat for the surface recombination rate, $S_p = 10^3$ m/s and compare.

**3-9-1.** Consider a Hall sample. The current is $I = 10$ mA. The sample length, width, and thickness are 100 µm, 5 µm, and 0.5 µm, respectively. The applied voltage is 1 V. The magnetic field is $B = 0.2$ T. The Hall factor is $r_H = 1.2$. The Hall voltage, $V_H = 2.5$ mV. Find the Hall constant, the Hall mobility, and the drift mobility.

**3-9-2.** Streamlines in Fig. 3.9.3 show the hole velocity. How will this figure change for $n$-type samples?

# 4

# DIODES AND CONTACTS

## 4-1. INTRODUCTION

This chapter deals with $p$-$n$ junctions, metal-semiconductor junctions, and ohmic contacts. Metal-semiconductor diodes (called **Schottky diodes**) and $p$-$n$ junction diodes have found numerous and important applications in electrical engineering. These devices also represent basic components that are used in more sophisticated three-terminal devices. As was mentioned in the Introduction to this book, two $p$-$n$ junctions can be combined together to form a Bipolar Junction Transistor (see Chap. 5). A $p$-$n$ junction can be used as a gate in a Field Effect Transistor (see Chap. 6) or as an integral part of a photodetector (see Chap. 8). And a vast majority of semiconductor devices include ohmic metal-semiconductor contacts. Hence, understanding how these basic elements work is required in order to understand the operation of all other semiconductor devices.

## 4-2. *p-n* JUNCTIONS at ZERO BIAS

A $p$-$n$ junction forms when $n$-type and $p$-type regions of the same semiconductor material are brought into intimate contact. In reality, such a contact cannot be achieved by just clamping together two pieces of a semiconductor material doped $n$-type and $p$-type, respectively, since, in this case, the surface of the contact will be too contaminated and too nonideal. Usually, $p$-$n$ junctions are obtained by either changing the material doping from $n$-type to $p$-type (or vice versa) during the sample growth, or by **ion implantation** of dopants that compensate the dopants in this semiconductor sample.

**188**

Ion implantation is a process during which a semiconductor sample is bombarded by highly energetic impurity atoms. These ions usually penetrate into a thin surface region of a semiconductor and heavily damage the crystal structure of this surface layer. The penetration depth depends on the implanted species and on the ion energy and may vary from only a few hundred angstroms to a micron or so. After the implantation, the sample is heated and kept for some time at a fairly high temperature. This procedure (called **annealing**) removes the damage. As a result, we introduce donors or acceptors (depending on implanted species) into the surface region. A typical concentration profile is very nonuniform (see Fig. 4.2.1). However, for a simplified analysis, we can replace this nonuniformly doped layer with a uniformly doped surface layer as shown in Fig. 4.2.1. As an example, the *p-n* junction shown in Fig. 4.2.1 may be obtained by implanting $N_{as} = 2 \times 10^{12}$ cm$^{-2}$ of boron atoms per unit area into an *n*-type Si crystal doped by phosphorus atoms with concentration $N_d = 10^{16}$ cm$^{-3}$. (Phosphorus acts as a donor and boron acts as an acceptor in Si; see Section 3.4). The implant depth may be on the order of $d_{imp} = 0.2$ µm. Hence, the volume concentration of the implanted acceptors would be equal to $N_a = N_{as}/d_{imp} = 10^{17}$ cm$^{-3}$, making the effective doping concentration in the surface layer $N_{aeff} = N_a - N_d = 9 \times 10^{16}$ cm$^{-3}$.

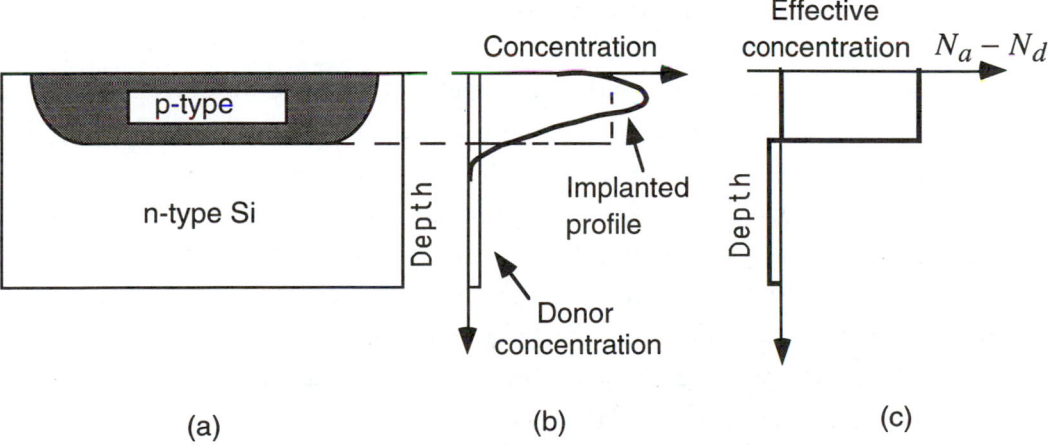

(a)                              (b)                              (c)

**Fig. 4.2.1.** (a) Si *p-n* junction obtained by implanting boron atoms into Si sample doped with phosphorus, (b) concentration profiles, and (c) simplified concentration profiles of donors and implanted acceptors.

In a $p$-$n$ junction, holes try to diffuse from the $p$-region, where their concentration is high, to the $n$-region, where their concentration is low. Since holes are charged positively, they leave behind, in the $p$-region, a layer of negatively charged acceptors. In a similar way, electrons try to diffuse from the $n$-region, where their concentration is high, to the $p$-region, where their concentration is low. Since electrons are charged negatively, they leave behind, in the $n$-region, a layer of positively charged donors (see Fig. I.4). These charges create a potential barrier that prevents more electrons from coming into the $p$-region and prevents more holes from coming into the $n$-region. This charged boundary layer (called a **space charge layer**) creates an electric field (called a **built-in electric field**) that maintains drift fluxes of the electrons and holes in the opposite direction to their diffusion fluxes. An equilibrium is reached when the Fermi level becomes constant throughout the entire system, and the overall fluxes (and, hence, the current densities) of electrons and holes are zero. Indeed, in equilibrium, the electron and hole current densities must be zero, and the quasi-Fermi levels of electrons and holes coincide with the position independent Fermi level, $E_F$:

$$j_n = \mu_n n \frac{\partial E_{Fn}}{\partial x} = \mu_n n \frac{\partial E_F}{\partial x} = 0$$

$$j_p = \mu_p p \frac{\partial E_{Fp}}{\partial x} = \mu_p p \frac{\partial E_F}{\partial x} = 0$$

where $j_n$, $j_p$, $n$, $p$, $\mu_n$, $\mu_p$, $E_{Fn}$, and $E_{Fp}$ are electron and hole current densities, carrier concentrations, mobilities, and quasi-Fermi levels, respectively [see eqs. (3-8-7) and (3-8-8) and the corresponding discussion in Section 3.8].

The energy band diagram of a $p$-$n$ junction is shown in Fig. 4.2.2. In the $n$-region, far from the $p$-$n$ junction, the electron and hole concentrations are $n_n \approx N_d$, $p_n = n_i^2/N_d$ where $n_i$ is the intrinsic carrier concentration. In the $p$-type region, far from the $p$-$n$ junction, the hole and electron concentrations are $p_p \approx N_a$, $n_p = n_i^2/N_a$. [Here, $N_d$ and $N_a$ are the concentrations of shallow (completely ionized) donors and acceptors.] Hence, the electron concentration changes from $N_d$ in the $n$-region to $n_i^2/N_a$ in the $p$-region, and the hole concentration changes from $n_i^2/N_d$ in the $n$-region to $N_a$ in the $p$-region (see Fig. 4.2.3). As shown in Fig. 4.2.2, far from the junction, the energy bands are flat. In the space charge region where the electron and hole concentrations change, the bands are bent. The electron and hole concentrations change by a factor of $\exp(1) \approx 2.718$ when

$E_c - E_F$ and $E_F - E_v$ change by a thermal energy $qV_{th} = k_B T$; see eqs. (3-3-2) and (3-3-3). ($V_{th} \approx 25.8$ mV and $qV_{th} \approx 25.8$ meV at $T = 300$ K.)

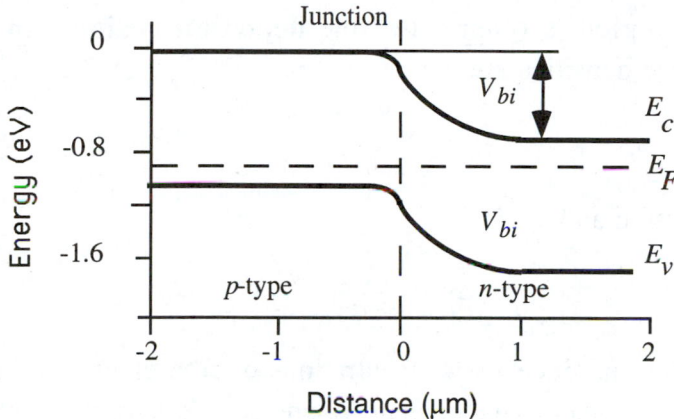

**Fig. 4.2.2.** Energy band diagram of a Si *p-n* junction. Acceptor density $N_a = 5 \times 10^{15}$ cm$^{-3}$, donor density $N_d = 1 \times 10^{15}$ cm$^{-3}$, $T = 300$ K.

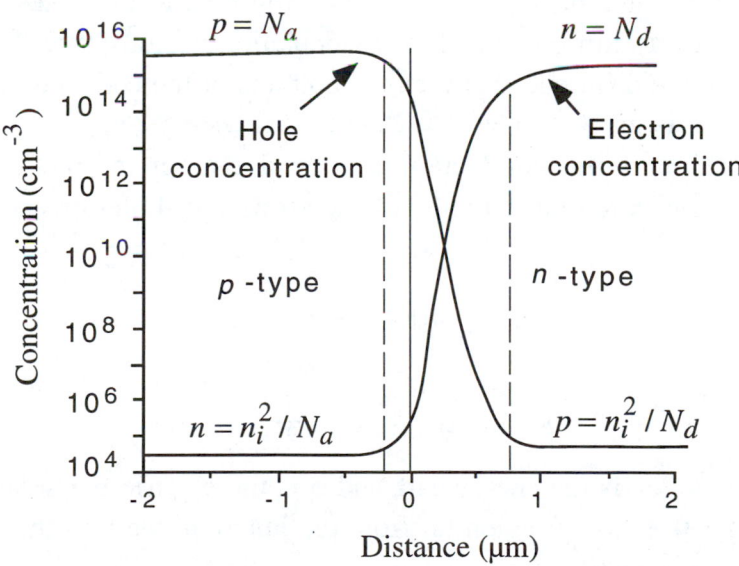

**Fig. 4.2.3.** Electron and hole carrier concentrations in a Si *p-n* junction. Acceptor density $N_a = 5 \times 10^{15}$ cm$^{-3}$, donor density $N_d = 1 \times 10^{15}$ cm$^{-3}$, $T = 300$ K. Dashed lines show the boundaries of the depletion region.

The total band bending is typically several tenths of an electron volt (depending on doping levels, the energy gap, and the effective densities of states in the conduction and valence bands), that is, much greater than $qV_{th}$. Hence, $n \ll N_d$ and $p \ll N_a$ in almost the entire space charge region where the bands are bent. Therefore, this region is often called the **depletion region.** In the depletion region, the charge densities are

$$\rho_n \approx qN_d \qquad (4\text{-}2\text{-}1)$$

in the $n$-type section and

$$\rho_p \approx -qN_a \qquad (4\text{-}2\text{-}2)$$

in the $p$-type section, since the concentrations of free electrons and holes in the depletion layer are much smaller than $N_d$ and $N_a$, respectively ($N_d \gg n \gg n_{po}$ and $N_a \gg p \gg p_{no}$, where $n_{po}$ and $p_{no}$ are the equilibrium electron and hole concentrations in the $p$- and $n$-type regions). This approximation is called the **depletion approximation**.

The depletion approximation is not valid in the boundary layers between the neutral sections and the depletion regions. However, as discussed below, these boundary layers are thin compared to the depletion region since the total band bending, that is, the difference between the bottoms of the conduction band in the $p$- and $n$-regions, is much greater than the thermal energy $qV_{th}$.

The bottom of the conduction band and the top of the valence band correspond to the potential energies of electrons and holes, respectively (see Section 3.7):

$$E_c = -q\phi + const \qquad (4\text{-}2\text{-}3)$$

$$E_v = -q\phi - E_g + const \qquad (4\text{-}2\text{-}4)$$

where $E_g = E_c - E_v$ is the energy gap, and $\phi$ is the electric potential. Choosing $E_c = 0$ and $\phi = 0$ in the $n$-region far from the junction, we find the electron and hole concentration profiles

$$n = N_d \exp\left(\frac{\phi(x)}{V_{th}}\right) = N_d \exp\left(-\frac{E_c(x)}{qV_h}\right) \qquad (4\text{-}2\text{-}5)$$

$$p = N_a \exp\left[-\frac{V_{bi} + \phi(x)}{V_{th}}\right] = N_a \exp\left[\frac{-qV_{bi} + E_c(x)}{qV_{th}}\right] \qquad (4\text{-}2\text{-}6)$$

where $V_{bi}$ is the **built-in voltage** (see Figures 4.2.2 and 4.2.3).

In the *n*-type region far from the junction $n = N_d$, $\phi = 0$, and, according to eq. (4-2-6), $p = N_a \exp(-V_{bi}/V_{th})$. Since $pn = N_d N_a \exp(-V_{bi}/V_{th}) = n_i^2$ we find that

$$V_{bi} = V_{th} \ln\left(\frac{N_d N_a}{n_i^2}\right) \qquad (4\text{-}2\text{-}7)$$

Since

$$n_i = (N_c N_v)^{1/2} \exp\left(-\frac{E_g}{2k_B T}\right) \qquad (4\text{-}2\text{-}8)$$

[see eq. (3-3-11)], eq. (4-2-7) can be rewritten as follows:

$$V_{bi} = E_g / q + V_{th} \ln\left(\frac{N_d N_a}{N_c N_v}\right) \qquad (4\text{-}2\text{-}9)$$

This equation shows that the built-in voltage is approximately proportional to the energy gap (see Fig. 4.2.4).

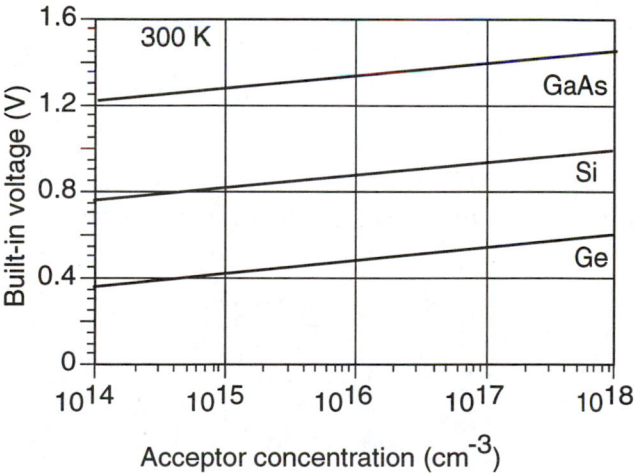

**Fig. 4.2.4.** Built-in voltage for $n^+$-$p$ junctions versus acceptor concentration. The donor density in the $n^+$ region is $10^{19}$ cm$^{-3}$.

The field distribution in the depletion region is determined from Poisson's equation

$$\frac{dF}{dx} = \frac{\rho}{\varepsilon_s}$$

(4-2-10)

where $\varepsilon_s$ is the dielectric permittivity of the semiconductor and $\rho = q(p - n + N_d - N_a)$ is the space charge density.

Using the depletion approximation, we obtain

$$\frac{dF}{dx} = \begin{cases} -\dfrac{qN_a}{\varepsilon_s} & \text{for} \quad -x_p < x < 0 \\[2mm] \dfrac{qN_d}{\varepsilon_s} & \text{for} \quad 0 < x < x_n \end{cases}$$

(4-2-11)

Here, $x = 0$ corresponds to the boundary between the $p$- and $n$- type regions, and $x_p$ and $x_n$ are the depletion widths on the two sides of the junction. An example of such a charge distribution is shown in Fig. 4.2.5.

Integrating eq. (4-2-11) and using the conditions $F(x = x_n) = 0$, $F(x = -x_p) = 0$, we obtain

$$F = \begin{cases} -F_m\left(1 + \dfrac{x}{x_p}\right) & \text{for} \quad -x_p < x < 0 \\[2mm] -F_m\left(1 - \dfrac{x}{x_n}\right) & \text{for} \quad 0 < x < x_n \end{cases}$$

(4-2-12)

(see Fig. 4.2.5). The maximum magnitude of the electric field, $F_m$, in the junction (reached at the $p$-$n$ interface, at $x = 0$) is given by

$$F_m = \frac{qN_d x_n}{\varepsilon_s} = \frac{qN_a x_p}{\varepsilon_s}$$

(4-2-13)

Since

$$F = -\frac{d\phi}{dx} = \frac{1}{q}\frac{dE_c}{dx}$$

where $\phi = -E_c/q$ is the electric potential, the potential distribution is found by integrating eq. (4-2-12):

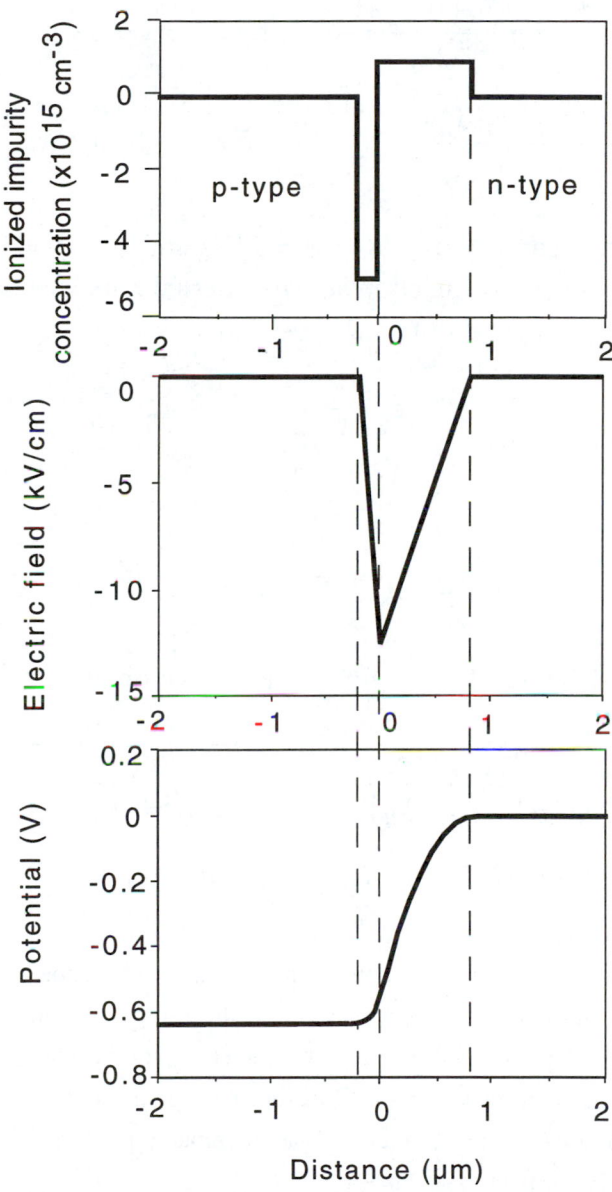

**Fig. 4.2.5.** Ionized dopant density (negative numbers correspond to ionized acceptors and positive to ionized donors), electric field, and potential profiles for a silicon *p-n* junction. Acceptor density $N_a = 5 \times 10^{15}$ cm$^{-3}$, donor density $N_d = 1 \times 10^{15}$ cm$^{-3}$, $T = 300$ K.

$$\phi = \begin{cases} -V_{bi} + \dfrac{qN_a\left(x + x_p\right)^2}{2\varepsilon_s} & \text{for} \quad -x_p < x < 0 \\[3ex] -\dfrac{qN_d\left(x - x_n\right)^2}{2\varepsilon_s} & \text{for} \quad 0 < x < x_n \end{cases} \qquad (4\text{-}2\text{-}14)$$

where we used the conditions $\phi(-x_p) = -V_{bi}$ and $\phi(x_n) = 0$. The potential distribution for a silicon p-n junction at zero external bias is shown in Fig. 4.2.5.

From eqs. (4-2-13) and (4-2-14), we find

$$\frac{x_n}{x_p} = \frac{N_a}{N_d} \qquad (4\text{-}2\text{-}15)$$

$$\frac{qN_d x_n^2}{2\varepsilon_s} + \frac{qN_a x_p^2}{2\varepsilon_s} = V_{bi} \qquad (4\text{-}2\text{-}16)$$

Solving eqs. (4-2-15) and (4-2-16) for $x_n$ and $x_p$, we obtain

$$x_n = \sqrt{\frac{2\varepsilon_s V_{bi}}{qN_d\left(1 + N_d / N_a\right)}} \qquad x_p = \sqrt{\frac{2\varepsilon_s V_{bi}}{qN_a\left(1 + N_a / N_d\right)}} \qquad (4\text{-}2\text{-}17)$$

For a $p^+$-n junction, where $N_a \gg N_d$, we have $x_n \gg x_p$ (see Fig. 4.2.5).

**Example 4-2-1.**
Calculate and plot the total depletion width, $x_n + x_p$, as a function of doping for a Si p-n junction with the donor concentration in the n-type region, $N_d$, equal to the acceptor concentration in the p-region, $N_a$ (such a junction is called a **symmetrical junction**). Temperature $T = 300$ K. The dielectric permittivity of Si is $1.05 \times 10^{-10}$ F/m. The intrinsic carrier concentration is approximately $10^{10}$ cm$^{-3}$. Vary the doping density from $10^{14}$ cm$^{-3}$ to $10^{17}$ cm$^{-3}$.

**Solution:**
For a symmetrical junction, $x_n + x_p = 2\sqrt{\varepsilon_s V_{bi}/(qN)}$ where $N = N_d = N_a$, and $V_{bi} = 2V_{th} \ln(N/n_i)$ where $V_{th} = k_B T/q$ is the thermal voltage (see Fig. 4.2.6).

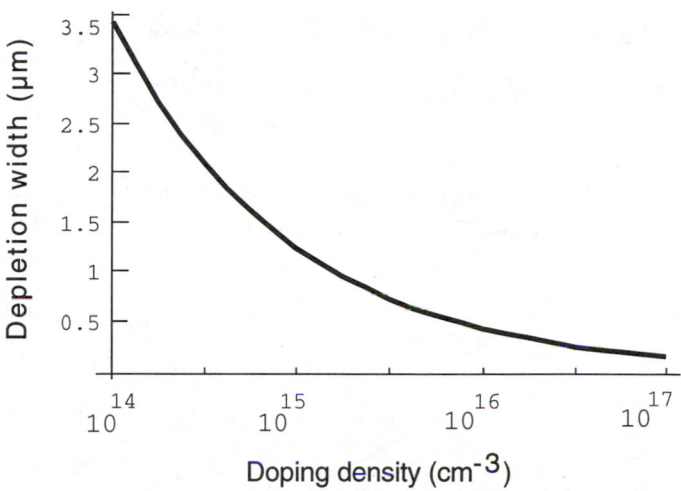

**Fig. 4.2.6.**  Depletion width of symmetrical Si *p-n* junction versus doping density.

The depletion approximation breaks down in the thin boundary layers between the depletion layer and neutral layers where the carrier concentrations are comparable with the doping concentrations.  We choose $\phi = 0$ in the neutral *n*-type region where $n \approx N_d$.  In other regions of the device, $n = N_d \exp(\phi/V_{th})$, and the space charge density $\rho = q(N_d - n) = N_d - N_d \exp(\phi/V_{th})$.  Hence, in the boundary layer in the *n*-type region, Poisson's equation becomes

$$\frac{d^2\phi}{dx^2} = -\frac{qN_d}{\varepsilon_s} + \frac{qN_d}{\varepsilon_s}\exp\left(\frac{\phi}{V_{th}}\right) \tag{4-2-18}$$

When  $n - N_d \ll N_d$  (i. e., $\phi \ll V_{th}$),  the  second  term  on  the  right-hand side of eq. (4-2-18) can be expanded in a Taylor series, $[\exp(\phi/V_{th}) \approx 1 + \phi/V_{th}]$, yielding

$$\frac{d^2\phi}{dx^2} \approx \frac{qN_d\phi}{\varepsilon_s V_{th}} = \frac{\phi}{L_{Dn}^2} \tag{4-2-19}$$

where the characteristic scale of the potential variation

$$L_{Dn} = \sqrt{\frac{\varepsilon_s V_{th}}{qN_d}} \tag{4-2-20}$$

is called the Debye length.  The width of the boundary layer is of the order of several Debye lengths.  For the $p$-type region, the width of the boundary region is determined by several Debye lengths, $L_{Dp} = \sqrt{\dfrac{\varepsilon_s V_{th}}{qN_a}}$.

**Example 4-2-2.**

Calculate and plot the Debye length in GaAs ($\varepsilon_s = 1.14 \times 10^{-10}$ F/m) versus doping density, $N$, at $T = 300$ K and T = 77 K for $10^{14}$ cm$^{-3}$ < $N$ < $10^{17}$ cm$^{-3}$.

**Solution:**

See Fig. 4.2.7.

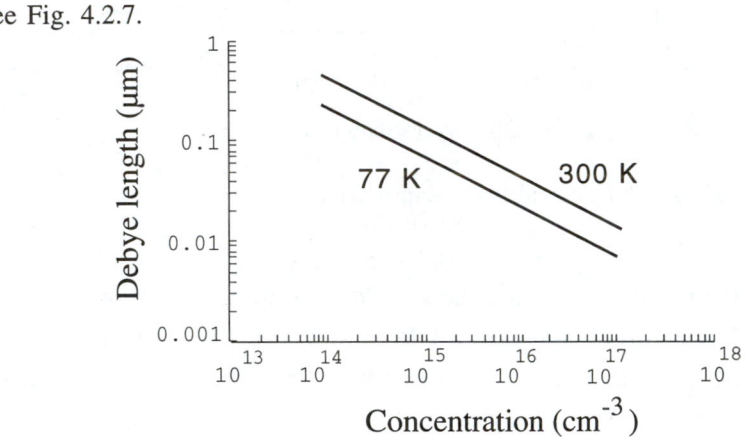

**Fig. 4.2.7.**  Debye length versus doping in GaAs.

The summary of  equations relevant to properties of $p$-$n$ junctions at zero bias is given in Table 4.2.1.

| Maximum electric field in a $p$-$n$ junction | $F_m = \dfrac{qN_d x_n}{\varepsilon_s} = \dfrac{qN_a x_p}{\varepsilon_s}$ |
|---|---|
| Built-in voltage | $V_{bi} = V_{th}\, \ln\!\left(N_d N_a / n_i^2\right)$ |
| Width of depletion regions | $x_n = \sqrt{\dfrac{2\varepsilon_s V_{bi}}{qN_d\left(1 + N_d / N_a\right)}}$ , $x_p = \sqrt{\dfrac{2\varepsilon_s V_{bi}}{qN_a\left(1 + N_a / N_d\right)}}$ |
| Debye lengths | $L_{Dn} = \sqrt{\dfrac{\varepsilon_s V_{th}}{qN_d}}$ , $L_{Dp} = \sqrt{\dfrac{\varepsilon_s V_{th}}{qN_a}}$ |

**Table 4.2.1.**  Summary of equations describing $p$-$n$ junctions at zero bias.

# 4-3. *p-n* JUNCTIONS under BIAS

### 4.3.1.   The law of the junction.

A negative voltage applied to the *p*-region (with respect to the *n*-region) pulls the electrons and holes away from the junction.  It increases the potential barrier between the *n*- and *p*-regions.  This polarity corresponds to a **reverse bias**.  A **forward bias** (positive potential applied to the *p*-region) decreases the potential barrier between the *n*- and *p*-regions and, hence, the diode resistance.  (Since under reverse bias or small forward bias, the resistance of the depletion region is much greater than the resistance of the neutral sections, almost all applied bias drops across the depletion region.)

At zero bias, the drift current (caused by this electric field) is exactly compensated by the diffusion current.  The net current density is zero:

$$j_n = q\left(n\mu_n F + D_n \frac{\partial n}{\partial x}\right) = q\mu_n n dE_{Fn}/dx = q\mu_n n dE_F/dx = 0 \qquad (4\text{-}3\text{-}1a)$$

$$j_p = q\left(p\mu_p F - D_p \frac{\partial p}{\partial x}\right) = q\mu_p p dE_{Fp}/dx = q\mu_p p dE_F/dx = 0 \qquad (4\text{-}3\text{-}1b)$$

and the Fermi level is independent of position [see eqs. (3-7-2) to (3-7-3) and (3-8-7) to (3-8-8)].  An external voltage violates this balance between the drift and diffusion currents.  However, for small voltages, the net current is still much smaller than the drift and diffusion currents in the depletion region evaluated separately.   This means that the carrier distribution is still close to the equilibrium state and, consequently, the electron and hole quasi-Fermi levels defined by

$$n = N_c \exp\left(\frac{E_{Fn} - E_c}{k_B T}\right), \quad p = N_v \exp\left(\frac{E_v - E_{Fp}}{k_B T}\right) \qquad (4\text{-}3\text{-}2)$$

remain nearly constant throughout the depletion region (see Fig. 4.3.1).  Numerical simulations of *p-n* junctions at reverse bias and low forward biases confirm that $E_{Fn}$ and $E_{Fp}$ in the depletion region remain practically the same as $E_{Fn}$ and $E_{Fp}$ in the *n*-type and *p*-type regions, respectively.

At zero bias, the potential barrier between the *n* and *p* regions is equal to $qV_{bi}$, and the Fermi level is constant throughout the diode (see Fig. 4.3.1).

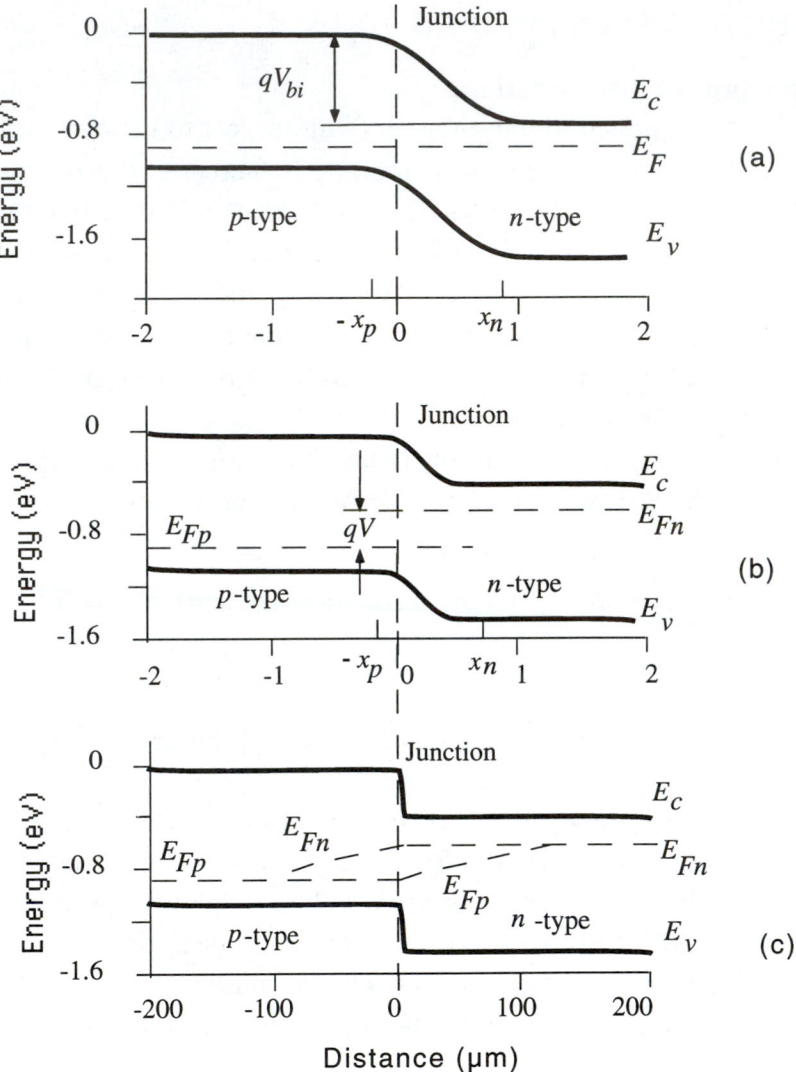

**Fig. 4.3.1.** Energy band diagram of a Si *p-n* junction for (a) zero bias and (b, c) forward bias of 0.24 V. Acceptor density $N_a = 5 \times 10^{15}$ cm$^{-3}$, donor density $N_d = 1 \times 10^{15}$ cm$^{-3}$, $T = 300$ K. $-x_p$ and $x_n$ are the boundaries between the depletion and neutral regions. Notice that $E_{Fn}$ and $E_{Fp}$ remain practically constant within the depletion region (see Fig. 4.3.1b). Outside the depletion region, $E_{Fn}$ and $E_{Fp}$ vary over distances much larger than the depletion layer width (see Fig. 4.3.1c). The characteristic lengths of the $E_{Fn}$ and $E_{Fp}$ variations are determined by the diffusion lengths of electrons and holes, $L_n$ and $L_p$, (see Section 3.8).

A forward bias, $V$, reduces the potential barrier to $V_{bi} - V$, and the difference between $E_{Fn}$ and $E_{Fp}$ in the depletion region becomes

$$qV = E_{Fn} - E_{Fp} \qquad (4\text{-}3\text{-}3)$$

(see Fig. 4.3.1b). Using eq. (4-3-3), we find the *pn* product in the depletion region and at its boundaries:

$$pn = n_i^2 \exp\left(\frac{E_{Fn} - E_{Fp}}{k_B T}\right) = n_i^2 \exp\left(\frac{V}{V_{th}}\right) \qquad (4\text{-}3\text{-}4)$$

Equation (4-3-4) is called the **law of the junction**.

As was done in Section 2.2, we will denote the electron and hole concentrations in the $n$ and $p$ type regions as $p_n$, $n_n$ and $p_p$, $n_p$, respectively. Let us consider a point in the neutral region at the $n$-side of the junction close to the depletion region boundary where $n_n \approx N_d$. From eq. (4-3-4) we obtain

$$p_n(x_n) \approx \frac{n_i^2}{N_d} \exp\left(\frac{V}{V_{th}}\right) \qquad (4\text{-}3\text{-}5)$$

Similarly, for the neutral region at the $p$-side of the junction close to the depletion region boundary, we find

$$n_p(-x_p) \approx \frac{n_i^2}{N_a} \exp\left(\frac{V}{V_{th}}\right) \qquad (4\text{-}3\text{-}6)$$

Hence, a small variation in bias can lead to a large variation in the minority carrier concentration. At room temperature ($T = 300$ K), the change in $V$ by $V_{th} = k_B T/q = 0.02584$ V leads to an increase in the minority carrier concentrations at the edges of the depletion region by a factor of $\exp(1) \approx 2.71828$.

**Example 4-3-1.**

The intrinsic carrier concentration in GaAs at 300 K is approximately $10^6$ cm$^{-3}$. Estimate the forward bias required to create an electron-hole density in the depletion region (at the point where $p = n$) of $10^{17}$ cm$^{-3}$.

**Solution:**

From the law of the junction $pn/n_i^2 = \exp(V/V_{th})$. Hence,

$$V = V_{th} \ln\left(\frac{pn}{n_i^2}\right) = 0.02584 \times \ln\left(10^{11}\right) = 0.654 \text{ (V)}$$

### 4.3.2. Minority carrier profiles in neutral regions.

As was shown in Section 3.8, the concentration, $p_n$, of minority carriers (holes) in the $n$-type neutral region is described by the diffusion equation

$$D_p \frac{\partial^2 p_n}{\partial x^2} - \frac{p_n - p_{no}}{\tau_{pl}} = 0 \tag{4-3-7}$$

[compare with eq. (3-8-26)]. A general solution of eq. (4-3-7) is given by

$$p_n(x) - p_{no} = A \exp\left(\frac{x - x_n}{L_p}\right) + B \exp\left(-\frac{x - x_n}{L_p}\right) \tag{4-3-8}$$

[compare with eq. (3-8-27)] where

$$L_p = \sqrt{D_p \tau_{pl}} \tag{4-3-9}$$

is called the **hole diffusion length** and the constants $A$ and $B$ are to be determined from the boundary conditions:

$$p_n(x_n) = p_{no} \exp\left(\frac{V}{V_{th}}\right) \tag{4-3-10}$$

$$p_n(x \rightarrow \infty) = p_{no} \tag{4-3-11}$$

Using these boundary conditions, we find from eq. (4-3-8)

$$p_n(x) - p_{no} = p_{no}\left[\exp\left(\frac{V}{V_{th}}\right) - 1\right] \exp\left(-\frac{x - x_n}{L_p}\right) \tag{4-3-12}$$

The distributions of electrons and holes in a typical $p$-$n$ junction are shown in Fig. 4.3.2 for zero bias and a forward bias of 0.24 V. As can be seen from Fig. 4.3.2, the forward bias sharply increases the concentrations of the minority carriers at the boundaries between the neutral regions and depletion region. On a scale comparable to the width of the depletion layer, the variation of the minority concentrations in the neutral regions is not noticeable. These concentrations decay on a scale determined by the diffusion lengths (see Fig. 4.3.2b). In the semilog scale used in Fig. 4.3.2b, the exponential variations of the minority carrier concentrations predicted by eq. (4-3-12) are represented by straight lines.

**Example  4-3-2.**

The slope of the semilog dependence of the hole concentration on distance in the *n*-type region of a *p-n* diode at room temperature ($T = 300$ K) is 1 decade/120 μm. The hole mobility is 250 cm²/V.  What is the hole lifetime?

**Solution:**

The slope in the semilog scale is equal to $1/[L_p \ln(10)]$.  Hence $L_p$ = 120/ln(10)=52.1 μm.  Using the Einstein relation, we find $D_p = \mu_p V_{th}$ = 250×0.02584 = 6.46 (cm²/s).  Hence, the hole lifetime,

$$\tau_{pl} = \frac{L_p^2}{D_p} = \frac{\left(52.1 \times 10^{-4}\right)^2}{6.46} = 4.20 \times 10^{-6} \, (s)$$

As an exercise, estimate the diffusion lengths of electrons in the *p*-type region and holes in the *n*-type region from Fig. 4.3.2b.

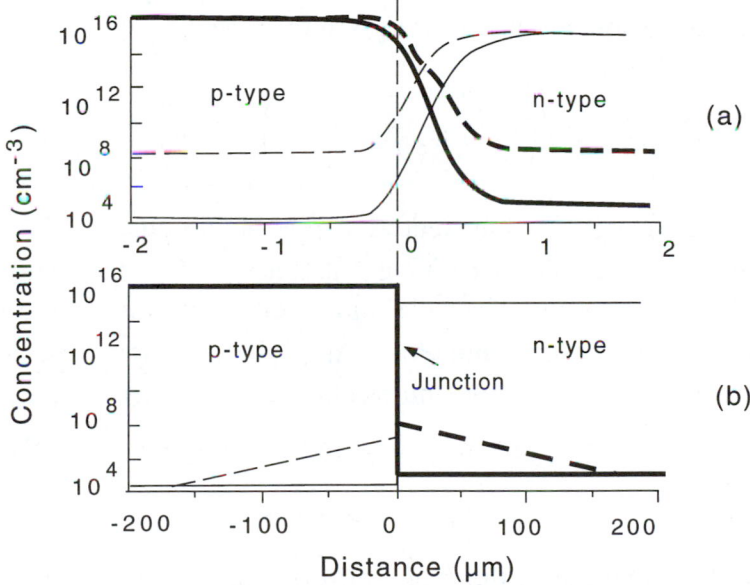

**Fig. 4.3.2.** Electron (thin lines) and hole (thick lines) concentrations in Si *p-n* junction for zero bias (solid lines) and forward bias of 0.24 V (dashed lines), (a) large and (b) small scale.  $N_a = 10^{16}$ cm⁻³, $N_d = 1 \times 10^{15}$ cm⁻³, $T = 300$ K.

## 4.3.3.    Ideal Diode equation.

The hole current density, $j_p$, in the neutral *n*-type region, is primarily caused by diffusion (see Problem 4-3-2), that is,

$$j_p \approx j_{pD} = -qD_p \frac{\partial p_n}{\partial x} = \frac{qD_p p_{no}}{L_p} \left[ \exp\left(\frac{V}{V_{th}}\right) - 1 \right] \exp\left(-\frac{x - x_n}{L_p}\right) \qquad (4\text{-}3\text{-}13)$$

The electron diffusion current density in the same region, $j_{nD}$, can be found from the quasi neutrality condition:

$$n_n - n_{no} \approx p_n - p_{no}$$

which leads to

$$j_{nD} = qD_n \frac{\partial n_n}{\partial x} \approx qD_n \frac{\partial p_n}{\partial x} = -\frac{D_n}{D_p} j_{pD} \qquad (4\text{-}3\text{-}14)$$

$$= -\frac{qD_n p_{no}}{L_p} \left[ \exp\left(\frac{V}{V_{th}}\right) - 1 \right] \exp\left(-\frac{x - x_n}{L_p}\right)$$

so that the total diffusion current density $j_D$ in the $n$-type neutral region becomes

$$j_D = j_{pD} + j_{nD} \approx \frac{qp_{no}}{L_p} (D_p - D_n) \left[ \exp\left(\frac{V}{V_{th}}\right) - 1 \right] \exp\left(-\frac{x - x_n}{L_p}\right) \qquad (4\text{-}3\text{-}15)$$

The total current density in the neutral $n$-region is equal to the sum of the diffusion current density and the drift current density. This total current density, $j$, is independent of position. The hole and electron diffusion current densities vary with distance due to recombination. As explained above, the characteristic scales of this variation are the hole and electron diffusion lengths, $L_p = \sqrt{D_p \tau_{pl}}$ and $L_n = \sqrt{D_n \tau_{nl}}$ for the $n$-type and $p$-type regions, respectively. Typically,

$$L_n \gg x_n + x_p \qquad L_p \gg x_n + x_p \qquad (4\text{-}3\text{-}16)$$

(see Fig. 4.3.2b). Therefore, the hole and electron diffusion current densities remain nearly constant throughout the depletion region, and the total current density is given by

$$j \approx j_{Dp}\big|_{x=x_n} + j_{Dn}\big|_{x=-x_p} \qquad (4\text{-}3\text{-}17)$$

where $j_{Dp}$ and $j_{Dn}$ are the hole and electron diffusion current densities in the neutral $n$- and $p$-type regions, respectively. (Here we have taken into account that the drift components of the minority carrier currents are small compared to their diffusion currents; see Problem 4-3-2). Hence, from eq. (4-3-13) and from a similar equation for the electrons in the *p*-type neutral region, we find

$$I = I_s \left[ \exp\left(\frac{V}{V_{th}}\right) - 1 \right] \tag{4-3-18}$$

where  $I = jS$ is the diode current, $S$ is the cross section, and $I_s$ is the **diode saturation current** given by

$$I_s = S\left( \frac{qD_p\, p_{no}}{L_p} + \frac{qD_n\, n_{po}}{L_n} \right) \tag{4-3-19}$$

Equation (4-3-18) is called the **Shockley equation** or the **Ideal Diode equation**. In many cases, this equation does not accurately describe the current-voltage characteristics of *p-n* junction diodes at forward biases. The reason is that, in addition to the diffusion current components considered above, two other, nonideal, current components (called **generation** and **recombination currents**) may play an important role (see Subsection 4.3.5). These currents depend on the concentrations, distribution, and energy levels of different impurities in the depletion region, which may trap or emit electrons and holes. Such impurities (called **traps**) are always present in a semiconductor material.

### 4.3.4.  Short *p+-n* diode.

The case of a short *p+-n* diode such that the length of the *n*-section, $X_n$, is much smaller than the hole diffusion length, $L_p$, is important for the analysis of the bipolar junction transistor (see Chap. 5). For this case, the solution of the diffusion equation (4-3-7) is given by

$$p_n(x) \approx p_{no} + p_{no}\left[ \exp\left(\frac{V}{V_{th}}\right) - 1 \right]\left( \frac{X_n - x}{X_n - x_n} \right) \tag{4-3-20}$$

(see Problem 4-3-3), and the hole distribution in the *n*-type region is a linear function of *x*. In this case, the  current density is given by

$$j \approx -qD_p \left. \frac{\partial p_n}{\partial x} \right|_{x=x_n} = \frac{qD_p p_{no}}{X_n - x_n}\left[ \exp\left(\frac{V}{V_{th}}\right) - 1 \right] \tag{4-3-21}$$

This current density is $L_p/(X_n - x_n)$ larger than for a long *p+-n* diode [compare with eq. (4-3-19)].

### 4.3.5.   Generation and recombination currents.

The **generation current** density becomes important at reverse bias voltages when the depletion region is devoid of carriers and, according to the law of the junction,

$$\frac{pn}{n_i^2} = \exp\left(-\frac{|V|}{V_{th}}\right) << 1 \qquad (4\text{-}3\text{-}22)$$

Under thermal equilibrium, the thermal generation of electron-hole pairs is balanced by their recombination ($G = R$).  Under a reverse bias, there are very few electron-hole pairs in the depletion region , and the recombination rate, $R$, is nearly zero.  However, the thermal generation processes continuously supply the electron-hole pairs to the depletion region.  This thermal generation rate can be estimated as

$$G_{thermal} = \frac{n_i}{\tau_{gen}} \qquad (4\text{-}3\text{-}23)$$

where $\tau_{gen}$ is the effective generation time of electron-hole pairs in the depletion region.  The total electron charge (equal to the hole charge) supplied into the depletion region per unit area per second is equal to $qG_{thermal}x_d = qn_ix_d/\tau_{gen}$ where $x_d = x_n + x_p$ is the width of the depletion region.

Figure 4.3.3 illustrates what happens to these charges: they are swept away by the junction electric field during the transit time $t_{tr} \approx x_d/v_s$, where $v_s$ is the carrier saturation velocity (since the electric field in the depletion region under reverse bias is very strong).  The carrier saturation velocity is on the order of $10^5$ m/s, and the depletion width is typically a few microns or less (see Fig. 4.2.6).  Hence, the transit time is on the order of 10 ps or so.  The generation times typically vary between a microsecond and a nanosecond (orders of magnitude higher that $t_{tr}$).  On the time scale of $\tau_{gen}$, the generated carriers are swept away almost instantaneously, and the generation current density is

$$j_{gen} = \frac{qn_ix_d}{\tau_{gen}} \qquad (4\text{-}3\text{-}24)$$

Substituting $V_{bi}$ with $V_{bi} - V$ in eqs. (4-2-17) and (4-2-18) for the depletion widths, $x_n$ and $x_p$, we find

$$x_n = \sqrt{\frac{2\varepsilon_s\left(V_{bi} - V\right)}{qN_d\left(1 + N_d / N_a\right)}} \tag{4-3-25}$$

$$x_p = \sqrt{\frac{2\varepsilon_s\left(V_{bi} - V\right)}{qN_a\left(1 + N_a / N_d\right)}} \tag{4-3-26}$$

Thus, $x_d = x_p + x_n$ and $j_{gen}$ are proportional to $(V_{bi} - V)^{1/2}$.

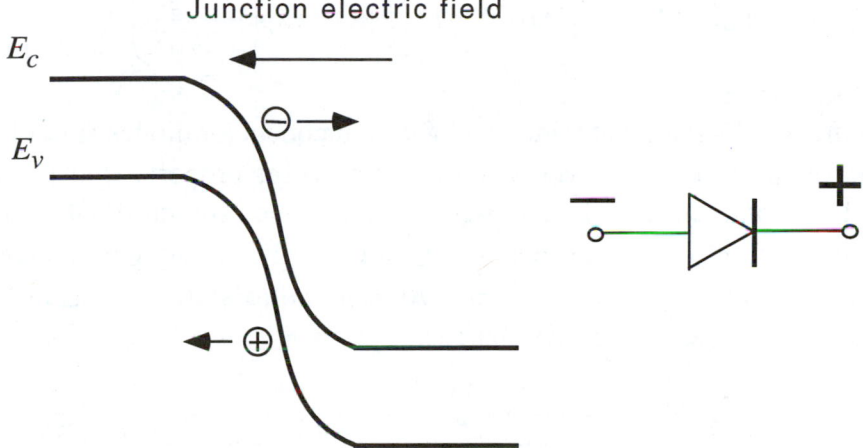

**Fig. 4.3.3.**  Band diagram of *p-n* diode under reverse bias.  Arrows show the direction of electron and hole drift in the junction electric field. Also shown is the diode circuit symbol with voltage polarity corresponding to reverse bias.

The total reverse current density, $j_R$, is given by:

$$j_R = j_s + j_{gen} \tag{4-3-27 a}$$

where the diffusion component is

$$j_s = \left(\frac{qD_p}{N_d L_p} + \frac{qD_n}{N_a L_n}\right)n_i^2 \tag{4-3-27 b}$$

[compare with eq. (4-3-19)].  Since $j_s$ is proportional to $n_i^2$ and $j_{gen}$ is proportional to $n_i$, the generation current will be dominant when $n_i$ is sufficiently small.   In practice, this is often the case for Si, GaAs, and wider gap semiconductors at room temperature and lower temperatures.

The intrinsic carrier density, $n_i$, was estimated in Section 3.3 assuming effective masses $m_n = 0.3\ m_e$ and $m_p = 0.6\ m_e$:

$$n_i\left(\mathrm{m}^{-3}\right) = 1.34 \times 10^{21} \times T^{3/2}(K) \exp\left(-\frac{E_g}{2k_BT}\right)$$

[see eq. (3-3-12) and Fig. 3.3.3].  Assuming the depletion region volume to be 1 $\mu\mathrm{m}^3$ and the generation time of 1 ns, we obtain

$$I_{gen}(\mathrm{A}) = 2.14 \times 10^{-7} \times T^{3/2}(\mathrm{K}) \times \exp\left(-\frac{E_g}{2k_BT}\right)$$

(see Fig. 4.3.4).

Figure 4.3.4 shows that wide band gap semiconductor diodes should behave almost as open circuits at reverse bias.  This useful property can be used in integrated circuits where reverse biased *p-n* junctions provide device isolation. Wide band gap semiconductors such as SiC or GaN (the energy gap of GaN is 3.4 eV) may eventually be used in nonvolatile solid-state memories where information is encoded by electric charges.

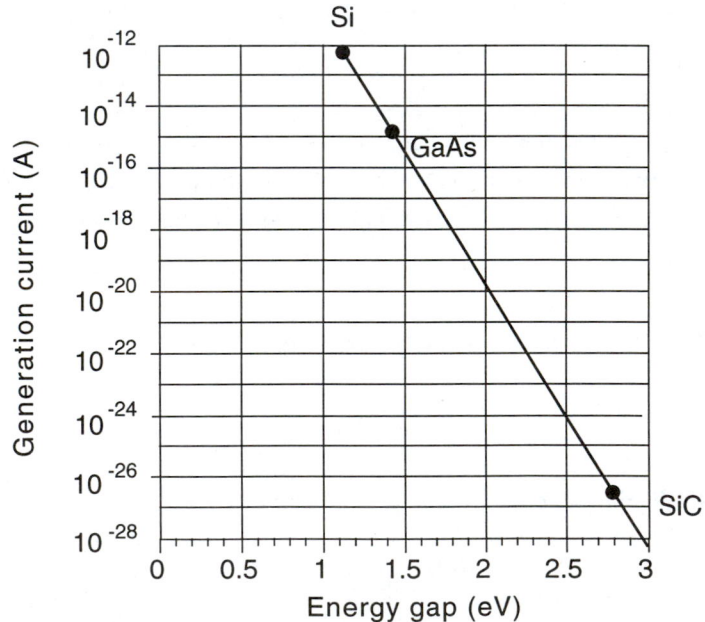

**Fig. 4.3.4.**   Generation current (per 1 $\mu\mathrm{m}^3$ of the depletion region volume) versus energy gap (at room temperature).

A very long storage time of such memories will be ensured by the vanishingly small reverse current in the depletion regions, which will isolate these charges. In practice, however, very small values of the reverse current are difficult to obtain because a parasitic surface leakage current determines the lower bound of the leakage current.

Under forward bias conditions ($V > 0$) excess electrons and holes are injected into the depletion region, where some of them recombine. The recombination current density is equal to the total electron charge per unit area recombining in the depletion region in one second:

$$j_{rec} = q \int_{-x_p}^{x_n} U_R \, dx \qquad (4\text{-}3\text{-}28)$$

(This electron charge is equal to the total hole charge per unit area recombining in the depletion region.) Here $U_R$ is the net recombination rate. For a simple model accounting for only one impurity (trap) energy level near the middle of the energy gap, this integral can be evaluated, leading to the following expression:

$$j_{rec} = j_{recs} \exp\left(\frac{V}{2V_{th}}\right) \qquad (4\text{-}3\text{-}29)$$

where

$$j_{recs} \approx \frac{\pi}{2} \frac{q n_i V_{th}}{\tau_{rec} F_{max}} \qquad (4\text{-}3\text{-}30)$$

[see van der Ziel (1976)]. However, in practical devices, many trap levels may play a role, and the following empirical expression for the forward bias recombination current is more accurate:

$$j_{rec} = j_{recs} \exp\left(\frac{V}{m_r V_{th}}\right) \qquad (4\text{-}3\text{-}31)$$

Here $m_r$ may (and often does) differ from 2. The total forward current density now becomes

$$j_F = j_s \exp\left(\frac{V}{V_{th}}\right) + j_{recs} \exp\left(\frac{V}{m_r V_{th}}\right) \qquad (4\text{-}3\text{-}32)$$

Since $j_s$ is proportional to $n_i^2$ [see eq. (4-3-27)] and $j_{recs}$ is proportional to $n_i$ [see eq. (4-3-30)], the recombination current is more important in large gap semiconductors, such as GaAs, SiC, or GaN where $n_i$ is small enough. At large voltages, the diffusion current density [proportional to $\exp(V/V_{th})$] becomes dominant since the recombination current density {proportional to $\exp[V/(m_r V_{th})]$ where $m_r \approx 2$} increases more slowly with the forward bias.

Often, instead of using eqs. (4-3-27) and (4-3-32), we use the following empirical diode equation:

$$I = I_{seff}\left[\exp\left(\frac{V}{\eta V_{th}}\right) - 1\right] \tag{4-3-33}$$

where $\eta$ is called the ideality factor and $I_{seff}$ is the effective saturation current. The deviation of $\eta$ from unity may be considered a measure of the importance of the recombination current.

In a real semiconductor diode, the parasitic series resistance, $R_S$, of the device contacts and the semiconductor neutral regions may play an important role, and eq. (4-3-34) has to be modified to include the voltage drop across this series resistance:

$$I = I_s\left[\exp\left(\frac{V - IR_s}{\eta V_{th}}\right) - 1\right] \tag{4-3-34}$$

or

$$V = IR_s + \eta V_{th}\ln\left(\frac{I}{I_s}\right) \tag{4-3-35}$$

An approximate analytical solution relating $I$ to $V$ can be obtained from eq. (4-3-35) [see Lee et al. (1993)].

Figure 4.3.5 shows how the value of the series resistance in this empirical diode equation affects the diode current-voltage characteristics. When the series resistance is high, the diode $I$-$V$ characteristic becomes almost linear at high bias with the slope determined by $R_s$. At small values of $R_s$, the diode in a certain voltage range behaves similarly to an ideal switch: almost no current below a certain "cut-in" voltage and a very small resistance at voltages higher than the cut-in voltage. This is why a diode can be used as a switch or as an element shifting the voltage level by the value of the cut-in voltage (typically 0.6 V or so for Si diodes at room temperature).

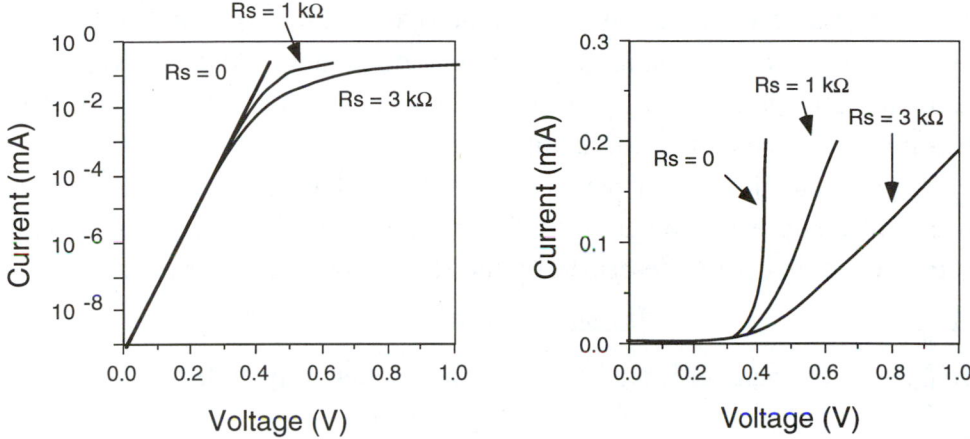

**Fig. 4.3.5.** Effect of series resistance on current-voltage characteristics of *p-n* diode. $\eta = 1$, $I_{seff} = 10^{-8}$ mA. The characteristics are shown in semilog and linear scales.

As discussed above, the reverse saturation current is a function of the reverse bias. Depending on the band gap, either the diffusion or the generation component of the reverse saturation current is dominant. The forward current is the sum of the diffusion and recombination currents. A relative contribution of these components depends on semiconductor material parameters, doping, applied bias, and temperature. All these important details are often ignored in a practical diode simulation because of the lack of detailed information about device physics and device parameters. Instead, experimental data are fitted by the empirical diode equation (4-3-35). In this case, the effective saturation current, $I_s$, should be scaled with device temperature in the same way as the intrinsic carrier density, $n_i^2$, when the diffusion current is dominant and as $n_i$ when the generation current is dominant. Typically, models implemented in circuit simulators (see Section 2.6) use the following expression:

$$I_s(T) = I_s(T_o) \left( \frac{T}{T_o} \right)^{\frac{\kappa}{\eta}} \exp\left( \frac{E_g}{k_B T_o} \right) \exp\left( -\frac{E_g}{k_B T} \right) \qquad (4\text{-}3\text{-}36)$$

Here $E_g$ is called the **activation energy** (with a default value set to be equal to the energy gap), $T$ is the device temperature, $T_o$ is a nominal device temperature (at which the device parameters are specified in the circuit simulator; usually $T_o$ = 300 K), and $\kappa$ is an empirical temperature exponent. $\kappa = 3$ and $\eta = 1$ for $p$-$n$ diodes when the diffusion saturation current is dominant. (As an exercise, please explain why $\kappa = 3$ in this case and suggest the values of $\kappa$ and $\eta$ appropriate when the recombination current is dominant.)

In practical devices, a parasitic parallel leakage current often plays an important role and has to be accounted for. In circuit simulation programs, this parasitic current is usually described by the following empirical equation:

$$I_{leakage} = G_{min} V \tag{4-3-37}$$

where $G_{min}$ is the parallel leakage conductance (taken to be independent of $V$).

**Example 4-3-3.**

The default value of $G_{min}$ in SPICE is $10^{-12}$ $1/\Omega$. The doping level of the $n$-region of a Si $p^+$-$n$ diode $N_d = 10^{15}$ cm$^{-3}$, the generation time $\tau_{gen} = 10^{-8}$ s, the intrinsic carrier density $n_i = 10^{10}$ cm$^{-3}$, the hole diffusion length $L_p = 100$ μm, the hole diffusion coefficient $D_p = 10$ cm$^2$/V, the length of the neutral $n$-type region $X_n - x_n$ is 10 μm, the dielectric permittivity $\varepsilon_s = 1.05 \times 10^{-10}$ F/m, the built-in voltage $V_{bi} = 0.6$ V, and the diode cross section $S = 10^{-2}$ cm$^{-2}$. Compare the saturation diffusion current, generation current, and parasitic leakage current at 10 V reverse bias (using $G_{min} = 10^{-12}$ $1/\Omega$).

**Solution:**

$$I_{leakage} = 10^{-12}\left(\frac{1}{\Omega}\right) \times 10 \text{ (V)} = 10^{-11} \text{ A}.$$

Since $p_{no} = \dfrac{n_i^2}{N_d} = \dfrac{10^{20}}{10^{15}} = 10^5 \left(\text{cm}^{-3}\right) \gg n_{po}$, and $X_n - x_n \ll L_p$ , using eq.

(4-3-21) and converting to the SI units, we find

$$I_{Ds} \approx \frac{qD_p p_{no} S}{X_n - x_n} = \frac{1.602 \times 10^{-19} \times 10 \times 10^{-4} \times 10^{11} \times 10^{-6}}{10 \times 10^{-6}} = 1.602 \times 10^{-12} \text{ (A)}$$

For a $p^+$-$n$ diode from eqs. (4-3-24) and (4-3-25), we find

$$x_n = \sqrt{\frac{2\varepsilon_s (V_{bi} - V)}{qN_d}} = \sqrt{\frac{2 \times 1.05 \times 10^{-10} \times [0.6 - (-10)]}{1.602 \times 10^{-19} \times 10^{21}}} = 3.73 \times 10^{-6} \text{ (m)}$$

$$I_{gen} = \frac{qn_i x_d S}{\tau_{gen}} = \frac{1.602 \times 10^{-19} \times 10^{10} \times 3.73 \times 10^{-6} \times 10^{-6}}{10^{-8}} = 5.97 \times 10^{-13} \text{ (A)}$$

In this example, the leakage current determined by the default SPICE parameter is the largest. This shows how important it is to understand the meaning of the models implemented in SPICE and the default values of the SPICE parameters.

The semi-empirical equations (4-3-37) and (4-3-38) give an example of an empirical simplified device model. Such models are primarily used for parameter extraction and/or circuit simulation (see Section 2.6). Since even the circuit modeling of the simplest semiconductor device – a *p-n* diode – has to rely on a semi-empirical model, it should come as no surprise that similar approaches are also used for more complicated devices, such as transistors!

Table 4.3.1 summarizes equations related to *p-n* junctions at arbitrary bias.

| Hole and electron diffusion lengths | $L_p = \sqrt{D_p \tau_{pl}} \qquad L_n = \sqrt{D_n \tau_{nl}}$ |
|---|---|
| Depletion widths under bias | $x_n = \sqrt{\dfrac{2\varepsilon_s (V_{bi} - V)}{qN_d(1 + N_d / N_a)}} \quad x_p = \sqrt{\dfrac{2\varepsilon_s (V_{bi} - V)}{qN_a(1 + N_a / N_d)}}$ |
| Ideal and empirical Diode Equations | $I = I_s \left[ \exp\left(\dfrac{V}{V_{th}}\right) - 1 \right], \quad I = I_s \left[ \exp\left(\dfrac{V}{\eta V_{th}}\right) - 1 \right]$ |
| Diode equation (with series resistance) | $I = I_s \left[ \exp\left(\dfrac{V - IR_s}{\eta V_{th}}\right) - 1 \right]$ |
| Parallel leakage current | $I_{leakage} = G_{min} V$ |
| Reverse diode current density | $j_R = j_s + j_{gen} \qquad \text{where} \qquad j_{gen} = q \displaystyle\int_{-x_p}^{x_n} \lvert U_R \rvert dx = \dfrac{qn_i x_d}{\tau_{gen}}$ $j_s = \left( \dfrac{qD_p}{N_d L_p} + \dfrac{qD_n}{N_a L_n} \right) n_i^2 \ \text{(long diode)}$ $j_s = \left[ \dfrac{qD_p}{N_d(X_n - x_n)} + \dfrac{qD_n}{N_a(X_p - x_p)} \right] n_i^2 \ (\text{short diode})$ |
| Temperature dependence of the effective saturation current | $I_s(T) = I_s(T_o)\left(\dfrac{T}{T_o}\right)^{\frac{\kappa}{\eta}} \exp\left(\dfrac{E_g}{k_B T_o}\right)\exp\left(-\dfrac{E_g}{k_B T}\right)$ |

**Table 4.3.1.** Summary of equations describing *p-n* junctions under bias.

## 4-4.  DEPLETION AND DIFFUSION CAPACITANCES

Figure 4.4.1 shows the distributions of electric charges in a *p-n* junction and in a parallel plate capacitor, as well as an equivalent circuit of the space charge layer. (Arrows in Fig. 4.4.1b represent the streamlines of an electric field in a parallel plate capacitor, which are similar to those in a *p-n* junction.)

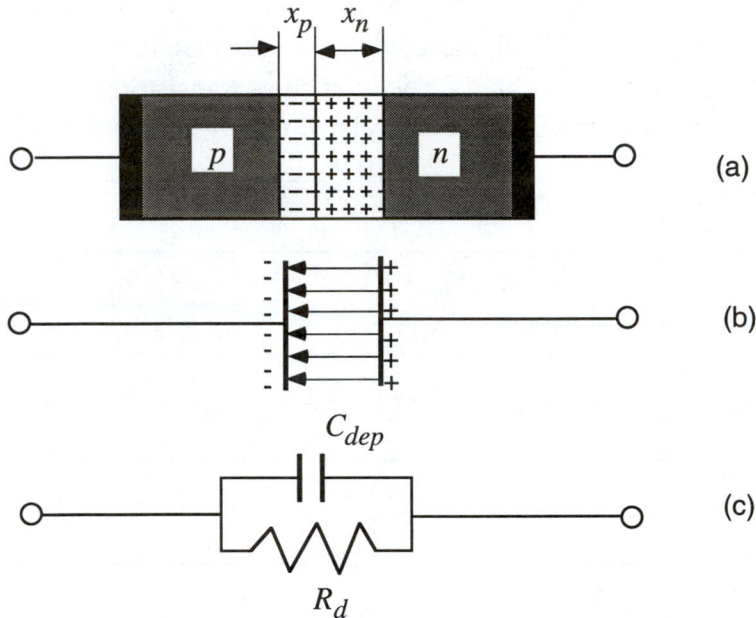

**Fig. 4.4.1.**  Distribution of electric charges (a) in a *p-n* junction and (b) in a parallel plate capacitor, and (c) an equivalent circuit of the space charge layer.  $C_{dep}$ is the equivalent differential capacitance of the depletion layer [see eq. (4-4-1)]; $R_d = dV/dI$ is the equivalent differential resistance of the *p-n* junction.

At reverse bias, zero bias, or small forward bias, the differential capacitance of a *p-n* diode is given by

$$C_{dep} = \left| \frac{dQ_d}{dV} \right|$$

(4-4-1)

where

$$|Q_d| = qN_d x_n S = qN_a x_p S \tag{4-4-2}$$

is the depletion charge, $V$ is the bias voltage, $S$ is the diode cross section, and $x_n$ and $x_p$ are the widths of the depletion layer in the $n$-region and $p$-region of the diode, respectively. When an external bias, $V$, is applied, $x_n$ and $x_p$ are given by eqs. (4-2-17) and (4-2-18). Using these equations, we obtain

$$dx_n = -x_n \frac{dV}{2\sqrt{V_{bi} - V}}, \quad dx_p = -x_p \frac{dV}{2\sqrt{V_{bi} - V}} \tag{4-4-3}$$

$$dQ_d = qN_d S dx_n = qN_a S dx_p = -qN_d x_n \frac{S dV}{2\sqrt{V_{bi} - V}} = -qN_a x_p \frac{S dV}{2\sqrt{V_{bi} - V}} \tag{4-4-4}$$

$$C_{dep} = S\sqrt{\frac{q\varepsilon_s N_{eff}}{2(V_{bi} - V)}} \tag{4-4-5}$$

where $N_{eff} = \dfrac{N_a N_d}{N_a + N_d}$.

Equation (4-4-5) shows that the depletion capacitance of an abrupt junction is proportional to $1/(V_{bi} - V)^{1/2}$. This equation can be rewritten in a more elegant form:

$$C_{dep} = \frac{\varepsilon_s S}{x_d} \tag{4-4-6}$$

where $x_d = x_n + x_p$ is the total width of the depletion region. Even though we derived this expression for an abrupt junction with uniform doping densities of acceptors and donors, it applies to $p$-$n$ junctions with arbitrary doping profiles [see, for example, Shur (1990)]. In particular, for a linearly graded profile ($N_d - N_a = ax$ where $a$ is a constant), the depletion width is given by

$$x_d = \left[\frac{12\varepsilon_s(V_{bi} - V)}{qa}\right]^{1/3} \tag{4-4-7}$$

where the built-in voltage, $V_{bi}$, for the linearly graded junction is given by

$$V_{bi} = 2V_{th} \ln\left(\frac{ax_d}{2n_i}\right) \qquad (4\text{-}4\text{-}8)$$

(see Problem 4-4-1) and we find

$$C_{dep} = S\left[\frac{qa\varepsilon_s^2}{12(V_{bi} - V)}\right]^{1/3} \qquad (4\text{-}4\text{-}9)$$

For a large variety of different doping profiles, including a uniformly doped and linearly graded doping profiles, the voltage dependence of the depletion capacitance can be approximated by the following empirical expression:

$$C_{dep} = \frac{C_{jo}}{(1 - V/V_{bi})^m} \qquad (4\text{-}4\text{-}10)$$

where $m$ is called the **grading coefficient**. [As can be seen from eqs. (4-4-5) and (4-5-10), $m = 1/2$ for an abrupt junction and $m = 1/3$ for a linearly graded junction.]

Differentiating eq. (4-4-6) with respect to $V$, we obtain:

$$N_d(x_d) = \frac{C_{dep}^3}{q\varepsilon_s S^2 \left|dC_{dep}/dV\right|} = \frac{2}{q\varepsilon_s S^2} \left|\frac{1}{d\left(1/C_{dep}^2\right)/dV}\right| \qquad (4\text{-}4\text{-}11)$$

Figure 4.4.2 shows the calculated $C_d$ versus $V$ dependence of a Si $p^+$-$n$ junction. Figure 4.4.3 shows the $1/C_{dep}^2$ versus $V$ dependence for the same device.

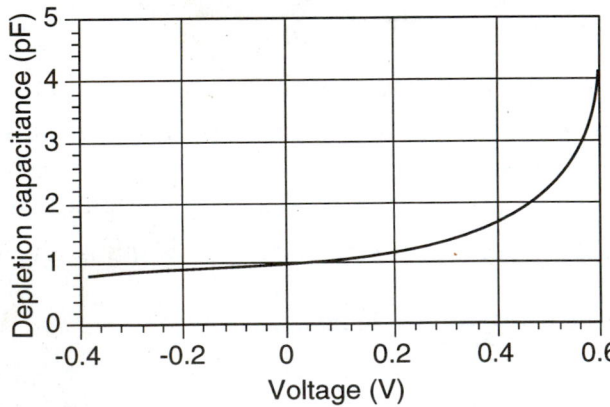

**Fig. 4.4.2.** Calculated depletion capacitance-voltage characteristic of Si $p^+$-$n$ junction. $N_a = 5 \times 10^{15}$ cm$^{-3}$. $N_d = 10^{15}$ cm$^{-3}$.

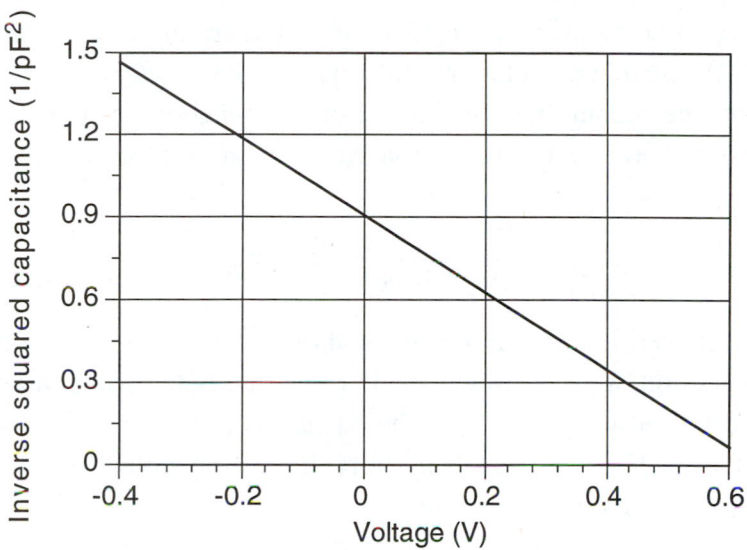

**Fig. 4.4.3.** Calculated $1/C_{dep}^2$ versus $V$ dependence for the same $p^+$-$n$ junction.

As can be seen from eq. (4-4-11), the slope of $1/C_{dep}^2$ versus $V$ dependence is related to the doping density. Therefore, eq. (4-4-11) may be used to deduce the doping profile from the measured capacitance-voltage characteristics at reverse and small forward bias. (As an exercise, estimate the diode cross section using the data shown in Figs. 4.4.2 and then estimate $N_d$ from the data shown in Fig. 4.4.3.)

The $p$-$n$ junction also has a capacitance related to the minority carriers injected into the neutral regions. Indeed, the minority carrier distribution in the neutral region cannot change instantaneously. A characteristic time of the change of minority carrier profile is on the order of the minority carrier lifetime. This time delay can be interpreted in terms of an equivalent $RC$ time constant, which is $C_{dif}R_d$, where the characteristic capacitance is called the diffusion capacitance and $R_d = dV/dI$ is the equivalent differential resistance of the $p$-$n$ junction. From eq. (4-3-33), we find that under the forward bias

$$R_d \approx \frac{\eta V_{th}}{I} \qquad (4\text{-}4\text{-}12)$$

In order to find the diffusion capacitance, we have to calculate the device impedance using the diffusion equations for the minority carriers. These equations have to be modified to include the time derivatives $\partial p/\partial t$ and $\partial n/\partial t$ (see Problem 4-4-2). Such a calculation confirms that the $C_{dif}R_d = C_{dif}/G_d$ product is determined by the recombination time. For a $p^+$-$n$ diode at low frequencies $\omega$ such that $\omega\tau_{pl} \ll 1$, where $\tau_{pl}$ is the hole recombination time, we obtain

$$C_{dif} = \frac{I\tau_{pl}}{2V_{th}} = \frac{1}{2}\frac{\tau_{pl}}{R_d} = \frac{1}{2}G_d\tau_{pl} \qquad (4\text{-}4\text{-}13)$$

As we showed in Section 4.3, for a short $p^+$-$n$ diode (with the length of the neutral $n$-section, $X_n - x_n$, much smaller than the hole diffusion length, $L_p$) $L_p$ should be substituted with $X_n - x_n$ in the equations describing the diode [see eqs. (4-3-20) and (4-3-21)]. For such a short diode, the hole lifetime, $\tau_{pl}$, in eq. (4-4-13) has to be replaced as follows:

$$\tau_{pl} = \frac{L_p^2}{D_p} \rightarrow \frac{(X_n - x_n)^2}{D_p} = 2t_{tr} \qquad (4\text{-}4\text{-}14)$$

where $t_{tr}$ is called the **hole transit time**. Based on the solution of the diffusion equation, one can show that $t_{tr}$ represents the time it takes for holes in a short diode to diffuse from the boundary of the depletion region to the contact (a more formal term would be a "**characteristic time scale**" of the hole diffusion process). That $t_{tr}$ is a characteristic time scale of the hole diffusion process is quite natural since it is the only combination of powers of $D_p$ and $X_n - x_n$ that has the dimension of time.

In the case when $t_{tr} \ll \tau_{pl}$, which is very important for practical diodes, eq. (4-4-13) should be replaced by

$$C_{dif} = G_d t_{tr} \qquad (4\text{-}4\text{-}15)$$

The diode response time is determined by the time constant $C_d R_d$. Since

$$\frac{t_{tr}}{\tau_{pl}} = \frac{(X_n - x_n)^2}{2L_p^2} \ll 1$$

and the differential conductance of a short diode is $L_p/(X_n - x_n)$ larger than for a long diode [see eq. (4-3-21) and related discussion], the time response of a short diode is $L_p/(X_n - x_n)$ faster than for a long diode.

Equations (4-4-13) [or (4-4-15) and (4-4-12)] predict that $C_{dif}$ increases exponentially with an increase in the forward bias. Equation (4-4-5) shows that the depletion capacitance, $C_{dep}$, increases as $1/\sqrt{V_{bi}-V}$ with an increase in the forward bias. However, at large forward bias, $V$, eqs. (4-4-5) and (4-4-13) are no longer valid because the densities of minority carriers are no longer proportional to $\exp(V/V_{th})$ and the very concept of the depletion layer becomes invalid. Once the difference between the built-in voltage, $V_{bi}$, and applied bias $V$ becomes comparable to the thermal voltage, the barrier between the $n$ and $p$ regions practically ceases to exist. At this point, the depletion widths, $x_n$ and $x_p$, calculated using the depletion approximation, become comparable to

$$L_{Dn} = \sqrt{\frac{\varepsilon_s k_B T}{q^2 N_d}} \text{ and } L_{Dp} = \sqrt{\frac{\varepsilon_s k_B T}{q^2 N_a}} \tag{4-4-16}$$

respectively, where $L_{Dn}$ and $L_{Dp}$ are the Debye lengths for the $n$- and $p$-region, respectively [see eq. (4-2-20)]. Any further increase in the applied voltage will drop almost entirely across the diode series resistance, $R_s$, which determines the diode current-voltage characteristic at very large forward biases (as shown in Fig. 4.3.5). This sets up a lower limit for the depletion width, on the order of $L_{Dp} + L_{Dn}$, and an upper limit for the diode differential capacitance

$$C_{d\max} = \frac{\varepsilon_s S}{L_{Dn} + L_{Dp}} \tag{4-4-17}$$

This limitation of the diode differential capacitance under a large forward bias (close to $V_{bi}$) can be taken into account by introducing an effective differential junction capacitance, $C_d$:

$$C_d = \frac{1}{1/\left(C_{dep} + C_{dif}\right) + 1/C_{d\max}} \tag{4-4-18}$$

Figure 4.4.4 shows the equivalent circuit of a $p$-$n$ junction diode. In addition to the differential junction capacitance, $C_d$, and the differential junction resistance, $R_d$, the equivalent circuit includes a series resistance, $R_s$ (that, in turn, includes the contact resistances and of the resistance in the neutral regions of the semiconductor), as well as a parasitic inductance, $L_s$, and the geometric capacitance of the sample

$$C_{geom} = \varepsilon_s S / L \tag{4-4-19}$$

where $L$ is the sample length.

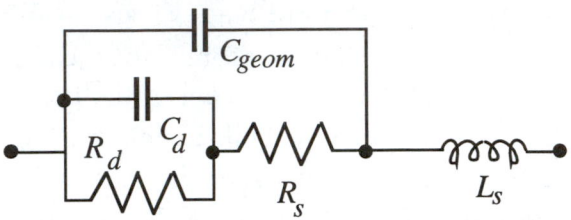

**Fig. 4.4.4.** The small signal equivalent circuit of a $p$-$n$ junction diode.

**Example 4-4-1.**

Plot the forward bias dependence of the different capacitance components for a Si $p^+$-$n$ diode at $T = 300$ K. The doping level of the $n$-region $N_d = 10^{15}$ cm$^{-3}$, the intrinsic carrier density $n_i = 10^{10}$ cm$^{-3}$, the hole diffusion length $L_p = 100$ μm, the hole diffusion coefficient $D_p = 10$ cm$^2$/V, the length of the neutral $n$-type region $X_n - x_n$ is 10 μm, the dielectric permittivity $\varepsilon_s = 1.05 \times 10^{-10}$ F/m, the built-in voltage $V_{bi} = 0.6$ V, and the diode cross section $S = 10^{-2}$ cm$^{-2}$.

**Solution:**

We use the SI units. The hole lifetime,

$$\tau_{pl} = \frac{L_p^2}{D_p} = \frac{\left(10^{-4}\right)^2}{10^{-3}} = 10^{-5}\,(\text{s}).$$

The Debye length, $L_{Dp} = \sqrt{\dfrac{\varepsilon_s V_{th}}{qN_a}} = \sqrt{\dfrac{1.05 \times 10^{-10} \times 0.02584}{1.602 \times 10^{-19} \times 10^{21}}} = 1.30 \times 10^{-7}\,(\text{m})$

The maximum junction capacitance,

$$C_{d\,max} = \frac{\varepsilon_s S}{L_{Dn}} = \frac{1.05 \times 10^{-10} \times 10^{-6}}{1.30 \times 10^{-7}} = 8.07 \times 10^{-10}\,(\text{F})$$

The differential conductance is calculated from eq. (4-3-21):

$$G_d = \frac{qD_p P_{no} S}{V_{th}(X_n - x_n)} \exp\left(\frac{V}{V_{th}}\right)$$

to yield the diffusion capacitance $C_{dif} = G_d t_{tr}$ where $t_{tr}$ is found from eq. (4-3-21). The depletion capacitance is calculated from eq. (4-3-25). The resulting plots are shown in Fig. 4.4.5.

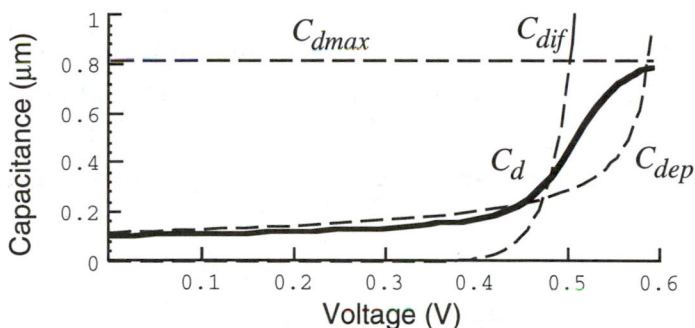

**Fig. 4.4.5.** Depletion ($C_{dep}$), diffusion ($C_{dif}$), maximum ($C_{dmax}$), and differential ($C_d$) capacitances of Si $p$-$n$ junction versus bias.

Practical measurements of the small signal junction capacitance are usually done at frequencies, $f \approx 2$ MHz or so, at which the parasitic inductance, $L_s$, may become very important. Hence, the measured impedance (calculated based on the equivalent circuit shown in Fig. 4.4.4) is given by

$$Z_{meas} \approx R_s + R_d / (1 + i\omega C_d R_d) + i\omega L_s \qquad (4\text{-}4\text{-}19)$$

When the differential resistance, $R_d$, becomes small enough at large forward bias, the second term on the right-hand side of eq. (4-4-19) becomes small, and the impedance changes from capacitive to inductive. Thus, the estimates of the differential capacitance, $C_d$, of a $p$-$n$ junction under large forward bias have to account for this inductance.

An equivalent circuit gives only an approximate description of a small signal response of a $p$-$n$ junction diode. The diode is a fairly complicated system that cannot be accurately described at all frequencies by the lumped element equivalent circuit shown in Fig. 4.4.4. At frequencies higher than the **diode cutoff frequency**

$$f_T = 1/(2\pi C_d R_s) \qquad (4\text{-}4\text{-}20)$$

the lumped element model may no longer be applicable. At such frequencies, the voltage drop across the diode capacitance is smaller than the voltage drop across the series resistance.

**Example 4-4-2.**

Estimate the series resistance, $R_s$, of the neutral $n$-type region and cutoff frequency, $f_T$, for the Si $p^+$-$n$ junction considered in Example 4-4-1 assuming an electron mobility of 1,000 cm$^2$/Vs.

**Solution:**

We use the SI units. $R_s \approx X_n/(q\mu_n N_d S) = 10^{-5}/(1.602\times10^{-19}\times0.1\times10^{21}\times10^{-6}) = 0.624$ ($\Omega$). From Fig. 4.4.4, $C_d \leq 800$ pF. Hence, $f_T = 1/(2\pi C_d R_s) \approx 3.19\times10^8$ (Hz).

The summary of equations related to the *p-n* junction capacitances is given in Table 4.4.1.

| | |
|---|---|
| Depletion capacitance | $C_{dep} = \dfrac{\varepsilon_s S}{x_d}$    where $x_d = x_n + x_p$ <br><br> $x_n = \sqrt{\dfrac{2\varepsilon_s(V_{bi} - V)}{qN_d(1 + N_d/N_a)}}$    $x_p = \sqrt{\dfrac{2\varepsilon_s(V_{bi} - V)}{qN_a(1 + N_a/N_d)}}$ |
| Depletion capacitance for different doping profiles | $C_{dep} = \dfrac{C_{jo}}{(1 - V/V_{bi})^m}$    where grading coefficient $m = 1/2$ for an abrupt junction, $m = 1/3$ for linearly graded junction |
| Diffusion capacitance of $p^+$-$n$ junction | $C_{dif} = \dfrac{I\tau_{pl}}{2V_{th}} = \dfrac{1}{2}\dfrac{\tau_{pl}}{R_d} = \dfrac{1}{2}G_d\tau_{pl}$ |
| Diffusion capacitance of a short $p^+$-$n$ junction | $C_{dif} = \dfrac{qD_p p_{no} S t_{tr}}{2(X_n - x_n)V_{th}}\exp\left(\dfrac{V}{V_{th}}\right) = G_d t_{tr}$ <br><br> where $t_{tr} = \dfrac{(X_n - x_n)^2}{2D_p}$ |
| Differential capacitance | $C_d = \dfrac{1}{1/(C_{dep} + C_{dif}) + 1/C_{d\max}}$ <br><br> where $C_{d\max} = \dfrac{\varepsilon_s S}{L_{Dn} + L_{Dp}}$ |
| Electron and hole Debye lengths | $L_{Dn} = \sqrt{\dfrac{\varepsilon_s k_B T}{q^2 N_d}}$    $L_{Dp} = \sqrt{\dfrac{\varepsilon_s k_B T}{q^2 N_a}}$ |
| Geometric capacitance | $C_{geom} = \varepsilon_s S/L$ |
| Measured impedance | $Z_{meas} \approx R_s + R_d/(1 + i\omega C_d R_d) + i\omega L_s$ |

**Table 4.4.1.** Equations related to the *p-n* junction capacitances.

## 4-5.  JUNCTION BREAKDOWN

Fig. 4.5.1 shows the current-voltage characteristics of a *p-n* junction diode simulated by PSpice$^{tm}$.  As can be seen from the figure, the diode current under a reverse bias is very small.  However, once the reverse bias exceeds a certain critical value, the reverse current rises very steeply.  Such behavior is very typical for any *p-n* junction diode.

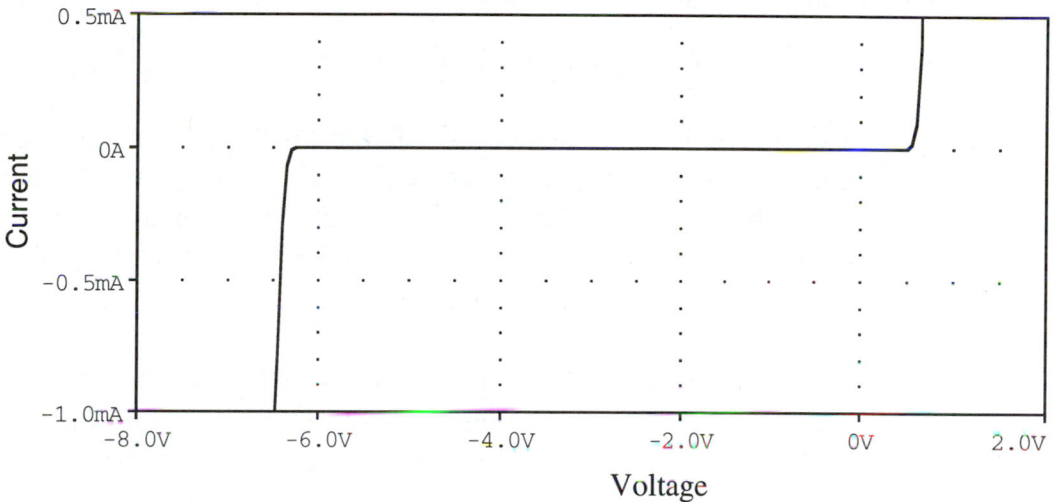

**Fig. 4.5.1.**  Diode *I-V* characteristics simulated by PSpice$^{tm}$.  Default parameters are used in the simulation (see Section 4.6) except for the diode series resistance (50 $\Omega$) and the breakdown voltage (6 V).

This sharp increase in the reverse current may be caused either by an **avalanche breakdown** or by **tunneling breakdown**.

The avalanche breakdown is caused by **impact ionization**.  In this process, an electron (or a hole) acquires enough energy from the electric field to break a bond and promote another electron from the valence band into the conduction band (see Section 3.8), creating an electron-hole pair. The electrons and holes created by impact ionization are accelerated by the electric field and may create more electron-hole pairs.  This will lead to an uncontrolled rise of current caused by the impact ionization (until the current is either limited by an external load or the diode burns out).

The generation rate of electron-hole pairs caused by impact ionization was discussed in Section. 3.8.  A simpler (but also a less accurate) description of the impact ionization process relies on the concept of the breakdown electric field, $F_{br}$. According to this concept, when the electric field is less than $F_{br}$, the impact ionization generation rate can be totally neglected. However, once the electric field anywhere within a device reaches or exceeds $F_{br}$, avalanche breakdowns occurs.

For a $p^+$-$n$ junction, equating the maximum electric field in the $p^+$-$n$ junction to the breakdown field, we obtain the following estimate of the critical voltage, $V_{abr}$, causing the avalanche breakdown:

$$V_{abr} = \varepsilon_s F_{br}^2 / (2qN_d) \tag{4-5-1}$$

The breakdown field at 300 K is of the order of 100 kV/cm for germanium, 300 kV/cm for silicon, 400 kV/cm for gallium arsenide, and 2300 kV/cm or more for silicon carbide. The breakdown voltage can be quite large for diodes with low-doped $n$-regions and, for some devices, can exceed several thousand volts. (See Fig. 4.5.2, which compares the breakdown voltages for one-sided abrupt Si, GaAs, and SiC $p$-$n$ junctions.)

Power electronic devices should have a high breakdown voltage (which decreases with doping) and a high conductance (which increases with doping). Equation (4-5-1) shows that, at a given breakdown voltage, the acceptable doping level increases as $F_{br}^2$. Thus, SiC power devices should be able to achieve at least $(2,300/300)^2 \approx 59$ times higher doping and much smaller resistance than Si devices.

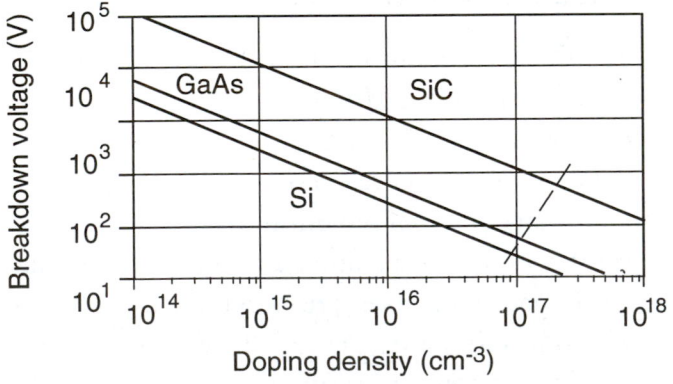

**Fig. 4.5.2.** Breakdown voltage versus doping density for one-sided abrupt Si, GaAs, and SiC $p$-$n$ junctions. To the right of the dashed line we expect that tunneling breakdown may become dominant.

As mentioned above, during an impact ionization process, an electron (or a hole) is accelerated by an applied electric field. This acceleration is interrupted by scattering events – electron (or hole) collisions with impurities, lattice vibrations, or crystal imperfections. These collisions limit the energy gained by electrons or holes from the electric field. Since these collisions are random processes, once in a while a certain electron (or a hole) gets "lucky" and travels long enough to gain enough energy (more than the energy gap, $E_g$) to create an electron-hole pair. Since the impact ionization process depends on carriers gaining energy from the electric field between scattering events, the probability of the impact ionization decreases when the scattering events become more probable. Hence, typically the breakdown field, $F_{br}$, increases with increasing temperature.

Figure 4.5.3 shows the *p-n* junction band diagram at high reverse bias. As can be seen from the figure, electrons may tunnel from occupied states in the valence band of the *p*-type region into empty states of the conduction band in the *n*-type region if the distance separating these two regions is small enough (which is the case in highly doped semiconductors). The tunneling probability is determined by a triangular potential barrier and can be calculated using equations derived in Section 1.3. Tunneling breakdown usually occurs in fairly highly doped semiconductors when the maximum electric field in the depletion layer approaches values of the order of $10^6$ V/cm.

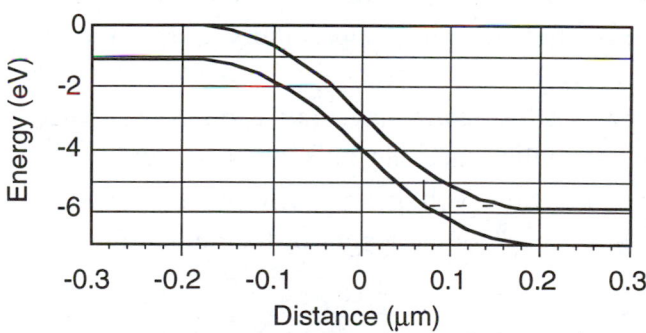

**Fig. 4.5.3.** Energy band diagram of a Si *p-n* junction. Acceptor density $N_a = 10^{17}$ cm$^{-3}$, donor density $N_d = 10^{17}$ cm$^{-3}$, $T = 300$ K. Dashed lines show the potential barrier that electrons in the valence band have to tunnel through to get to the conduction band. As an exercise, roughly estimate the reverse bias voltage using the band diagram shown on this figure.

As can be seen from Fig. 4.5.3, the height of the potential barrier for tunneling is approximately equal to the energy gap, $E_g$. The width of this potential barrier, $d_{tun}$, can be estimated as

$$d_{tun} \approx \frac{E_g}{qF_{\max}}$$

(4-5-2)

where $F_{\max}$ is the maximum electric field in the junction. [$F_{\max}$ is proportional to $1/(V_{bi} + |V|)^{1/2}$ where $V_{bi}$ is the built-in voltage and $V$ is the reverse bias voltage.]  Equation (4-5-2) shows that tunneling becomes more likely with an increase in the reverse bias.

The critical voltage for tunneling breakdown, $V_{tbr}$, may be estimated by defining the onset of breakdown as the reverse voltage, which gives a 10-fold increase in the reverse current over the saturation current, $I_s$, that is,

$$I(V_{brt}) \approx 10SI_s$$

(4-5-3)

According to Sze (1981), the critical voltage for tunneling breakdown is usually less than approximately $4E_g/q$.  When the breakdown voltage is higher than $6E_g/q$, the breakdown is typically caused by the avalanche effect.  Both mechanisms may play a role for breakdown voltages between $4E_g/q$ and $6E_g/q$.

When power dissipation in a $p$-$n$ diode is large, a so-called thermal breakdown may lead to a runaway increase in the reverse current and to an S-type negative differential resistance.  This mechanism is especially important in devices made from relatively narrow gap materials (such as Ge), where the device temperature may increase appreciably even at relatively low current densities.

Often, surface breakdown may play a dominant role, since the surface breakdown field is smaller than the bulk electric field.  Therefore, $p$-$n$ diodes designed to withstand high voltages often have a conical shape (as shown in Fig. 4.5.4) in order to increase the length of the depletion region at the surface.

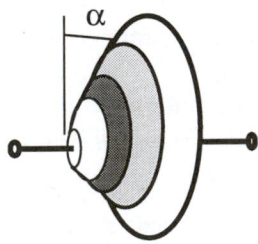

**Fig. 4.5.4.**  A power semiconductor diode.  The depletion region is shaded.

The cone angle, $\alpha$, may need to be as small as just a few degrees for the largest breakdown voltage possible!

In circuit simulation programs (such as SPICE, see Section 4.6), the reverse breakdown current is usually described by the following empirical equation:

$$I_{breakdown} = I_{br} \exp\left(-\frac{V + V_{br}}{V_{th}}\right) \tag{4-5-4}$$

This equation is used no matter what breakdown mechanism takes place. Perhaps, in certain cases, such a crude approach may be justified if the exact temperature dependence of the breakdown current does not matter that much. This is not a very good justification, but it is hard to come up with a better argument in favor of eq. (4-5-4). Perhaps, some of the readers of this book will take time and effort to implement a better junction breakdown model into SPICE.

A more detailed discussion of the breakdown mechanisms may be found, for example, in the text by Shur (1990).

Figure 4.5.1 shows the current-voltage characteristics of a diode experiencing a reverse breakdown. As can be seen from the figure, the current rises rapidly in the breakdown regime, while the voltage across the diode remains nearly constant. Hence, diodes operating in the breakdown regime can be used for maintaining a certain voltage (approximately equal to the breakdown voltage) across the diode. A diode designed for this purpose (called a reference diode or a Zener diode) finds many applications in electronic circuits (see Problem 4-5-2). The following circuit symbol is used for a Zener diode: ⟶⧏⟶.

Table 4.5.1 gives a summary of results related to the breakdown in $p$-$n$ junctions.

| Impact ionization breakdown voltage for a $p^+$-$n$ junction | $V_{abr} = \dfrac{\varepsilon_s F_{br}^{\,2}}{2qN_d}$ |
|---|---|
| Empirical equation for breakdown current used by SPICE | $I_{breakdown} = I_{br} \exp\left(-\dfrac{V + V_{br}}{V_{th}}\right)$ |
| Breakdown voltages for avalanche and tunneling | Higher than $6E_g/q$ for avalanche breakdown, less than approximately $4E_g/q$ for tunneling breakdown |

**Table 4.5.1.**  Important results related to the breakdown in $p$-$n$ junctions.

## *4-6.  TUNNEL DIODES

As we discussed in Section 4.5, a heavily doped $p$-$n$ diode can experience tunneling breakdown [see Fig. 4.5.2, which shows (by the dashed line) the barrier separating electrons in the valence band of the $p$-type region from the conduction band of the $n$-type region]. In very heavily doped diodes (with degenerate $p$-type and $n$-type regions), this barrier may become so thin that tunneling becomes important at forward bias as well. Under such conditions, the current-voltage characteristic of the diode changes dramatically (see Fig. 4.6.1) and has a region where the current actually decreases with an increase in the forward bias (a region of negative differential resistance $dV/dI < 0$). A $p$-$n$ diode operating in this regime is called a **tunnel diode**.

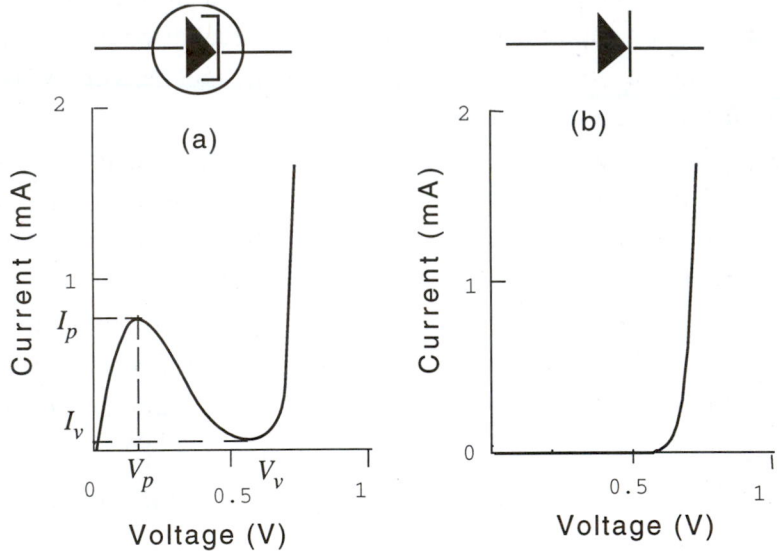

**Fig. 4.6.1.** Calculated current-voltage characteristics of (a) GaAs tunnel diode and (b) conventional diode. At $T = 300$ K, typical values of the peak and valley voltages, $V_p$ and $V_v$, are about 200 mV and 600 mV for GaAs tunnel diodes and about 100 mV and 300 mV for Ge tunnel diodes, respectively. Also shown are circuit symbols for tunnel diode and conventional diode.

Figure 4.6.2 shows the band diagrams of a tunnel diode for several bias voltages. An electron from the conduction band in the $n$-type region may tunnel into an empty energy state in the valence band in the $p$-type region with the same

energy (since the energy must be conserved in a tunneling process).   Figure
4.6.2b shows the potential barrier for separating an electron with the energy $E_F^{(n)}$
from an empty energy state in the valence band with the same energy.   It also
shows the band of states available for tunneling that exists when a relatively small
bias $V \le V_p$ where $V_p$ is the peak voltage of the current-voltage characteristics
(see Fig. 4.6.1a) applied to the diode.   If the barrier for tunneling is narrow
enough (which is the case in a highly doped diode where the depletion region is
very, very thin), the tunneling current is much larger than the normal diode
forward current.   When the forward bias is increased more, no states become
available for tunneling (see Fig. 4.6.2c).   The tunneling current drops, and the
subsequent rise in the diode current shown in Fig. 4.6.1a is related to the normal
diode current-voltage characteristic.

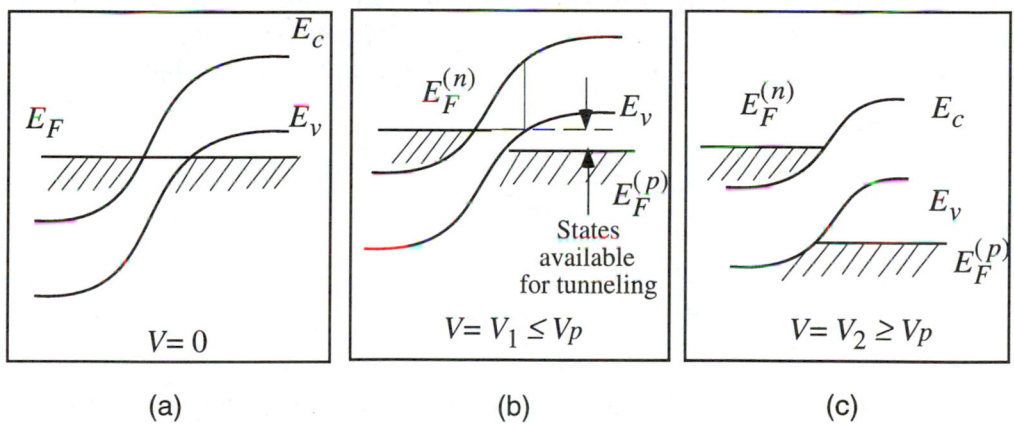

(a)                                (b)                                (c)

**Fig. 4.6.2.**  Band diagrams of tunneling diode.  Dashed areas correspond
to occupied states.

The total current, $I$,  in  a tunnel diode can be approximated by

$$I = I_{tun} + I_{diode} + I_{excess} \qquad\qquad (4\text{-}6\text{-}1)$$

[see Sze (1981), p. 529] where

$$I_{diode} \approx I_s \exp\left[\left(\frac{V}{\eta V_{th}}\right) - 1\right] \qquad\qquad (4\text{-}6\text{-}2)$$

is the normal diode current.   Here $I_s$ is the saturation current, $\eta$ is the ideality
factor, and $V_{th}$ is the thermal voltage.   For the tunneling current, $I_{tun}$, we will
use the following interpolation:

$$I_{tun} \approx \frac{V}{R_o} \exp\left[-\left(\frac{V}{V_o}\right)^m\right] \tag{4-6-3}$$

$I_{excess}$ is the additional tunneling current related to parasitic tunneling processes via impurity energy levels in the energy gap. This current usually determines the minimum (valley) current, $I_v$ (see Fig. 4.6.1). We will interpolate the excess current by the following equation:

$$I_{excess} \approx \frac{V}{R_v} \exp\left[\left(\frac{V - V_v}{V_{ex}}\right)\right] \tag{4-6-4}$$

where $R_v$ and $V_{ex}$ are the interpolation parameters and $V_v$ is the valley voltage (see Fig. 4.6.1). This interpolation formula makes it relatively easily to extract the interpolation parameters $R_o$, $V_o$, $m$, $R_v$, and $V_{ex}$ from measured data. $R_o$ is the equal to the diode resistance at low voltages. Parameters $V_o$ and $m$ are related to the peak voltage, $V_p$, and the maximum differential conductance, $g_{d\max} = (dI_{tun}/dV)_{\max}$, as follows:

$$\frac{V_p}{V_o} = \left(\frac{1}{m}\right)^{1/m} \tag{4-6-5}$$

$$|g_{d\max}|R_o = \frac{m}{\exp\left(\frac{1+m}{m}\right)} \tag{4-6-6}$$

[Please derive eqs. (4-6-5) and (4-6-6) by finding the maxima of the tunneling current, $I_{tun}$, and of the tunneling current derivative, $dI_{tun}/dV$, from eq. (4-6-2).]

Figure 4.6.3 shows $|g_{d\max}R_o|$ and $V_p/V_o$ as functions of $m$.

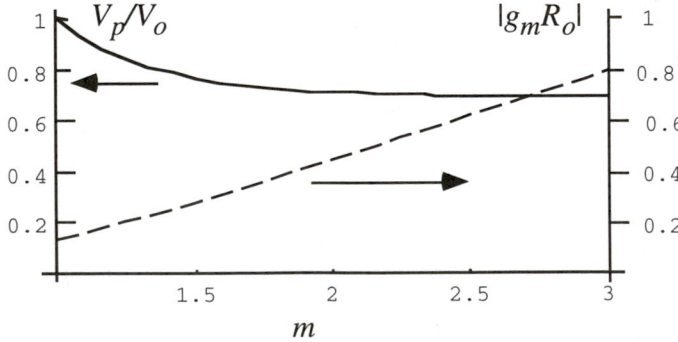

**Fig. 4.6.3.** $|g_{d\max}R_o|$ and $V_p/V_o$ as functions of $m$.

**Example  4-6-1.**

Extract  parameters  $R_o$,  $V_o$,  and  $m$  from  the  tunnel  diode  current-voltage characteristics shown in Fig. 4.6.1a.  (Neglect the excess tunneling current.)

**Solution:**

We first determine the slope of the $I$-$V$ characteristic at $V \rightarrow 0$. This slope is equal to 10 mA/V, corresponding to $R_o = 100\ \Omega$. Then we find the maximum negative differential conductance from maximum slope in the negative differential resistance region $|g_{d\max}| \approx 2.5$ mA/V (see Fig. 4.6.4). Hence $|g_{d\max}R_o| \approx 0.25$. From Fig. 4.6.3 we determine $m = 1.5$. For $m = 1.5$, we find from Fig. 4.6.3 $V_p/V_o \approx 0.7$. Since from Fig. 4.6.1, $V_p \approx 0.14$ V, we estimate $V_o \approx 0.2$ V.

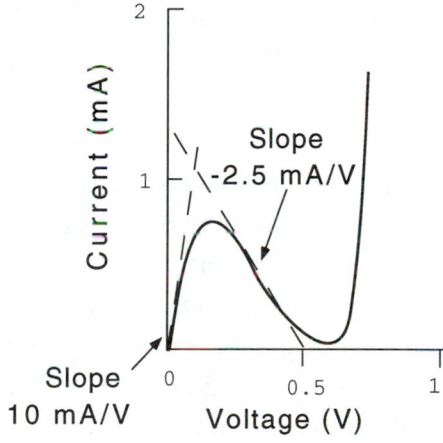

**Fig. 4.6.4.**  Parameter extraction for tunnel diode.

Equations (4-6-1) to (4-6-6) represent an empirical approach to the description of a tunnel diode.  The physics-based approach should rely on the theory of tunneling developed in Section 1.3 and should account for the number of the states available for tunneling.  This can be done using the densities of states and occupation functions considered in Section 2.4.  Also, we will have to determine the barrier height, $V_{bi} - V$, where $V_{bi}$ is the built-in voltage and $V$ is the applied bias.  In Section 4.2, we derived the expression for $V_{bi}$ for a nondegenerate diode, and we can obtain the equation for $V_{bi}$ for a degenerately doped diode using the expressions for the electron and hole Fermi levels given in Section 3.3.  So we have the knowledge, but do we have the time to allow ourselves the intellectual pleasure of a long derivation?  If we don't, we could look up the results in a more advanced book by Sze (1981), p. 516.

The most interesting property of a tunnel diode is its negative differential resistance in the voltage range $V_p \leq V \leq V_v$ (see Fig. 4.6.1).  This allows us to use a tunnel diode as an active element (an oscillator) in microwave and millimeter wave circuits. The tunneling process is very fast, since  the distances traveled by carriers are very short.

Another application of the tunnel diode is in switching circuits.  In order to understand how this works let us consider the tunnel diode with the $I$-$V$ characteristic shown in Fig. 4.6.1 connected in series with a 200 $\Omega$ resistance and a voltage source.  The operating point (or points) are found by equating the current through the diode to the current through the resistor:

$$I = \frac{V_s - V}{R} \qquad\qquad (4\text{-}6\text{-}7)$$

The voltage drop across the diode, $V$, is a fairly complicated function of $I$ [see eqs. (4-6-1) to (4-6-4)], and eq. (4-6-7) is difficult to solve analytically. However, it can be easily solved graphically.  Figure 4.6.5 shows the diode current-voltage characteristic with the superimposed dependencies $(V_s - V)/R_s$ (called **load lines**) for a 0.5 V voltage source (dashed line) and a 1 V voltage source (solid line).  The intercept points (shown by open circles in Fig. 4.6.5) correspond to possible operating points.  For the 0.5 V voltage source, we have but one operating point.  However, for the 1 V voltage source, we have three, and the problem arises which one we should choose.  In order to answer this question, we have to consider a **small signal equivalent circuit** of the tunnel diode (see Fig. 4.6.6).   [A small signal equivalent circuit of an electronic device is the circuit of lumped circuit elements, such as resistances, capacitances, etc. that has the same (or nearly the same) response to a small amplitude ac signal as the device.]

In this equivalent circuit, the diode is represented by a parallel combination of the differential resistance $R_d = dV/dI$ and the differential capacitance $C_d = dQ/dV$.  Once again equating the currents through the diode and the resistor, we obtain

$$C_d \frac{du}{dt} + i = \frac{u_s - u}{R} \qquad\qquad (4\text{-}6\text{-}8)$$

where $u$ is the small signal *ac* voltage across the diode and $u_s$ is a small signal voltage source.  Substituting $i = u/R_d$ into eq. (4-6-7), we obtain

$$\frac{du}{dt} = \frac{u_s}{RC_d} - \frac{u}{\tau}$$

(4-6-9)

where

$$\tau = \frac{C_d}{1/R_d + 1/R}$$

(4-6-10)

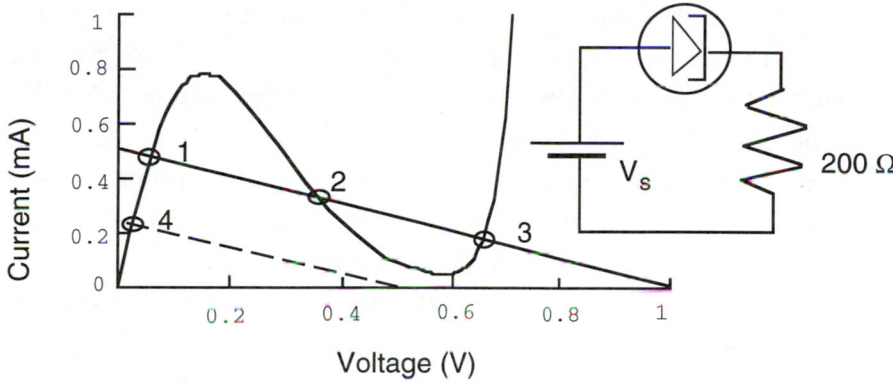

**Fig. 4.6.5.**  Operating points of tunnel diode for $V_s = 1$ V and $V_s = 0.5$ V.

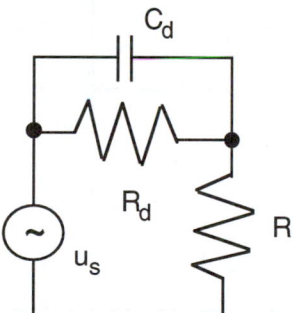

**Fig. 4.6.6.**  Small signal circuit of tunnel diode.

Integrating differential equation (4-6-10), we find

$$u(t) = \frac{R_d u_s}{R + R_d} + A \exp\left(-\frac{t}{\tau}\right)$$

(4-6-11)

The first term in the right-hand side of eq. (4-6-11) describes the voltage division of the ac source voltage. The second term describes the time evolution of $u$ (the constant $A$ has to be determined from the initial condition). As we can see from this equation, when $\tau < 0$ (which is the case for the operating point 2 in Fig. 4.6.5 corresponding to the negative differential resistance region), any small deviation of the initial condition from the exact value $R_d u_s/(R_d + R)$ will exponentially grow with time. When $\tau > 0$, any such deviation will decay with time. Hence, we conclude that operating points 1 and 3 in Fig. 4.6.5 are stable and operating point 2 is unstable. A circuit with two stable operating points is called **bistable**, a circuit with one stable operating point is called **monostable**. Depending on the tunnel diode bias history, it can wind up in either stable operating point. If we start from small bias (small currents), such as the current corresponding to operating point 4 in Fig. 4.6.5, and gradually increase the current, we will end up in operating point 1. However, if we increase the bias voltage so that the current reaches the peak value, the device will switch into operating point 3. Hence, the device can be switched between operating points 1 and 3 with a voltage pulse. Such a switch has many applications in digital electronic circuits.

Table 4.6.1 summarizes equations describing current voltage characteristics of tunnel diodes.

| Current through tunnel diode | $I = I_{tun} + I_{diode} + I_{excess}$ |
|---|---|
| Tunneling current | $I_{tun} \approx \dfrac{V}{R_o} \exp\left[-\left(\dfrac{V}{V_o}\right)^m\right]$ |
| Normal diode current | $I_{diode} \approx I_s \exp\left[\left(\dfrac{V}{\eta V_{th}}\right) - 1\right]$ |
| Excess tunneling current | $I_{excess} \approx \dfrac{V}{R_v} \exp\left[\left(\dfrac{V - V_v}{V_{ex}}\right)\right]$ |

**Table 4.6.1.** Equations for *I-V* characteristics of tunnel diodes.

## 4-7.  CIRCUIT MODELING OF *p-n* JUNCTION DIODES

Practical design, fabrication, and applications of any semiconductor device rely on computer simulation of device fabrication and computer-aided design of semiconductor devices and circuits.

In this section, we describe a diode model implemented in the circuit simulator program called SPICE, since this program is used by a vast majority of electrical engineers.  Hence, it is very important to understand and realistically choose parameters describing diodes in SPICE.

Originally, SPICE was developed at the University of Berkeley.  Now there are many commercial versions of SPICE.  Some of them, such as PSpice[tm] from Microsim, Inc. or AIM-Spice developed by the international team from the Korean Institute of Advanced Science and Technology, Norwegian Institute of Technology, and the University of Virginia, run on microcomputers.  In this book, we will use both PSpice[tm] and AIM-Spice (see Appendix A8) in order to run all circuit examples.  A description of PSpice[tm] is given by Tuinenga (1990) and a description of AIM-Spice is given by Lee et al. (1993).

These two books and other books on SPICE contain information on how to prepare a circuit description for the circuit simulation using SPICE and how to specify different types of circuit analyses.  The goal of this section is to relate the SPICE *p-n* junction diode model to the equations derived in this chapter.  This is very useful for both understanding default SPICE parameters and for an intelligent choice of parameters for SPICE simulations.

AIM-Spice has two *p-n* diode models.  The model for a conventional *p-n* diode is called the Level 1 model.  This model is identical to the *p-n* diode model in other versions of SPICE.  The second model is a Heterostructure Diode model.  This model is unique to AIM-Spice and is beyond the scope of this book.   The parameters of the Level 1 model are given in Table 4.7.1.

As can be seen from the table, the default parameters related to the temperature dependence of the saturation current correspond to Si *p-n* diodes (since $E_g = 1.11$ eV) and to the case when the diffusion current is dominant (since this current scales with temperature proportionally to $n_i^2$).  This is not always the case in real diodes (see Section 4.3).

Some of the SPICE parameters (primarily those dealing with diode noise) are beyond the scope of this book.  Except for very special cases (such as noise analysis), these parameters can be kept at their default values.

| SPICE parameter | SPICE parameter name | Unit | Spice default | Chap. 2 notation | Relevant equation |
|---|---|---|---|---|---|
| IS | Saturation current | A | 1.0e-14 | $I_s$ | (4-3-35) |
|  |  |  |  |  | (4-3-36) |
| RS | Series resistance | $\Omega$ | 0 | $R_s$ | (4-3-35) |
| N | Ideality factor | - | 1 | $\eta$ | (4-3-35) |
| TT | Transit time | s | 0 | $t_{tr}$ | (4-4-31) |
|  |  |  |  |  | (4-4-32) |
| CJO | Zero bias capacitance | F | 0 | $C_{jo}$ | (4-4-13) |
| VJ | Built-in voltage potential | V | 1 | $V_{bi}$ | (4-4-13) |
| M | Grading coefficient | - | 0.5 | $m$ | (4-4-13) |
| EG | Energy gap | eV | 1.11 | $E_g$ | (4-3-37) |
| XTI | Saturation current temperature exponent | - | 3.0 | $\kappa$ | (4-3-37) |
| KF | Flicker noise coefficient | - | 0 | - | - |
| AF | Flicker noise exponent | - | 1 | - | - |
| FC | Coefficient for forward-bias depletion capacitance | - | 0.5 | $m$ | - |
| BV | Reverse breakdown voltage | V | infinite | $V_{br}$ | (4-5-4) |
| IBV | Current at breakdown | A | $10^{-3}$ | $I_{br}$ | (4-5-4) |
| TNOM | Temperature at which parameters are specified | $^\circ$C | 27 | $T_o$ (K) ($T_o$=TNOM +273) | (4-3-37) |

**Table 4.7.1.** Parameters of *p-n* diode SPICE model.

Figures 4.7.1 to 4.7.4 show the results of the PSpice[tm] simulation of a Si *p-n* diode. In this simulation, $t_{tr} = 2$ ns, $C_{jo} = 1$ ps, $R_s = 100$ $\Omega$.

Figure 4.7.1 shows the decrease in the slope of the semilog current-voltage characteristics at higher temperatures (for relatively small forward biases). The semilog plots deviate from the straight lines at high forward bias when the voltage drop, $IR_s$, across the diode series resistance becomes important. This trend can be also seen from Fig. 4.7.2: at high forward bias, the diode current voltage characteristics become nearly linear with the slope determined by the series resistance.

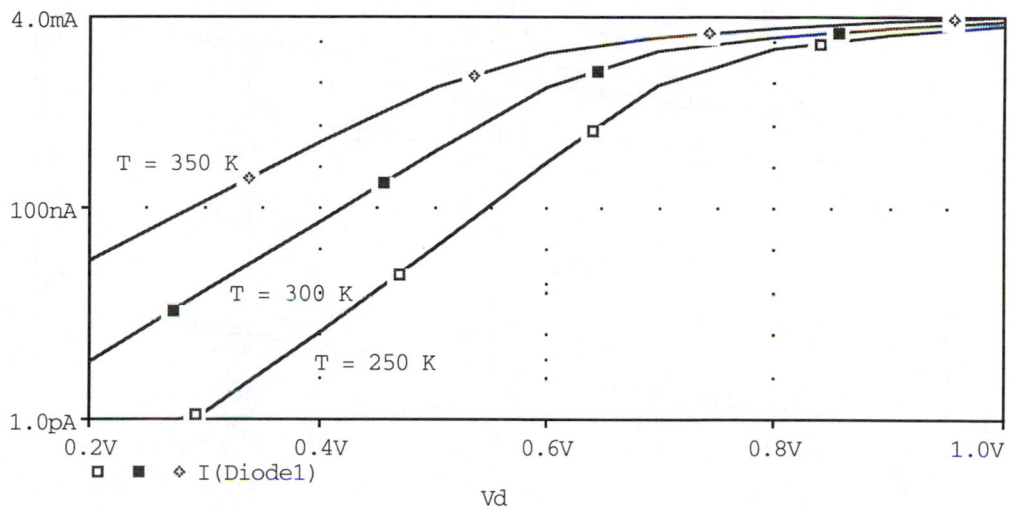

**Fig. 4.7.1.** Forward *I-V* characteristics of Si *p-n* junction for three temperatures: 250 K, 300 K, and 350 K (semilog scale).

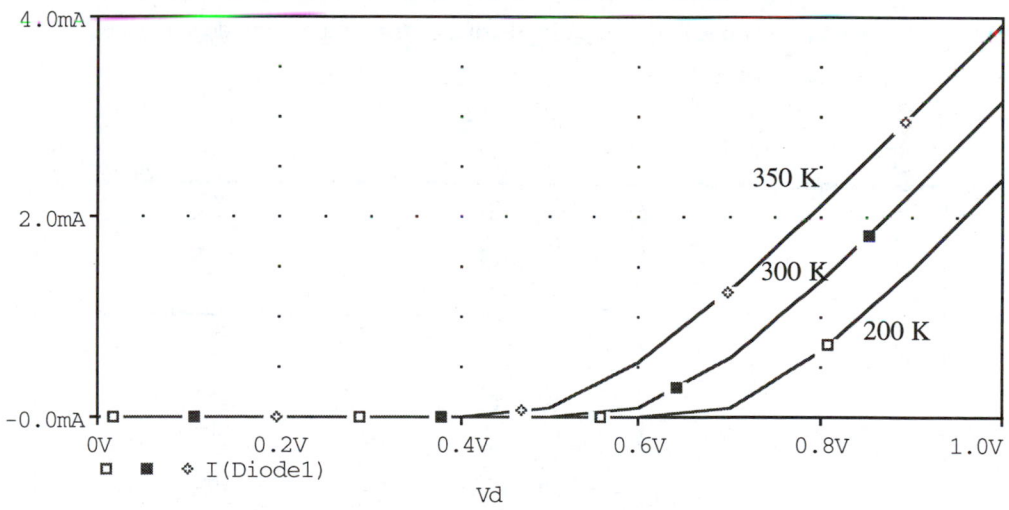

**Fig. 4.7.2.** Forward *I-V* characteristics of Si *p-n* diode for  temperatures 200 K, 300 K, and 350 K (linear scale).

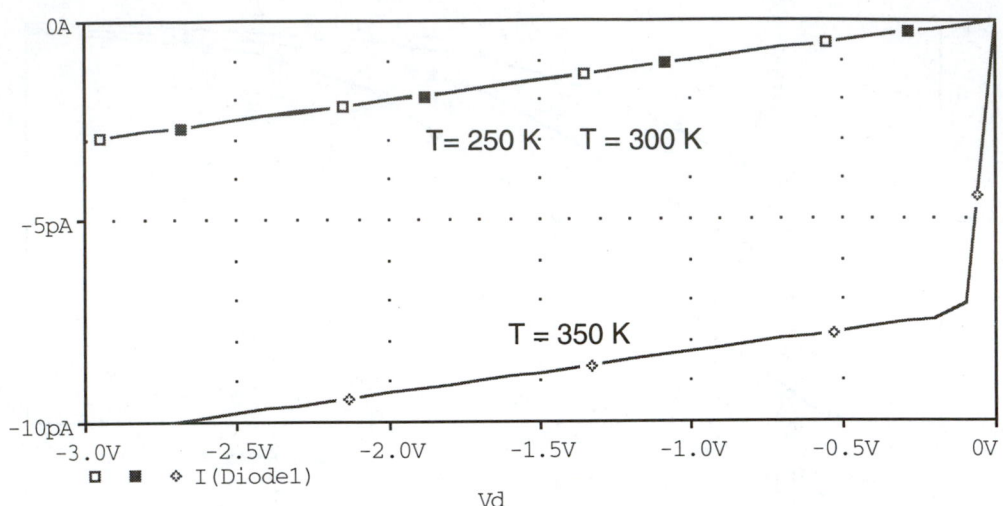

**Fig. 4.7.3.** Reverse *I-V* characteristics of Si *p-n* junction for three temperatures: 250 K, 300 K, and 350 K. Symbols are the same as for Fig. 4.7.1. Notice that at 250 K and 300 K the diode current is the same. This is caused by the default value of the minimum parallel conductance, $G_{min} = 10^{-12}$ 1/$\Omega$, and does not necessarily reproduce the temperature dependence in practical devices.

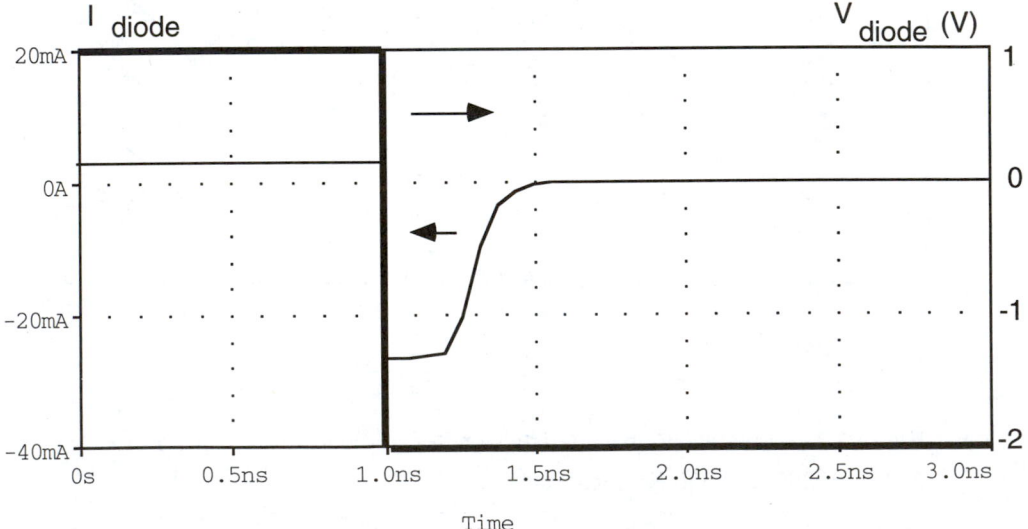

**Fig. 4.7.4.** Transient response of Si *p-n* junction for input voltage waveform shown in the figure.

An important diode characteristic is the **turn-on  voltage,** which is defined as the voltage at which the diode current reaches a certain fixed value, let us say 0.5 mA.  Figure 4.7.2 shows that the turn-on voltage decreases with an increase in temperature.  (The reason for this decrease is that at higher temperatures, electrons and holes have a higher thermal energy, and it is easier for them to overcome the potential barrier separating the *p* and *n* regions; see Fig. 4.2.2.)

Figure 4.7.3 illustrates features of the SPICE diode model that are not described by the standard diode equations.  The SPICE model includes a small constant conductance, $G_{min}$, shunting the diode.  This conductance is responsible for a constant slope in the reverse diode characteristics.  Also, when the diode saturation current becomes very small, the current through this conductance can become much larger than the diode current.  (This corresponds to the curves for 300 K and 250 K shown in Fig. 4.7.3.)

Figure 4.7.4 gives an example of a more sophisticated simulation.  It shows the response of a *p-n* diode to a voltage step with voltage changing from forward to reverse bias.  Such a transient is very important in the circuits where a *p-n* junction diode is used as a switch with resistance changing from a very small value under a forward bias to a very large value under a reverse bias.  Such switching often takes place in power electronic devices.  As can be seen from Fig. 4.7.4, when the voltage polarity is changed instantaneously, the diode current first changes sign.  The reason for this change in sign can be clearly understood when we consider an equivalent circuit of the diode at the forward bias (see Fig. 4.7.5b and compare with Fig. 4.4.3).  The voltage across the diode does not

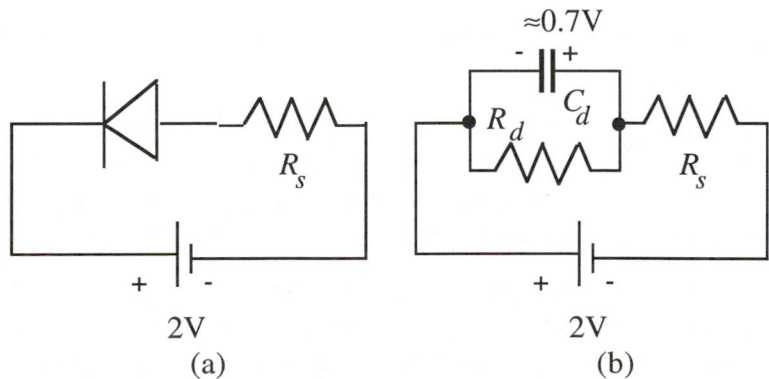

**Fig. 4.7.5.**  (a) The diode under reverse bias and (b) the equivalent circuit of this diode at $t = 0^+$.

change at $t = 0$ since the voltage across the diode capacitance cannot change instantaneously. The current through the series resistance, $R_s = 100 \, \Omega$, at $t = +0$ can be found using the Kirchoff's voltage law and Ohm's law:

$$I(t = +0) \approx -\frac{0.7 + 2}{100} \approx -27 (\text{mA}) \qquad (4\text{-}7\text{-}1)$$

(The negative sign corresponds to a reverse current.) The current remains nearly constant for as long as the concentration of minority carriers near the junction exceeds the equilibrium concentration of the minority carriers since, under such conditions, the voltage across the diode will still correspond to the forward bias. The time scale of this process is determined by the transit time, $t_{tr}$ (2 ns in our example), since it takes that long for the minority carriers to diffuse from the boundary of the depletion region to the contacts [see eqs. (4-4-31) and (4-4-32)]. (As was discussed in Section 4.4, this characteristic time scale is the minority carrier recombination time for diodes with neutral regions longer than the minority carrier diffusion length.) Then the current approaches the equilibrium value corresponding to the negative bias. The time constant of the latter process is determined by the diode $RC$ constant where $C$ is the diode depletion capacitance and $R = R_s$. (Of course, SPICE will compute the diode response for us even if we may not yet have a full understanding of the device physics.) Clearly, we have two ways to design a fast switching diode: either to reduce the minority carrier recombination time (perhaps by adding impurities, which serve as recombination centers) or to make the neutral region shorter. Still another approach is to try to choose such doping profiles that the internal electric fields will aid the removal of the minority carriers during the switching process. Such an approach is used in a so-called **step-recovery** diode [see Sze (1981) for more details].

The implementation of device models in SPICE allows us to analyze problems that may be far beyond our limited capabilities for analytical calculations. But in order to use this program [as well as other Computer Aided Design (CAD) tools for semiconductor device and circuit simulation] one has to understand basic semiconductor physics and basic device models. Otherwise, as it happens so often, computer simulations will follow one of the basic rules: "Garbage in – garbage out," no matter how sophisticated our modeling tools may be.

# 4-8. SCHOTTKY DIODES

### 4-8-1. Metal-semiconductor contact at zero bias.

Electrons in the conduction band of a crystal can be viewed as sitting in a potential box formed by the crystal boundaries (see Fig. 4.8.1). This potential box for electrons is usually deeper in a metal than in a semiconductor. If a metal and a semiconductor are brought together into a close proximity, some electrons from the metal will move into the semiconductor and some electrons from the semiconductor will move into the metal. However, since the barrier for the electron escape from the metal is higher, more electrons will transfer from the semiconductor into the metal than in the opposite direction. At thermal equilibrium, the metal will be charged negatively, and the semiconductor will be charged positively, forming a dipole layer that is very similar to that in a $p^+$-$n$ junction (see Section 4.2). The Fermi level will be constant throughout the entire metal-semiconductor system, and the energy band diagram in the semiconductor will be similar to that for an $n$-type semiconductor in a $p^+$-$n$ junction (see Fig. 4.8.2).

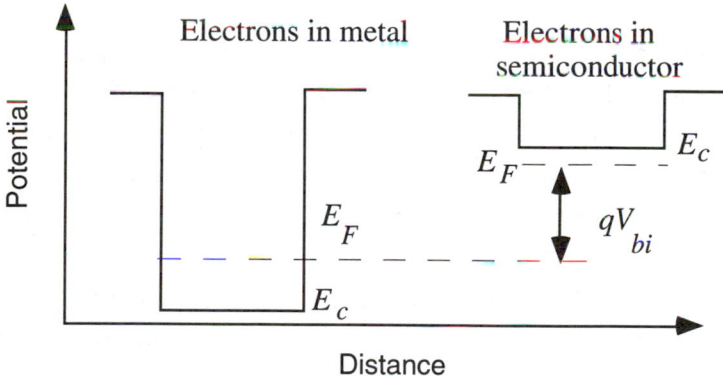

**Fig. 4.8.1.** Schematic energy diagram for electrons in conduction bands of a metal and of a semiconductor.

Energies $\Phi_m$ and $\Phi_s$ shown in Fig. 4.8.2 are called the metal and the semiconductor **work functions**. The **work function** is equal to the difference between the **vacuum level** (which is defined as a free electron energy in vacuum) and the Fermi level. The **electron affinity** of the semiconductor, $X_s$ (also shown in Fig. 4.8.2), corresponds to the energy separation between the vacuum level and the conduction band edge of the semiconductor.

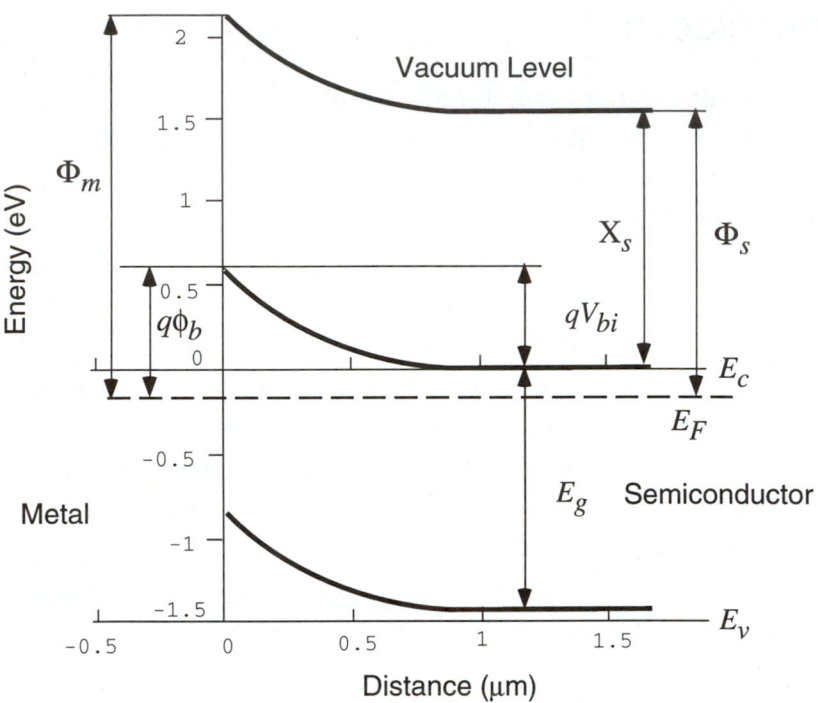

**Fig. 4.8.2.** Simplified energy diagram of GaAs metal-semiconductor barrier $q\phi_b$ is the barrier height (0.75 eV), $X_s$ is the electron affinity in the semiconductor, $\Phi_s$ and $\Phi_m$ are the semiconductor and the metal work functions, and $V_{bi}$ (0.591 V) is the built-in voltage. Donor concentration in GaAs is $10^{15}$ cm$^{-3}$.

A metal-semiconductor diode is called a **Schottky diode**. In the idealized picture of the Schottky junction shown in Fig. 4.8.2, the energy barrier between the semiconductor and the metal is

$$q\phi_b = \Phi_m - X_s \tag{4-8-1}$$

Since $\Phi_m > \Phi_s$ the metal is charged negatively. The positive net space charge in the semiconductor leads to a band bending

$$qV_{bi} = \Phi_m - \Phi_s \tag{4-8-2}$$

where $V_{bi}$ is called the **built-in voltage**, in analogy with the corresponding quantity in a *p-n* junction. Note that $qV_{bi}$ is also identical to the difference between the Fermi levels in the metal and the semiconductor when separated by a large distance (no exchange of charge); see Fig. 4.8.1.

However, eq. (4-8-1) and Fig. 4.8.2 are not quite correct.  In reality, a change in the metal work function, $\Phi_m$, is not equal to the corresponding change in the barrier height ,$\phi_b$, as predicted by eq. (4-8-1).  In actual Schottky diodes, $\phi_b$ increases with an increase in $\Phi_m$ but only by  0.1 to 0.3 eV  when $\Phi_m$  increases by 1 to 2 eV.  Even though a detailed and accurate understanding of Schottky barrier formation remains a challenge, many properties of Schottky barriers may be understood independently of the exact mechanism determining the barrier height. In other words, we can simply determine the effective barrier height from experimental data.  Usually, as a crude and empirical rule of thumb, we can assume that the Schottky barrier height for an $n$-type semiconductor is close to 1/2 and 2/3 of the energy gap.

In a Schottky diode, the semiconductor band diagram looks very similar to that of an $n$-type semiconductor in a $p^+$-$n$ diode (compare Fig. 4.3.1a and 4.8.2). Hence, the variation of the space charge density, $\rho$, the electric field, $F$, and the potential, $\phi$, in the semiconductor near the metal-semiconductor interface can be found using the depletion approximation:

$$\rho = qN_d \tag{4-8-3}$$

$$F = -\frac{qN_d(x_n - x)}{\varepsilon_s} \tag{4-8-4}$$

$$\phi = -\frac{qN_d(x_n - x)^2}{2\varepsilon_s} = -V_{bi}\left(1 - \frac{x}{x_n}\right)^2 \tag{4-8-5}$$

(Here  $x = 0$  corresponds to the metal-semiconductor interface.)  The depletion layer width, $x_n$, at zero bias is given by

$$x_n = \sqrt{\frac{2\,\varepsilon_s V_{bi}}{qN_d}} \tag{4-8-6}$$

**4-8-2.  Schottky diode under bias.**

Forward bias corresponds to a positive voltage applied to the metal with respect to the semiconductor.  Just as for a $p^+$-$n$ junction, the depletion width under small forward bias and reverse bias may be obtained by substituting $V_{bi}$ with $V_{bi} - V$, where $V$ is the applied voltage.  As illustrated in Fig. 4.8.3, the application of a forward bias decreases the potential barrier for electrons moving from the semiconductor into the metal. Hence, the current-voltage characteristic of a Schottky diode can be described by a **diode equation**, similar to that for a *p-n* junction diode [see eq. (4-3-34)]:

$$I = I_s \left[ \exp\left( \frac{V - IR_s}{\eta V_{th}} \right) - 1 \right]$$

(4-8-7)

where $I_s$ is the saturation current, $R_s$ is the series resistance, $V_{th} = k_B T/q$ is the thermal voltage, and $\eta$ is the ideality factor ($\eta$ typically varies from 1.02 to 1.6).

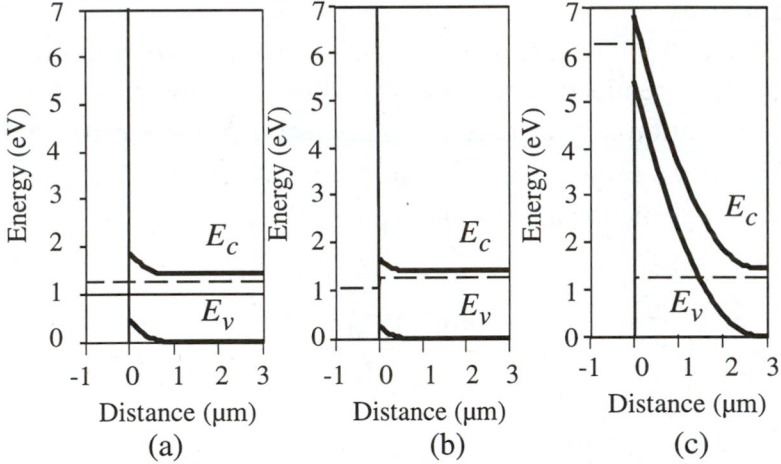

(a) (b) (c)

**Fig. 4.8.3.** Band diagrams for a GaAs Schottky barrier diode at (a) zero bias, (b) 0.2 V forward bias, and (c) 5 V reverse bias. Dashed line shows the position of the Fermi level in the metal ($x < 0$) and in the semiconductor ($x > 0$).

**Example 4-8-1.**

Estimate the doping density in the GaAs layer from the data in Fig. 4.8.3.

**Solution:**

From Fig. 4.8.3, the depletion width, $x_n$, at zero bias is approximately 0.7 μm. The depletion width in Fig. 4.8.2 for $N_d = 10^{15}$ cm$^{-3}$ is approximately 1 μm. Since the depletion width varies approximately as $1/N_d^{1/2}$, we conclude that $N_d \approx 2 \times 10^{15}$ cm$^{-3}$. Somewhat more accurately, we can estimate $x_n$ from the data for reverse bias using eq. (4-8-6) where we substitute $V_{bi}$ with $V_{bi} - V$ where $V$ is negative for reverse bias. Check that this estimate gives approximately the same result.

### 4-8-3. Thermionic emission.

The diode saturation current, $I_s$, is typically much larger for Schottky barrier diodes than in $p$-$n$ junction diodes since the Schottky barrier height is smaller than the barrier height in $p$-$n$ junction diodes. In a $p$-$n$ junction, the height of the barrier separating electrons in the conduction band of the $n$-type region from the bottom of the conduction band in the $p$-region is on the order of the energy gap. A typical Schottky barrier height is only about two thirds of the energy gap or less, as mentioned above. Also, the mechanism of the electron conduction is different. One can show that the saturation current density in a Schottky diode with a relatively low doped semiconductor is given by

$$j_{ss} = A^* T^2 \exp\left(-\frac{\phi_b}{k_B T}\right) \tag{4-8-8}$$

where $A^*$ is called the Richardson constant. For a conduction band minimum with spherical surfaces of equal energy (such as the $\Gamma$ minimum in GaAs),

$$A^* = \alpha \frac{m_n q k_B^2}{2\pi^2 \hbar^3} \approx 120\, \alpha \frac{m_n}{m_e} \left(\frac{A}{cm^2 K^2}\right) \tag{4-8-9}$$

where $m_n$ is the effective mass and $\alpha$ is an empirical factor on the order of unity. The Schottky diode model described by eqs. (4-8-8) and (4-8-9) is called the **thermionic emission model**. For Schottky barrier diodes fabricated on the {111} surfaces of Si, $A^* = 96$ A/(cm$^2$K$^2$). For GaAs, $A^* = 4.4$ A/(cm$^2$K$^2$).

The basic assumption of the thermionic model is that electrons have to pass over the barrier in order to cross the boundary between the metal and the semiconductor. Hence, to find the saturation current, we have to estimate the number of electrons passing over the barrier and their velocities. The number of electrons, $N(E)dE$, having energies between $E$ and $E + dE$ is proportional to the product of the Fermi-Dirac distribution function, $f(E)$, and the number of states in this energy interval, $g(E)dE$, where $g(E)$ is the density of states:

$$N(E)dE = g(E)f(E)dE \tag{4-8-10}$$

$[N(E) = dn(E)dE$ where $n(E)$ is the number of electrons in the conduction band with energies higher than $E$; see Fig. 2.4.2.] At high energies, the Fermi-Dirac occupation function is very close to the Boltzmann distribution function [see eqs. (2-4-4) and (2-4-5)]:

$$f_n(E) = \frac{1}{1 + \exp\left(\dfrac{E - E_F}{k_B T}\right)} \approx \exp\left(\frac{E_F - E}{k_B T}\right) \qquad (4\text{-}8\text{-}11)$$

The next step should be to multiply the number of the electrons, $N(E)dE$, in the energy interval from $E$ to $E + dE$ by the velocity of these electrons. We have to account for different directions of the electron velocities and integrate over energies higher than the barrier height in order to determine the flux of the electrons coming from the semiconductor into the metal. Finally, we deduct the flux of the electrons coming from the metal into the semiconductor. The difference between these two fluxes will be proportional to the current density predicted by the thermionic model. [Such a derivation is done in many books including Shur (1990)]. However, we can take a much simpler route if we are interested in understanding the physics of the thermionic model. To this end, let us consider a Schottky diode under a strong reverse bias when $V$ is negative and $-V \gg \eta k_B T$. Then $I = -I_s$ [see eq. (4-8-7)], and the band diagram looks like that shown in Fig. 4.8.3c. In this case the energy difference between the Fermi level in the semiconductor and the top of the barrier is so large that practically no electrons are available to come from the semiconductor into the metal. However, the Fermi level in the metal is much closer to the top of the barrier, and electrons still come from the metal into the semiconductor. The flux of these electrons constitutes the saturation current. In order to estimate this flux, we should recall that the density of states is a relatively slow function of energy [$g(E)$ is proportional to $(E - E_c)^{1/2}$; see eq. (2-4-19)] compared to the distribution function, which decreases by $\exp(1) \approx 2.718$ each time $E$ increases by $k_B T$. Hence, the largest contribution into the electron flux will come from the electrons that are a few $k_B T$ above the barrier. The number of such electrons will be proportional to the effective density of states in the semiconductor

$$N_c = 2\left(\frac{m_n k_B T}{2\pi\hbar^2}\right)^{3/2} \qquad (4\text{-}8\text{-}12)$$

[see eq. (2-4-21)] and to $\exp(-\phi_b/k_BT)$.   Their velocity in the direction perpendicular to the metal semiconductor interface is proportional to the thermal velocity

$$v_{thnx} = \sqrt{\frac{k_BT}{m_n}} \qquad (4\text{-}8\text{-}13)$$

[see eq. (3-6-6)].   Hence, the saturation current density is given by

$$j_{ss} = CqN_c \exp\left(-\frac{\phi_b}{k_BT}\right)v_{thnx} = 2C\left(\frac{m_n k_BT}{2\pi\hbar^2}\right)^{3/2}\sqrt{\frac{k_BT}{m_n}}\exp\left(-\frac{\phi_b}{k_BT}\right)$$

$$= \sqrt{\frac{1}{2\pi^3}}C\frac{m_n k_B^2 T^2}{\hbar^3}\exp\left(-\frac{\phi_b}{k_BT}\right) \qquad (4\text{-}8\text{-}14)$$

where $C$ is a numerical constant of the order of unity.   With a proper choice of $C$, this equation coincides with eqs. (4-8-8) and (4-8-9).

In a relatively lightly doped semiconductor with low mobility, the current through the Schottky barrier may be limited more by diffusion and drift processes in the space charge region than by the barrier at the metal-semiconductor interface. Even in this special case, eq. (4-8-7) is still applicable but eq. (4-8-8) is no longer valid.

## 4-8-4.  Thermionic-field emission.

In relatively highly doped semiconductors, the depletion region becomes so narrow that electrons can tunnel through the barrier near the top (see Fig. 4.8.4b).   This process is called **thermionic-field emission**.   In order to understand thermionic-field emission, we have to recall once again that the number of electrons with energies above a given energy $E$ decreases exponentially with energy as $\exp[-E/(k_BT)]$.   On the other hand, the barrier transparency increases exponentially with the decrease in the barrier width.   Hence, as the doping increases and the barrier becomes thinner, the dominant electron tunneling path occurs at lower energies than the top of the barrier (see Fig. 4.8.4b).

In degenerate semiconductors, especially in semiconductors with a small electron effective mass such as GaAs, electrons can tunnel through the barrier near or at the Fermi level, and the tunneling current is dominant.   This mechanism is called **field emission** (see Fig. 4.8.4c).

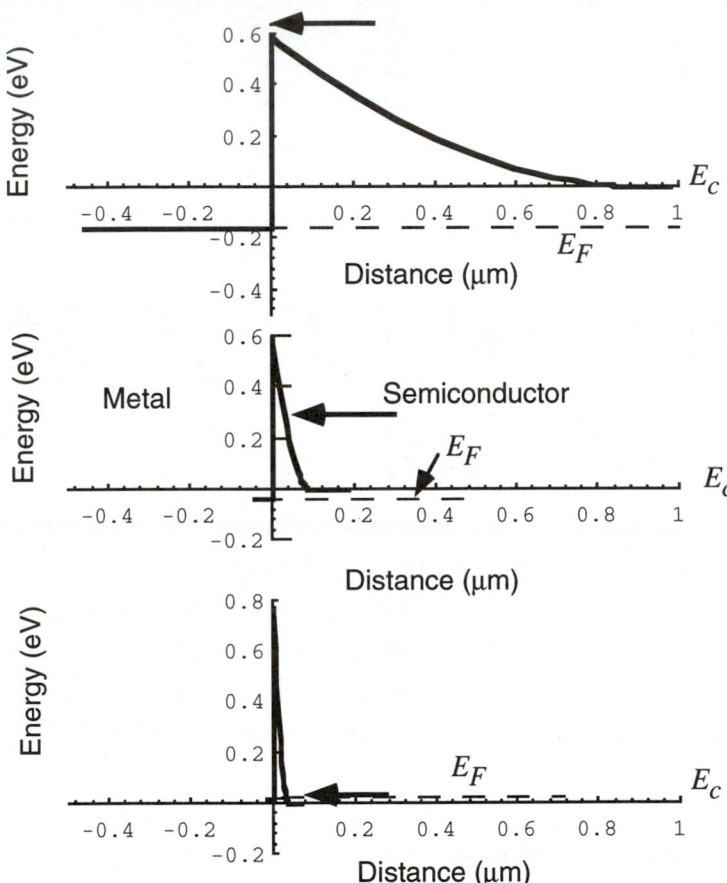

**Fig. 4.8.4.** Band diagrams of Schottky barrier junctions for GaAs for doping levels $N_d = 10^{15}$ cm$^{-3}$ (top graph), $N_d = 10^{17}$ cm$^{-3}$ (middle graph), and $N_d = 10^{18}$ cm$^{-3}$ (bottom graph). Arrows indicate electron transfer across the barrier under forward bias. At very low doping levels, electrons go over the barrier closer to the top of the barrier (this process is called thermionic emission). At moderated doping levels, electrons tunnel across the barrier closer to the top of the barrier (this process is called thermionic-field emission). In highly doped degenerate semiconductors, electrons near the Fermi level tunnel across a very thin depletion region (this process is called field emission).

The current-voltage characteristic of a Schottky diode in the case of thermionic-field emission can be calculated using the same approach as for the thermionic model, except that in thermionic-field emission case, we have to evaluate the product of the tunneling transmission coefficient and the number of electrons at a given energy as a function of energy and integrate over the states in the conduction band.  Such a calculation [see Rhoderick and Williams (1988)] yields the following expression for the current density in the thermionic-field emission regime under forward bias:

$$j = j_{stf} \exp\left(\frac{qV}{E_o}\right) \tag{4-8-15}$$

where

$$E_o = E_{oo} \coth\left(\frac{E_{oo}}{k_B T}\right) \tag{4-8-16}$$

$$E_{oo} = \frac{qh}{4\pi}\sqrt{\frac{N_d}{m_n \varepsilon_s}} = 1.85 \times 10^{-11} \left[\frac{N_d\left(\text{cm}^{-3}\right)}{\left(m_n/m_o\right)\left(\varepsilon_s/\varepsilon_o\right)}\right]^{1/2} \text{(eV)} \tag{4-8-17}$$

$$j_{stf} = \frac{A^* T \sqrt{\pi E_{oo}\left(\phi_b - qV - E_c + E_{Fn}\right)}}{k_B \cosh\left(E_{oo}/k_B T\right)} \exp\left[-\frac{E_c - E_{Fn}}{k_B T} - \frac{\left(\phi_b - E_c + E_{Fn}\right)}{E_o}\right] \tag{4-8-18}$$

In GaAs Schottky diodes, the thermionic-field emission becomes important for $N_d > 10^{17}$ cm$^{-3}$ at  300 K  and for  $N_d > 10^{16}$ cm$^{-3}$ at  77 K.  In silicon, the corresponding values of $N_d$ are several times larger.

**Example 4-8-2.**

Using eqs. (4-8-16) to (4-8-18), calculate and plot forward $j$-$V$ characteristics for GaAs Schottky diodes doped at $10^{15}$, $10^{17}$, and $10^{18}$ cm$^{-3}$.  Estimate ideality factors $\eta$.  Comment on expected reverse $j$-$V$ characteristics.

**Solution:**

The forward $j$-$V$ characteristics are shown in Fig. 4.8.5.

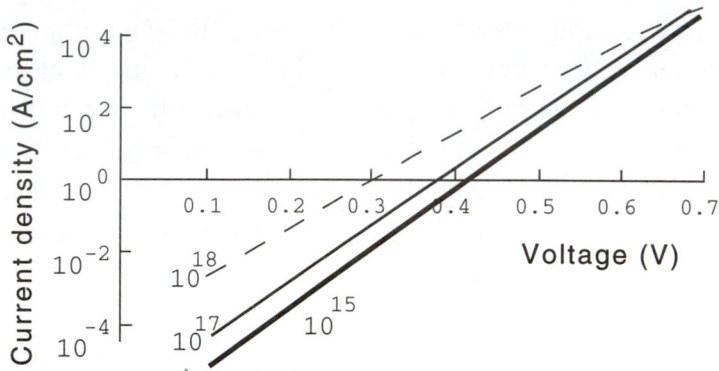

**Fig. 4-8-5.** Forward $j$-$V$ characteristics of GaAs Schottky diodes doped at $10^{15}$, $10^{17}$, and $10^{18}$ cm$^{-3}$ (curves are marked accordingly) at $T = 300$ K.

We notice a larger current density (smaller differential resistance) at higher doping levels caused by tunneling. For $10^{18}$ cm$^{-3}$ doping, the change in bias from 0.3 to 0.6 V changes the current density by 4 orders of magnitude. Hence, $\eta = 0.3/[4 \ln(10)\ 0.02584] \approx 1.26$. A similar calculation yields $\eta \approx 1.13$ for $10^{17}$ cm$^{-3}$ and $\eta \approx 1$ for $10^{15}$ cm$^{-3}$. Larger values of $\eta$ for higher doping levels is an indication of tunneling. Extrapolating the calculated curves to $V = 0$, we estimate the saturation current densities, $j_s$, of $8 \times 10^{-5}$, $9 \times 10^{-6}$, and $3 \times 10^{-7}$ A/cm$^2$, for $10^{18}$ cm$^{-3}$, for $10^{17}$ cm$^{-3}$, and $10^{15}$ cm$^{-3}$, respectively. At reverse bias, $j \approx j_s$, independent of bias, once $-V \gg V_{th}$.

The resistance of the Schottky barrier in the field emission regime is quite low. Therefore metal-$n^+$ contacts are used as ohmic contacts. The specific contact resistance, $\rho_c$, decreases with the increase in the doping level of the semiconductor. (This resistance may vary from $10^{-3}$ $\Omega$cm$^2$ to $10^{-7}$ $\Omega$cm$^2$ or even smaller depending on semiconductor material, doping level, contact metal, and ohmic contact fabrication technology.)

### 4-8-5. Small signal equivalent circuit and applications.

The small signal equivalent circuit of a Schottky barrier (see Fig. 4.8.6) is very similar to that of a $p$-$n$ diode (see Fig. 4.4.4). It includes a parallel combination of the differential resistance of the Schottky barrier

$$R_d = \frac{dV}{dI} \tag{4-8-19}$$

and the differential capacitance of the space charge region:

$$C_{dep} = S\sqrt{\frac{qN_d\varepsilon_s}{2(V_{bi} - V)}} \tag{4-8-20}$$

[compare with eq. (4-4-5)]. The equivalent circuit also includes the series resistance, $R_S$, which accounts for the contact resistance and the resistance of the neutral semiconductor region between the ohmic contact and the depletion region, the equivalent series inductance, $L_s$, and the device geometric capacitance:

$$C_{geom} = \varepsilon_s S / L \tag{4-8-21}$$

where $L$ is the device length and $S$ is the device cross section

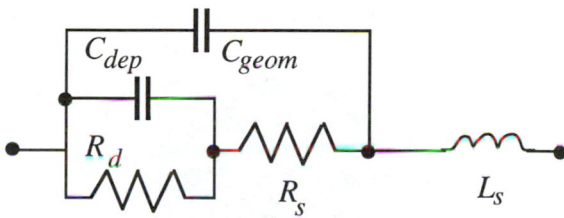

**Fig. 4.8.6.** The small signal equivalent circuit of a Schottky diode.

The major difference between the equivalent circuit of a Schottky diode and that of a *p-n* junction is the absence of the diffusion capacitance in the case of the Schottky diode. The diffusion capacitance of a *p-n* junction is associated with a time delay caused by the electron-hole recombination. A Schottky diode is a majority carrier device, where recombination is usually not important. Hence, Schottky diodes have a much faster response under forward bias conditions than *p-n* junction diodes. Therefore, Schottky diodes are used in applications where the speed of a response is important, for example, in microwave **detectors, mixers,** and **varactors** [see Maas (1993)].

In most cases, $R_S \ll R_d$ and $C_{dep} \gg C_{geom}$ and the frequency response of a Schottky diode is limited by

$$f_T = \frac{1}{2\pi R_s C_{dep}}$$
(4-8-22)

Small Schottky diodes respond to very high frequencies (up to five terahertz – $5 \times 10^{12}$ Hz or higher). Such frequencies, $f$, correspond to photon energies, $\hbar\omega = 2\pi\hbar f$, causing transitions between energy levels of gas molecules. Therefore, Schottky diodes are used as detectors determining the presence and concentration of certain gases in the upper atmosphere. As an example, Fig. 4.8.7 shows the distribution of chlorine monoxide in the upper atmosphere at approximately 20 km height from the sea level. [This distribution was measured by Upper Atmosphere Research Satellite (UARS) and reported by Dr. Joe Waters, Jet Propulsion Lab.] Chlorine monoxide is the gas destroying the ozone layer, and its distribution roughly coincides with the ozone "hole" in the upper atmosphere. The Schottky diodes used in this measurement were fabricated in the Semiconductor Device Laboratory at the University of Virginia by Professors Crowe and Mattauch and their associates.

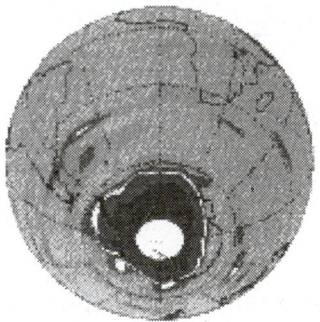

**Fig. 4.8.7.** Distribution of chlorine monoxide in the upper atmosphere at 20 km height from sea level measured by Upper Atmosphere Research Satellite. Dark area indicates a high concentration of chlorine monoxide. (After G. Zarpette, Sensing Climate Change, Spectrum, p. 23, July 1993, copyright © IEEE 1993.)

A Schottky diode mixer uses a nonlinearity of the forward current-voltage characteristic of a Schottky diode. As a consequence, when two ac signals with radian frequencies $\omega_1$ and $\omega_2$ are applied to the Schottky diode, signals with frequencies $\omega_1 + \omega_2$ and $\omega_1 - \omega_2$ appear as well (see Problem 4-8-6).

A varactor is a device having a capacitance that varies under bias. A Schottky diode varactor uses a nonlinear capacitance of a Schottky diode [see eq.

(4-8-20)].  A Schottky diode varactor operates under reverse bias, since this allows us to obtain a larger capacitance variation with a change in the applied voltage.

Just like Schottky barrier detectors, Schottky barrier mixers and varactors operate up to very high frequencies (up to several terahertz) and find numerous applications in radioastronomy, manufacturing, and environment control, and microwave, millimeter wave, and optoelectronic circuits.

### 4-8-6.  SPICE simulation of Schottky diodes.

The circuit simulator, SPICE, discussed in Section 4.7 does not include a separate model for a Schottky diode.  However, circuits with Schottky diodes may be still simulated using a judicious choice of the parameters for the SPICE diode model. The SPICE diode parameters that must be adjusted in order to describe a Schottky diode are given in Table 4.8.1.

| SPICE parameter | SPICE parameter name | Unit | Schottky diode value | Chap. 4 notation |
|---|---|---|---|---|
| TT | Transit time | s | 0 | $t_{tr}$ |
| EG | Energy gap | eV | 0.65 | $\phi_b$ |
| XTI | Saturation current temperature exponent | – | $2\eta$ | $\kappa$ |

**Table 4.8.1.**  Parameters that must be adjusted to describe a Schottky diode in SPICE model (Level 1 model of AIM-Spice).

In a *p-n* junction, the transit time, $t_{tr}$, is related to the diffusion time  (see Section 4.4).  However, in a Schottky diode, a transit time delay is associated with the electron drift across the depletion region when the width of the depletion region changes.  This time constant is given by

$$t_{tr} = \Delta d / v_T \tag{4-8-23}$$

where $\Delta d$ is the change of the depletion region width during transient, and $v_T$ is the electron thermal velocity.  The value of $t_{tr}$ is usually quite small (a fraction of a picosecond) and, in most cases, choosing $t_{tr} = 0$ as in Table 4.8.1 is appropriate.

In order to explain the suggested values of other SPICE parameters, we have to reproduce eq. (4-3-37) used by SPICE in order to describe the temperature dependence of the saturation current:

$$I_s(T) = I_s(T_o)\left(\frac{T}{T_o}\right)^{\frac{\kappa}{\eta}} \exp\left(\frac{E_g}{k_B T_o}\right) \exp\left(-\frac{E_g}{k_B T}\right) \qquad (4\text{-}3\text{-}37)$$

The current-voltage characteristics of a Schottky diode and a *p-n* diode calculated by SPICE are compared in Fig. 4.8.8.

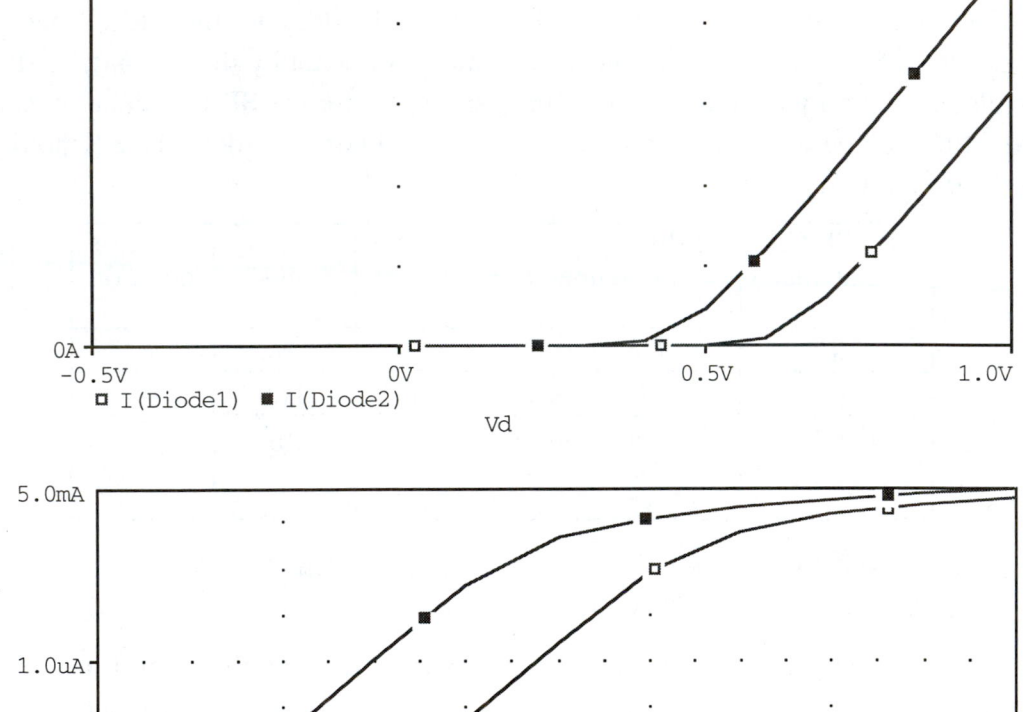

**Fig. 4.8.8.** Current-voltage characteristics of a Schottky diode (diode 2) and a *p-n* diode (diode 1) calculated by SPICE. Saturation currents are $10^{-11}$ A and $10^{-14}$ A, respectively.

As can be seen from eq. (4-8-8), the activation energy for the Schottky diode saturation current density is determined by the barrier height, $\phi_b$. Hence, the value for the SPICE parameter EG (in eV) should be chosen to be equal to $\phi_b$ (in eV). Since the pre-exponential term in eq. (4-8-8) is proportional to $T^2$, the value of *XTT* [the value of $\kappa$ in eq. (2-3-37)] suggested in Table 4.8.1 is equal to $2\eta$. The key differences from a *p-n* junction diode are a much larger saturation current density in a Schottky diode and a considerably smaller turn-on voltage.

The summary of the most important results relevant to properties of Schottky junctions is given in Table 4.8.2.

| | |
|---|---|
| Space charge density in the depletion layer | $\rho = qN_d$ |
| Field distribution in the depletion layer | $F = -\dfrac{qN_d(x_n - x)}{\varepsilon_s}$ |
| Potential distribution in the depletion layer | $\phi = -\dfrac{qN_d(x_n - x)^2}{2\varepsilon_s} = -V_{bi}\left(1 - \dfrac{x}{x_n}\right)^2$ |
| Depletion layer width | $x_n = \sqrt{\dfrac{2\varepsilon_s(V_{bi} - V)}{qN_d}}$ |
| Empirical diode equation | $I = I_s\left[\exp\left(\dfrac{V}{\eta V_{th}}\right) - 1\right]$ |
| Parallel leakage current | $I_{leakage} = G_{min}V$ |
| Diode equation (with series resistance) | $I = I_s\left[\exp\left(\dfrac{V - IR_s}{\eta V_{th}}\right) - 1\right]$ |
| Reverse diode current density (saturation current density) | $J_{ss} = A^{*}\overset{*}{T}^{2}\exp\left(-\dfrac{\phi_b}{V_{th}}\right)$ where $A^* = \alpha\dfrac{m_n qk_B^2}{2\pi^2\hbar^3} \approx 120\alpha\dfrac{m_n}{m_e}\left(\dfrac{A}{cm^2 K^2}\right)$ |
| Temperature dependence of the saturation current | $I_s(T) = I_s(T_o)\left(\dfrac{T}{T_o}\right)^2 \exp\left(\dfrac{\Phi_b}{k_B T_o}\right)\exp\left(-\dfrac{\Phi_b}{k_B T}\right)$ |

**Table 4.8.2.**  Important equations describing Schottky diodes.

## 4-9.  OHMIC CONTACTS

All electronic devices have to be connected to the outside world in order to form an electronic circuit in combination with other circuit elements.  In the case of a *p-n* diode, for example, contacts have to be provided to both *p*-type and *n*-type regions of the device in order to connect the diode to an external circuit.  These contacts have to be as unobtrusive as possible, so that the current flowing through a semiconductor device and, hence, through the contacts, leads to the smallest parasitic voltage drop possible.  Whatever voltage drop does occur across the contact has to be proportional to the current so that the contacts do not introduce uncontrollable and unexpected nonlinear elements into the circuit.  Since such contacts satisfy Ohm's law, they are usually called **ohmic contacts**.

As was discussed in Section 4.8, a contact between a metal and a semiconductor is typically a Schottky barrier contact.  However, if the semiconductor is very highly doped, the Schottky barrier depletion region becomes very thin, as illustrated in Fig. 4.8.4.  At very high doping levels, a thin depletion layer becomes quite transparent for electron tunneling (see Fig. 4.7.4).  This suggests that a practical way to make a good ohmic contact is to make a very highly doped semiconductor region between the contact metal and the semiconductor.

It may have been better to use a metal with a work function, $\Phi_m$, which is equal to or smaller than the work function of a semiconductor, $\Phi_s$, (see Problem 4-9-1).  However, for most semiconductors, it is difficult to find such a metal acceptable for practical contacts.

Current-voltage characteristics of a Schottky barrier diode (calculated using equations given in Section 4.8) and of an ohmic contact are compared in Fig. 4.9.1.  As was mentioned above, a good ohmic contact should have a linear current-voltage characteristic and a very small resistance that is negligible compared to the resistance of the active region of the semiconductor device.  An ohmic contact with the *I-V* characteristic shown in Fig. 4.9.2 does not satisfy fully these conditions since the voltage drop across this contact is not negligibly small compared with the voltage drop across the Schottky diode at moderate current densities above 0.1 kA/cm$^2$.

As was discussed in Section 4.8, the barrier between a metal and a semiconductor is usually smaller for semiconductors with smaller energy gaps.  Hence, another way to decrease the contact resistance is to place a layer of a narrow gap highly doped semiconductor material between the active region of the device and the contact metal.  Some of the best ohmic contacts to date have been made this way.

A quantitative measure of the contact quality is the **specific contact resistance**, $\rho_c$, which is the contact resistance of a unit area contact.  Depending on the semiconductor material and on the contact quality, $\rho_c$ can vary anywhere from $10^{-3}$ $\Omega$cm$^2$ to $10^{-7}$ $\Omega$cm$^2$ or even less.

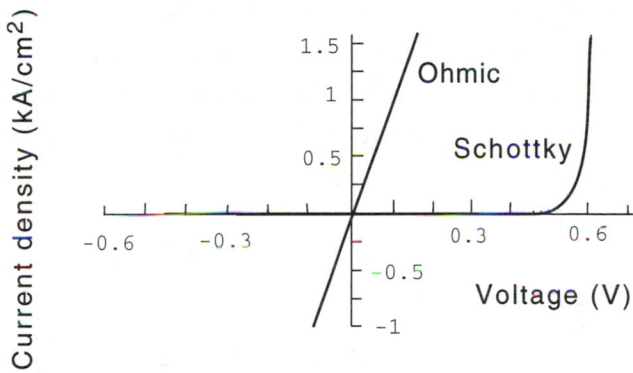

**Fig. 4.9.1.**  Current-voltage characteristics of ohmic and Schottky barrier metal-semiconductor contacts to GaAs.  (Schottky contact is to GaAs doped at $10^{15}$ cm$^{-3}$.)  Ohmic contact resistance is $10^4$ $\Omega$cm$^2$.

Most semiconductor devices have either a sandwich structure or a planar structure, as illustrated in Fig. 4.9.2.  The contact resistance of each contact in a sandwich structure contact is given by

$$R_c = \frac{\rho_c}{S} \qquad (4\text{-}9\text{-}1)$$

A typical current density in a sandwich type device can be as high as $10^4$ A/cm$^2$. Hence, the specific contact resistance of $10^{-5}$ $\Omega$cm$^2$ would lead to a voltage drop on the order of 0.1 V.  This may be barely acceptable.  A larger specific contact resistance of $10^{-4}$ $\Omega$cm$^2$ or so would definitely lead to problems, as we can see from Fig. 4.9.1.

For a more typical, planar structure, the contact resistance is proportional to the contact width but no longer proportional to the contact area.  In this case, the current density is larger near the contact edge.  For planar devices, an important measure of the contact quality is the contact resistance related to the unit width, $R_{cm}$, which can vary anywhere from 0.1 $\Omega$mm to 5 $\Omega$mm or more depending on a semiconductor material and the contact quality. The contact resistance is given by

$$R_c = \frac{R_{cm}}{W} \qquad (4\text{-}9\text{-}2)$$

A typical current in a planar semiconductor device used in a monolithic integrated circuit can be on the order of 100 mA/mm. Hence, the voltage drop across a contact with 1 $\Omega$mm contact resistance can be on the order of 0.1 V, quite noticeable compared to typical voltages involved that are on the order of 1 to 3 V. Still, this voltage drop across the contact is often tolerable. However, some devices may have a higher current still, up to more than 1 A/mm, leading to even higher voltage drops across the contacts.

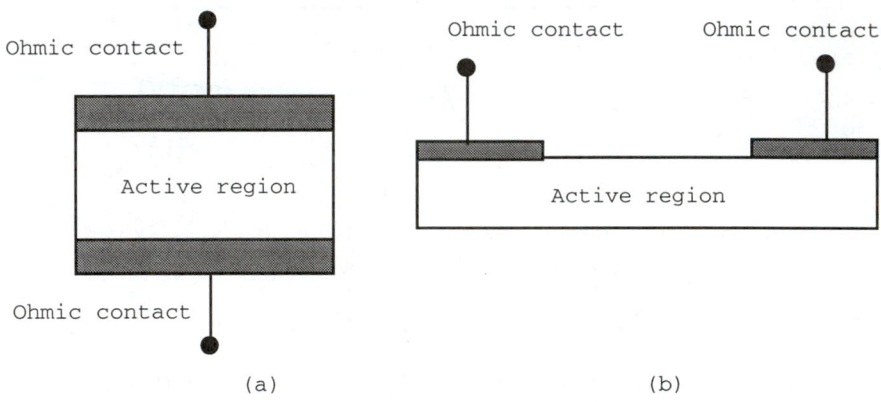

Fig. 4.9.2. (a) Sandwich and (b) planar device structures.

These estimates show that a semiconductor material can become viable for applications in electronic devices only when good ohmic contacts with low contact resistances become available. Often, poor ohmic contacts become a major stumbling block for applications of new semiconductor materials.

Figure 4.9.3 shows a characterization pattern used in integrated circuit technology for determining $R_c$. This pattern is called a **Transmission Line Model** (TLM) pattern because the current distribution under the planar ohmic contact is described by the same equation as a transmission line.

The resistance, $R_{n,n+1}$, between two adjacent ohmic contacts in the TLM pattern is given by

$$R_{n,n+1} = 2R_c + R_{sq}\frac{L_{n,n+1}}{W} \qquad (4\text{-}9\text{-}3)$$

where $L_{n,n+1}$ is the distance between the contacts, $R_{sq}$ is the resistance of the semiconductor film per square

$$R_{sq} = \frac{1}{\sigma t} \qquad (4\text{-}9\text{-}4)$$

$t$ is the film thickness, and $\sigma$ is the film conductivity. Hence, plotting the resistances $R_{n,n+1}$ between the contacts in this pattern as a function of distances between the contacts, we can determine the contact resistance and the resistance of the semiconductor film per square as shown in Fig. 4.9.4.

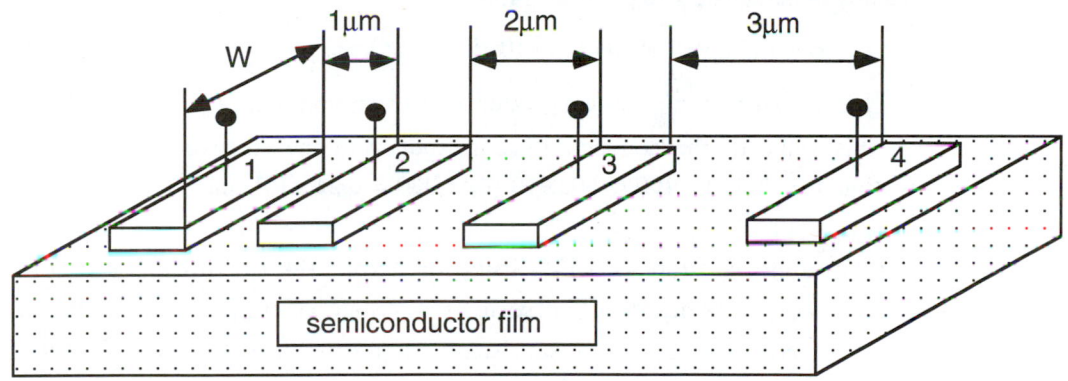

**Fig. 4.9.3.**   TLM characterization pattern for determining contact resistances.

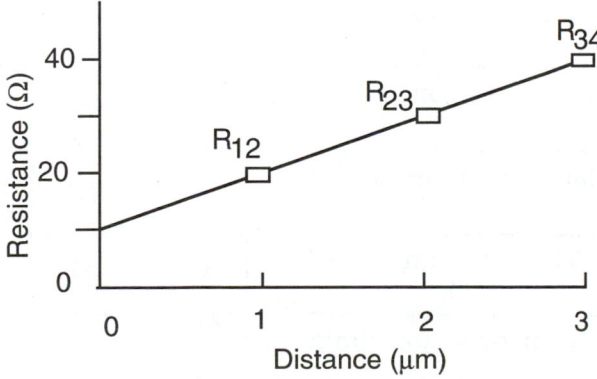

**Fig. 4.9.4.**  TLM resistances versus distance between the contacts.

**Example 4-9-1.**

The plot in Fig. 4.9.4 is for a pattern with contact width $W = 50$ μm. The film thickness is 0.2 μm. From the data shown in the figure, determine the contact resistance, the resistance of the semiconductor film per square, the film conductivity, and the carrier concentration at room temperature (assume an $n$-type Si film).

**Solution:**

The resistance intercept is $2R_c = 10$ Ω. Hence, the contact resistance related to a 1 mm width, $R_{cm} = 0.25$ Ωmm. The slope $R_{sq}/W = 10$ Ω/μm. Hence, the film resistance per square, $R_{sq} = 10 \times 50 = 500$ Ω per square. [The meaning of this notation (Ω per square) is that any square of this film has the same resistance independently of the square size.] The film conductivity

$$\sigma = 1/(R_{sq}t) = 1/(500 \times 0.2 \times 10^{-4}) = 100 \ 1/\Omega cm \qquad (4\text{-}9\text{-}5)$$

For an $n$-type silicon film at room temperature, the electron mobility $\mu_n \approx 1,000$ cm$^2$/Vs, this value of the conductivity corresponds to the electron carrier concentration $n = \sigma/(q\mu_n) = 100/(1.602 \times 10^{-19} \times 1,000) = 6.24 \times 10^{17}$ cm$^{-3}$.

Circuit simulators, such as AIM-Spice or PSpice[tm], do not have special models to describe ohmic contacts. The series resistance of ohmic contact has to be estimated using eq. (4-9-1) for a sandwich structure and eq. (4-9-2) for a planar structure and should be implemented as a part of the device series resistance (see Fig. 4.3.3, which shows the effect of a series resistance on current-voltage characteristics of a $p$-$n$ diode).

Table 4.9.1 summarizes equations related to ohmic contacts.

| Contact resistance for a sandwich structure | $R_c = \dfrac{\rho_c}{S}$ |
|---|---|
| Contact resistance for a planar structure | $R_c = \dfrac{R_{cm}}{W}$ |
| Resistance of TLM structure | $R_{n,n+1} = 2R_c + R_{sq}\dfrac{L_{n,n+1}}{W}$ |
| Resistance of semiconductor film per square | $R_{sq} = \dfrac{1}{\sigma t}$ |

**Table 4.9.1.** Important equations related to ohmic contacts.

## *4-10.  HETEROJUNCTION DIODES

Figure 4.10.1 shows a schematic diagram of a **Molecular Beam Epitaxy (MBE)** system. In this system, a deep vacuum is maintained inside the MBE chamber, and a heated semiconductor substrate is bombarded by ion beams produced by ion sources. These ions form a high-quality semiconductor film on the substrate. Closing or opening shutters can control the film composition and/or doping levels literally within one atomic distance.

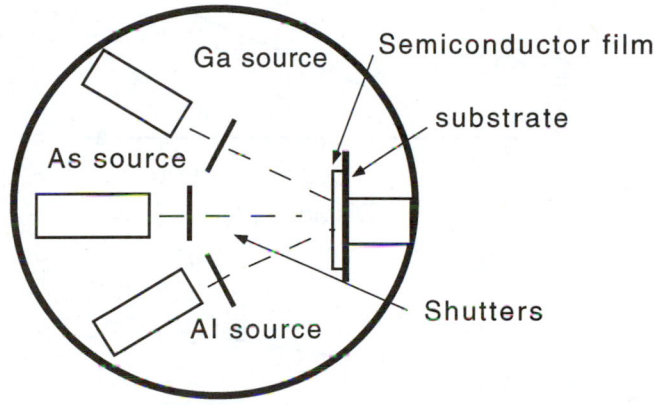

**Fig. 4.10.1.** Schematic diagram of a Molecular Beam Epitaxy (MBE).

An ability to change the material composition as a function of distance provides new opportunities for governing the behavior of electrons and holes. For example, we can grow an $Al_xGa_{1-x}As$ film with the molar fraction, $x$, linearly varying with distance from zero to 0.45. The bottom of the conduction band and the top of the valence band vary with $x$ as follows:

$$E_c(eV) = 0.7482x$$

$$E_v(eV) = -1.42 - 0.5x$$

where the zero energy level is chosen to coincide with the bottom of the conduction band in GaAs (see Fig. 4.10.2.) As we recall, the bottom of the conduction band represents the electron potential energy. The derivative of this energy with respect to distance determines the force acting on the electrons in the conduction band. The top of the valence band represents the hole potential energy. The derivative of this energy with respect to distance determines the force acting on the holes in the valence band. The signs of these forces are such that they push electrons in the conduction band down the slope of the conduction band and up the slope of the valence band. This means that forces acting on electrons and holes in the semiconductor film with the composition profile shown in Fig. 4.10.2 are such that both types of charge carriers are pushed in the same direction!

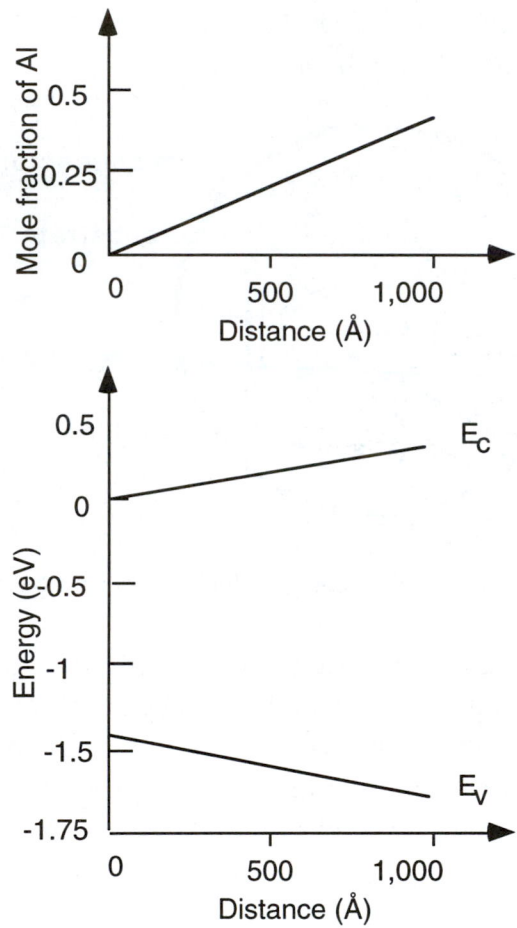

**Fig. 4.10.2.** Energy band diagram for AlGaAs film with graded composition.

This is very different from the forces acting on the charge carriers in an electric field. An electric field tilts the edges of the valence and conduction bands in such a way that they remain parallel to each other. An electric field pushes the electrons and holes in opposite directions.  Hence, the change in chemical composition – gradual or abrupt – provides new opportunities for governing the carrier motion in semiconductor devices.  We already considered in Chapter 1 the quantum well formed by a thin narrow energy gap layer sandwiched between two layers of wide band gap semiconductor material.  The abrupt changes in the conduction and valence bands in the layered structure are created by chemical forces produced by the contact between the two dissimilar materials forming the heterostructure.

**Heterostructure devices** include diodes, transistors, solar cells, light emitting diodes, lasers, and other optoelectronic devices.

When two different semiconductors are joined together, the atoms at the heterointerface have to form chemical bonds.  If a semiconductor film grown on a dissimilar semiconductor substrate is very thin, it can be deformed enough to adjust to the lattice constant of the substrate.  Otherwise, the strain may become too large leading to film imperfections.  One alternative is to grow pairs of semiconductor materials with a very close lattice constant match for forming a nearly ideal heterostructure.  Examples of such pairs include AlGaAs-GaAs, $In_{0.53}Ga_{0.47}As$-InP, GaN-AlN.  The other alternative is to grow heterostructure films  sufficiently thin so that the strain caused by the lattice constant mismatch can be still tolerated.  (Such heterostructures are called **pseudomorphic**.)

The first model of a heterojunction was proposed by R. A. Anderson in his Master's thesis.  In his paper published in *IBM Journal of Research and Development*, vol. 4., p. 283 (1960), he suggested that the **conduction band discontinuity** at the heterointerface between two semiconductors is determined by the difference of the electron affinities, $X_1$ and $X_2$, in these two materials:

$$\Delta E_c = X_1 - X_2 \tag{4-10-1}$$

(see Fig. 4.10.3).  Then the **valence band discontinuity** is  given by

$$\Delta E_v = \Delta E_g - \Delta E_c = \Delta E_g - X_1 + X_2 \tag{4-10-2}$$

where $\Delta E_g$ is the energy gap discontinuity.  This model is very much similar to the primitive model of a Schottky barrier, which predicts that the Schottky barrier height is equal to the difference between the work functions of the metal and the semiconductor forming the Schottky barrier.  And just like this simple Schottky model, the Anderson model of heterointerface discontinuities does not withstand the scrutiny of an experimental verification.  A better model (proposed by J. Terzoff in 1984) relates the conduction band discontinuity to the difference in the Schottky barrier heights in vacuum, $q\phi_{b1}$ and $q\phi_{b2}$,  for the two semiconductor materials:

$$\Delta E_c = q\left(\phi_{b1} - \phi_{b2}\right) \qquad (4\text{-}10\text{-}3)$$

Then , the valence band discontinuity is found as

$$\Delta E_v = \Delta E_g - \Delta E_c \qquad (4\text{-}10\text{-}3a)$$

A typical band diagram of a heterojunction is shown in Fig. 4.10.3.

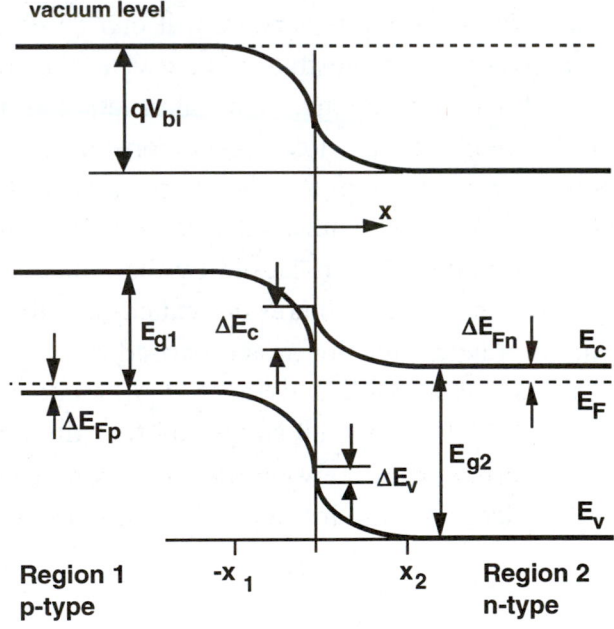

**Fig. 4.10.3.**  Energy band diagram of a *p-n* heterojunction (lower part) shown together with the variation in the vacuum level (upper part) (from Lee et al., 1993, copyright © Prentice Hall, 1993, reproduced by permission of Prentice Hall, Inc., Englewood Cliffs, NJ).

The built-in voltage, $V_{bi}$, may be found from this figure:

$$qV_{bi} = E_{g1} - \Delta E_{Fn} - \Delta E_{Fp} + \Delta E_c \qquad (4\text{-}10\text{-}4)$$

Here, $E_{g1}$ is the energy gap of the narrower gap material (region 1), which we assume to be doped $p$-type; $\Delta E_{Fp}$ is the difference between the Fermi level and the top of the valence band in this material; and $\Delta E_{Fn}$ is the difference between the bottom of the conduction band in the wide gap $n$-type material (region 2) and the Fermi level far from the heterointerface.

In a conventional $p$-$n$ junction, the potential difference, $V_{bi} - V$, between the $p$-region and the $n$-region is divided as follows:

$$V_{bi1} - V_1 = \xi_h(V_{bi} - V) \tag{4-10-5}$$

$$V_{bi2} - V_2 = (1 - \xi_h)(V_{bi} - V) \tag{4-10-6}$$

where $V_{bi1} - V_1$ and $V_{bi2} - V_2$ are the potential drops in the depletion regions of semiconductor 1 and 2, respectively,

$$\xi_h = \frac{N_d}{N_d + N_a} \tag{4-10-7}$$

(see Section 4.2). In a heterojunction, we have to account for the difference in the dielectric constants $\varepsilon_1$ and $\varepsilon_2$ of the $p$-type and $n$-type regions. At the heterointerface,

$$\varepsilon_1 F_1 = \varepsilon_2 F_2 \tag{4-10-8}$$

where $F_1$ and $F_2$ are the electric fields at the heterointerface in the $p$-type and $n$-type regions, respectively. Therefore, for heterojunction, we have to replace $\xi_h$ with

$$\xi = \frac{\varepsilon_2 N_d}{\varepsilon_2 N_d + \varepsilon_1 N_a} \tag{4-10-9}$$

The depletion widths in the $p$-type region, $x_{d1}$, and in the $n$-type region, $x_{d2}$, are also given by the equations similar to those for a conventional $p$-$n$ junction:

$$x_{d1} = \sqrt{\frac{2\varepsilon_1 \xi(V_{bi} - V)}{q N_a}} \qquad x_{d2} = \sqrt{\frac{2\varepsilon_2 (1 - \xi)(V_{bi} - V)}{q N_d}} \tag{4-10-10}$$

Hence, the depletion capacitance per unit area is

$$c_d = \sqrt{\frac{q\varepsilon_1\varepsilon_2 N_a N_d}{2(V_{bi} - V)(\varepsilon_1 N_a + \varepsilon_2 N_d)}} \qquad (4\text{-}10\text{-}11)$$

(As for a conventional $p$-$n$ junction, the diffusion capacitance in a heterostructure diode may become important or even dominant under forward bias.)

Depending on relative doping levels of the $p$ and $n$ regions and on the applied bias, conduction and valence band discontinuities may lead to the appearance of additional barriers for the electron and hole transport. A detailed analysis leads to the following modified diode equation for heterojunction diodes:

$$j = \left( \frac{q D_n n_{po}}{L_n \zeta_n} + \frac{q D_p p_{no}}{L_p \zeta_p} \right) \left[ \exp\left( \frac{V}{V_{th}} \right) - 1 \right] \qquad (4\text{-}10\text{-}12)$$

where

$$\zeta_n = \tanh\left( \frac{X_p - x_{d1}}{L_n} \right) + \frac{D_n}{v_n L_n} \exp\left( \frac{\Delta E_n}{q V_{th}} \right), \quad \zeta_p = \tanh\left( \frac{X_n - x_{d2}}{L_p} \right) \quad (4\text{-}10\text{-}13)$$

$\Delta E_n = \Delta E_c - V_{bi1} - V_1$ is the effective additional barrier for the electron transport, $v_n = v_n = \sqrt{\dfrac{k_B T}{2\pi m_n}}$ is the mean effective electron thermal velocity, $L_n$ and $L_p$ are diffusion length of minority electrons and holes, respectively, and $X_n$ and $X_p$ and $x_{d1}$ and $x_{d2}$ are the lengths of the $n$ and $p$ regions and the depletion widths in these regions, respectively [see Lee et al. (1993)].

Figure 4.10.4 shows the $I$-$V$ characteristic of a typical heterostructure diode calculated using this model (which is the AIM-Spice Level 2 diode model).

As was pointed out by Lee et al. (1993), in a real heterostructure diode, the current level is often much higher than that predicted above. The reason may be a self-heating of the heterojunction under bias or, perhaps, an inhomogeneity over the device cross section, which will cause the current to flow where the barrier is lowest. In addition, the above equations do not account for tunneling, which may be important in strongly doped diodes. For a circuit modeling, an apparent fix is to reduce the value of $\Delta E_n$ and to account for self-heating by increasing the device temperature, $T$.

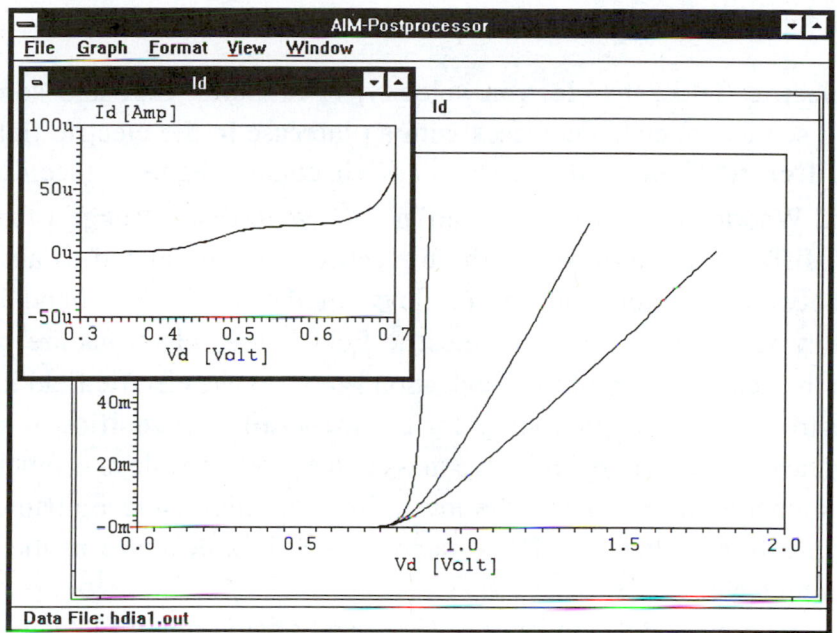

**Fig. 4.10.4.** AIM-Spice current-voltage relationship of an ideal heterostructure $p^+$-$n$ junction. The plateau between 0.5 and 0.6 V indicated in the small window is caused by the presence of the spike in the conduction band edge (see Fig. 4.10.3). (From Lee et al., 1993, copyright © Prentice Hall, 1993, reproduced by permission of Prentice Hall, Inc., Englewood Cliffs, NJ.)

Table 4.10.1 summarizes important equations related to heterojunctions.

| Conduction band discontinuity | $\Delta E_c = q(\phi_{b1} - \phi_{b2})$ |
|---|---|
| Valence band discontinuity | $\Delta E_v = \Delta E_g - \Delta E_c$ |
| Heterojunction built-in voltage | $qV_{bi} = E_{g1} - \Delta E_{Fn} - \Delta E_{Fp} + \Delta E_c$ |
| Modified diode equation for heterojunction diodes | $j = \left( \dfrac{qD_n n_{po}}{L_n \zeta_n} + \dfrac{qD_p p_{no}}{L_p \zeta_p} \right)\left[ \exp\left( \dfrac{V}{V_{th}} \right) - 1 \right]$ $\zeta_n = \tanh\left( \dfrac{X_p - x_{d1}}{L_n} \right) + \dfrac{D_n}{v_n L_n} \exp\left( \dfrac{\Delta E_n}{qV_{th}} \right)$ $\zeta_p = \tanh\left( \dfrac{X_n - x_{d2}}{L_p} \right)$ |

**Table 4.10.1.** Important equations related to heterojunctions.

## *4-11.  GUNN DIODES

In high electric fields, the electron velocity, $v$, in GaAs, InP, and some other compound semiconductors decreases with an increase in the electric field, $F$, so that the **differential mobility** $\mu_d = dv/dF$ becomes negative (see Chap. 3). Ridley and Watkins in 1961 and Hilsum in 1962 were first to suggest that such a **negative differential mobility** in high electric fields is related to an electron transfer between different minima (valleys) of the conduction band in GaAs (**intervalley transfer**).  When the electric field is low, electrons are primarily located in the central valley of the conduction band.  As the electric field increases, many electrons gain enough energy for the **intervalley transition** into higher satellite valleys.  The electron effective mass in the satellite valleys is much greater than in the central valley.  Also, the intervalley transition is accompanied by an increased electron scattering.  These factors result in a decrease of the electron velocity in high electric fields.

The velocity-field dependence in GaAs and several other related compound semiconductors can be approximated by the following equation:

$$v(F) = v_s\left[1 + \frac{F/F_s - 1}{1 + A(F/F_s)^t}\right] \tag{4-11-1}$$

where $F_s = v_s/\mu$ is the saturation field and $\mu$ is the low field mobility.  The saturation velocity, $v_s$, and constants $A$ and $t$ depend on the value of $\mu$.  For GaAs

$$v_s = 0.6(1+\mu) - 0.2\mu^2 \ (10^5 \text{ m/s})$$

$$A = \frac{0.6}{\exp[10(\mu - 0.2)] + \exp[-35(\mu - 0.2)]} + 0.01 \tag{4-11-2}$$

$$t = 4\left[1 + \frac{320}{\sinh(40\mu)}\right]$$

where $\mu$ is in $m^2/Vs$.  [For $0.2 \text{ m}^2/\text{Vs} \leq \mu \leq 1 \text{ m}^2/\text{V}$ such an interpolation is accurate for GaAs within 10% to 15%; see Shur (1987).]  Figure 4.11.1 shows the $v(F)$ curves for GaAs calculated using eqs. (4-11-1) and (4-11-2) for $\mu = 0.85$ $m^2/Vs$ and $\mu = 0.5 \text{ m}^2/\text{Vs}$.  In both cases we observe a range of electric fields where the differential mobility $\mu_d = dv/dF$ is negative.

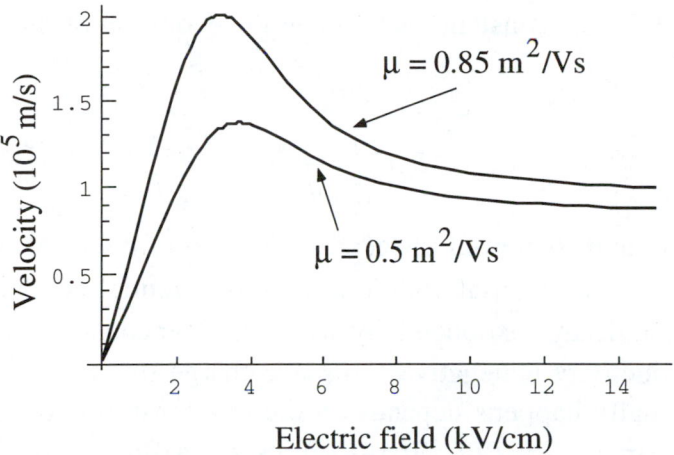

**Fig. 4.11.1.**  Velocity-field curves for GaAs.

A simplified equivalent circuit of a uniformly doped semiconductor may be presented as a parallel combination of the differential resistance (see Fig. 4.11.2):

$$R_d = \frac{L}{q\mu_d n_o S} \qquad (4\text{-}11\text{-}3)$$

and the differential capacitance:

$$C_d = \frac{\varepsilon S}{L} \qquad (4\text{-}11\text{-}4)$$

Here $S$ is the cross section of the sample, $L$ is the sample length, and $n_o$ is the electron concentration.

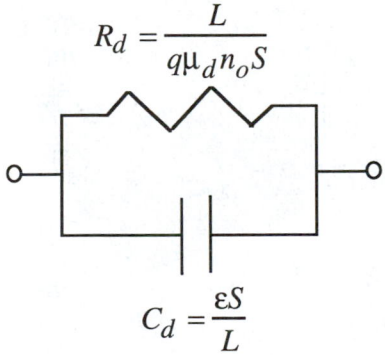

**Fig. 4.11.2.**  Equivalent circuit for a uniform piece of semiconductor.

The equivalent $RC$ time constant determining the evolution of the space charge is given by

$$\tau_{md} = R_d C_d = \frac{\varepsilon}{q\mu_d n_o} \tag{4-11-5}$$

($\tau_{md}$ is called is the **differential dielectric relaxation time** or Maxwell dielectric relaxation time.) In a material with a positive differential conductivity, a space charge fluctuation decays exponentially with this time constant. However, if the differential conductivity is negative, the space charge fluctuation may grow with time. What actually happens depends on the relationship between $\tau_{md}$ and the electron transit time, $t_{tr} = L/v$. If $(-\tau_{md}) \gg t_{tr}$, a fluctuation of the electron concentration occurring near the negatively biased terminal (cathode) grows very little during its transit time toward the positively biased terminal (anode). However, when $(-\tau_{md}) \ll t_{tr}$, a space charge fluctuation grows tremendously during a small fraction of the transit time. In this case, it develops into a high field region (called a **high field domain**), which propagates from the cathode toward the anode with the velocity that is approximately equal to the electron saturation velocity, $v_s$. This process is illustrated in Fig. 4.11.3, which shows a high field domain formation and propagating toward the anode.

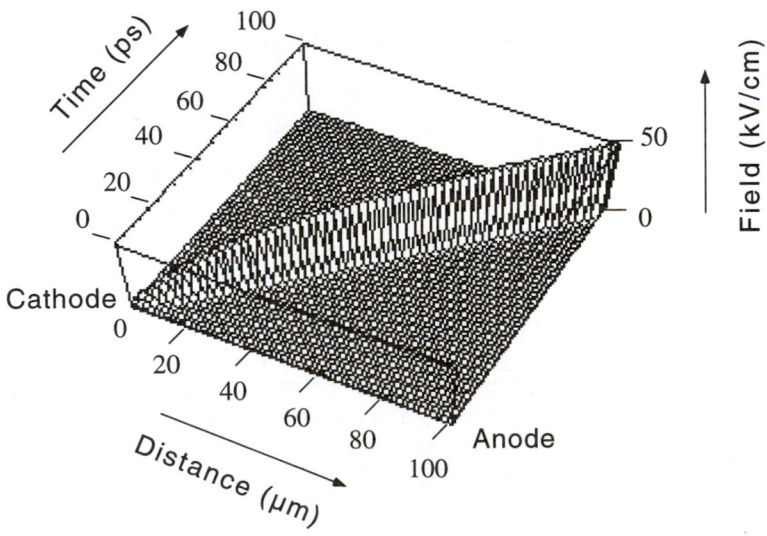

**Fig. 4.11.3.** Nucleation and propagation of a high field domain.

The condition $(-\tau_{md}) << t_{tr}$ leads to the following criterion of a high field domain formation:

$$n_o L >> \frac{\varepsilon v_s}{q \, |\mu_d|} \qquad (4\text{-}11\text{-}6)$$

For $v \simeq 10^5$ m/s, $|\mu_d| \simeq 0.15$ m$^2$/vs, $\varepsilon = 1.14 \times 10^{-10}$ F/m, we obtain $n_o L >> 1.5 \times 10^{11}$ cm$^{-2}$. This condition (first introduced by Professor Herbert Kroemer in 1965) is called the **Kroemer criterion**.

John Gunn was first to observe high field domains in GaAs in 1963. Ever since, these GaAs two-terminal devices are often called Gunn diodes (though many people prefer a different term – transferred electron devices – which emphasizes the role of the electron intervalley transfer.)

Each domain nucleates near the cathode, propagates through the sample (during the transit time $L/v_s$), and is destroyed near the anode. After the domain annihilation at the anode, the next domain nucleates near the cathode and the process repeats itself. The current density in a sample with a domain is equal to $j_s = q n_o v_s$ (see Fig. 4.11.4). When the domain is extinct, the current density is equal to the peak current density, $j_p = q n_o v_p$ where $v_p$ is the peak velocity. During the domain annihilation and the formation of a new domain the current density rises from $j_s$ to $j_p$. .

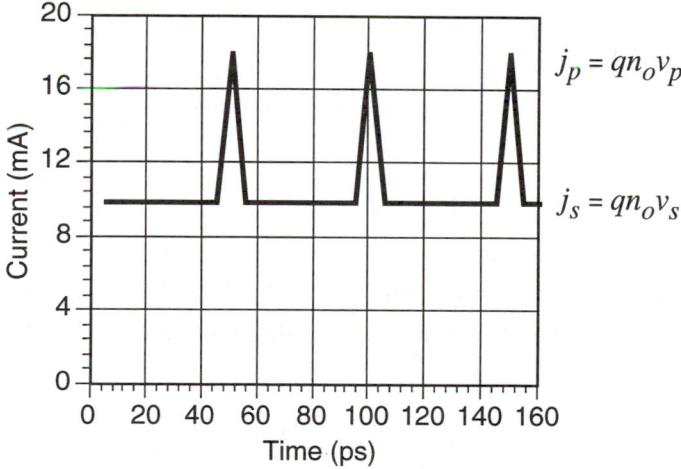

**Fig. 4.11.4.** Current waveform of a Gunn diode. The oscillation period is $\approx L/v_s$ where $L$ is the diode length, and $v_s$ is the electron saturation velocity.

The domain formation time decreases with an increase in $n_o$ (since $|\tau_{md}|$ decreases) and may vary from a few picoseconds to a few hundred picoseconds. The domain shape is also dependent on $n_o$. When $n_o$ is large (much larger than $\sim 10^{15}$ cm$^{-3}$ for GaAs), a small relative change in the electron concentration is sufficient to produce a large space charge, supporting a high domain field. In this case the domain shape is nearly symmetrical. In the opposite limiting case ($n_o \ll 10^{15}$ cm$^{-3}$) the domain leading edge is totally depleted of carriers, creating a positive space charge density $qn_o = qN_d$ (see Fig. 4.11.5). The electron density in the accumulation layer (the domain trailing edge) is limited by diffusion processes and is much larger than $n_o$.

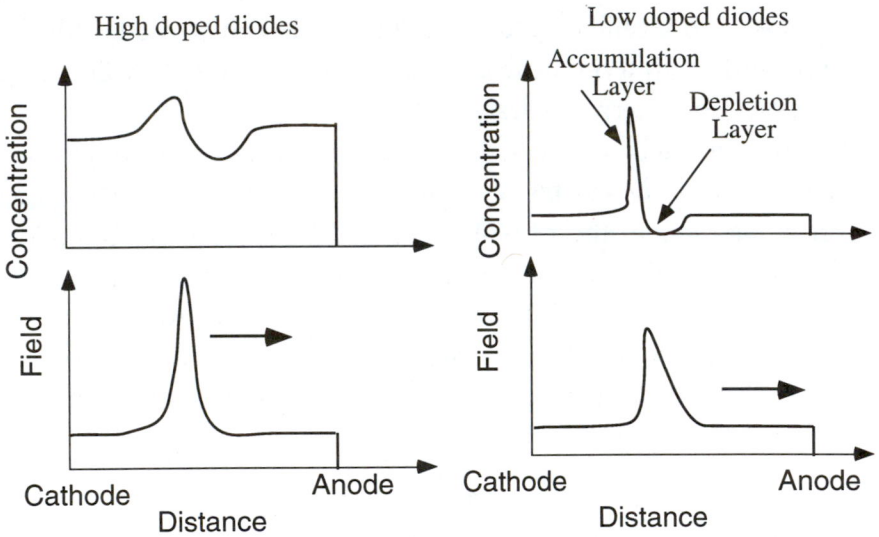

**Fig. 4.11.5.** Concentration and field profiles in high and low doped Gunn diodes.

The resulting field distribution is very similar to the field distribution in a $p^+$-$n$ junction moving with a domain velocity close to the electron saturation velocity, $v_s$. In the nearly depleted leading edge, the drift and diffusion currents are small and the current density is determined by the displacement current $j = \varepsilon \partial F / \partial t$ so that the total current is given by

$$I = q v_s n_o(x_d) S(x_d) \tag{4-11-7}$$

where $x_d = v_s t$ is the coordinate of the moving domain (see Problem 4-11-3 and Fig. 4.11.6). This means that the current waveform should reproduce the $n_o(x)S(x)$ profile. This domain property may be used for implementing numerous analog and logic functions.

Stable propagating domains exist in relatively long samples that operate as microwave oscillators at relatively low frequencies (10 to 40 GHz or less). In short samples, operating as microwave oscillators at higher frequencies (50 GHz and above), the distribution of the electric field is quite different. In such samples, growing **accumulation layers** are nucleated (see Fig. 4.11.7). The propagation velocity of accumulation layers may be even larger than the peak velocity in GaAs. In both regimes, the Gunn diodes can operate as microwave or millimeter wave oscillators. (The frequency of oscillations increases approximately in inverse proportion to the diode length.)

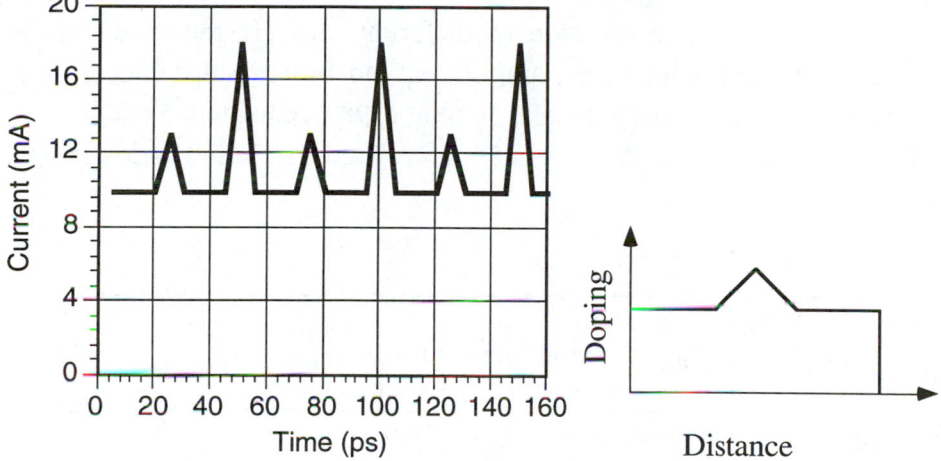

**Fig. 4.11.6.** Current waveform of a Gunn diode with a nonuniform doping profile shown in the figure. Large current spikes are caused by the domain formation and extinction (compare with Fig. 4.11.4). Smaller current spikes reproduces the shape of the doping profile.

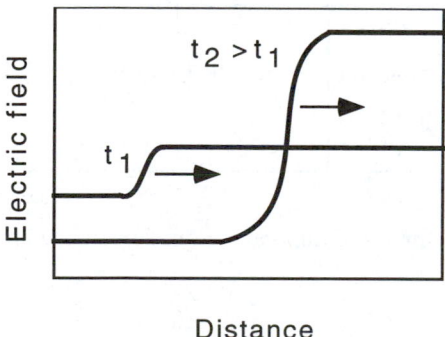

**Fig. 4.11.7.** Propagation of an accumulation layer in a Gunn diode.

The output power decreases approximately as $1/f^2$ (or as $L^2$ where $L$ is the diode length) and varies from tens of kilowatts in a pulsed mode of operation at frequencies of the order of a few gigahertz to a few milliwatts at frequencies on the order of 200 GHz. (These power levels are high enough for the operation of a police radar – one of typical applications of a Gunn diode.) The reason for this dependence is that the output power is proportional to $U^2/R_L$ where $R_L$ is the load impedance and $U$ is the applied bias (which is proportional to $F_p L$ where $F_p$ is the peak field).

Gunn oscillators can also produce the second or even the third harmonic of the fundamental frequency generating power at frequencies higher than 200 GHz.

A more detailed discussion of different transferred electron devices (including microwave generators, amplifiers, and functional devices) and their characteristics was given, for example, by Shur (1987), Chapters 5 and 6.

The summary of important equations related to Gunn diodes is given in Table 4.11.1.

| Velocity-field dependence in GaAs and related compounds | $v(F) = v_s\left[1 + \dfrac{F/F_s - 1}{1 + A(F/F_s)^t}\right]$ |
|---|---|
| Condition of high domain formation (Kroemer criterion) | $n_o L \gg \dfrac{\varepsilon v_s}{q\,\lvert\mu_d\rvert}$ |
| Dependence of the Gunn diode frequency on device length | Roughly proportional to $1/L$ |
| Dependence of the Gunn diode output power on device length | Roughly proportional to $1/L^2$ |
| Current waveform in a Gunn diode with nonuniform product of the doping density and cross section | $I = q v_s n_o(x_d) S(x_d)$   where $x_d = v_s t$ |

**Table 4.11.1.** Summary of important equations related to Gunn diodes.

## *4-12.  IMPATT AND TRAPATT DIODES

The Gunn diodes discussed in Section 4.11 use bulk negative differential resistance related to the intervalley transition in GaAs and other compound semiconductors. Another family of power microwave devices, which includes **IMPATT** and **TRAPATT diodes**, relies on the phase shifts between the device current and applied voltage, creating a "dynamic" negative differential resistance.

To understand the mechanism of the dynamic negative resistance, let us consider the small signal sinusoidal current, $i(t) = u(t)/R_d$, flowing through a differential resistance, $R_d$, when a small signal sinusoidal voltage, $u(t)$, is applied. Figure 4.12.1 shows the small signal current and voltage waveforms for $R_d > 0$ and $R_d < 0$. As can be seen from the figure, when $R_d < 0$, the phase of $u(t)$ and $i(t)$ are opposite. Hence, shifting the phase of the current waveform by $\pi$ has the same effect as having a negative differential resistance.

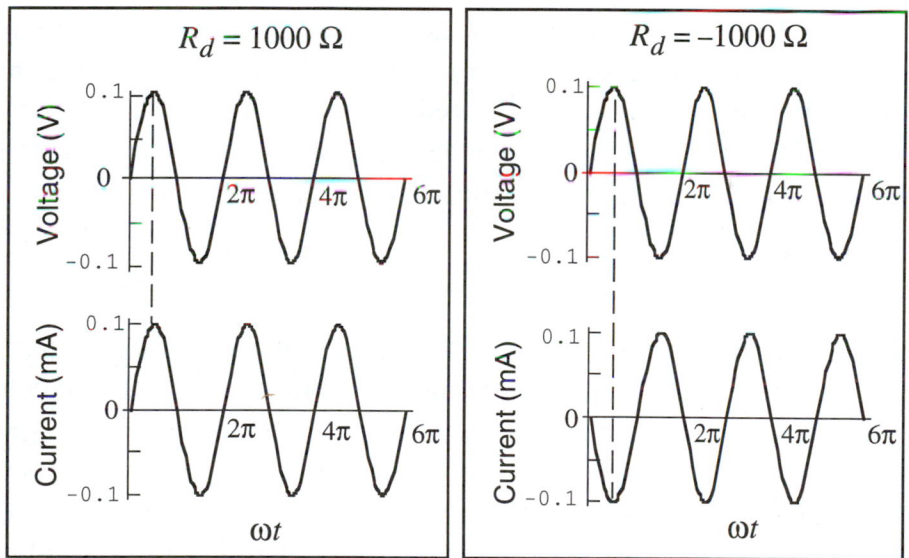

**Fig. 4.12.1.**  Current and voltage waveforms for $R_d > 0$ and $R_d < 0$.

Figure 4.12.2 shows a schematic diagram of an **IMPact ionization Avalanche Transit Time (IMPATT) diode** along with a typical doping profile and the electric field distribution. Avalanche breakdown (generating electron-hole pairs) and carrier drift across a special "drift" section of the device create a phase delay responsible for a dynamic negative resistance. (Under reverse bias, electron-hole pairs are generated at the $p^+$-$n$ interface where the electric field is the highest.)

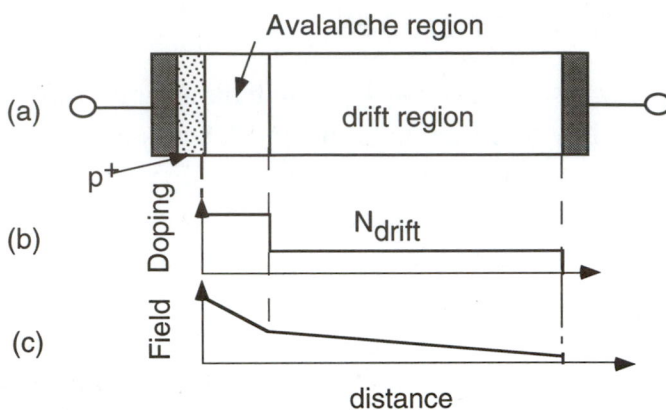

**Fig. 4.12.2.** (a) Schematic diagram of an IMPATT diode along with (b) a typical doping profile and (c) electric field distribution.

These carriers move in the electric field creating an electric current at the $p^+$-$n$ interface. In order to understand how this helps create a phase delay between an ac voltage and an ac current, let us consider the situation when the dc voltage applied is equal to the critical voltage, $V_{br}$, needed to initiate avalanche breakdown. During half of each period, the ac voltage brings the total voltage above $V_{br}$, leading to the impact ionization. The impact ionization generation rate is given by

$$G = \alpha_{no} n v_n \exp\left[\left(-\frac{F_{in}}{F}\right)^{m_{in}}\right] \tag{4-12-1}$$

where $F_{in}$ is the characteristic field of the impact ionization, $m_{in}$ is a constant (typically between 1 and 2), $\alpha_{no}$ is a constant, and $n$ and $v_n$ are the electron concentration and velocity, respectively [see eqs. (3-8-12) and (3-8-13)]. Since the electron current density, $j = q v_n n$, and $\partial n / \partial t \approx G$, the rate of current increase, $\partial j / \partial t$, is proportional to the generation rate. $G$ changes in phase with the ac voltage, and the phase shift between $j$ and $\partial j / \partial t$ is $\pi/4$ (90°) [just as between $\sin(\omega t)$ and $\cos(\omega t)$ functions]. Therefore, the current phase shift with respect to the ac voltage created by the impact ionization is approximately $\pi/2$. The purpose of the drift region is to create an additional 90° phase delay. This delay occurs since the current flows even after the impact ionization stops during the transit time, $T_{tr}$, of the electrons across the drift region. The corresponding phase shift is equal to $\omega T_{tr}$ where $\omega$ is the oscillation frequency. If $\omega T_{tr} \approx \pi/2$, the total phase delay of the electron current with respect to the applied voltage is close to $\pi$ ; see Fig. 4.12.3.

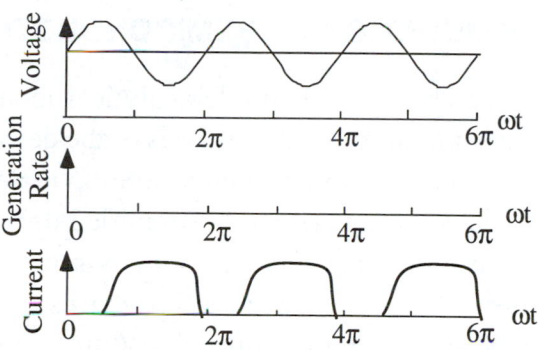

**Fig. 4.12.3.** Voltage, generation rate, and current waveforms in IMPATT diode.

Since $\omega T_{tr} \approx \pi/2$, the drift transit time is equal to one half of the period of oscillations, and the operating frequency, $f$, of an IMPATT diode is given by

$$f = \frac{v_{drift}}{2L} \tag{4-7-1}$$

where $L$ is the device length and $v_{drift}$ is the drift velocity of carriers.

In 1958, Read was the first to propose the idea of an IMPATT device and, therefore, this device is frequently called a **Read diode**. (At approximately the same time, this device was developed independently in Russia by A. Tager.) Different modifications of the Read diode have been proposed, and almost any reverse biased $p$-$n$ junction diode may operate as an IMPATT diode.

At relatively low frequencies, IMPATT diodes can also operate in a completely different and more efficient regime. In this regime, during a part of the cycle, the drift region of the diode is filled by a high-density electron-hole plasma that is "trapped" within the device. In this state the device impedance is very small and, hence, the voltage across the device is low. The electrons and holes are removed from the device via electron and hole transit to the contacts. This regime is called the **TRApped Plasma Avalanche Triggered Transit (TRAPATT) regime** and the diodes operating in this regime are called TRAPATT diodes. TRAPATT diodes typically operate at frequencies below 10 GHz and have high output power in pulsed regime of operation (in order to avoid excessive device heating). Output powers higher than a kilowatt for several TRAPATT diodes connected in series and operating at 1 GHz or so have been obtained.

IMPATT and TRAPATT diodes are very powerful microwave and millimeter wave sources (but they are relatively noisy because of the stochastic nature of impact ionization).

## *4-13. COMPUTER SIMULATION OF SEMICONDUCTOR DEVICES

Understanding device physics and simple analytical models are very helpful in semiconductor device design and in the analysis of the device operation. However, many, if not most, device structures require more accurate computer simulations. As an example, we show in Fig. 4.13.1a a simplified device structure of a silicon semiconductor diode. In this device, both contacts are on the top of a semiconductor wafer, which is very typical for devices used in integrated circuits. If we want to use the *p-n* diode model developed in Section 4.3, what should we take as the device area? How should we estimate the diode series resistance? Fortunately, commercial and public domain two-dimensional device simulators help us answer such questions. Most such simulators are two dimensional, that is, they solve the semiconductor equations for the device cross section (such as shown in Fig. 4.13b) and provide the distributions of electric potential, field, and carrier concentrations in the device cross section. The current is computed per unit of the device width and has to be multiplied by the device width for the comparison with experimental data.

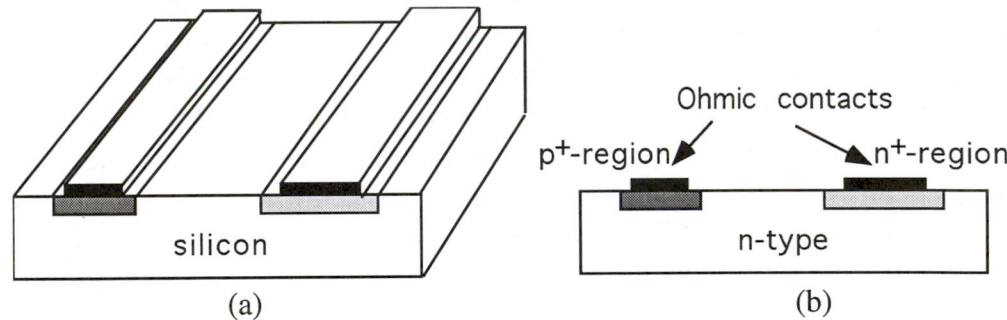

**Fig. 4.13.1.** (a) Silicon *p-n* diode with both contacts on top of the wafer and (b) device cross-section.

Most modern device simulators have their roots in PISCES – the pioneering device simulator developed by the Stanford group. Dutton and Yu of Stanford gave a detailed description of PISCES and other topics related to state-of-the-art device simulation [see Dutton and Yu (1993)]. Using their book and simulator's manuals, one can learn how to use this software for simple devices, such as shown in Fig. 4.13.1, in just a few hours (the simulation of complex device structures requires certain skills and experience).

As an example, we give here an input file used by the ATLAS-II device simulator. The statements in this file are chosen in such a way that they are fully compatible with the Stanford PISCES. The comment statements included in this file should make it fairly clear how to modify this example to simulate a diode with different dimensions, different doping levels, or different geometry. (Since nearly all device simulators come with numerous examples, a user can often modify an example file in a trivial way to simulate her or his device structure.)

```
Title  Si p+ -n -n+ diode simulation
$ A line starting with $ is a comment line which does not affect
$ the simulation results
$
$ Semiconductor equations are solved numerically using
$ a mesh of nodes covering the device cross section
$
$ Specifying the mesh and dimensions
mesh      rect nx=41 ny=41
$ The above line specifies the rectangular mesh shown in the figure
$ below.  The mesh has 411 nodes in x-direction and 40 nodes in y-direction
x.m       n=1  loc=0.0
x.m       n=41 loc=10.
y.m       n=1  loc=0.0
y.m       n=41 loc=2.0
$
$ The x.m and y.m statements specify the location of the nodes in microns
$  In this case, the sample dimensions are 10x2 micron.
$
$ Specifying regions of different materials
region    num=1 iy.l=1 iy.h=41 ix.l=1 ix.h=41 silicon
$ The above statement describes a silicon region. This rectangle is
$ specified by giving the node numbers of its corners. This region is
$ assigned number 1.  In a more general case, we may specify many
$ regions of different materials
$
$ Specifying electrodes
elec      num=1 ix.l=6  ix.h=12 iy.l=1 iy.h=1
elec      num=2 ix.l=29 ix.h=35 iy.l=1 iy.h=1
$ The above statements specify ohmic contacts (default electrodes)
$ Schottky contacts may be specified as well if required
$
$ Doping distribution
doping    uniform conc=1.0e16 n.type
doping    uniform conc=5.0e18 p.type x.l=1 x.r=3.0 y.t=0 y.b=0.3
doping    uniform conc=5.0e18 n.type x.l=6 x.r=9.0 y.t=0 y.b=0.3
$ The above statements specify doping (in cm^-3) in rectangles
$ with the corner coordinates given by x.l, xr, y.t, y.b in microns
$ If no coordinates are specified the doping is for the entire device
$
$ Material and model specification
models    srh
$ The above statement specifies the Shockley-Read recombination mechanism
```

```
$ Since no parameters are specified here, the default values of
$ silicon parameters will be used in the simulation
$
$ Solving procedures
symb    newton carr=2
method
$ The above statements specify the method of solution (Newton)
$ and the bipolar nature of this problem (carr = 2)
$
$ Calculate forward I-V characteristic
log outf=diode.log
$ The above statement specifies the output file (diode.log)
solve   init
$ The above statement generates the initial solution for zero bias
$ voltages which is then used as the initial condition for subsequent
$ simulations
solve   v1=0.05 vstep=0.05 nstep=39 elect=1
$ The above statement generates the solutions for the bias voltage
$ v1 at the first contact for v1=0.05 vstep=0.05 for 39 steps and
$ zero (default) bias at the second contact
end
$ The above statement signifies the end of the input file
```

As explained in the input file, the simulation uses a mesh of nodes. (ATLAS-II as well as Stanford PISCES use triangular meshes.) A small part of this mesh is shown in Fig. 4.13.2.

p⁺-region

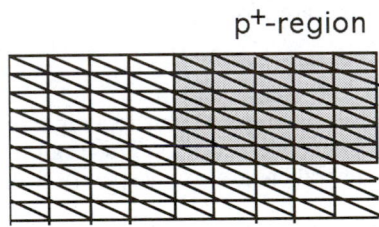

**Fig. 4.13.2.**  A part of the mesh near upper left corner of diode cross section used in the numerical device simulation ($p^+$-region is shaded.)

Sophisticated and user-friendly graphics packages included as a part of modern device simulators allow us to generate colors for the two-dimensional distributions of electric potential, field, carrier concentrations, and so on, or even animated color movies illustrating the variation of these distributions with time.

Figure 4.13.3 shows the diode current voltage characteristic computed by ATLAS-II using the input file given above for a 10 μm wide device.  The thin straight line in Fig. 4.13.3 shows the results of an analytical calculation using eq. (4-3-21).  In this analytical calculation, we assumed the hole diffusion coefficient $D_p = \mu_p k_B T/q$, hole mobility $\mu_p = 400$ cm$^2$/Vs, thermal voltage, $k_B T/q = 0.26$ V,

$p_{no} = n_i^2/N_d$ where the intrinsic carrier concentration $n_i = 1.5 \times 10^{10}$ cm$^{-3}$, donor concentration $N_d = 10^{16}$ cm$^{-3}$. We assumed the effective diode length, $X_n$ - $x_n$, to be equal to 5.5 μm (the distance between the centers of the $p^+$ and $n^+$ regions), and the device area to be 20 μm$^2$ (the product of the 10 μm device width and the 2 μm width of the $p^+$ region). Over four decades of current (from approximately 10$^{-7}$ mA to 10$^{-3}$ mA) this crude analytical model agrees fairly well with the ATLAS-II simulation. At smaller currents, the computed values are higher because of the contribution from the recombination current. At larger currents, the computed values of the current are smaller because of the high concentrations of minority carriers and because of the effects caused by the series resistance. It is remarkable that it is possible to simulate these complicated phenomena by writing just a few simple statements into the input file of a two-dimensional device simulator.

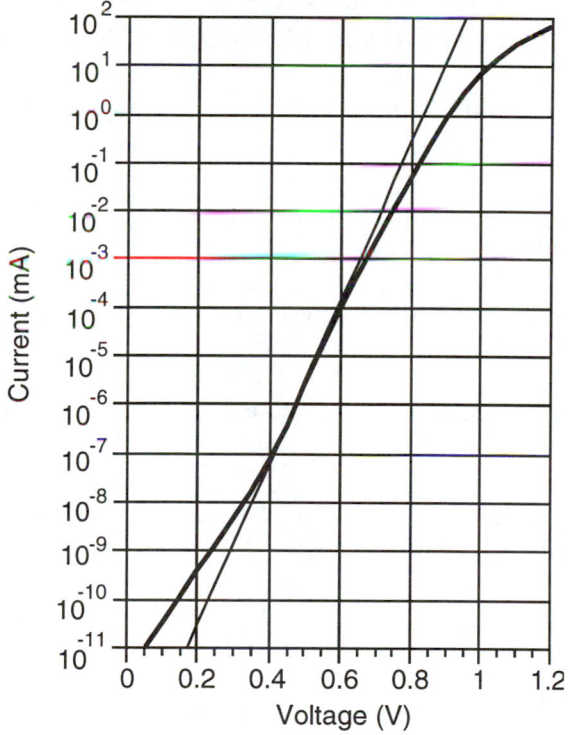

**Fig. 4.13.3.** Comparison between numerical simulation by ATLAS-II (thick curve) and analytical model (thin line) for current-voltage characteristics of silicon $p$-$n$ diode.

# REFERENCES

R. W. DUTTON AND Z. YU, Technology CAD. *Computer Simulation of IC Processes and Devices*, Kluwer, Boston, MA (1993)

K. LEE, M. SHUR, T. A. FJELDLY, AND T. YTTERDAL, *Semiconductor Device Modeling for VLSI*, Prentice Hall, Englewood Cliffs, NJ (1993)

S. A. MAAS, *Microwave Mixers*, Second Edition, Artech House, MA (1993)

E. H. RHODERICK AND R. H. WILLIAMS, Metal-Semiconductor Contacts, Second Edition, Claredon Press, Oxford University Press, New York (1988)

M. SHUR, *GaAs Devices and Circuits*, Plenum, New York (1987)

M. SHUR, *Physics of Semiconductor Devices*, Prentice Hall, Englewood Cliffs, NJ (1990)

S. M. SZE, *Physics of Semiconductor Devices.* Second Edition, John Wiley & Sons, New York (1981)

S. M. SZE, *Semiconductor Devices. Physics and Technology*, John Wiley & Sons, New York (1985)

S. M. SZE, Editor, *VLSI Technology*, Second Edition, McGraw Hill, New York (1988)

P. W. TUINENGA, *SPICE. A Guide to Circuit Simulation & Analysis Using PSpice®*, Prentice Hall, Englewood Cliffs, NJ (1988)

A. VAN DER ZIEL, *Solid State Physical Electronics*, Prentice Hall, Englewood Cliffs, NJ (1976)

# BIBLIOGRAPHY

ATLAS-II. 2D DEVICE SIMULATION FRAMEWORK. User's Manual, Silvaco International, Santa Clara, CA, March (1994)

M. E. LEVINSHTEIN AND G. S. SIMIN, *Getting to Know Semiconductors*, World Scientific, Singapore (1992)

*Popular introduction to the physics of semiconductor devices.*

B. G. STREETMAN, *Solid State Electronic Devices*, Fifth Edition, Prentice Hall, Englewood Cliffs, NJ (1995)

*Concise and clear undergraduate text on semiconductor devices.*

D. A. NEIMAN, *Semiconductor Physics and Devices. Basic Principles*, Irwin, Boston (1992)

*Undergraduate text on semiconductor devices with many examples and problems.*

G. W. NEUDECK, *The p-n Junction Diode*, Modular Series on Solid State Devices, Vol. 2, R. F. Pierret and G. W. Neudeck, Editors, Addison-Wesley, Reading, MA (1983)

*A good undergraduate text describing properties of p-n junctions.*

R. M. WARNER, JR, AND B. L. GRUNG, *Semiconductor Device Electronics*, Holt, Rinehart, and Winston, Philadelphia (1991)

*This undergraduate book contains a detailed theory of p-n junctions.*

## REVIEW QUESTIONS

**1.** A Si sample is uniformly doped with phosphorus at $10^{16}$ cm$^{-3}$. We make a *p-n* junction by implanting boron atoms into the surface layer. The implanted depth is 0.1 μm.

What should the surface concentration of the boron implant be to obtain a *p-n* junction with equal concentrations of holes in the *p*-region and electrons in the *n*-region? (Assume the 100% implant activation.)

$\qquad$ 2 points

What is the built-in voltage at room temperature? ($n_i = 10^{10}$ cm$^{-3}$)

$\qquad$ 1 point

**2.** The figure below shows four pieces of different materials in vacuum: metal, dielectric, and semiconductors (*n*-type and *p*-type). The dielectric permittivity of

both the semiconductor and the dielectric is $1 \times 10^{-10}$ F/m. The external electric field is 100 kV/cm. Sketch the field distribution inside these materials.
(The *n*-type and *p*-type semiconductors are doped at $3 \times 10^{14}$ cm$^{-3}$ with shallow impurities.)

▢   <u>5 points</u>

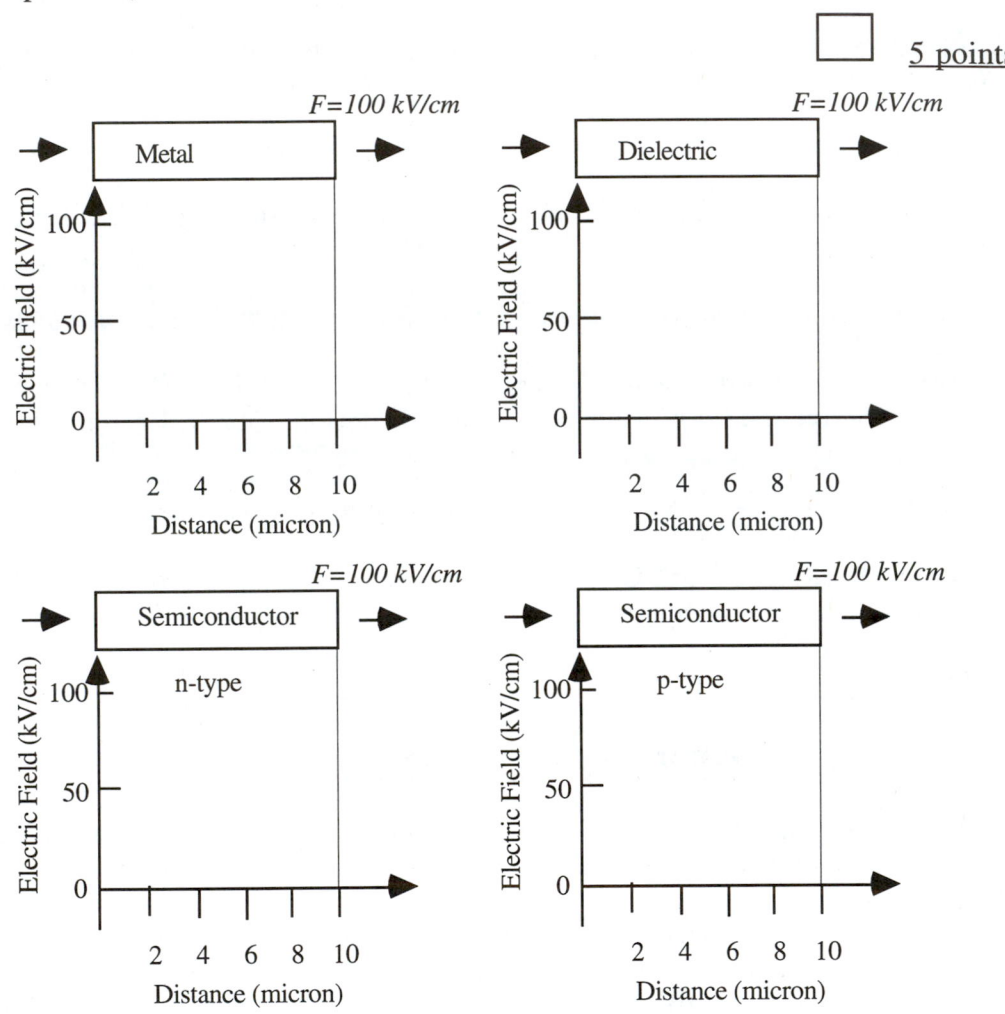

**3.** Consider two $p^+$-$n$ junction diodes with the electron concentrations of $1 \times 10^{15}$ cm$^{-3}$ and $4 \times 10^{15}$ cm$^{-3}$ in the *n*-type regions, respectively. What are the ratios of the maximum electric fields and depletion widths for these diodes?

▢   <u>2 points</u>

**4.** Consider a Si *p-n* junction with doping of $10^{15}$ cm$^{-3}$ in both the *p*-type region and the *n*-type region. The intrinsic carrier concentration is $10^{10}$ cm$^{-3}$. Temperature, $T = 300$ K. Silicon dielectric permittivity is $1.05 \times 10^{-10}$ F/m. The

critical field of the avalanche breakdown is 300 kV/cm.
a. Sketch a band diagram for a 1 V reverse bias. Show the scales.

☐ 3 points

b. Find the depletion capacitance, $C$, of this diode at a reverse bias of 1 V assuming a diode cross section of $10^{-2}$ cm$^2$.

☐ 1 point

c. What is the breakdown voltage of this $p$-$n$ junction diode?

☐ 1 point

**5.** Consider two ideal $p$-$n$ junction diodes that differ only by the lengths of the neutral regions. The first diode has neutral regions that are much longer than the diffusion lengths of minority carriers. The second diode has neutral regions that are much shorter than the diffusion lengths of minority carriers. Which diode is expected to have a higher current at the same forward bias? Why?

☐ 2 points

**6.** What is the Law of the Junction?

☐ 1 point

**7.** Write down the diode equation accounting for the diode series resistance.

☐ 1 point

**8.** The figures below show the TLM pattern and the TLM pattern resistances versus the distance between the ohmic contacts. The pattern width is 50 μm. The film thickness is 0.1 μm. The electron concentration is $10^{17}$ cm$^{-3}$.

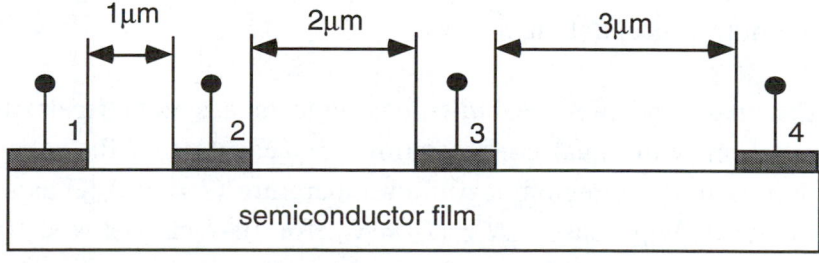

Cross section of TLM characterization pattern.

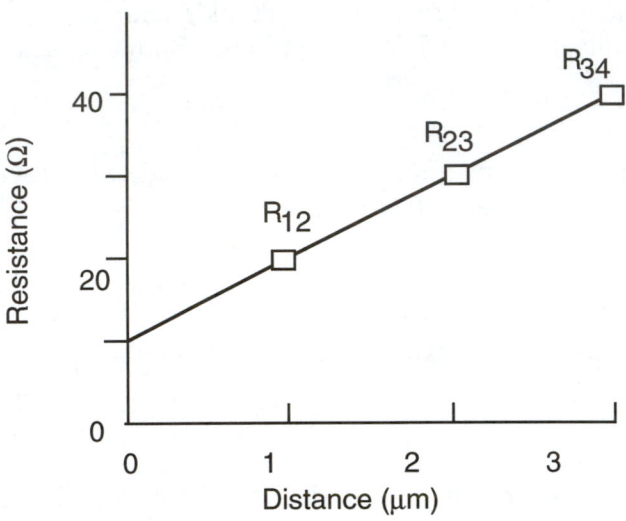

TLM resistances versus distance between the contacts.

a. What is the electron mobility?

☐ 1 point

b. What is the ohmic contact resistance for a 1 mm width contact?

☐ 1 point

# PROBLEMS

**4-2-1.** Derive eq. (4-2-7) for the built-in voltage, $V_{bi}$, from eqs. (4-2-5) and (4-2-6) by considering the $p$-type region far from the junction where the hole concentration, $p = N_a$.

**4-2-2.** Calculate and plot the built-in voltage of a symmetrical silicon $p$-$n$ junction with equal concentrations of acceptors in the $p$-region and donors in the $n$-region at room temperature ($T = 300$ K) as a function of the doping density, $N = N_a = N_d$, for $10^{13}$ cm$^{-3}$ < $N$ < $10^{20}$ cm$^{-3}$. The energy gap of Si is 1.12 eV. The intrinsic carrier concentration is approximately $10^{10}$ cm$^{-3}$ at $T = 300$ K.

**4-2-3.** Plot the depletion width for a symmetrical silicon $p$-$n$ junction with equal concentrations of acceptors in the $p$-region and donors in the $n$-region, $N = N_a = N_d = 10^{15}$ cm$^{-3}$, at zero bias as a function of temperature, $T$, for 100 K < $T$ < 300 K. The dielectric permittivity of Si is $1.05 \times 10^{-10}$ F/m. The energy gap of Si is 1.12 eV. The intrinsic carrier concentration is approximately $n_{io} \approx 10^{10}$ cm$^{-3}$ at $T_o = 300$ K.

Assume that $n_i = n_{io} \left( \dfrac{T}{T_o} \right)^{3/2} \exp\left[ \dfrac{E_g}{2k_B}\left( \dfrac{1}{T_o} - \dfrac{1}{T} \right) \right]$.

**4-2-4.** Derive the equation relating the ratio of the depletion width of a $p^+$-$n$ silicon junction (at zero bias) over the Debye length to the doping density and the built-in voltage. Assume zero bias voltage.

**4-2-5.** Calculate and plot electron and hole concentrations versus distance for a $p^+$-$n$ silicon junction with the acceptor density in the $p$-type region, $N_a = 10^{16}$ cm$^{-3}$ and with the donor density in the $n$-type region, $N_d = 10^{14}$ cm$^{-3}$ at room temperature and zero bias voltage. The dielectric permittivity of Si is $1.05 \times 10^{-10}$ F/m. The energy gap of Si is 1.12 eV. The densities of states in the conduction and valence bands at room temperature are $3.22 \times 10^{19}$ cm$^{-3}$ and $1.83 \times 10^{19}$ cm$^{-3}$, respectively.

**4-2-6.** Sketch the field distributions in a $p$-$n$ silicon junction with the doping profiles shown. Assume zero applied bias and room temperature ($n_i \approx 10^{10}$ cm$^{-3}$). The silicon dielectric permittivity is $1.05 \times 10^{-10}$ F/m.

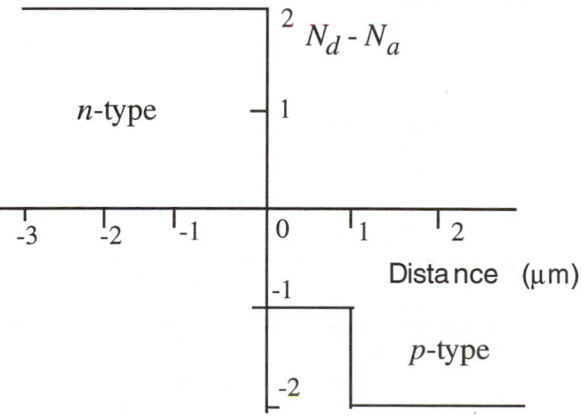

Doping versus distance.

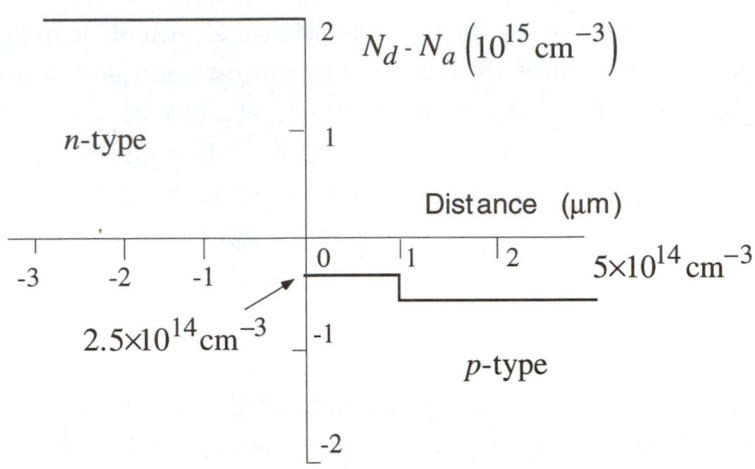

Doping versus distance.

**4-2-7.** Sketch the band diagram and the position of the quasi-Fermi levels within the depletion region in a Si $p$-$n$ junction doped at $N_d = N_a = 10^{15}$ cm$^{-3}$ at 300 K. Repeat for three bias voltages: $V = 0$, $V = 0.2$ V (forward bias), and $V = -1$ V (reverse bias). The densities of states in the conduction and valence bands at 300 K are $3.22 \times 10^{19}$ cm$^{-3}$ and $1.83 \times 10^{19}$ cm$^{-3}$, respectively. The energy gap is 1.12 eV. The silicon dielectric permittivity is $1.05 \times 10^{-10}$ F/m.

**4-2-8.** For the band diagram shown in the figure, calculate and plot the electron and hole carrier concentrations at 300 K as functions of position. The densities of states in the conduction and valence bands at 300 K are $3.22 \times 10^{19}$ cm$^{-3}$ and $1.83 \times 10^{19}$ cm$^{-3}$, respectively.

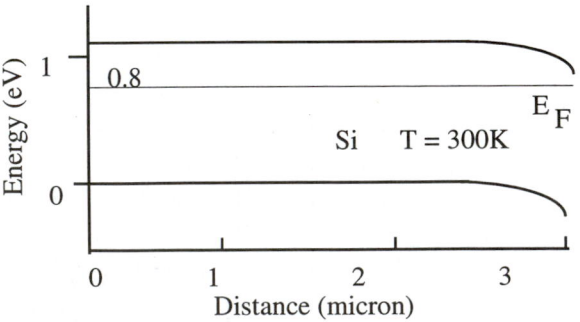

**4-3-1.** From the *I-V* characteristics of the *p-n* junction shown in the figure, estimate the diode saturation current, the ideality factor, and series resistance.

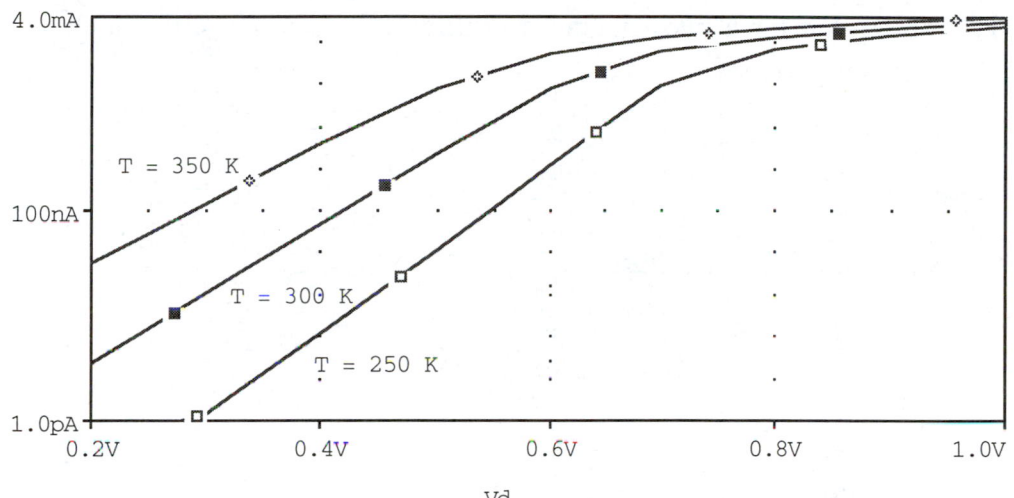

Using the following empirical equation for the temperature dependence of the saturation current, estimate the energy gap of the semiconductor material of the diode:

$$I_s(T) = I_s(T_o)\left(\frac{T}{T_o}\right)^{\frac{\kappa}{\eta}} \exp\left(\frac{E_g}{k_B T_o}\right) \exp\left(-\frac{E_g}{k_B T}\right)$$

**4-3-2.** Plot the hole diffusion current densities versus distance in the *n*-type region of a *p-n* diode at the forward bias of 0.24 V. The hole mobility, $\mu_p = 300$ cm$^2$/Vs. The intrinsic carrier concentration $n_i = 10^{10}$ cm$^{-3}$. Deduce other parameters from Fig. 4.3.2.

**4-3-3.** Derive eq. (4-3-20) for a short $p^+$-$n$ diode.

**4-3-4.** Design a Si $p^+$-$n$ junction diode with a reverse saturation current smaller than 1 nA at room temperature at $-5$ V and forward current at 0.5 V of approximately 1 mA at $T = 300$ K. Specify the device dimensions and indicate doping levels. **Hint:** Assume reasonable values of parameters. Use eq. (4-3-34) and choose $x_n \ll X_n \ll L_p$ where $X_n$

is the length of the $n$-region, $x_n$ is the width of the depletion region, and $L_p$ is the hole diffusion length. (You may estimate $L_p$ to be on the order of 50 μm.)

**4-4-1.** Show that for a linearly graded profile, $(N_d - N_a = ax)$, the depletion width is given by

$$x_d = \left[ \frac{12\varepsilon_s (V_{bi} - V)}{qa} \right]^{1/3}$$

where the built-in voltage,

$$V_{bi} = 2V_{th} \ln\left( \frac{ax_d}{2n_i} \right)$$

**\*4-4-2.** Solve the diffusion equation

$$D_p \frac{\partial^2 p_n}{\partial x^2} - \frac{p_n - p_{no}}{\tau_{pl}} = \frac{\partial p_n}{\partial t}$$

for the hole concentration, $p_n$, in the $n$-type region of a $p$-$n$ diode assuming that the applied voltage, $V(t)$, and the current density, $j(t)$, are given by large signal stationary terms, $V_o$ and $j_o$, and small signal time dependent terms:

$$V(t) = V_o + v^* \exp(i\omega t)$$

$$j(t) = j_o + j^* \exp(i\omega t)$$

where $v^* \ll V_o$ and $j^* \ll j_o$. Seek the solution in the following form:

$$p_n(x,t) = p_{ns}(x) + p^*(x)\exp(i\omega t)$$

and assume the following boundary conditions:

$$p^*(x \to \infty) = 0$$

$$p_n(x = x_n) = p_{no} \exp\left[ \frac{V_o + v^* \exp(i\omega t)}{V_{th}} \right]$$

The expansion of the exponent on the right-hand side of the last equation into a Taylor series with respect to $v*\exp(i\omega t)$ yields the second boundary condition for $p*$: $p*(x=x_n)=\dfrac{p_{no}v*}{V_{th}}\exp\left(\dfrac{V_o}{V_{th}}\right)$.

Compare the obtained solution with the steady-state solution and calculate the device impedance. Compare the results with eq. (4-4-13).

**4-4-3.** At what bias voltage will the diffusion capacitance of a Si $p^+$-$n$ diode with the $n$-type region doped at $N_d = 10^{15}$ cm$^{-3}$ exceed the depletion capacitance by a factor of 3 at room temperature? The length of the $n$-type region is 5 $\mu$m (much smaller than the hole diffusion length). Assume the intrinsic carrier concentration $n_i = 10^{10}$ cm$^{-3}$, $\varepsilon_s = 1.04\times10^{-10}$ F/m, $V_{bi} = 0.7$ V. To simplify your calculations, assume that the length of the neutral region, $X_n$, is much larger than the depletion width, $x_n$. Still, be prepared to solve (graphically or numerically) a simple algebraic transcendental equation.

**4-5-1.** Assuming a breakdown field of 300 kV/cm, plot the breakdown voltage versus donor concentration in the $n$-type region of a $p^+$-$n$ junction for $10^{13}$ cm$^{-3} < N_d < 10^{17}$ cm$^{-3}$. Assume the dielectric permittivity $\varepsilon_s = 1.05\times10^{-10}$ F/m and built-in voltage $V_{bi} = 0.7$ V.

**4-7-1.** Using default PSpice$^{tm}$ diode parameters, $R_s = 10$ k$\Omega$, $C_{jo} = 10$ pF, simulate the output voltage waveform for the following circuit and the input voltage waveform:

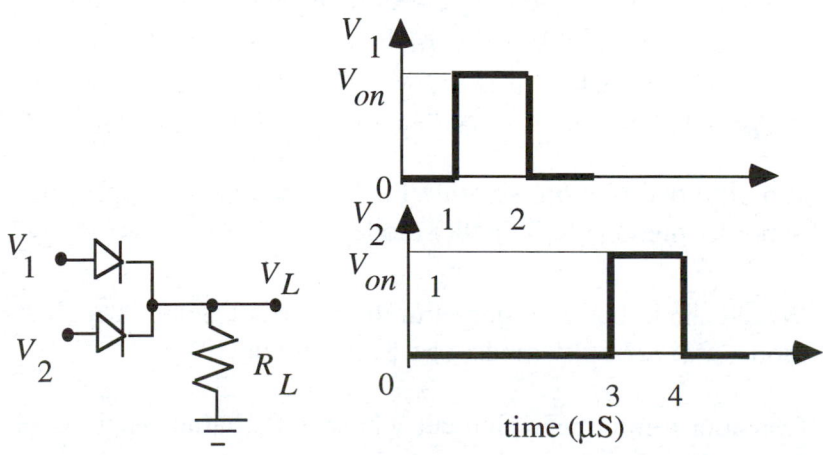

(Choose reasonable values of $R_L$ and $V_{on}$ so that the circuit will perform an "OR" function.)

**4-8-1.** Choose reasonable values of the current density $j_{seff}$ in the diode equation

$$j = j_{seff}\left[\exp\left(\frac{V}{\eta V_{th}}\right) - 1\right]$$

for the $I$-$V$ characteristics of (a) a Si $p^+$-$n$ junction diode (with the $n$-type region doped at $10^{15}$ cm$^{-3}$ with the hole lifetime of 1 $\mu$s), (b) a Si Schottky diode, (c) a GaAs Schottky diode at room temperature. [Assume that the saturation current of a Si $p$-$n$ junction can be described using eq. (4-3-19).]

**4-8-2.** The measured capacitance of the reversed biased silicon Schottky barrier diode is given by

$$\frac{1}{C^2} = \frac{1}{C_o^2} - kV$$

where $C_0 = 1$ pF and $k = 2$ pF$^{-2}$V$^{-1}$. The diode is uniformly doped. Dielectric permittivity $\varepsilon_s \approx 1.05\times10^{-10}$ F/m. Device area is $S = 10^{-4}$ cm$^{-2}$. The effective density of states density in the conduction band $N_c = 2.8\times10^{19}$ cm$^{-3}$. Find the Schottky barrier height.

**4-8-3.** Draw the energy band diagram of an $n$-type GaAs Schottky diode at (a) zero bias, (b) 0.2 V forward bias, and (c) 2 V reverse bias. The Schottky barrier height is 0.7 V. The concentration of shallow donors in GaAs is $10^{16}$ cm$^{-3}$. The dielectric permittivity is $1.14\times10^{-10}$ F/m.

**4-8-4.** Which diode is more suitable for high frequency applications: a GaAs Schottky diode, a Si Schottky diode, or a Si $p$-$n$ diode? Why?

**4-8-5.** Which diode has a larger saturation current density: a GaAs Schottky diode, a Si Schottky diode, or a Si $p$-$n$ diode? Why?

**4-8-6.** Consider a nonlinear element whose $I$-$V$ characteristic is given by $I =$

$aV^2$ where $V = V_O \sin(\omega_1 t) + V_O \sin(\omega_2 t)$ is the applied voltage. This element is connected in series with a resistor, $R_S$. Calculate the voltage, $V_S$, across $R_S$ and show that it contains components with frequencies $\omega_2 - \omega_1$ and $\omega_2 + \omega_1$. (This problem illustrates the operation of a semiconductor mixer.)

**4-9-1.** Consider a semiconductor sample with two contacts shown in the figure.

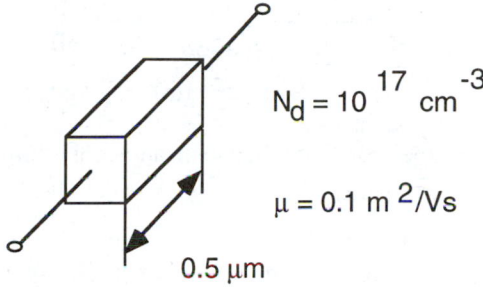

$$N_d = 10^{17} \text{ cm}^{-3}$$

$$\mu = 0.1 \text{ m}^2/\text{Vs}$$

0.5 μm

How small should the specific contact resistance be to ensure that the voltage drop across the ohmic contacts is less than 10% of the applied voltage?

**4-9-2.** The measured dependence of the resistance between ohmic contacts for the transmission-line-model pattern shown in the figure is shown as a function of the distance between the contacts. The thickness of the doped $n$-type layer is 0.1 μm. The device width is 5 μm. Assume an electron mobility of 1,000 cm$^2$/Vs. Find the electron concentration in the channel and the contact resistance per millimeter width assuming a uniform doping level in the channel.

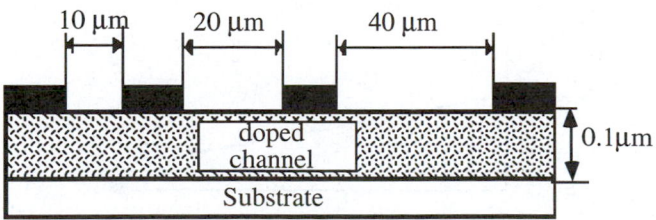

10 μm    20 μm    40 μm

doped
channel

0.1μm

Substrate

Transmission-line-model pattern. Substrate is undoped.

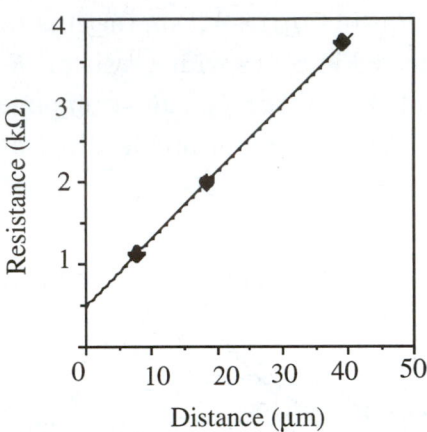

Dependence of the resistance between adjacent ohmic contacts on the distance between the contacts.

**4-10-1.** Sketch a qualitative band diagram of an $n$-GaAs-$p$-Al$_{0.3}$Ga$_{0.7}$As diode at zero bias.

**4-10-2.** For a semiconductor sample with the band diagram shown in the figure, sketch approximate dependencies of electron and hole concentrations on distance at $T = 600$ K. The densities of states in the conduction and valence bands are $4.7 \times 10^{17}$ cm$^{-3}$ and $7 \times 10^{18}$ cm$^{-3}$, respectively.

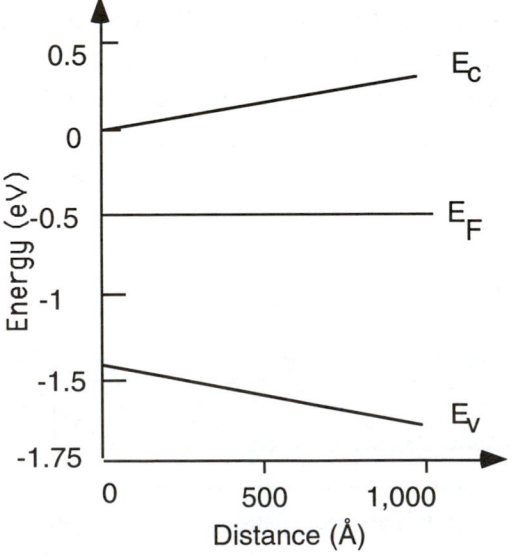

**4-11-1.** Design a GaAs Gunn diode operating at 35 GHz in a quenched mode (the quenched mode of operation is the regime in which the frequency is larger than the transit frequency and is determined by the resonant frequency of the microwave cavity). (Choose a doping-length product of $3 \times 10^{12}$ cm$^{-2}$ and a reasonable operating current of a few mA.)

**4-11-2.** Consider the simplified two-valley model of the conduction band in GaAs shown in the figure. The effective mass of density of states in the low valley is 0.067 $m_e$ where $m_e$ is the free electron mass. The effective mass of density of states in the upper valley is 1 $m_e$. The electron mobility in the low valley $\mu_1 = 10,000$ cm$^2$/Vs. The electron mobility in the upper valley $\mu_2 = 200$ cm$^2$/Vs. Assuming that the electron temperature, $T_e$, is given by

$$T_e = T_o + \frac{2}{3}\frac{q\mu_1 F^2 \tau_e}{k_B}$$

where $T_o = 300$ K is the sample temperature, $q$ is the electronic charge, $k_B$ is the Boltzmann constant, $F$ is the electric field, and $\tau_e = 5 \times 10^{-12}$ s is the energy relaxation time, calculate the dependence of the electron drift velocity, $v$, on the electric field for $0 < F < 10$ kV/cm.

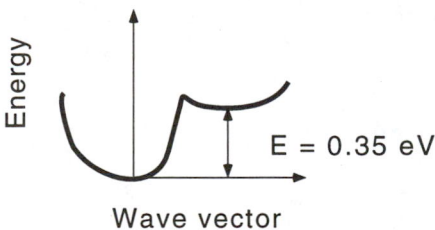

Wave vector

**4-11-3.** Derive eq. (4-11-7). **Hint:** The field distribution in the domain leading edge can be approximated as $F = F_{max} - \dfrac{qN_d(x_d)(x - x_d)}{\varepsilon}$ where $F_{max}$ is the maximum domain field, $x$ is the space coordinate, $x_d = v_s t$ is the domain coordinate, $t$ is time, $v_s$ is the electron saturation velocity, $N_d \approx$

$n_o$ is the doping density, $n_o$ is the equilibrium electron concentration, and $\varepsilon$ is the semiconductor dielectric permittivity.

**4-12-1.** Choose the length of the drift region for an IMPATT diode operating at 30 GHz assuming a carrier saturation velocity of $8 \times 10^4$ m/s.

# 5

# Bipolar Junction Transistors

## 5-1.  PRINCIPLE OF OPERATION

### 5.1.1.  Bipolar Junction Transistor operation.

A transistor is a three-terminal electronic device.  It has an input current loop and an output current loop (see Fig. 5.1.1).  Transistors are designed in such a way that a small change in the input transistor current or voltage can result in a large change in the output current or voltage.  Thus a transistor may provide a voltage and/or current **gain**.  Alternatively, the input signal can practically shut off the output current.  This property is used in all digital electronic circuits.

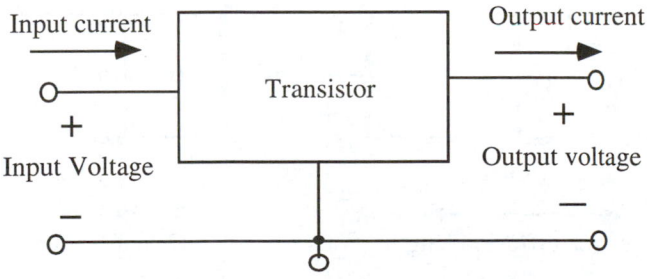

**Fig. 5.1.1.**  Transistor input and output.

Each year, the U.S. Patent office issues hundreds of patents for new transistor designs.  However, most of these transistors belong to one of two types

– **Bipolar Junction Transistors (BJTs)** or **Field Effect Transistors (FETs)**. The Bipolar Junction Transistor was invented at Bell Laboratories by Brattain, Bardeen, and Shockley in 1947. A Field Effect Transistor was patented by Lilienfeld in 1930. But only after the work of Shockley in the early fifties, did it become a useful electronic device. Nowadays, silicon FETs dominate semiconductor electronics. BJTs also occupy an important niche. A new technology that combines both types of transistors (called **BICMOS**) has emerged as a technology that is often advantageous for fast electronic circuits. Bipolar transistors are called **bipolar** because they use both electrons and holes. In contrast, Field Effect Transistors are **unipolar** devices using either electrons (more often) or holes.

A BJT consists of two $p$-$n$ junctions. Since BJTs are so closely related to $p$-$n$ junction diodes, we will consider bipolar transistors first. This chapter gives basic information about BJTs and describes basic BJT models. Field Effect Transistors, which are even more important, will be dealt with in the next two chapters.

A bipolar junction transistor (BJT) is composed of two back-to-back $p$-$n$ junctions (see Fig. 5.1.2). These devices can be made in two polarities shown in the figure. They are called **$n$-$p$-$n$** and **$p$-$n$-$p$** transistors, respectively.

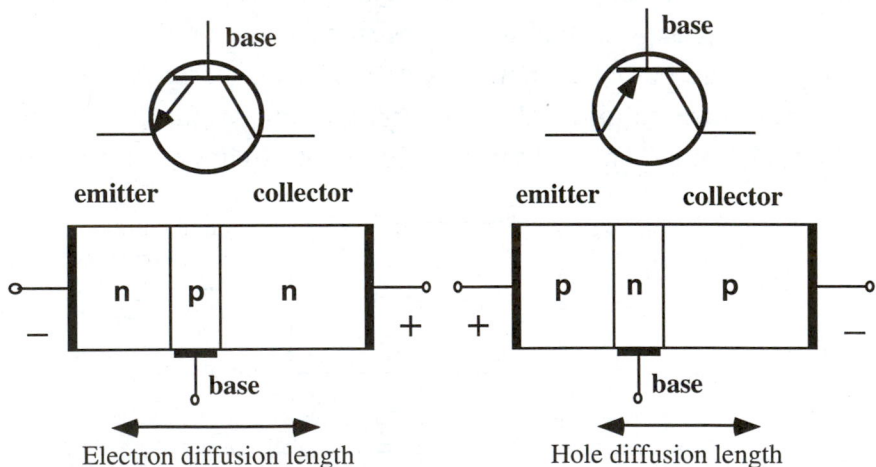

**Fig. 5.1.2.** Schematic diagrams and circuit symbols of $n$-$p$-$n$ and $p$-$n$-$p$ BJTs. Signs indicate voltage polarities with respect to the base. As schematically shown, in a typical BJT, the base width is much smaller than the diffusion length of minority carriers in the base.

In a typical regime of BJT operation, one of the *p-n* junctions is forward-biased and the other one is reverse-biased. (This regime of operation is called the **active forward mode** for reasons that will be explained later.) The forward-biased junction is called the **emitter-base** junction. The reverse-biased junction is called the **collector-base junction**. The three BJT contacts are called **emitter**, **base**, and **collector**, respectively.

To be specific, let us consider an *n-p-n* transistor. (Of course, all our arguments will equally apply to *p-n-p* transistors if we simply interchange words "electron" and "hole" and change voltage signs accordingly.)

Qualitative charge and field distributions and an energy band diagram for an *n-p-n* transistor are shown in Fig. 5.1.3.

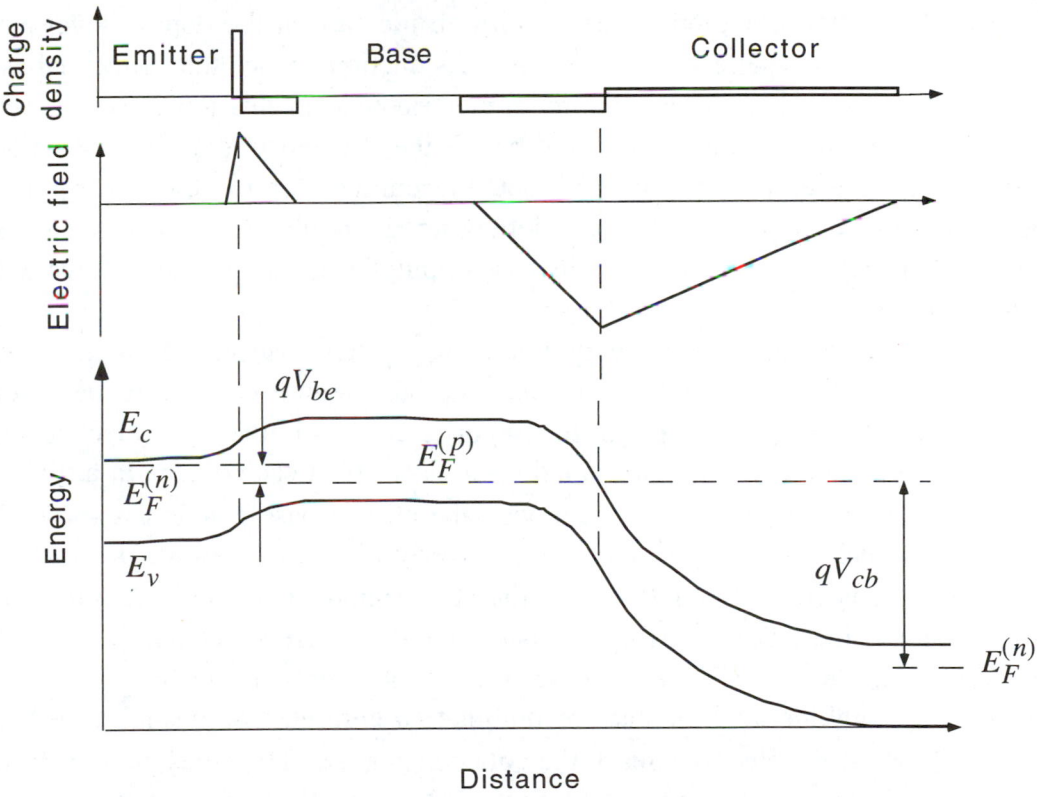

**Fig. 5.1.3.** Schematic space charge density and electric field distributions and energy band diagram for *n-p-n* BJT. Also shown are positions of quasi-Fermi levels and bias voltages.

Positive and negative charge densities in Fig. 5.1.3 correspond to ionized donors and ionized acceptors, respectively. Figure 5.1.3 is not drawn to scale in order to show the important features of the charge and field distributions more clearly. However, the figure reproduces the relationship between the doping levels in BJT regions. The emitter region is doped higher than the base region and the base is doped higher than the collector. Figure 5.1.3 is drawn for a typical regime of a BJT operation when the emitter-base junction is forward-biased and the collector-base junction is reverse biased.

The relative doping levels correspond to the magnitudes of the charge densities in the depletion regions of the emitter-base and collector base junctions. As we discussed in Section 2.2, the electric field outside of a depletion region is zero (or very small if the current flows across the $p$-$n$ junction). The maximum field in the depletion region increases with an increase in the doping level and with an increase in the reverse bias across the junction. Since the collector-base junction is reverse-biased, the maximum electric field in this junction is larger than in the emitter-base junction even though the collector region has a smaller doping than the emitter region. In fact, the emitter doping level is not just larger, but much larger than the base doping level, and the collector level is not just smaller, but much smaller than the base doping level, for reasons that we will discuss below.

If the base region were very long, the emitter-base and collector-base junctions would be independent. In fact, the base region is made to be much shorter than the diffusion length of the minority carriers in the base (electrons in a $n$-$p$-$n$ BJT). Therefore, the two junctions greatly affect each other. In a typical regime of operation (forward active mode), the emitter-base junction is forward-biased, and the collector-base junction is reverse-biased. As in a $p$-$n$ junction diode, electrons from the emitter enter the base region. If the base region were much longer than the electron diffusion length, these electrons would all recombine in the base. Since the base is very short, only a tiny fraction of the electrons recombine, and the base recombination current, $I_{rb}$, is small (see Fig. 5.1.4). Most of the electrons reach the collector region. The holes coming from the base into the emitter region constitute only a small fraction of the emitter current since (1) the base doping is very small compared to the emitter doping and (2) the hole mobility and diffusion coefficient are smaller than the electron mobility and diffusion coefficient. Hence, we can describe the BJT operation in the forward active mode as follows. If the base current, $I_b$, is increased, the emitter-base forward bias, $V_{be}$, must increase as well, as follows from the diode equation (see Section 4.3) describing the emitter-base junction. $I_b$ includes the hole current component, $I_{pe}$, which is proportional to $D_p \dfrac{n_i^2}{N_{de}} \exp\left(\dfrac{qV_{be}}{k_B T}\right)$ (as follows from the diode equation, see Section 4.1.3). Here $D_p$ is the hole diffusion coefficient, $n_i$ is the intrinsic carrier concentration, $N_{de}$ is the donor concentration in the emitter region, and $T$ is temperature. This increase in $V_{be}$ causes a much larger increase in the electron component of the emitter current, $I_{ne}$, since $I_{ne}$ is proportional to $D_n \dfrac{n_i^2}{N_{ab}} \exp\left(\dfrac{qV_{be}}{k_B T}\right)$ and the acceptor concentration in the base, $N_{ab} \ll N_{de}$. Also, the electron diffusion coefficient, $D_n \gg D_p$. Hence, a small variation of the base current causes a large variation of the emitter and collector currents.

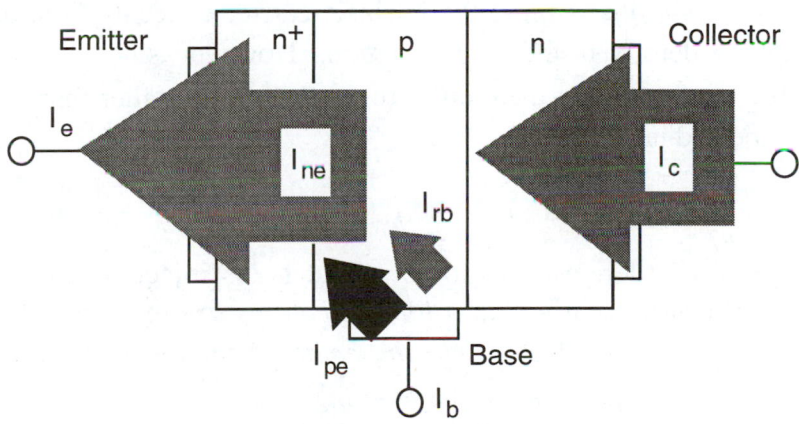

**Fig. 5.1.4.** Principle of BJT operation. Light shaded arrows represent electronic current components and recombination current in the base region. Darkly shaded arrow represents hole current component of the emitter current. Electron fluxes are in the direction opposite to $I_e, I_c, I_{rb}$. Hole flux is in the same direction as $I_{pe}$.

### 5.1.2.   Minority carrier profiles.

Since the emitter-base junction is forward-biased, the minority carrier (electron) concentration in the base at the boundary of the emitter-base depletion region is given by (see Section 4.3):

$$n_{be} = n_{bo} \exp\left(V_{be} / V_{th}\right) \gg n_{bo} \qquad (5\text{-}1\text{-}1)$$

where $n_{bo} = n_{ib}^2/N_{ab}$, $n_{ib}$ is the intrinsic carrier concentration in the base region, $N_{ab}$ is the acceptor doping density in the base region, $V_{th} = k_B T/q$ is the thermal voltage, and $V_{be}$ is the emitter-base voltage ($V_{be} \gg V_{th}$). Hence, under forward bias, minority carriers are supplied (we say "**injected**") into the base region from the emitter region.

As we understood in Section 4.3 analyzing a $n^+$-$p$ junction, these minority carriers – electrons – diffuse away from the junction and recombine with the majority carriers – holes. Of course, holes are injected into the emitter region, just like electrons are injected into the base. The minority carrier (hole) concentration in the emitter at the boundary of the emitter-base depletion region is given by

$$p_{eb} = p_{eo} \exp(V_{be}/V_{th}) \gg p_{eo} \qquad (5\text{-}1\text{-}2)$$

where $p_{eo} = n_{ie}^2/N_{de}$, $n_{ie} = n_{ib}$ is the intrinsic carrier concentration, and $N_{de}$ is the donor doping density in the emitter region. However, since $N_{de} \gg N_{ab}$ the number of holes injected into the emitter region is much smaller than the number of electrons injected into the base:

$$p_{eb} \ll n_{be} \qquad (5\text{-}1\text{-}3)$$

Hence, to the first order, we can forget about holes injected into the emitter region. The important point is that *a lot of electrons are injected into the base,* but *only a few holes are injected into the emitter* because of the much higher doping in the emitter region than in the base region.

When electrons are injected into the base, they bring along a negative charge. Attracted by this charge, holes enter the base. Their positive charge practically compensates the negative electron charge so that the neutral base region remain almost neutral (or as we say, "quasineutral"). In mathematical terms, it means that

$$|\Delta p - \Delta n| \ll \Delta p \qquad (5\text{-}1\text{-}4)$$

where $\Delta p$ and $\Delta n$ are excess hole and electron concentrations in the base. Hence, we have additional electron-hole pairs in the base region. Such a system can be described by the solution of the continuity equation for minority carriers (in our case, electrons) alone [see eqs. (3-7-28) and (3-8-30)]. The total current density in each point in the base can be found as the sum of the electron diffusion current density and the recombination current density. (As discussed above, practically all recombination must take place in the base, where both electrons and holes are available.  In the emitter region we have lots of electrons but practically no holes so that no appreciable recombination takes place.)

    If the separation between the emitter-base and collector-base junctions were much greater than the electron diffusion length in the base, the two junctions would not affect each other.  All electrons injected by the emitter into the base would recombine with the majority carriers (holes) in the neutral region of the base before ever reaching the space charge (i. e., depletion region) of the collector base junction.  However, in a BJT, the base region is always made much shorter than the diffusion length of the minority carriers.  Since in a typical regime of operation, the collector-base junction is reverse-biased, the electron concentration in the base at the depletion zone edge of the collector-base junction, $n_{bc}$, is much smaller than the equilibrium concentration:

$$n_{bc} = n_{bo} \exp(V_{bc} / V_{th}) << n_{bo} \tag{5-1-5}$$

Here $V_{bc}$ is the base-collector voltage $(V_{bc} < 0)$.

    Since the width, $W$, of the base is small compared to the diffusion length, $L_{nb}$, of electrons in the base, the carrier recombination in the base is small.  To first order, the electron concentration in the base varies linearly with distance. Mathematically it can be proven by solving the diffusion equation for the minority carriers (electrons) in the base:

$$D_n \frac{d^2 n_b}{dx^2} - \frac{n_b - n_{bo}}{\tau_{nl}} = 0 \tag{5-1-6}$$

[compare with eq. (3-8-30)].  Here $D_n$ is the electron diffusion coefficient, and $\tau_{nl}$ is the electron lifetime.  The solution of this second-order linear differential equation is given by

$$\Delta n = A \exp\left(\frac{x}{L_{nb}}\right) + B \exp\left(-\frac{x}{L_{nb}}\right) \tag{5-1-7}$$

[compare with eq. (3-8-31)].  Here $\Delta n = n_b - n_{bo}$ and $L_{nb} = (D_n \tau_{nl})^{1/2}$ is the electron diffusion length.  Since $W << L_{nb}$, we can expand the exponential functions in eq. (5-1-7) into a Taylor series keeping only constant and linear terms:

$$\Delta n = A + B + (A - B)\frac{x}{L_{nb}} \tag{5-1-8}$$

We assume that the widths of the depletion regions at the emitter-base and collector-base interfaces are much smaller than the base width (which is typically the case) and choose $x = 0$ to correspond to the position of the emitter-base junction. Then the boundary conditions are

$$\Delta n(0) = n_{be} - n_{bo} \tag{5-1-9}$$

$$\Delta n(W) \approx -n_{bo} \tag{5-1-10}$$

Hence, we find that

$$A + B = n_{be} - n_{bo}$$
$$A + B + (A - B)\frac{W}{L_{nb}} = -n_{bo} \tag{5-1-11}$$

Solving these two algebraic equations with respect to $A$ and $B$, we finally obtain

$$\Delta n = n_{be}\left(1 - \frac{x}{W}\right) - n_{bo} \tag{5-1-12}$$

[This linear variation of $n_b$ with distance can be also understood without solving eq. (5-1-6). Indeed, if we neglect electron recombination in the base, the second term in the right hand side of eq. (5-1-6) can be neglected. This means that the second derivative of $n_b$ with respect to $x$ is zero; hence, the first derivative is constant, and $n_b$ varies linearly with distance.]

The electron diffusion current in the base

$$I_n = SqD_n\frac{dn_b}{dx} \tag{5-1-13}$$

remains practically constant throughout the base. (Here, $S$ is the device cross section.) In other words, practically all the electrons injected into the base diffuse straight into the base-collector depletion region since very few electrons recombine with holes in the thin base region. Once they reach this depletion region, they are grabbed by the strong electric field of the reverse-biased collector-base junction and carried away toward the neutral collector region. As a result, the collector current is practically equal to the emitter current.

Substituting eq. (5-1-1) into eq. (5-1-12) and the resulting equation into eq. (5-1-13), we find the emitter current, $I_e$, and the collector current, $I_c$, as functions of the emitter-base voltage:

$$I_e \approx I_c \approx |I_n| \approx |I_{ne}| = \frac{SqD_n n_{bo}}{W} \exp\left(\frac{V_{be}}{V_{th}}\right) \qquad (5\text{-}1\text{-}14)$$

where $I_{ne}$ is the electron component of the emitter current, which is practically equal to the electron diffusion current in the base. Since $I_e$ and $I_c$ are proportional to $\exp(V_{be}/V_{th})$, an increase of the emitter-base voltage by $V_{th}$ (approximately 26 mV at room temperature) leads to an increase of $I_e$ and $I_c$ by a factor of $\exp(1) \approx 2.718$.

**Example 5-1-1.**

Relate the emitter current to the total concentration of acceptors, $n_G$, in the base (called the **Gummel number**).

**Solution:**

Substituting $n_{bo} = \dfrac{n_i^2}{N_{ab}}$ into eq. (5-1-14) we obtain

$$I_e \approx \frac{SqD_n n_i^2}{n_G} \exp\left(\frac{V_{be}}{V_{th}}\right) \text{ since } n_G = N_{ab}W.$$

The collector and emitter currents are approximately equal because the base current is small. A transistor satisfies the generalized Kirchoff's current law

$$\mathbf{I}_e + \mathbf{I}_c + \mathbf{I}_b = 0 \qquad (5\text{-}1\text{-}15)$$

where the directions of $\mathbf{I_e}$, $\mathbf{I_c}$, and $\mathbf{I_b}$ are chosen to enter the BJT (see Fig. 5.1.5).

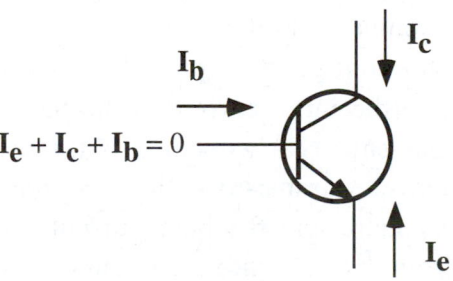

**Fig. 5.1.5.** Generalized Kirchoff's law for a BJT.

### 5.1.3.  BJT as an amplifier.

A BJT can operate as an amplifier.  This means that we apply an input signal represented by an ac voltage or current and obtain a more powerful output signal.  Since a BJT has three terminals, one terminal has to be shared between the input and output loops, as shown in Fig. 5.1.1.  This can be done in three different ways and, therefore, we have three possible circuit configurations of the bipolar junction transistor – the common base, common emitter, and common collector configurations (see Fig. 5.1.6).

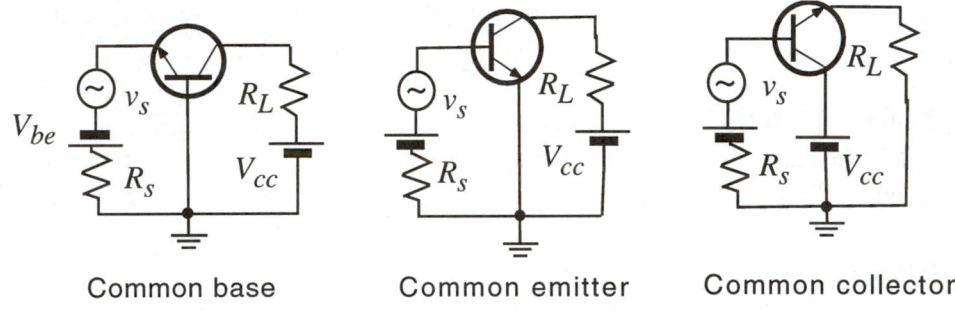

**Common base**              **Common emitter**              **Common collector**

**Fig. 5.1.6.** Circuit configurations for a bipolar junction transistor.

Let us first discuss the common-base configuration.  A small variation of the emitter-base voltage, $v_e$, caused by the signal, leads to a nearly equal variation of the emitter and the collector currents  $(\Delta I_c \approx - \Delta I_e)$.  If $R_L >> R_s$, the voltage variation   $\Delta V_L = \Delta I_c R_L$ across $R_L$ can be much greater than $v_s$.  (With an appropriate value of $R_L$, the maximum value of $\Delta V_L$ is limited by the collector voltage supply, $V_{cc}$.)  Hence, we can have an ac voltage gain of the order of $R_L/R_s$, an *ac* current gain, $\Delta I_c/\Delta I_e \approx - 1$, and an ac power gain nearly the same as the ac voltage gain.  The voltage and power gains occur because electrons are injected from the emitter region through the base into the collector region where the voltage drop is large compared to the small forward emitter-base voltage.

The common emitter configuration provides not only a voltage gain but a current gain as well, since the input current is the base current and the output current is the collector current.  (No wonder it is the most popular BJT circuit configuration.)  In modern-day transistors, the base recombination current, $I_{rb}$ (see Fig. 5.1.4) is fairly small, and the base current, $I_b$, is primarily the hole current, $I_{pe}$, that flows from the base into the emitter region:

$$I_b \approx I_{pe} \approx SqD_p \left.\frac{\partial p_e}{\partial x}\right|_{x=0} \approx \frac{SqD_p p_{eo}}{X_e} \exp\left(\frac{V_{be}}{V_{th}}\right) \tag{5-1-16}$$

[compare with eq. (5-1-14) for the electronic component of the emitter current, $I_{ne}$]. Here, $X_e$ is the width of the emitter region and $p_{eo} = n_{ie}^2/N_{de}$ is the equilibrium minority carrier concentration, where $N_{de}$ is the donor density and $n_{ie}$ is the intrinsic carrier concentration in the emitter, and we assumed that $X_e \ll L_{pe}$, where $L_{pe}$ is the diffusion length of holes in the emitter region. Since the emitter region is much more heavily doped than the base region, $p_{eo} \ll n_{bo}$ and $I_b \approx I_{pe} \ll I_{ne} \approx I_e$.

Dividing $I_e$ given by eqs. (5-1-14) by $I_b$ given by eq. (5-1-16), we obtain an estimate for the common-emitter current gain, $\beta_o = \partial I_c/\partial I_b \approx \partial I_e/\partial I_b$

$$\beta_o = \frac{D_n N_{de} X_e}{D_p N_{ab} W} \tag{5-1-17}$$

Since $N_{de} \gg N_{ab}$, and $D_n/D_p = \mu_n/\mu_p > 1$ (approximately 2.5 for silicon), $\beta_o$ is much greater than one. (Since $\mu_n/\mu_p > 1$, Si $n$-$p$-$n$ BJTs usually have a higher current gain than Si $p$-$n$-$p$ BJTs.)

Even though $I_{pe}$ is the dominant component of the base current, it is instructive to consider the recombination component of the base current. Let us now revisit the solution of the diffusion equation for electrons in the base. However, we shall now include the second and third order terms into the Taylor series expansion used in eq. (5-1-8):

$$\Delta n = A + B + (A-B)\frac{x}{L_{nb}} + (A+B)\frac{x^2}{2L_{nb}^2} + (A-B)\frac{x^3}{6L_{nb}^3} \tag{5-1-18}$$

Once again using the boundary conditions specified by eqs. (5-1-9) and (5-1-10), we obtain

$$\Delta n = \left[n_{be}(0) - n_{bo}\right]\left(1 - \frac{xW}{2L_{nb}^2} + \frac{x^2}{2L_{nb}^2} + \frac{x^3}{6L_{nb}^3}\right) - \frac{x}{W}n_{be}(0) \tag{5-1-19}$$

Hence, $dn_{be}/dx$ and the electron diffusion current vary along the base. In fact, the electron diffusion current at the boundary of the collector-base depletion region is smaller than  the electron diffusion current at the boundary of the emitter-base depletion region by $1 - \dfrac{W^2}{2L_{nb}^2}$ times. The difference is the component of the base current related to the electron recombination in the base. Hence, the common-emitter current gain limited by the recombination in the base is on the order of

$$\beta_R = \frac{2L_{nb}^2}{W^2} \qquad (5\text{-}1\text{-}20)$$

The overall maximum common-emitter current gain can now be estimated as

$$\beta = \left(\frac{1}{\beta_o} + \frac{1}{\beta_R}\right)^{-1} = \left(\frac{D_p N_{ab} W}{D_n N_{de} X_e} + \frac{W^2}{2L_{nb}^2}\right)^{-1} \qquad (5\text{-}1\text{-}21)$$

[In eq. (5-1-21), we added up inverse common emitter current gains. From a mathematical point of view this is similar to what we did by adding up inverse mobilities to find an overall carrier mobility in Chap. 3. In both cases, such an approach reproduces correctly limiting cases when one of the components is dominant and provides a reasonable interpolation in between.]

**Example 5-1-2.**
Estimate $\beta_o$ and $\beta_R$ for the following parameter values, typical for Si BJTs: $N_{de}$ =$10^{19}$cm$^{-3}$, $N_{ab}$ =$2 \times 10^{17}$cm$^{-3}$, $\mu_n$ = 900 cm$^2$/Vs, $\mu_p$ = 300 cm$^2$/Vs, $W$ = 0.1 μm, $X_e$ = 0.4 μm, $L_{nb}$ = 10 μm.

**Solution:**
From eqs. (5-1-17) and (5-1-20), we find $\beta_o$ = 600, $\beta_R$ = 20,000.

Another important BJT characteristic related to the short circuit common-emitter current gain is the **common-base current gain**, $\alpha$, defined as

$$\alpha = \frac{\partial I_c}{\partial I_e} \qquad (5\text{-}1\text{-}22)$$

This definition can be rewritten as

$$\alpha = \frac{\partial I_c}{\partial I_e} = \frac{\partial I_c}{\partial I_{ne}} \frac{\partial I_{ne}}{\partial I_e} \approx \alpha_T \gamma \qquad (5\text{-}1\text{-}23)$$

where $\gamma = \dfrac{\partial I_{ne}}{\partial I_e}$ is the **emitter injection efficiency** and

$$\alpha_T = \frac{\partial I_c}{\partial I_{ne}} \tag{5-1-24}$$

is called the **base transport factor**. For the uniformly doped base, we obtain using eq. (5-1-19),

$$\alpha_T = 1 - \frac{W^2}{2L_{nb}} \tag{5-1-25}$$

However, modern BJTs have a nonuniform doping in the base, with the acceptor doping density decreasing towards the collector-base junction. (Such transistors are called "**graded base**" transistors.) The nonuniform doping leads to a non-uniform hole concentration along the base. Hence, the holes diffuse from the emitter side of the base toward the collector side of the base. This creates an excess positive charge at the collector side of the base and an excess negative charge at the emitter side of the base. This, in turn, leads to a built-in electric field, which pushes minority carriers (electrons) injected into the base toward the collector. As a consequence, the minority carriers take less time to traverse the base and their recombination is less effective than for a uniformly doped base. For graded base transistors, the base transport factor is given by

$$\alpha_T = 1 - \frac{W^2}{2fL_{nb}} \tag{5-1-26}$$

where factor $f$ can be as large as 2 (see Problem 5-1-2).

Using the definition of the common emitter and common base current gains and the generalized Kirchoff's current law given by eq. (5-1-15), we can relate $\alpha$ and $\beta$:

$$\beta = \frac{\alpha}{1-\alpha}, \quad \alpha = \frac{\beta}{1+\beta} \tag{5-1-27}$$

Equation (5-1-21) implies that in order to obtain a high common emitter gain, the emitter region should be doped as high as possible and the base should be made as narrow as possible. Usually, the emitter doping is chosen to be very high ($10^{19}$ - $10^{20}$ cm$^{-3}$). However, at very high emitter doping levels, the Bohr radii of impurity atoms overlap, creating an impurity band of states. This leads to a narrower energy gap in highly doped semiconductors. For Si, this energy gap narrowing can be estimated as

$$\Delta E_g(\text{eV}) = 2 \times 10^{-11} N_a^{0.5}\left(\text{cm}^{-3}\right) \tag{5-1-28}$$

The band gap narrowing leads to a sharp (exponential) increase in the intrinsic carrier concentration in the emitter region, and hence to a sharply increased hole injection into the emitter region and reduced gain.

The base also should not be made too thin. The effective base thickness, $W_{eff}$, is smaller than $W$ because of the extension of the emitter-base and collector-base depletion regions into the base (see Fig. 5.1.3):

$$W_{eff} = W - X_{be} - X_{bc} \tag{5-1-29}$$

where $X_{be}$ and $X_{bc}$ are the depletion widths for the emitter-base and collector-base junctions, respectively. (Actually, we should replace $W$ with $W_{eff}$ in all the equations given above in this section.) If the base doping is chosen to be too low, the two depletion regions in the base can even overlap and $W_{eff}$ can shrink to zero with the increase in the collector-base reverse bias. This effect (called base **punch-through**) will manifest itself as a sharp increase in the emitter and collector currents, that is, as a device breakdown. One possible way to prevent such catastrophic consequences is to choose the collector doping fairly low, much smaller than the base doping, so that the widest collector-base depletion region will extend primarily into the collector region.

Even more important limitations of the doping level in the base and of the base width can be understood by considering the BJT shown in Fig. 5.1.7. As can be seen from the figure, the base current has to flow in the lateral direction from the central section of the device. The resistance, $R_{bs}$, associated with this current path is called the **base spreading resistance**. The voltage drop, $I_b R_{bs}$, leads to a nonuniform injection along the base. The emitter current density will be higher at the edges, since there the voltage difference between the emitter and base region is larger. Since the emitter current density is a very strong (exponential) function of the emitter-base bias, the result is what is called "the **emitter current crowding**" at the edges. Hence, the base spreading resistance should be minimized or at least controlled. This means that the base should not be made too thin or too low doped.

As a consequence of these trade-offs, the actual BJT doping profile is chosen to be something like the profile shown in Fig. 5.1.8. (As we see from the figure, the actual doping profile is very nonuniform. As we discussed above, such a non-uniform doping in the base improves the transistor characteristics.)

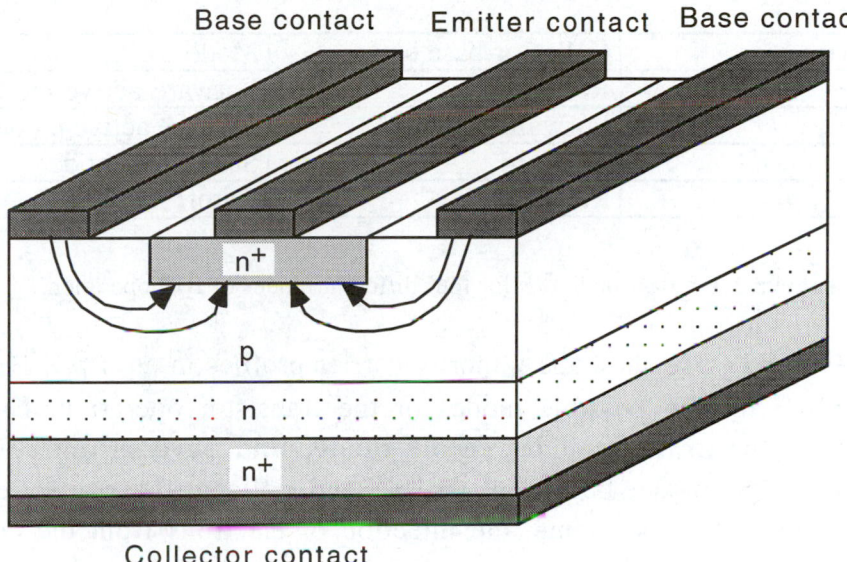

**Fig. 5.1.7.** Schematic diagram of a BJT. Arrows show streamlines of the base current.

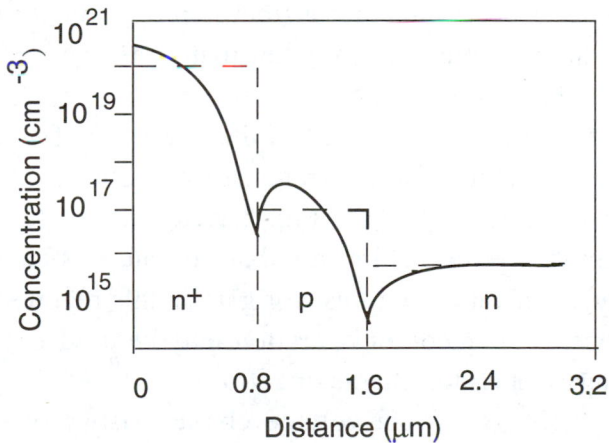

**Fig. 5.1.8.** Example of realistic (solid line) and idealized (dashed lines) BJT doping profile.

### 5.1.4.  BJT modes of operation.

Above, we mostly consider BJT operation when the emitter-base junction is forward biased and collector-base junction reverse-biased. This mode of operation is called the **forward active mode**. All possible bias conditions and modes of BJT operation are listed in Table 5.1.1.

| Emitter-base bias | Collector-base bias | Mode |
|---|---|---|
| Forward | Reverse | Forward active mode |
| Reverse | Forward | Reverse active mode |
| Forward | Forward | Saturation mode |
| Reverse | Reverse | Cutoff mode |

**Table 5.1.1.** Bias conditions for four different modes of BJT operation.

In Fig. 5.1.9 we show the minority carrier profiles in an $n$-$p$-$n$ BJT base corresponding to four possible modes of the transistor operation: **forward active mode, saturation mode, cutoff mode, and reverse active mode**. In the saturation mode, both emitter-base and collector-base junctions are forward biased. In this regime, the injection of electrons from the collector region is in the opposite direction compared to the injection of the electrons from the emitter region. The minority carrier profiles injected from the emitter and collector region, respectively, are shown in Fig. 5.1.9b by dashed lines. The sum of the two profiles (shown by a solid line) has a smaller slope than for the forward active mode. Hence, the current (that is proportional to the slope of the electron profile in the base region) is smaller than in the active forward mode of operation (compare Fig. 5.1.9a and 5.1.9b).

In the cutoff region (see Fig. 5.1.9c) there are very few minority carriers in the base and, hence, the transistor currents are very small.

In the reverse active mode (see Fig. 5.1.9d) more holes are injected into the collector region than electrons into the base because of the higher doping in the base. This leads to a very low transistor gain in this regime, which is of little practical importance. This mode of operation may be used for the extraction of transistor parameters (transistor characterization).

Figure 5.1.10 shows current-voltage characteristics of a BJT simulated using PSpice[tm] (see Section 5.2). The PSpice[tm] default parameters were used in this simulation, except for the collector series resistance, $R_c$, which was 1 k$\Omega$.

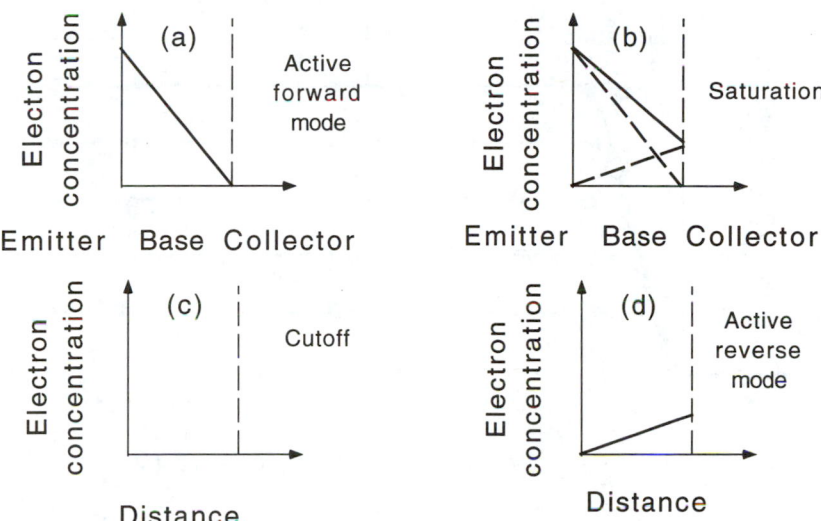

**Fig. 5.1.9.** Minority carrier profiles in an *n-p-n* BJT base corresponding to four possible modes of the transistor operation: (a) forward active mode, (b) saturation mode, (c) cutoff mode, and (d) reverse active mode. Dashed lines in Fig. 5-1-10 b show the minority carrier profiles injected from emitter and collector junctions.

At large emitter-collector voltages, $V_{ce}$, when the collector-base junction is reverse-biased and emitter-base junction is forward biased, the transistor operates in the active forward mode. As can be seen from Fig. 5.1.10, in this regime, the collector current is proportional to the base current ($I_c \approx \beta I_b$). When $V_{ce}$ is decreased to values comparable to the built-in voltage, the bias is no longer large enough to forward bias the emitter-base junction and reverse bias the collector-base junction. Both junctions become forward-biased, and the BJT operates in the saturation mode. As can be seen from Fig. 5.1.11, in this mode, the collector current rapidly decreases with the decrease in the collector-emitter bias. (The slope of this decrease is primarily determined by the collector parasitic series resistance.) Since a BJT can operate in both active forward and saturation mode, it can be used as a switch. Indeed, if the BJT emitter is grounded and the collector is attached via a load resistance, $R_L$, to a power supply, $V_{cc}$ (see Fig. 5.1.11), the collector-emitter voltage is given by

$$V_{ce} = V_{cc} - I_c R_L \qquad\qquad (5\text{-}1\text{-}30)$$

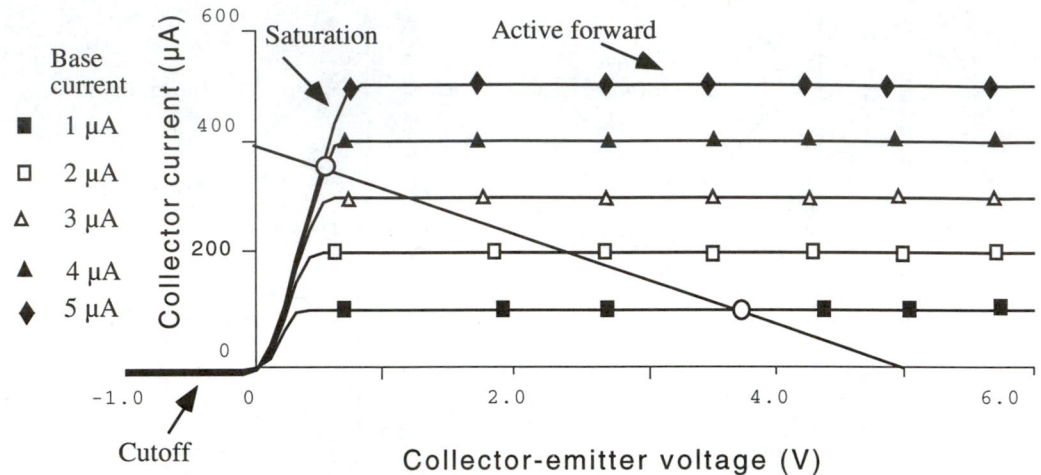

**Fig. 5.1.10.** BJT *I-V* output characteristics. Also shown is a load line for the load resistance $R_L$ = 12.5 kΩ and two operating points corresponding to the base currents of 1 µA and 5 µA, respectively.

In Fig. 5.1.10, this dependence is represented by a straight line (called a **load line**). The intersection of the load line with the transistor *I-V* characteristics yields operating points. As can be seen from the figure, for a small base current (1 µA), the operating point corresponds to a high collector-emitter voltage, low collector current state. For a larger base current (5 µA), the operating point corresponds to a low collector-emitter voltage, high collector current state. Hence, switching between these two states can be achieved by a small change in the base current.

For applications in digital circuits utilizing BJTs, these devices may be fabricated with multiple collector or emitter contacts allowing for many outputs.

As discussed above, the operation of a BJT is based on the exponential variation of the injected carrier density with the change in the height of the potential barrier between the emitter and base regions controlled by the base-emitter voltage, $V_{be}$. This leads to the exponential dependence of the emitter and collector currents on the emitter-base voltage [see eq. (5-1-5)] and to a very high **transconductance**, $g_m$, in a forward active mode of operation, especially for high values of the collector current, $I_c$ [see eq. (5-1-22)]: $g_m = \dfrac{\partial I_c}{\partial V_{be}} = \dfrac{q I_c}{k_B T}$.

Let us compare the BJT transconductance with the corresponding parameter for a field effect transistor, which utilizes a capacitive modulation of charge in a conducting channel (see Chap. 6).

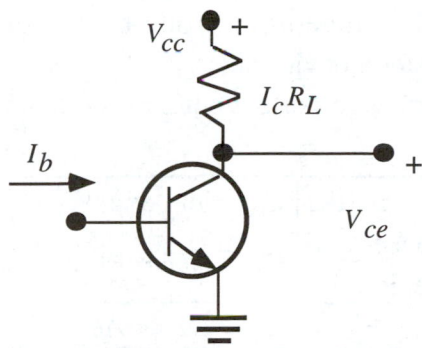

**Fig. 5.1.11.** Transistor connected to  power supply via  resistor.  The bias voltage, $V_{cc}$, is divided between the voltage drop across the transistor, $V_{ce}$,  and the voltage drop, $I_c R_L$, across the resistor.

For a 1 mA collector current, for example, at room temperature, the BJT transconductance $g_m$ = 1 mA/ 0.025 V $\approx$ 40 mS.  A typical FET transconductance is on the order of 1 to 6 mS for a 20 μm wide device.  High transconductance and current swing make the bipolar junction transistor a device of choice for many high-speed and high-power applications in both discrete and integrated circuits.

On the negative side, the BJT technology is power hungry.  This limits the maximum number of transistors one can put on a semiconductor chip.  In other words, the power requirements limit the **integration scale**.  The very strong exponential dependence of the BJT current on temperature may also be a negative factor.  If a certain transistor in an integrated circuit would have a higher temperature because of an unavoidable parameter variation or because of poor cooling conditions in this particular spot, the current through this particular transistor will be considerably higher, leading to a larger power dissipation in this particular spot and to a further temperature increase.

Field Effect Transistors (FETs), do not behave that way and, hence, FET technologies are more suitable for **Very Large Scale Integrated Circuits (VLSI)** – the bread and butter of modern electronics.  Recently, however, a combined BJT-FET technology became a popular choice for fast integrated circuits.  In this technology, called **BiCMOS** (from **Bi**polar **C**omplementary **M**etal **O**xide **S**emiconductor Field Effect Transistors)**,** BJTs and FETs are fabricated next to each other in the same integrated circuit so that relatively high-power high-speed BJTs can be used sparingly, as required, in an electronic circuit for the largest impact on the overall circuit performance.  However, even this technology faces a stiff competition from more conventional field effect technology utilizing very small device sizes.

Table 5.1.2 summarizes important equations related to BJT operation.

| Electron concentrations at the boundaries of emitter-base and collector-base depletion region | $n_{be} = n_{bo} \exp(V_{be} / V_{th})$ <br> $n_{bc} = n_{bo} \exp(V_{bc} / V_{th})$ |
|---|---|
| Quasineutrality condition in base | $\lvert \Delta p - \Delta n \rvert \ll \Delta p$ |
| Diffusion equation for electrons in the base | $D_n \dfrac{d^2 n_b}{dx^2} - \dfrac{n_b - n_{bo}}{\tau_{nl}} = 0$ |
| Electron diffusion current in the base | $I_n = SqD_n \dfrac{dn_b}{dx}$ |
| Emitter and collector currents for the active forward mode | $I_e \approx I_c \approx \lvert I_n \rvert \approx \dfrac{SqD_n n_{bo}}{W} \exp\left(\dfrac{V_{be}}{V_{th}}\right)$ |
| Base current for the active forward mode | $I_b \approx I_{pe} \approx \dfrac{SqD_p p_{eo}}{X_e} \exp\left(\dfrac{V_{be}}{V_{th}}\right)$ |
| Kirchoff's current law | $\mathbf{I_e + I_c + I_b = 0}$ |
| Minority (electron) carrier distribution in the base | $\Delta n = \left[ n_{be}(0) - n_{bo} \right] \left( 1 - \dfrac{x}{W} - \dfrac{xW}{2L_{nb}^2} + \dfrac{x^2}{2L_{nb}^2} \right)$ |
| Overall maximum common-emitter current gain | $\beta = \left( \dfrac{1}{\beta_o} + \dfrac{1}{\beta_R} \right)^{-1} = \left( \dfrac{D_p N_{ab} W}{D_n N_{de} X_e} + \dfrac{W^2}{2L_{nb}^2} \right)^{-1}$ |
| Emitter injection efficiency | $\gamma = \partial I_{ne} / \partial I_e$ |
| Base transport factor | $\alpha_T = \dfrac{\partial I_{ne}}{\partial I_{nc}} = 1 - \dfrac{W^2}{2fL_{nb}}$ |
| Common-base current gain | $\alpha = \dfrac{\partial I_c}{\partial I_e} = \dfrac{\partial I_c}{\partial I_{ne}} \dfrac{\partial I_{ne}}{\partial I_e} \approx \alpha_T \gamma$ |
| Effective base width | $W_{eff} = W - X_{be} - X_{bc}$ |
| Common emitter and common-base current gains | $\beta = \dfrac{\alpha}{1 - \alpha} \qquad \alpha = \dfrac{\beta}{1 + \beta}$ |
| BJT transconductance | $g_m = \dfrac{\partial I_c}{\partial V_{be}} = \dfrac{qI_c}{k_B T}$ |

**Table 5.1.2.** Important equations related to BJT operation.

## 5-2.  BJT MODELING

### 5.2.1.   Ebers-Moll  Model.

Figure 5.2.1 shows a simple large signal equivalent circuit of a bipolar junction transistor, which includes two diodes and two current-controlled current sources describing the injection of minority carriers from the emitter-base junction into the collector-base junctions and vice versa.  (Current sources representing the recombination and generation currents in the emitter-base and collector-base junctions are also added to this circuit.)

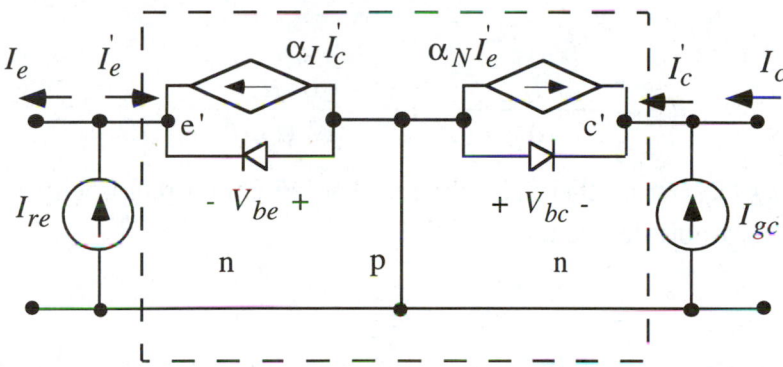

**Fig. 5.2.1.**  Equivalent circuit for an *n-p-n* bipolar transistor.

The model for the "intrinsic" transistor (shown in the dashed box in Fig. 5.2.1) is called the **Ebers-Moll  model**.  This model expresses the intrinsic emitter and collector currents in terms of the *p-n* diode currents.  Using Kirchoff's current law at nodes e' and c', we obtain

$$I_e' = -\alpha_I I_c' - I_{eo}\left[\exp\left(\frac{V_{be}}{V_{th}}\right) - 1\right] \tag{5-2-1}$$

$$I_c' = -\alpha_N I_e' - I_{co}\left[\exp\left(\frac{V_{bc}}{V_{th}}\right) - 1\right] \tag{5-2-2}$$

where $\alpha_N$ and $\alpha_I$ are the "normal" and "inverse" common base current gains, $I_{eo}$ is the emitter-base saturation current (with the collector-base circuit open), and $I_{co}$ is the collector-base saturation current (with the emitter-base circuit open). Substituting $I_c'$ from eq. (5-2-2) into eq. (5-2-1) and solving for $I_e'$ we obtain

$$I_e' = a_{11}\left[\exp\left(\frac{V_{be}}{V_{th}}\right) - 1\right] + a_{12}\left[\exp\left(\frac{V_{bc}}{V_{th}}\right) - 1\right] \tag{5-2-3}$$

where

$$a_{11} = -\frac{I_{eo}}{1-\alpha_N\alpha_I} \qquad a_{12} = \frac{\alpha_I I_{co}}{1-\alpha_N\alpha_I} \tag{5-2-4}$$

Substituting $I_e'$ from eq. (5-2-1) into eq. (5-2-2) and solving for $I_c'$ we obtain

$$I_c' = a_{21}\left[\exp\left(\frac{V_{be}}{V_{th}}\right)-1\right] + a_{22}\left[\exp\left(\frac{V_{bc}}{V_{th}}\right)-1\right] \tag{5-2-5}$$

where

$$a_{22} = -\frac{I_{co}}{1-\alpha_N\alpha_I} \qquad a_{12} = \frac{\alpha_N I_{eo}}{1-\alpha_N\alpha_I} \tag{5-2-6}$$

Here $V_{th} = k_B T/q$ is the thermal voltage. The so-called reciprocity relationship for an ideal two-port device requires

$$a_{21} = a_{12} \tag{5-2-7}$$

This leaves us with only three independent parameters of the Ebers-Moll model – $a_{11}$, $a_{12}$, and $a_{22}$, which are related to the material parameters, the transistor dimensions, and the doping levels via the solution of the continuity equations for minority carriers in the base. Below we provide the equations for $a_{11}$, $a_{12}$, and $a_{22}$ for reference purposes. For $n$-$p$-$n$ transistors and for the sign convention for the currents $I_e'$ and $I_c'$ used in Fig. 5.2.1, we have

$$a_{11} \approx -qSn_i^2\left[\frac{D_n}{N_{ab}W_{eff}} + \frac{D_p}{N_{de}X_e}\right] \tag{5-2-8}$$

$$a_{12} = a_{21} \approx \frac{qSD_n n_i^2}{N_{ab}W_{eff}} \tag{5-2-9}$$

$$a_{22} \approx -qSn_i^2\left[\frac{D_n}{N_{ab}W_{eff}} + \frac{D_p}{N_{dc}X_c}\right] \tag{5-2-10}$$

Here, $W_{eff}$ is the width of the neutral base region

$$W_{eff} = W - X_{be} - X_{bc} \qquad (5\text{-}2\text{-}11)$$

where $X_{be}$ and $X_{bc}$ are the widths of the depletion regions extended into the base at the emitter-base and collector-base junctions, respectively [see eq. (5-1-30)]; $X_e$ and $X_c$ are the widths of the emitter and collector regions (assumed to be much less than the corresponding minority carrier diffusion lengths); $N_{de}$, $N_{ab}$ and $N_{dc}$ are the doping densities of the emitter, the base, and the collector, respectively; and $D_n$ and $D_p$ are the diffusion constants for minority electrons and minority hole. The other symbols are defined in Section 5.1. Equations (5-2-3) to (5-2-5) are obtained using the equations for the minority carrier distribution in the base derived in Section 5.1 and assuming that $W_{eff} \ll L_{nb}$ (see Problem 5-1-1).

The emitter-base voltage may be expressed in terms of the emitter current of the intrinsic transistor, $I_e'$, using eq. (5-2-1). From eqs. (5-2-4) and (5-2-6), we find the normal (or forward) common-base current gain:

$$\alpha_N = -a_{12}/a_{11} \qquad (5\text{-}2\text{-}12)$$

and the **common-base collector reverse saturation current**:

$$I_{co} = \frac{a_{12}a_{21}}{a_{11}} - a_{22} \qquad (5\text{-}2\text{-}13)$$

For the active forward mode, $V_{bc}$ is negative, $|V_{bc}|$ is much larger than $V_{th}$, and eq. (5-2-1) leads to

$$I_c = \alpha I_e + I_{cbo} \qquad (5\text{-}2\text{-}14)$$

where $I_{cbo}$ is the collector current with open emitter. (Here, we neglect the recombination current, $I_{re}$, and the generation current, $I_{gc}$, such that $I_e \approx -I_e'$, $\alpha_N \approx \alpha$ and $I_{co} \approx I_{cbo}$.) However, eq. (5-2-12) may account for the recombination current, $I_{re}$ (via the dependence of $\alpha$ on the collector current), and, hence, the current gains $\alpha_N$ and $\alpha$ and the saturation currents $I_{co}$ and $I_{cbo}$ do not have to be exactly equal.

In a similar manner, eq. (5-2-3) describes the inverse mode of operation when the collector-base junction is forward-biased and the emitter-base junction is reverse-biased.

As was mentioned above, the Ebers-Moll model for the intrinsic transistor has three parameters: $a_{11}$, $a_{12}$ ($a_{21} = a_{12}$), and $a_{22}$.   A more convenient parameter set (which can be expressed through $a_{11}$, $a_{12}$, and $a_{22}$) is $\alpha_N$, $I_{eo}$ and $I_{co}$. These three (only three!) parameters give a good idea about a BJT operating in the active forward mode at a moderate **injection level** (when the concentration of the minority carriers injected into the base from the emitter region is still much smaller than the doping concentration in the base).  However, modeling of real BJTs operating under realistic conditions in actual circuits is much more of a challenge.  First, as we understood analyzing $p$-$n$ junctions in Chapter 4, the simple model of a $p$-$n$ junction, such as used in the Ebers-Moll model, is inaccurate at both low and high injection levels.  It is also inaccurate at reverse biases.  The Ebers-Moll model is adequate for a basic understanding of device physics, for order of magnitude estimates and, perhaps, for guiding us toward reasonable empirical and semiempirical expressions.  If a simple theory cannot accurately describe a $p$-$n$ junction diode, no wonder that the basic Ebers-Moll model (based on that theory) has to be improved as well.  First, the current sources accounting for the recombination in the forward-biased emitter-base depletion region and for the electron-hole pair generation in the reverse-biased collector-base depletion region have to be added, as shown in Fig. 5.2.1.  This fixes the basic model at low injection levels.  At low injection levels, the recombination current in the forward-biased emitter-base depletion region

$$I_{reb} = I_{rebo} \exp\left(\frac{qV_{be}}{m_{reb}k_BT}\right) \tag{5-2-15}$$

provides a dominant contribution to the base current.  (Here, $m_{reb}$ is the ideality factor for the recombination current.  Typically, $m_{reb} \approx 2$.)  Hence, at low injection levels, the common-emitter current gain is limited by

$$\beta_{reb} = \beta_{rebo} \exp\left[\frac{qV_{be}}{k_BT}\left(1 - \frac{1}{m_{reb}}\right)\right] = \frac{A_R}{I_c^{1-1/m_{reb}}} \tag{5-2-16}$$

The expression for the overall common-emitter current gain [see eq. (5-1-21)] is fixed using our rule of adding up inverse gains limited by different mechanisms:

$$\beta = \left(\frac{1}{\beta_o} + \frac{1}{\beta_R} + \frac{1}{\beta_{reb}}\right)^{-1} = \left(\frac{D_pN_{ab}W_{eff}}{D_nN_{de}X_e} + \frac{W_{eff}^2}{2L_{nb}^2} + \frac{A_R}{I_c^{1-1/m_{reb}}}\right)^{-1} \tag{5-2-17}$$

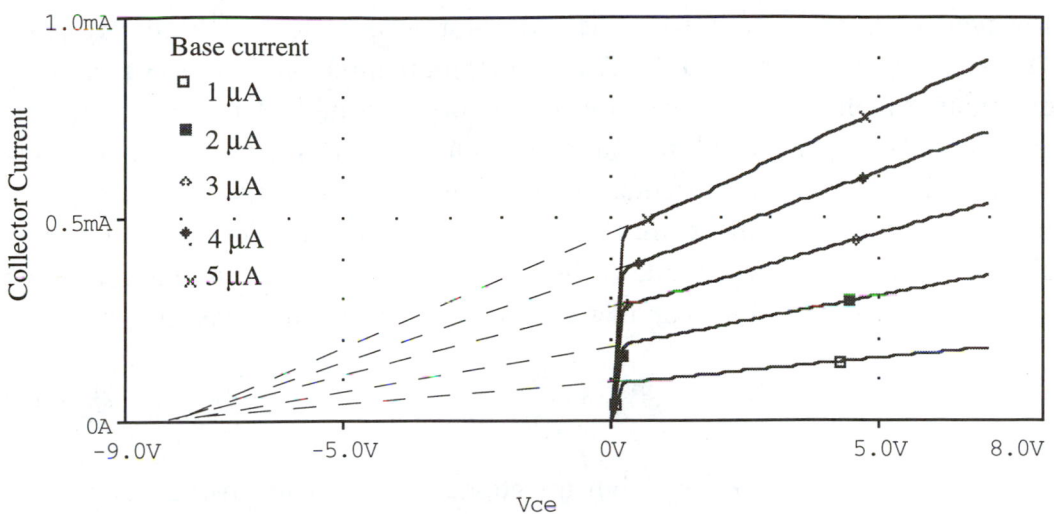

**Fig. 5.2.2.** BJT common-emitter characteristics calculated using PSpice$^{tm}$ default parameters (except for the Early voltage, $V_A$).

## 5.2.2.　Gummel-Poon model.

These characteristics shown in Fig. 5.2.2 are calculated using an PSpice$^{tm}$ **Gummel-Poon** BJT model. This model (proposed by Gummel and Poon in 1970) is more accurate and realistic than the Ebers-Moll model. The Gummel-Poon model accounts for high injection effects in a semiempirical way. This model is also implemented in other versions of SPICE (including AIM-Spice). However, this additional accuracy comes at a price. The Gummel-Poon model has more than 40 parameters describing a BJT! Just describing how to extract all these parameters from the experimental data required writing a separate book [see Getreu (1970)].

Using eq. (5-2-14) (linking the collector and emitter currents via the common base current gain, $\alpha$) and Kirchoff's law for a BJT [see eq. (5-1-15)], we can find the relationship between $\alpha$ and the common-emitter current gain, $\beta$. Substituting

$$I_e = I_c + I_b \tag{5-2-21}$$

into eq. (5-2-14) we obtain

$$I_c = \alpha I_c + \alpha I_b + I_{cbo} \tag{5-2-22}$$

However, even this improved model does not account for the **high injection effects**, for effects related to Auger recombination, which becomes very important at high concentrations of electron-hole pairs in the base region (see the discussion of Auger recombination in Section 3.8), and for the base spreading resistance.   In practical circuits, BJTs usually operate under high injection conditions, and all the effects that limit the common-emitter current gain at high injection levels are important.   The common-emitter current gain at high injection level decreases with an increase in the collector current as follows:

$$\beta_{hi} \approx \frac{A_{hi}}{I_c^{m_{hi}}}$$

(5-2-18)

where indices "*hi*" stand for "high injection."   Once again, using our rule of adding up inverse gains limited by different mechanisms, we obtain the following expression for the common-emitter current gain:

$$\beta = \left( \frac{1}{\beta_o} + \frac{1}{\beta_R} + \frac{1}{\beta_{reb}} + \frac{1}{\beta_{hi}} \right)^{-1}$$

(5-2-19)

Since $W_{eff}$ decreases with an increase in the reverse collector bias [see eq. (5-2-6)], the common-emitter current gain increases with the reverse bias. Typically, the doping levels in the collector-base region are such that this increase (called the **Early effect**) is relatively small, and the $I_c$-$V_{ce}$ dependence for the common emitter configuration can be approximated as

$$I_c = I_c^{(0)} \left( 1 + \frac{V_{ce}}{V_A} \right)$$

(5-2-20)

where $V_A$ is called the **Early voltage** (see Fig. 5.2.2).  The Early voltage is one of the most important BJT parameters.

**Example 5-2-1.**

Determine the Early voltage from the data shown in Fig. 5.2.2.

**Solution:**

From the $V_{ce}$-intercepts of the extrapolated $I$-$V$ curves, we find $V_A = 8$ V.

Hence,

$$I_c = \frac{\alpha}{1-\alpha} I_b + \frac{I_{cbo}}{1-\alpha} = \beta I_b + I_{ceo} \qquad (5\text{-}2\text{-}23)$$

where the **common-emitter current gain** is

$$\beta = \frac{\alpha}{1-\alpha} \qquad (5\text{-}2\text{-}24)$$

and the **common-emitter leakage current** is

$$I_{ceo} = \frac{I_{cbo}}{1-\alpha} \qquad (5\text{-}2\text{-}25)$$

Figure 5.2.3 shows the dependence of the common-emitter current gain, $\beta$, on the collector current calculated using the Ebers-Moll and Gummel-Poon models implemented in AIM-Spice.

As can be seen from the figure, both Ebers-Moll and Gummel-Poon models give very close results at small collector currents (i. e., at low injection levels). However, only the Gummel-Poon model correctly describes the decrease of the common-emitter current gain at large collector currents.

Figure 5.2.4 shows the common-base BJT current voltage characteristics calculated using the PSpice[tm] BJT model used for the calculation of the common collector current-voltage characteristics shown in Fig. 5.2.2.

We notice that the output conductance in Fig. 5.2.4 seems to be very small, in spite of the fact that the same Early voltage of 8 V was used in the simulation.

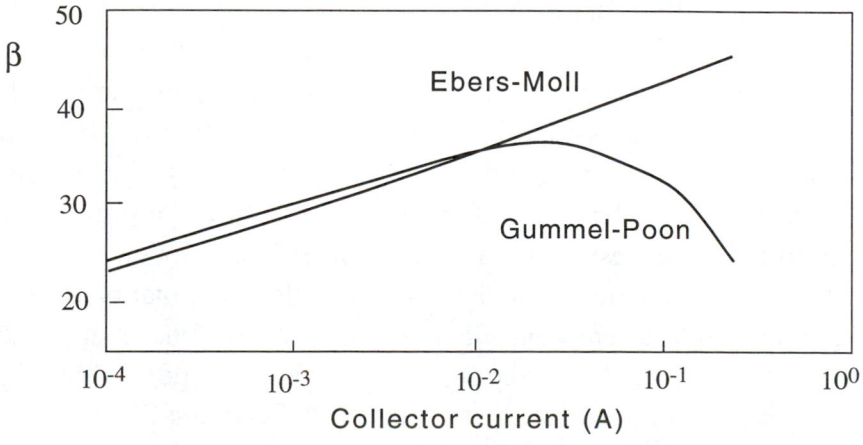

**Fig. 5.2.3.** Common-emitter current gain versus collector current .

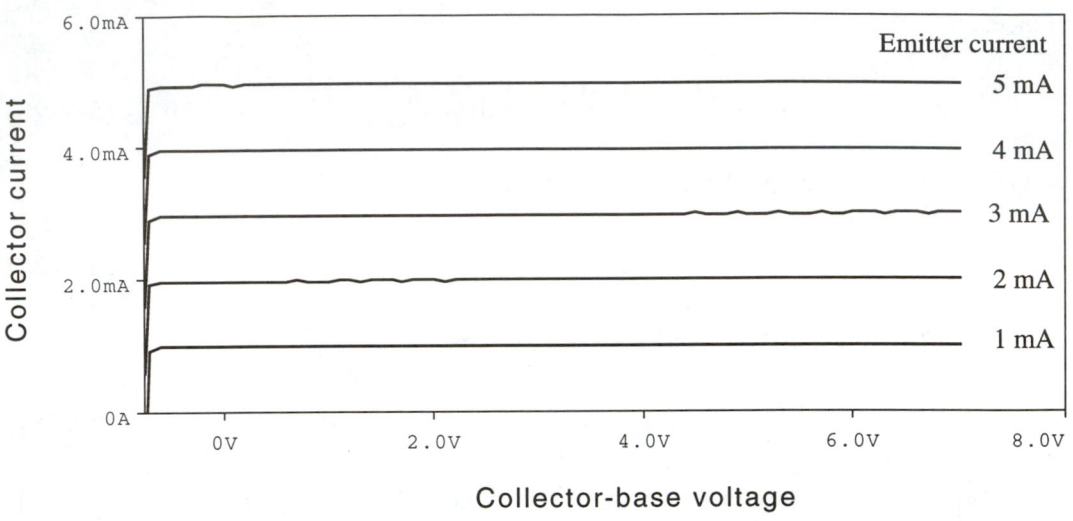

**Fig. 5.2.4.** Common-base characteristics of a BJT calculated using PSpice[tm] using default parameters (except for the Early voltage, $V_A = 8$ V).

The reason is that even relatively large changes in the common-emitter current gain, $\beta$, cause only a relatively small change in the common-base current gain, $\alpha$. From eq. (5-2-25), we find that

$$\alpha = \frac{\beta}{1+\beta} \qquad (5\text{-}2\text{-}26)$$

For example, if $\beta$ changes from 100 to 200, $\alpha$ changes only from 0.99 to 0.995.

### 5.2.3. BJT simulation using SPICE.

Let us now discuss the parameters of the Gummel-Poon model used in SPICE. As mentioned above, the number of these parameters is quite large. However, these parameters can be classified into groups corresponding to the emitter-base junction, collector-base junction, and so on. In this way, their default and typical values can be readily understood. Figure 5.2.5 shows a simplified equivalent circuit of an intrinsic transistor used in the Gummel-Poon model.

Each transistor junction is modeled as a parallel combination of two diodes and of the equivalent current-controlled current source. One of these diodes is called an "ideal" diode and the other one is a "leakage" diode, which represents leakage and recombination current components (see Section 4.3).

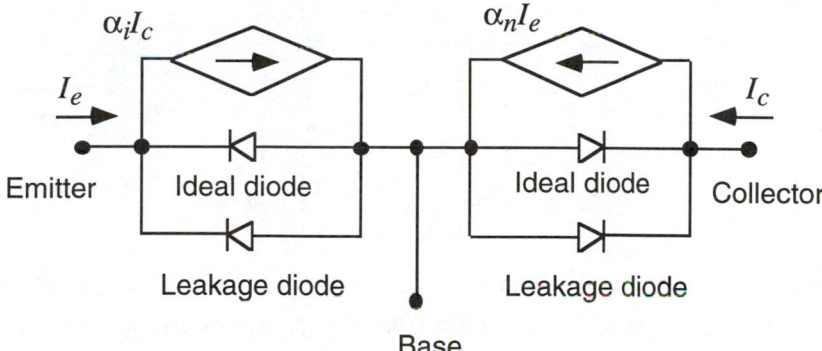

**Fig. 5.2.5.** Simplified equivalent circuit for an *n-p-n* bipolar transistor used in the Gummel-Poon model.

The ideal diode *I-V* characteristics for the emitter-base junction are described by the diode equation

$$I = \frac{I_s}{\beta_F}\left[\exp\left(\frac{V}{n_F V_{th}}\right) - 1\right] \tag{5-2-27}$$

where the default value of $n_F$ is unity [which reduces eq. (5-2-27) to an ideal diode equation], $\beta_F$ is called the **forward common-emitter current gain**, and the saturation current

$$I_s(T) = I_s(T_o)\left(\frac{T}{T_o}\right)^{\frac{\kappa}{\eta}}\exp\left(\frac{E_g}{k_B T_o}\right)\exp\left(-\frac{E_g}{k_B T}\right) \tag{5-2-28}$$

(see Section 4.3).

The ideal diode *I-V* characteristics for the collector-base junction are described by a diode equation similar to eq. (5-2-27)

$$I = \frac{I_s}{\beta_R}\left[\exp\left(\frac{V}{n_R V_{th}}\right) - 1\right] \tag{5-2-29}$$

where the default value of $n_R$ is unity [which reduces eq. (5-2-29) to an ideal diode equation], $\beta_R$ is called the **reverse common-emitter current gain**. (Notice that the Gummel-Poon model uses the same saturation current for both emitter-base and collector-base junctions, which may introduce a certain error.)

The *I-V* characteristics of the leakage diodes are described by

$$I = I_{se} \left[ \exp\left( \frac{V}{n_E V_{th}} \right) - 1 \right] \qquad (5\text{-}2\text{-}30)$$

$$I = I_{sc} \left[ \exp\left( \frac{V}{n_C V_{th}} \right) - 1 \right] \qquad (5\text{-}2\text{-}31)$$

for the emitter-base and collector-base junctions, respectively. In addition, if a bipolar junction transistor is fabricated as a part of an integrated circuit on a semiconductor substrate, the leakage current of the depletion regions formed at the boundaries with the substrate may become important. This current is also modeled by a diode equation (see Sections 4.3 and 4.7):

$$I = I_{ss} \left[ \exp\left( \frac{V}{n_S V_{th}} \right) - 1 \right] \qquad (5\text{-}2\text{-}32)$$

If the substrate current is important then the four nodes – collector, base, emitter, and substrate -- have to be specified. (The default value of the substrate potential is ground.) Appendix A4 (Table A4.1) summarizes the parameters related to the *I-V* characteristics of the emitter-base, collector-base, and substrate diodes used in the Gummel-Poon model.

The equivalent depletion emitter-base and collector-base diode capacitances are modeled in exactly the same way as the depletion capacitance of a *p-n* junction (see Sections 4.4 and 4.6). If a bipolar junction transistor is fabricated as a part of an integrated circuit on a semiconductor substrate, the capacitance of the depletion regions formed at the boundaries with the substrate may become important. This capacitance is also modeled in the same way as for a *p-n* junction. Appendix A4 (Table A4.2) summarizes the Gummel-Poon SPICE parameters characterizing the depletion capacitances.

The next set of device parameters are the parasitic resistances: emitter series resistance, $R_E$, collector series resistance, $R_C$, and the base spreading resistance, $R_B$. Resistances $R_E$ and $R_C$ can be considered independent of current levels. However, the base spreading resistance may be a function of the base current. Indeed, as can be seen from Fig. 5.1.8 (where arrows show streamlines of the base current) the base current creates a voltage drop across the base in the lateral direction. As a consequence, the base regions closer to the emitter periphery become more forward-biased than the regions in the center of the base.

(This effect is called emitter current crowding.).  At high bias, the majority carriers injected into the base reduce the base resistance.  In the Gummel-Poon Spice model, these effects are described by the following equation:

$$R_{Beff} = R_{Bm} + \left(R_B - R_{Bm}\right)\frac{I_{RB}}{I_B + I_{RB}} \qquad (5\text{-}2\text{-}33)$$

where $R_{Beff}$ is the effective base spreading resistance, $R_B$ is the maximum (zero base current) base spreading resistance, $R_{Bm}$ is the minimum base spreading resistance, $I_B$ is the base current, and $I_{RB}$ is the base current at which

$$R_{Beff}\left(I_{RB}\right) = \frac{R_B + R_{Bm}}{2} \qquad (5\text{-}2\text{-}34)$$

Parameters characterizing the parasitic resistances of a BJT in the Gummel-Poon SPICE model are summarized in Appendix A4 (Table A4.3).

The remaining three groups of the Gummel-Poon parameters deal with nonideal transistor effects, transistor time response, and with the temperature dependencies of the BJT parameters.  These three groups of parameters are summarized in Tables A4.4, A4.5, and A4.6 of Appendix A4, respectively.

Most transistor parameters characterizing nonideal effects, that is, deviations from the Ebers-Moll model, deal with high injection effects, which become important at high current densities.  For example, the effective forward common-emitter current gain (called "**forward beta**," $\beta_{Feff}$) decreases with an increase in the collector current at large collector currents (see the discussion above and Fig. 5.2.3).  In the Gummel-Poon SPICE model:

$$\beta_{Feff} = \frac{\beta_F}{1 + I_c / I_{KF}} \qquad (5\text{-}2\text{-}35)$$

A similar equation is used for the reverse beta:

$$\beta_{Reff} = \frac{\beta_R}{1 + I_c / I_{KR}} \qquad (5\text{-}2\text{-}36)$$

A more detailed description of the Gummel-Poon model can be found, for example, in books by Getreu (1970), Shur (1990), and Lee et al. (1993).

For a crude simulation of BJT characteristics at low frequencies, it may be sufficient to use the Ebers-Moll model described in the beginning of the section. The  Gummel-Poon BJT model used in PSpice$^{tm}$ reduces to the Ebers-Moll model when the default parameters are used in the simulation. Hence, in order to apply the Ebers-Moll model for low-frequency simulations, we have to specify only the SPICE parameters stated in Table A4.1, using the default values for all the remaining SPICE parameters. Table A4.1 expresses the SPICE parameters given in Table A4.1 in terms of the parameters of the Ebers-Moll model.

**Example 5-2-2.**

Estimate the parameters of the Ebers Moll model and use PSpice$^{tm}$ to calculate the $I_c$-$V_{ce}$ characteristics (where $I_c$ is the collector current and $V_{ce}$ is the collector-emitter voltage) for the base currents $I_b = 1$ μA, 2 μA, 3μA, and 4 μA. (Neglect the change in the effective base width due to the extension of the depletion regions into the base.) Transistor parameters are: emitter width $X_e = 0.5$ μm, base width $W = 0.5$ μm, collector width $X_c = 3$ μm, device area $S = 1\times10^{-8}$ m$^2$, emitter doping $N_{de} = 5\times10^{18}$ cm$^{-3}$, base doping $N_{ab} = 2\times10^{17}$ cm$^{-3}$, collector doping $N_{dc} = 5\times10^{15}$ cm$^{-3}$. Intrinsic carrier concentration at room temperature $n_i = 10^{10}$ cm$^{-3}$. Temperature $T = 300$ K. Electron mobility in the base region $\mu_n = 0.1$ m$^2$/Vs. Hole mobility in the emitter region $\mu_p = 0.04$ m$^2$/Vs.

**Solution:**

As pointed out above, the Gummel-Poon BJT model used in PSpice$^{tm}$ reduces to the Ebers-Moll model when the default parameters are used in the simulation. Using Appendix A4, we find that in order to apply the Ebers-Moll model for low-frequency simulations, we have to specify only the following SPICE parameters: saturation current, IS, ideal maximum forward beta, BF, ideal maximum reverse beta, BR, and substrate saturation current, ISS (see Table 5.2.1).

| Spice parameter | Spice parameter name | Unit | Spice default | Chap. 5 notation | Ebers-Moll parameter |
|---|---|---|---|---|---|
| IS | Saturation current | A | 1.0e-16 | $I_s$ | $-a_{11}$ |
| BF | Ideal maximum forward beta | | 100 | $\beta_F$ | $-\dfrac{a_{12}}{a_{11}+a_{12}}$ |
| BR | Ideal maximum reverse beta | | 1 | $\beta_R$ | $-\dfrac{a_{12}}{a_{22}+a_{12}}$ |
| ISS | Substrate saturation current | A | 1.0e-16 | $I_{ss}$ | 0 |

**Table 5.2.1.** SPICE parameters in terms of the parameters of the Ebers-Moll model. (Other SPICE parameters specified should have default values.)

The parameters of the Ebers-Moll model are related to physical transistor parameters via eqs. (5-2-8) to (5-2-10) except that in SPICE the effective base width, $W_{eff}$, is replaced with the geometrical base width, $W$, since the Early effect can be accounted for in SPICE simulations by choosing appropriate values for the Early voltage. The diffusion coefficients are found using the Einstein relation: $D_n = k_BT\mu_n/q = 2.6\times10^{-3}$ m²/s, $D_p = k_BT\mu_p/q = 1.04\times10^{-3}$ m²/s. Hence, $a_{11} = -4.23\times10^{-15}$ A, $a_{22} = -1.53\times10^{-14}$ A, $a_{12} = 4.165\times10^{-14}$ A, $\beta_F = 64.1$, $\beta_R = 1.37$, $I_s = 4.23\times10^{-15}$ A. The results are shown in Fig. 5.2.6.

PSpice$^{tm}$ input file:

```
BJT TEST
Q1 0 2 3 BJT
Q2 0 6 3 BJT
Q3 0 7 3 BJT
Q4 0 8 3 BJT
.model BJT NPN BF=64.1 BR=1.37 IS=4.23e-15 ISS=0
Vce 0 3 1
.DC Vce -1 7 0.1
Ib1 0 2 1u
Ib2 0 6 2u
Ib3 0 7 3u
Ib4 0 8 4u
.probe
.end
```

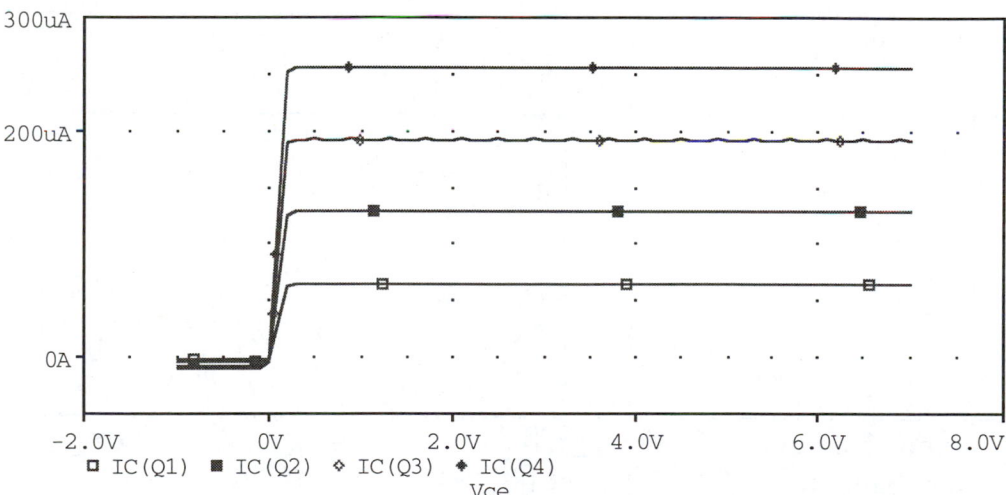

**Fig. 5.2.6.** Output characteristics ($I_c$ versus $V_{ce}$) of Bipolar Junction Transistor simulated using Ebers-Moll model implemented in PSpice$^{tm}$.

Important equations related to BJT modeling are summarized in Table 5.2.2.

| Ebers-Moll equation for emitter current | $I_e' = a_{11}\left[\exp\left(\dfrac{V_{be}}{V_{th}}\right) - 1\right] + a_{12}\left[\exp\left(\dfrac{V_{bc}}{V_{th}}\right) - 1\right]$ |
|---|---|
| Ebers-Moll equation for collector current | $I_c' = a_{21}\left[\exp\left(\dfrac{V_{be}}{V_{th}}\right) - 1\right] + a_{22}\left[\exp\left(\dfrac{V_{bc}}{V_{th}}\right) - 1\right]$ |
| Parameters of Ebers-Moll model | $a_{11} \approx -qSn_i^2\left[\dfrac{D_n}{N_{ab}W} + \dfrac{D_p}{N_{de}X_e}\right]$ <br><br> $a_{12} = a_{21} \approx \dfrac{qSD_n n_i^2}{N_{ab}W_{eff}}$ <br><br> $a_{22} \approx -qSn_i^2\left[\dfrac{D_n}{N_{ab}W} + \dfrac{D_p}{N_{dc}X_c}\right]$ |
| $I_c$ versus $I_e$ (common-base configuration) | $I_c = \alpha I_e + I_{cbo}$ |
| $I_c$ versus $I_b$ (common-emitter configuration) | $I_c = \beta I_b + I_{ceo}$ <br><br> $\beta = \dfrac{\alpha}{1-\alpha} \qquad I_{ceo} = \dfrac{I_{cbo}}{1-\alpha}$ |
| Early voltage | $I_c = I_c^{(0)}\left(1 + \dfrac{V_{ce}}{V_A}\right)$ |
| Common-emitter short circuit current gain | $\beta = \left(\dfrac{1}{\beta_o} + \dfrac{1}{\beta_R} + \dfrac{1}{\beta_{reb}}\right)^{-1}$ <br><br> $\beta = \left(\dfrac{D_p N_{ab} W_{eff}}{D_n N_{de} X_e} + \dfrac{W_{eff}^2}{2L_{nb}^2} + \dfrac{A_R}{I_c^{1-\frac{1}{m_{reb}}}} + \dfrac{I_c^{m_{hi}}}{A_{hi}}\right)^{-1}$ |
| Common-base short circuit current gain | $\alpha = \dfrac{\beta}{1+\beta}$ |

**Table 5.2.2.** Summary of important equations related to BJT modeling (see also Table 5.1.1).

## 5-3. BREAKDOWN IN BIPOLAR JUNCTION TRANSISTORS

### 5.3.1. Avalanche breakdown.

The **avalanche breakdown** of the collector-base junction is a typical mechanism of a BJT breakdown. Another important mechanism is the "**punch-through**" breakdown of the base, which occurs when that the collector-base depletion region merges with the emitter-base depletion region. (We already discussed this breakdown mechanism briefly in Subsection 5.1.3.) The mechanism of the avalanche breakdown in bipolar junction transistors is similar to that in $p$-$n$ diodes (see Section 4.5). However, the critical voltages of the avalanche breakdown in bipolar junction transistors are different for different transistor circuit configuration.

For a common-base configuration, the **avalanche breakdown voltage**, $BV_{cb}$, can be found by equating the maximum electric field at the collector-base interface to the **breakdown field**, $F_{br}$. Using the depletion approximation, just as was done in Section 4.5 for a $p$-$n$ junction, we obtain:

$$BV_{cb} \approx \frac{\varepsilon_s F_{br}^2}{2q}\left(\frac{1}{N_{ab}} + \frac{1}{N_{dc}}\right) \approx \frac{\varepsilon_s F_{br}^2}{2qN_{dc}} \tag{5-3-1}$$

[compare with eq. (4-5-1)]. The breakdown field, $F_{br}$, is approximately 300 kV/cm for Si, 400 kV/cm for GaAs, and 2,300 kV/cm or more for SiC.

A more accurate analysis of the avalanche breakdown can be done by introducing a **multiplication factor** $M_{cb}$ into the expression for the collector current for the common-base configuration:

$$I_c = M_{cb}\left(I_{cbo} + \alpha I_e\right) \tag{5-3-2}$$

where $I_{cbo}$ is the common-base collector current with an open emitter (see Section 5.2), $\alpha$ is the common-base current gain, and

$$M_{cb} = \frac{1}{1 - \left(V_{cb}/BV_{cb}\right)^{m_b}} \tag{5-3-3}$$

Here $V_{cb}$ is the collector-base voltage. Usually, the common-base breakdown voltage, $BV_{cb}$, is expected to increase with temperature since the enhanced atomic thermal motion leads to a more efficient carriers scattering, and, hence, fewer carriers are "lucky" enough to reach energies in excess of the energy gap needed for the creation of an electron-hole pair via impact ionization (see Section 3.8). The constant $m_b$ in eq. (5-3-3) depends on the doping profile in the collector region and on temperature. Typically, $m_b$ is between 2 and 5 for silicon transistors.

Let us now consider a common-emitter configuration. When the base is open, the collector current under avalanche multiplication conditions is obtained by substituting $I_e = I_c$ into eq. (5-3-2) and solving the resulting equation with respect to $I_c$:

$$I_c = I_{cbo} \frac{M_{cb}}{1 - \alpha M_{cb}} \tag{5-3-4}$$

Hence, for the common-emitter configuration, the breakdown occurs when

$$\alpha M_{cb} = 1 \tag{5-3-5}$$

Since the common-base current gain, $\alpha$, is close to unity, this condition is certainly much easier to satisfy than the corresponding condition for the common-base configuration, which is $M_{cb} \to \infty$. Usually, at the breakdown, the reverse collector-base voltage, $V_{cb}$, is much greater than the forward voltage bias across the emitter-base junction, $V_{be}$. Hence, at the breakdown $V_{ce} \approx V_{cb}$. Using this condition, we find from eq. (5-3-5)

$$BV_{ce} = BV_{cb}(1 - \alpha)^{1/m_b} \tag{5-3-6}$$

**Example 5-3-1.**
Find the ratio $BV_{ce}/BV_{cb}$ for a transistor with $\alpha \approx 0.99$ and $m_b = 3$.

**Solution:**
From eq. (5-3-6), we find directly $BV_{ce} \approx 0.215 \, BV_{cb}$.

The smaller value of $BV_{ce}$ is related to the current gain in the common-emitter configuration (see Fig. 5.3.1). Each hole created as a result of an avalanche breakdown leads to the injection of $1/(1 - \alpha)$ electrons. (As we will see in Section 5.5, a worse frequency response is another negative feature of the common-emitter configuration.)

As was shown in Section 5.2, the current, $I_{ceo}$, in the common-emitter configuration with an open base is much larger than the current, $I_{cbo}$, in the common-base configuration with an open emitter [see eq. (5-2-25)]:

$$I_{ceo} = \frac{I_{cbo}}{1 - \alpha} \tag{5-3-7}$$

Substituting $I_{cbo}$ from eq. (5-3-7) into eq. (5-3-4), we find that

$$I_c = M_{ce} I_{ceo} \qquad (5\text{-}3\text{-}8)$$

where

$$M_{ce} = M_{cb} \frac{1-\alpha}{1-\alpha M_{cb}} \qquad (5\text{-}3\text{-}9)$$

is the multiplication factor for the common-emitter configuration. Figure 5.3.2 shows the calculated dependence of the collector current on the bias voltage for the two configurations. This figure clearly demonstrates a higher leakage and a lower breakdown voltage for the common-emitter configuration.

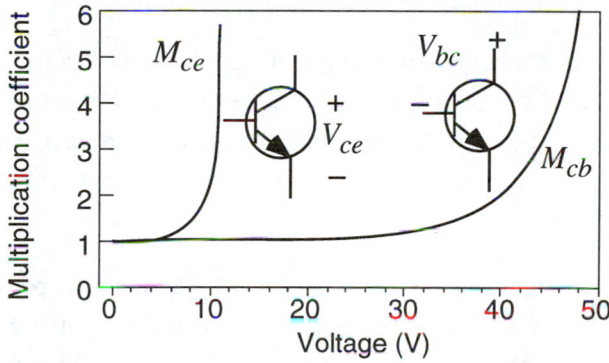

**Fig. 5.3.1.** Comparison of calculated multiplication factors for common-base (open emitter) and common-emitter (open base) configurations. Parameters used in the calculation: $BV_{cb} = 50$ V, $\alpha = 0.99$, $m_b = 3$.

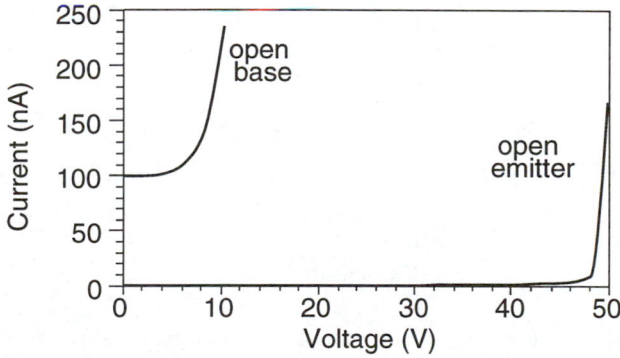

**Fig. 5.3.2.** Comparison of collector leakage currents for common-base (open emitter) and common-emitter (open base) configurations. Parameters used in the calculation: $BV_{cb} = 50$ V, $\alpha = 0.99$, $m_b = 3$, $I_{cbo} = 1$ nA.

### 5.3.2. Punch-through breakdown.

As was discussed in Subsection 5.1.3, a punch-through breakdown occurs when the reverse collector-base voltage becomes so large that the collector-base and the emitter-base depletion regions merge. Since the collector-base junction is reverse-biased, most of the voltage drop at punch-through occurs across the collector-base junction. Using the depletion approximation for the depletion width in the base at the collector-base interface, $d_{bc}$, and equating $d_{bc}$ to the base width, $W$, we obtain the following expression for the punch-through voltage, $V_{pth}$:

$$V_{pth} \approx \frac{qN_{ab}W^2}{2\varepsilon_s}\left(1 + \frac{N_{ab}}{N_{dc}}\right) \approx \frac{qN_{ab}^2 W^2}{2\varepsilon_s N_{dc}}. \tag{5-3-10}$$

Here we assume that the built-in voltage of the base-collector junction is much smaller than $V_{pth}$. According to eqs. (5-3-1) and (5-3-10), the breakdown voltages of the avalanche and the punch-through breakdowns increase when the collector doping, $N_{dc}$, is decreased.

> **Example 5-3-2.**
>
> For silicon, the breakdown field, $F_{br} \approx 3 \times 10^7$ V/m, the dielectric permittivity, $\varepsilon_s \approx 1.05 \times 10^{-10}$ F/m. At what base thickness will a silicon BJT with the base doping level $N_{ab} = 10^{17}$ cm$^{-3}$ have equal breakdown voltages $BV_{cb}$ and $V_{pth}$?
>
> **Solution:**
>
> The ratio of the avalanche breakdown voltage for the common-base configuration, $BV_{cb}$ [see eq. (5-3-1)], and the punch-through voltage, $V_{pth}$, is given by
>
> $$\frac{BV_{cb}}{V_{pth}} \approx \left(\frac{\varepsilon_s F_{br}}{qn_G}\right)^2 \tag{5-3-11}$$
>
> where $n_G = N_{ab}W$ is called the **Gummel number**. For the silicon BJT of this example, eq. (5-3-14) yields
>
> $$\frac{BV_{cb}}{V_{pth}} \approx \left(\frac{2 \times 10^{12}}{n_G(\text{cm}^{-2})}\right)^2 \tag{5-3-12}$$
>
> For the base doping level $N_{ab} = 10^{17}$ cm$^{-3}$, $V_{pth}$ is equal to $BV_{cb}$ for the base width $W = 0.2$ μm. For thinner bases, punch-through breakdown occurs at voltages smaller than $BV_{cb}$. For thicker bases, avalanche breakdown occurs at voltages smaller than $V_{pth}$.

The breakdown mechanisms limit the maximum voltages that a BJT can handle. The maximum collector current that practical power BJTs can handle is also limited. The emitter and collector currents are proportional to $D_n \dfrac{n_i^2}{N_{ab}} \exp\left(\dfrac{qV_{be}}{k_B T}\right)$ (see Subsection 5.1.1). The intrinsic carrier concentration, $n_i$, is proportional to $\exp\left(-\dfrac{E_g}{2k_B T}\right)$ where $E_g$ is the energy gap. Hence, the emitter and collector currents increase with temperature as $\exp\left(\dfrac{qV_{be} - E_g}{k_B T}\right)$.

Therefore, if the emitter and collector currents are kept constant, the temperature increase leads to a decrease of the emitter-base voltage with a negative temperature coefficient of approximately 1.5 mV/°C for a typical silicon transistor (see Example 5-3-3). If the emitter-base voltage is kept constant, the emitter and collector currents increase with temperature. This leads to a further temperature increase caused by Joule heating, which, in turn, leads to higher currents until the device is destroyed.

### Example 5-3-3.

Assuming that the collector current is kept constant, estimate the temperature coefficient of the emitter-base voltage, $dV_{be}/dT$, for a typical silicon transistor.

### Solution:

The temperature dependence of the collector current density is given by

$$j_c \approx j_e \approx \frac{qD_p n_i^2}{N_{ab}L_p} \exp\left(\frac{qV_{be}}{k_B T}\right) = \frac{qD_p \sqrt{N_c N_v}}{N_{ab}L_p} \exp\left(-\frac{E_g}{k_B T}\right) \exp\left(\frac{qV_{be}}{k_B T}\right) \tag{5-3-13}$$

since it is determined by the current-voltage characteristic of the emitter-base junction. Here $N_c$ and $N_v$ are the densities of states in the conduction and valence band. Solving for $V_{be}$, we obtain $V_{be} = qE_g + \dfrac{k_B T}{q} \ln\left(\dfrac{j_c}{j_{cT}}\right)$ where $j_{cT} = \dfrac{qD_p \sqrt{N_c N_v}}{N_{ab}L_p}$. We may neglect a relatively weak dependence of the log term on temperature. For Si, $E_g = 1.12$ eV and $V_{be}$ (at $T = 300$ K is typically 0.6 V. Hence, the temperature coefficient for the emitter-base voltage is equal to $\ln(j_c/j_{cT})$ $\approx -(1.12 - 0.6)/300 \approx -1.4$ mV/degree K.

Transistor manufacturers often specify the so-called **Safe Operating Area** (SOA) in the $I_c$-$V_{ce}$ plane. This area is limited by the lines corresponding to different failure modes (see Fig. 5.3.3). SOA is an important characteristic of a power bipolar junction transistor.

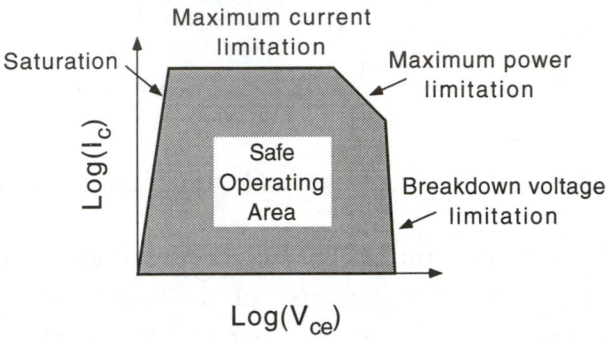

**Fig. 5.3.3.** Safe operating area (after Shur, 1990, copyright © Prentice Hall, 1990, reproduced by permission of Prentice Hall, Inc., Englewood Cliffs, NJ).

The equations related to BJT breakdown are summarized in Table 5.3.1.

| | |
|---|---|
| Collector-base breakdown voltage | $$BV_{cb} \approx \frac{\varepsilon_s F_{br}^2}{2q}\left(\frac{1}{N_{db}} + \frac{1}{N_{dc}}\right) \approx \frac{\varepsilon_s F_{br}^2}{2qN_{dc}}$$ |
| Increase in collector current for voltages higher than $BV_{cb}$ | $$I_c = M_{cb}\left(I_{cbo} + \alpha I_e\right)$$ |
| Multiplication factor due to avalanche breakdown in the collector-base junction | $$M_{cb} = \frac{1}{1-\left(V_{cb}/BV_{cb}\right)^{m_b}}$$ |
| Collector current under avalanche multiplication (open base) | $$I_c = M_{ce}I_{ceo} \quad \text{where} \quad M_{ce} = M_{cb}\frac{1-\alpha}{1-\alpha M_{cb}}$$ |
| Collector-emitter breakdown voltage | $$BV_{ce} = BV_{cb}\left(1-\alpha\right)^{1/m_b}$$ |
| Base punch-through voltage | $$V_{pth} \approx \frac{qN_{ab}W^2}{2\varepsilon_s}\left(1+\frac{N_{ab}}{N_{dc}}\right) \approx \frac{qN_{ab}^2 W^2}{2\varepsilon_s N_{dc}}$$ |

**Table 5.3.1.** Summary of equations related to BJT breakdown.

## 5-4.  SMALL SIGNAL EQUIVALENT CIRCUITS

In Section 5.1 we discussed the operating point of a BJT (see Fig. 5.1.10).  In Fig. 5.4.1, we show the common-emitter output characteristics of the same transistor with a superimposed small signal variation of the base current.  The operating point corresponds to a base current of 2 μA.  The amplitude of the base current variation is 1 μA so that the base current changes sinusoidally in time from 1 μA to 3 μA.  A small variation of the base current leads to a relatively large variation of the collector current (from 100 μA to 300 μA with the operating point at 200 μA) corresponding to a current gain of 100.  In other words, the transistor operates as an amplifier for an ac signal.

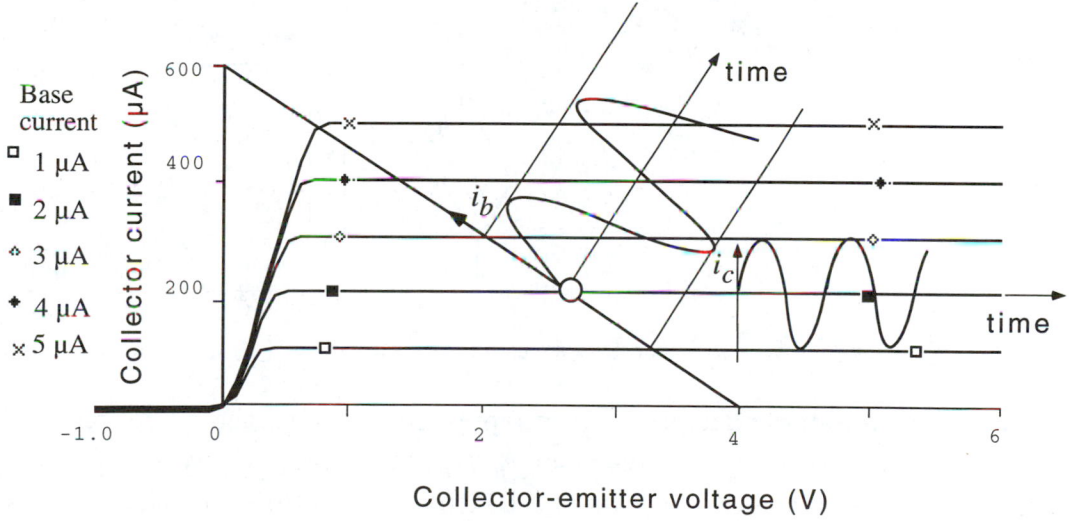

**Fig. 5.4.1.**  Small signal (linear) regime of transistor operation.

Since the base current variations are small, the transistor gain and other parameters remain practically the same for these variations of the base current. Under such conditions, the transistor response to a small signal (i. e., an ac response) may be described in terms of a linear two-port network.  In order to describe the *ac* response, we will  denote the input ac current and voltage as $i_1$ and $v_1$, the output *ac* current and voltage as $i_2$ and $v_2$, the total input current and voltage as $I_1$ and $V_1$, and the total output current and voltage as $I_2$ and $V_2$ (see Fig. 5.4.2).  Each port of a linear two-port network may be represented by its Norton or Thevenin equivalent (the Norton equivalent is a series combination of an impedance and of a voltage source; the Thevenin equivalent is a parallel combination of an admittance and of a current source). An **h-parameters** equivalent circuit shown in Fig. 5.4.2 represents the input port of a BJT by its Norton equivalent and the output port by its Thevenin equivalent. A set of four h-parameters fully describes a small signal response:

$$v_1 = h_{11}i_1 + h_{12}v_2 \qquad\qquad (5\text{-}4\text{-}1)$$

$$i_2 = h_{21}i_1 + h_{22}v_2 \qquad\qquad (5\text{-}4\text{-}2)$$

The **h-parameters** are most frequently used for transistor characterization [at least at relatively low frequencies (below 100 MHz or so)] and are specified by transistor manufacturers in their data sheets.

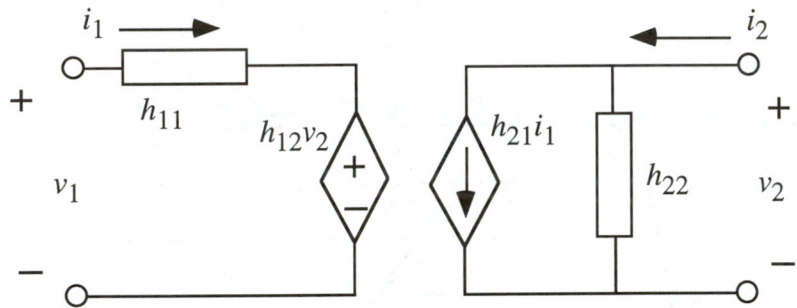

**Fig. 5.4.2.**  Two-port linear network using  h-parameters.

A z-**parameters** equivalent circuit represents the input and output ports of a BJT by their Thevenin equivalents. A **y-parameters** equivalent circuit represents the input and output ports of a BJT by their Norton equivalents. A **g-parameters** equivalent circuit represents the input port of a BJT by its Thevenin equivalent and the output port by its Norton equivalent.

**Example  5-4-1.**
Write equations for g-parameters similar to equations (5-4-1) and (5-4-2).

**Solution:**

$$i_1 = g_{11}v_1 + g_{12}i_2 \qquad\qquad (5\text{-}4\text{-}3)$$

$$v_2 = g_{21}v_1 + g_{22}i_2 \qquad\qquad (5\text{-}4\text{-}4)$$

The small signal parameters may be determined from different short-circuit or open-circuit measurements at the input and output ports:

$$h_{11} = \left.\frac{v_1}{i_1}\right|_{v_2=0}, \qquad h_{12} = \left.\frac{v_1}{v_2}\right|_{i_1=0} \qquad (5\text{-}4\text{-}5)$$

$$h_{21} = \left.\frac{i_2}{i_1}\right|_{v_2=0}, \qquad h_{22} = \left.\frac{i_2}{v_2}\right|_{i_1=0} \qquad (5\text{-}4\text{-}6)$$

The parameter $h_{11}$ ($h_i$) is called the **short-circuit input impedance**, $h_{12}$ is called the **open-circuit reverse voltage ratio** ($h_r$), $h_{21}$ is called the **short-circuit forward current ratio** ($h_f$), and $h_{22}$ is called the **open-circuit output admittance** ($h_o$). In the alternate notation ($h_i$, $h_r$, $h_f$, and $h_o$,), a second subscript is often used to denote the transistor configuration. For example, the $h$-parameters for the common-emitter transistor circuit configuration are denoted as $h_{ie}$, $h_{re}$, $h_{fe}$, and $h_{oe}$. The total of $h$-parameters is 12 (four for each transistor configuration). In their data sheets, transistor manufacturers usually provide only common-emitter $h$-parameters. Expressions relating other $h$-parameters to the common-emitter $h$-parameters are given in Table. 5.4.1.

The $h$-parameters can also be related to the parameters of the Ebers-Moll or the Gummel-Poon model. From eqs. (5-4-5) and (5-5-6), we obtain that at low frequencies, $h_{fe}$ is equal to $\beta$ and $h_{fb}$ is equal to $-\alpha$. For the other $h$-parameters, such relationships are more complicated. Therefore, the circuit analysis of BJTs often relies on other equivalent circuits whose parameters are more directly related to the physical device parameters. The equivalent circuit called the **hybrid-$\pi$ equivalent circuit** is often used for the analysis of the small signal response of the common-emitter configuration (see Fig. 5.4.3).

| Common-emitter $h$-parameters | Common-base $h$-parameters | Common-collector $h$-parameters |
|---|---|---|
| $h_{ie}$ | $h_{ib} = h_{ie}/(h_{fe} + 1)$ | $h_{ic} = h_{ie}$ |
| $h_{re}$ | $h_{rb} = h_{ie}h_{oe}/(h_{fe} + 1) - h_{re}$ | $h_{rc} = 1$ |
| $h_{fe}$ | $h_{fb} = -h_{fe}/(h_{fe} + 1)$ | $h_{fc} = -h_{fe} - 1$ |
| $h_{oe}$ | $h_{ob} = h_{oe}/(h_{fe} + 1)$ | $h_{oc} = h_{oe}$ |

**Table 5.4.1.** Different $h$-parameters in terms of $h_{ie}$, $h_{re}$, $h_{oe}$, and $h_{fe}$ (from Shur, 1990, copyright © Prentice Hall, 1990, reproduced by permission of Prentice Hall, Inc., Englewood Cliffs, NJ).

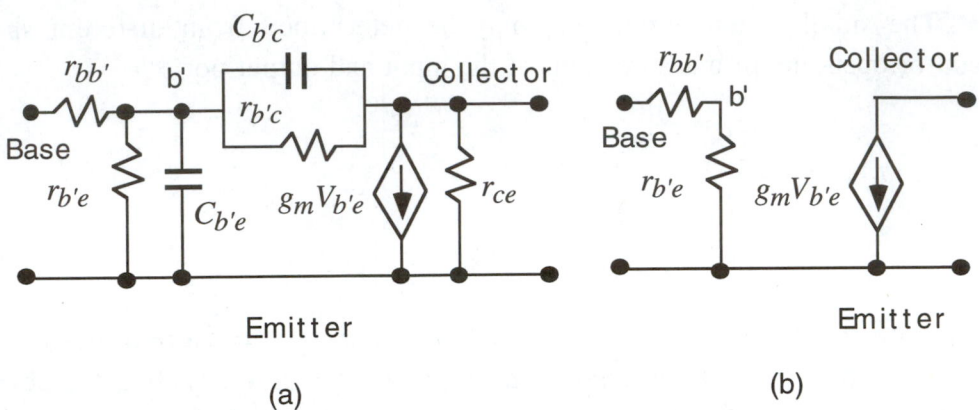

**Fig. 5.4.3.** (a) Full and (b) simplified hybrid-π equivalent circuits.

The transconductance, $g_m$, in Fig. 5.4.3 is related to the dynamic (differential) resistance, $r_e$, of the forward-biased emitter-base junction in the following way:

$$g_m = \frac{\partial I_c}{\partial V_{b'e}} = \alpha \frac{\partial I_e}{\partial V_{b'e}} \approx \frac{\alpha}{r_e} \approx \frac{I_c}{V_{th}} \tag{5-4-7}$$

where $V_{th} = k_B T/q$ is the thermal voltage. (Compare this with the differential resistance of the forward-biased *p-n* junction diode; see Section 4.4.). The resistance $r_{bb'}$ is the base spreading resistance. The resistance $r_{b'c}$ and the capacitance $C_{b'c}$ in the hybrid-π equivalent circuit (Fig. 5.4.3a) represent the **dynamic (differential) resistance** and the capacitance of the reverse-biased collector-base junction. (The collector-base capacitance $C_{b'c}$ is usually denoted as $C_{ob}$ in manufacturer data sheets.) Typically, the resistances $r_{ce}$ and $r_{b'c}$ are very large and, at low frequencies, the effects of the capacitances $C_{b'c}$ and $C_{b'e}$ can be neglected. In this case, the π-equivalent circuit may be simplified (see Fig. 5.4.3b), and the resistance $r_{b'e}$ be related to $r_e$ or $g_m$. Using the equivalent circuit shown in Fig. 5.4.3b, we find that

$$i_c \approx g_m v_{b'e} \tag{5-4-8}$$

$$v_{b'e} \approx i_b r_{b'e} \tag{5-4-9}$$

Substituting eq. (5-4-10) into eq. (5-4-9), we obtain

$$r_{b'e} \approx \frac{i_c}{i_b} \frac{1}{g_m} \approx \frac{h_{fe}}{g_m} = \frac{\beta}{g_m} \tag{5-4-10}$$

Another useful equivalent circuit is a **T-equivalent circuit** ( Fig. 5.4.4).

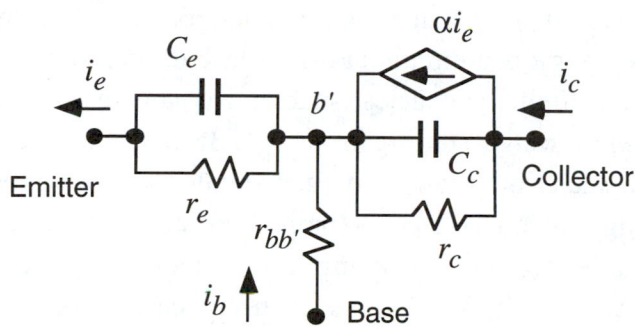

**Fig. 5.4.4.** T-equivalent circuit.

The **emitter capacitance**, $C_e$, in Fig. 5.4.4 is approximately equal to the sum of the diffusion capacitance of the emitter-base junction, $C_{edif}$, and the depletion capacitance, $C_{ed}$. The resistance $r_c$ in the T-equivalent circuit (Fig. 5.4.4) describes the Early effect (see Section 5.2).

Equations relating the parameters of the T- and the $\pi$-equivalent circuits at low frequencies to the $h$-parameters are given in Tables 5.4.2 and 5.4.3.

| Parameter | Expressed through $h$-parameters for common-emitter configuration | Expressed through $h$-parameters for common-base configuration |
|---|---|---|
| $r_e$ | $h_{re}/h_{oe}$ | $h_{ib}-h_{rb}\left(1+h_{fb}\right)/h_{ob}$ |
| $r_{bb'}$ | $h_{ie}-h_{re}\left(1+h_{fe}\right)/h_{oe}$ | $h_{rb}/h_{ob}$ |
| $r_c$ | $\left(1+h_{fe}\right)/h_{oe}$ | $\left(1-h_{rb}\right)/h_{ob}$ |
| $\alpha$ | $h_{fe}/\left(1+h_{fe}\right)$ | $-h_{fb}$ |
| $\beta$ | $h_{fe}$ | $-h_{fb}/\left(1+h_{fb}\right)$ |

**Table 5.4.2.** Parameters of T- and the $\pi$-equivalent circuits expressed through $h$-parameters.

| $h$-parameter | Relation to parameters of hybrid $\pi$-equivalent circuit |
|---|---|
| $h_{oe}$ | $1/\left(r_{b'c}+r_{b'e}\right)+1/r_{ce}+g_m r_{b'e}/\left(r_{b'c}+r_{b'e}\right)$ |
| $h_{ie}$ | $r_{bb'}+r_{b'e}r_{b'c}/\left(r_{b'e}+r_{b'c}\right)$ |
| $h_{fe}$ | $g_m r_{b'e}r_{b'c}/\left(r_{b'c}+r_{b'e}\right)$ |

**Table 5.4.3.** $h$-parameters and parameters of hybrid $\pi$-equivalent circuit.

The definitions of the small signal parameters show that these parameters can be introduced using open-circuit and short-circuit conditions at the input and output. However, at high frequencies (such as frequencies corresponding to the microwave frequency range starting from 1 GHz and above), it may be very difficult to realize such conditions. At such frequencies, interconnects between devices can no longer be considered as lumped resistances since their dimensions become comparable to the wavelength of electromagnetic field. These interconnects behave as transmission lines. In this case, small signal parameters called **s-parameters** (or scattering parameters) are used to describe the transistor response. We include the definitions of the s-parameters for reference purposes.

s-parameters are based on the concept of a **matched load**. A matched load is defined as a load that, when attached to a transmission line, does not introduce reflections. The matched load impedance is equal to the complex conjugate of the characteristic impedance of a transmission line, which is typically equal to 50 $\Omega$. The signal that would have been delivered to a matched load is called the incident signal. However, the actual load (such as the input loop of a BJT) is not always matched to the impedance of the signal source (be it a transmission line at high frequency or just a usual voltage source with a certain source resistance at low frequencies). The difference between the actual signal and the incident signal is called the reflected signal. Thus, the input and output voltages are presented as

$$i_1 = i_{1i} + i_{1r} \qquad i_2 = i_{2i} + i_{2r}$$
$$v_1 = v_{1i} + v_{1r} \qquad v_2 = v_{2i} + v_{2r}$$

(5-4-11)

We now denote the incident signals for the input and output ($i_{1i}$, $v_{1i}$ and $i_{2i}$, $v_{2i}$) as $a_1$ and $a_2$, respectively, and the reflected signals for the input and output ($i_{1r}$, $v_{1r}$ and $i_{2r}$, $v_{2r}$) as $b_1$ and $b_2$, respectively. Choosing the incident signals as independent variables and the reflected signals as dependent variables, we introduce the scattering parameters:

$$b_1 = s_{11}a_1 + s_{12}a_2$$
$$b_2 = s_{22}a_1 + s_{21}a_2$$

(5-4-12)

In this definition, it does not really matter whether we choose currents or voltages as $a$'s and $b$'s since the dimensionless $s$-parameters are defined only through their ratios.  The s-parameters can be expressed through $a$'s and $b$'s using eq. (5-4-12):

$$s_{11} = \frac{b_1}{a_1}\bigg|_{a_2 = 0} \tag{5-4-13}$$

is called the **input reflection ratio**,

$$s_{12} = \frac{b_1}{a_2}\bigg|_{a_1 = 0} \tag{5-4-14}$$

is called the **reverse transmission ratio**,

$$s_{21} = \frac{b_2}{a_1}\bigg|_{a_2 = 0} \tag{5-4-15}$$

is called the **forward transmission ratio**, and

$$s_{22} = \frac{b_2}{a_2}\bigg|_{a_1 = 0} \tag{5-4-16}$$

is called the **output reflection ratio**.  These parameters can be measured or calculated using SPICE [see, for example, Tuinenga (1992)].  Once the $s$-parameters are known they can be used for describing the transistor behavior at high frequencies [see, for example, Sze (1981)].

The summary of important equations related to small signal parameters of a BJT is given in Table 5.4.4.

| Definition of $h$-parameters | $v_1 = h_{11}i_1 + h_{12}v_2 \quad i_2 = h_{21}i_1 + h_{22}v_2$ |
|---|---|
| Dynamic resistance of the forward-biased emitter-base junction | $r_e = \dfrac{\partial V_{b'e}}{\partial I_e} \approx \dfrac{V_{th}}{I_e}$ |
| Transconductance | $g_m = \dfrac{\partial I_c}{\partial V_{b'e}} = \alpha\dfrac{\partial I_e}{\partial V_{b'e}} \approx \dfrac{\alpha}{r_e} \approx \dfrac{I_c}{V_{th}}$ |
| Resistance in the hybrid-$\pi$ equivalent circuit | $r_{b'e} \approx \dfrac{i_c}{i_b}\dfrac{1}{g_m} \approx \dfrac{h_{fe}}{g_m} = \dfrac{\beta}{g_m}$ |

**Table 5.4.4.** Summary of important equations related to BJT small signal parameters.

## 5-5. SMALL SIGNAL OPERATION AND CUTOFF FREQUENCIES

We can now analyze BJT operation in a small signal amplifier based on BJT equivalent circuits considered in Section 5.4.    In a typical circuit, the transistor operating point is chosen using **biasing resistors** (resistors $R_{B1}$ and $R_{B2}$ in Fig. 5.5.1).  For the circuit shown in Fig. 5.5.1, the base dc voltage (with respect to the ground) is equal to $V_{cc}R_{B1}/(R_{B1} + R_{B2})$ if the common-emitter gain, $\beta$, is large.  The operating point also depends on the **emitter series resistance**, $R_E$. The purpose of this resistance is to make the operating point more stable with respect to temperature variations.   (How this is achieved is analyzed in all introductory electronics textbooks.) The circuit shown in Fig. 5.5.1 is a one-stage amplifier, which could be a part of a larger circuit.  In this figure, a part of the circuit providing the input signal is represented by its Thevenin equivalent (represented by a source resistance, $R_s$, and a voltage source, $V_s$) and the part of the circuit receiving the output signal is represented by the load impedance. **Coupling capacitances** ($C_1$ and $C_2$ in Fig. 5.5.1) are used to connect the transistor stage to the rest of the circuit and to isolate the dc bias and ac signal.  A bypass capacitor, $C_E$, shunts the resistance, $R_E$, for the ac signal.

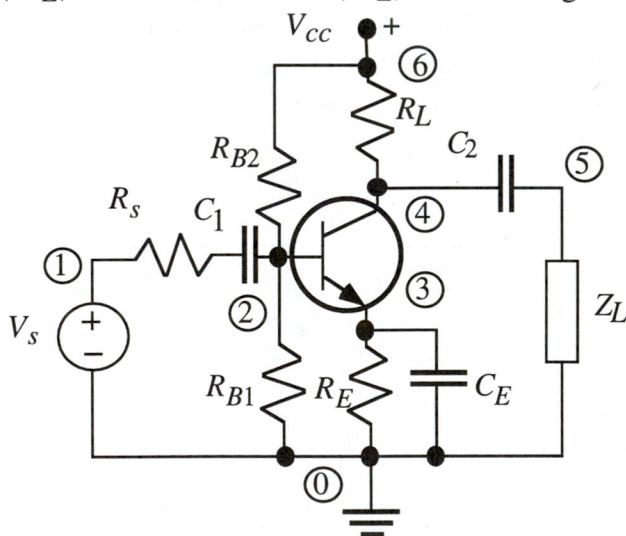

**Fig. 5.5.1.** Transistor amplifier in a common-emitter configuration. The operating point is chosen using biasing resistances, $R_{B1}$ and $R_{B2}$. Node numbers are used in the SPICE simulation later on in this section.

Here we only consider midfrequencies, which are high enough so that the coupling and bypass capacitances have very small impedances and can be substituted by shorts but are still much smaller than the BJT cutoff frequencies so that the internal transistor capacitances can be considered as open circuits. In this midfrequency range, the small signal transistor operation may be analyzed using one of the equivalent circuits discussed in Section 5.4.

Once a small signal equivalent circuit of a BJT is established, an elementary circuit theory can be used to calculate the current, voltage, and power gains as well as the input and output impedances of the BJT (see Fig. 5.5.2). Table 5.5.1 relates these gains and impedances to the $h$-parameters. Substituting the $h$-parameters for different transistor biasing configurations into the expressions given in Table 5.5.1, we can compare common-emitter, common-base, and common-collector configurations. The results of such a comparison for a typical BJT are summarized in Table 5.5.2.

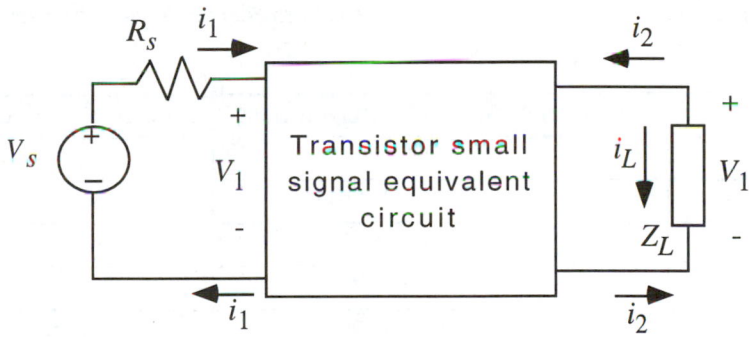

**Fig. 5.5.2.** Small signal equivalent circuit of transistor amplifier with source and load at midband.

As can be seen from Table 5.5.2, the common-emitter and common-collector configurations have a high current gain. The common-emitter configuration has a high voltage gain, the largest power gain, and medium input and output impedances. The common-collector configuration has a low voltage gain, large input impedance, and small output impedance. The common-base configuration has a high voltage gain, low input impedance, and high output impedance. For many applications, the common-emitter configuration has apparent advantages. However, among the trade-offs to consider when choosing

|  | Definition | Relation to $h$-parameters |
|---|---|---|
| Input impedance | $Z_i = v_1/i_1$ | $Z_i = h_i - h_f h_r /(h_o + Y_L)$ |
| Output admittance | $Y_o = i_2/v_2$ | $Y_o = h_o - h_f h_r /(h_i + R_s)$ |
| Voltage gain | $A_v = v_2/v_1$ | $A_v = A_i Z_L/Z_i$ |
| Voltage gain | $A_{vs} = v_2/v_s$ | $A_{vs} = A_v Z_i/(Z_i + R_s)$ |
| Current gain | $A_i = i_L/i_1 = -i_2/i_1$ | $A_i = h_f /(1 + h_o Z_L)$ |
| Current gain | $A_{is} = -i_2/i_s$ | $A_{is} = A_i R_s/(Z_i + R_s)$ |
| Power gains | $A_p = A_v A_i, A_{ps} = A_{vs} A_{is}$ | |

**Table 5.5.1.** Voltage, current, and power gains and input and output impedances (from Shur, 1990, copyright © Prentice Hall, 1990, reproduced by permission of Prentice Hall, Inc., Englewood Cliffs, NJ). $v_s$ is ac source voltage, $i_s = v_s/R_s$, $Z_L$ is load impedance, $R_s$ is source series resistance.

|  | Common emitter | Common base | Common collector |
|---|---|---|---|
| $h_i$ ($\Omega$) | 1250 | 16.4 | 1250 |
| $h_r$ | $4 \times 10^{-4}$ | $1.1 \times 10^{-5}$ | 1 |
| $h_f$ | 75 | -0.987 | -76 |
| $h_o$ ($\mu$mho) | 25 | 0.329 | 25 |
| $Z_i$ (k$\Omega$) | 1.19 | 0.0164 | 146 |
| $Y_o$($\mu$mho) | 15.77 | 0.334 | 24350 |
| $A_i$ | -71.43 | 0.986 | 72.38 |
| $A_{is}$ | -44.72  (33 dB) | 0.977 | 0.978 |
| $A_v$ | -119.76  (41.6 dB) | 120.1 | 0.991 |
| $A_{vs}$ | -44.72  (33 dB) | 0.977 | 0.978 |

**Table 5.5.2.** $h$–parameters and $A_i$, $Z_i$, $A_v$, $Y_o$, $A_{vs}$, and $A_{is}$ for Motorola 2N2219A transistor (from Shur, 1990, copyright © Prentice Hall, 1990, reproduced by permission of Prentice Hall, Inc., Englewood Cliffs, NJ). $R_s = 2$ k$\Omega$, $R_L = 2$ k$\Omega$, $I_c = 10$ mA (dc), $V_{ce} = 10$ V (dc), $f = 1$ kHz. (Data from *Small-Signal Transistor Data*, Motorola Semiconductor Products, Inc., Phoenix, AZ, 1983). [In decibels, $A_v$(dB)$=20\log_{10}|V_{out}/V_{in}|$, $A_i$(dB)$=20\log_{10}|i_{out}/i_{in}|$, $A_p$(dB)$=10\log_{10}|P_{out}/P_{in}|$.]

the biasing configuration for a BJT are a smaller breakdown voltage and a higher leakage current for the common-emitter configuration.

Let us now consider limitations of BJT operating frequencies imposed by time delays related to $RC$ time constants of a BJT equivalent circuit. Using the T-equivalent circuit for the common-base configuration shown in Fig. 5.4.4, we obtain the following expression for the ac emitter current (see Problem 5-5-1):

$$i_e = v_{b'e}(1 + j\omega C_e r_e)/r_e = i_{eo}(1 + j\omega C_e r_e) \tag{5-5-1}$$

where $i_{eo}$ is the ac current through resistance $r_e$. Hence, the common-base current gain $\alpha_\omega$ at the frequency $\omega$ is given by

$$\alpha_\omega = i_c / i_e = i_c /[i_{eo}(1 + j\omega C_e r_e)] = \alpha/(1 + j\omega C_e r_e) \tag{5-5-2}$$

where $\alpha$ is the common-base current gain at $\omega = 0$. Equation (5-5-2) may be rewritten as

$$\alpha_\omega = \alpha/(1 + j\,\omega/\omega_\alpha) \tag{5-5-3}$$

where

$$\omega_\alpha = 2\pi f_\alpha = 1/(C_e r_e) \tag{5-5-4}$$

is called the **alpha cutoff frequency**.

In a similar fashion, we can introduce the **beta cutoff frequency** by considering the common-emitter current gain. Using a simplified hybrid-$\pi$ equivalent circuit with the output shorted (see Fig. 5.5.3), we first express the ac input (base) current in terms of $v_{b'e}$:

$$i_b = v_{b'e}[g_{b'e} + j\omega(C_{b'e} + C_{b'c})] \tag{5-5-5}$$

where $g_{b'e} = 1/r_{b'e} = h_{fe}/r_e$ (see Section 5.4). Since $i_c = g_m v_{b'e}$, we obtain the short-circuit emitter current gain

$$\beta_\omega = \frac{i_c}{i_b} = \frac{g_m}{g_{b'e} + j\omega(C_{b'e} + C_{b'c})} = \frac{\beta}{\left(1 + j\omega/\omega_\beta\right)} \tag{5-5-6}$$

where

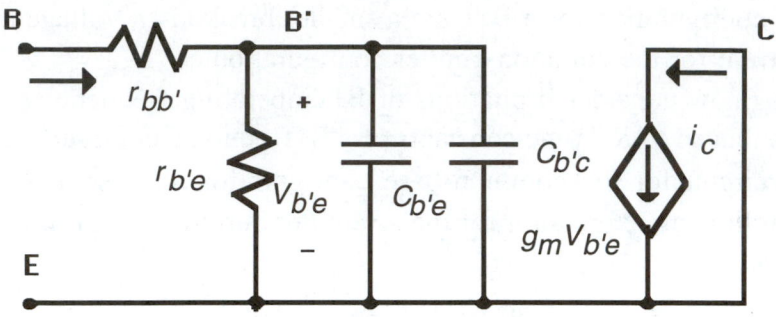

**Fig. 5.5.3.** Simplified hybrid-$\pi$ equivalent circuit for calculation of $\beta_\omega$.

$$\omega_\beta = 2\pi f_\beta = g_{b'e}/(C_{b'e} + C_{b'c}) \qquad (5\text{-}5\text{-}7)$$

is the **beta cutoff frequency**. Equation (5-5-7) may be rewritten as

$$f_\beta = g_m/\left[2\pi h_{fe}(C_{b'e} + C_{b'c})\right] \qquad (5\text{-}5\text{-}8)$$

Since, in most cases, $C_{b'e} \gg C_{b'c}$,

$$f_\beta \approx f_\alpha/h_{fe} \qquad (5\text{-}5\text{-}9)$$

The **cutoff frequency**, $f_T$, is defined as the frequency at which the magnitude of the short-circuit common-emitter current gain equals unity, that is, $|\beta_\omega| = 1$. From eqs. (5-5-6) and (5-5-9), we obtain

$$f_T = f_\beta\sqrt{\beta^2 - 1} \approx f_\beta h_{fe} \approx \frac{g_m}{2\pi(C_{b'e} + C_{b'c})} \qquad (5\text{-}5\text{-}10)$$

where $\beta = h_{fe}$ is the low-frequency, short-circuit, common-emitter current gain. Since usually $C_e \gg C_{b'c}, f_T \approx f_\alpha$.

At frequencies much larger than $f_\beta$ but much smaller than $f_T$,

$$\beta_\omega \approx \beta\omega_\beta/(j\omega) \qquad (5\text{-}5\text{-}11)$$

and, hence,

$$f_T \approx f_\beta h_{fe} \approx f|\beta_\omega| \qquad (5\text{-}5\text{-}12)$$

This equation is often used for deducing $f_T$ from the measured values of $\beta_\omega$ at high frequencies.

The above expressions are useful for a crude estimate of cutoff frequencies. However, we should remember that a BJT is a complex device, and a simple linear equivalent circuit consisting of lumped elements – resistances, capacitances, and controlled current sources – may not be accurate enough, especially at high frequencies. For a more accurate calculation of $f_T$ we have to account for additional time delays that are not represented by in the simple equivalent circuit. One such delay is associated with the collector depletion layer transit time

$$\tau_{cT} \approx \frac{x_{dcb}}{v_{sn}} \tag{5-5-13}$$

where $x_{dcb}$ is the width of the collector-base depletion region and $v_{sn}$ is the electron saturation velocity (for $n$-$p$-$n$ transistors). Another important delay is associated with the collector charging time, $\tau_c$, related to the collector series resistance, $r_{cs}$,

$$\tau_c = r_{cs} C_{b'c} \tag{5-5-14}$$

Finally, a parasitic capacitance, $C_p$, should be added to the collector capacitance, $C_{b'c}$. Hence, the cutoff frequency can be estimated as

$$f_T \approx \frac{1}{2\pi\tau_{eff}} \tag{5-5-15}$$

where

$$\tau_{eff} = \tau_e + \tau_c + \tau_{cT} \tag{5-5-16}$$

is the effective delay time and

$$\tau_e = \frac{C_{b'e} + C_{b'c} + C_p}{g_m} \approx \frac{\left(C_{b'e} + C_{b'c} + C_p\right) V_{th}}{I_c} \tag{5-5-17}$$

Since the emitter capacitance is primarily the emitter-base diffusion capacitance, $C_{edif}$, of the forward-biased emitter-base junction, the base transit time,

$W^2/(2D_n)$ (see Section 4.4), is included in the $C_{b'e}/g_m$ term.

  In most transistors, $\tau_e$ is an important or even the dominant contribution to the total delay time.  As can be seen from eq. (5-5-17), this time can be reduced by increasing the collector current.  However, at very large collector currents, high injection leads to the decrease of $f_T$ (see Fig. 5.5.4).

  An analysis of the above shows that narrow emitter stripes (small device areas) (since this reduces the emitter capacitance), large emitter and collector currents [see eq. (5-5-17)], very thin base regions (which lead to a smaller emitter-base diffusion capacitance), and low parasitic capacitances help achieve a high cutoff frequency.  A typical Si BJT may have a cutoff frequency on the order of 50 to 250 MHz.  High-frequency Si BJTs have a cutoff frequency on the order of 10 to 30 GHz.  Very high cutoff frequencies (over 300 GHz) have been achieved in **Heterojunction Bipolar Transistors** (see Section 5.6).

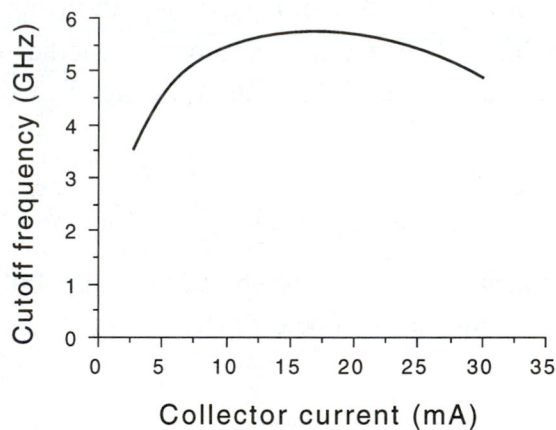

**Fig. 5.5.4.**  Dependence of $f_T$ on the collector current for a Motorola high-frequency silicon *n-p-n* transistor 2N6603 (from *Small-Signal Transistor Data*, Motorola Semiconductor Products, Inc., Phoenix, AZ, 1983).

  The frequency at which the power gain of the transistor is equal to unity under optimum matching conditions for the input and output impedances is called the **maximum oscillation frequency**, $f_{max}$.  Using a simplified $\pi$-equivalent circuit where we neglect $r_{b'c}$ and $r_{ce}$, we obtain

$$f_{max} = \sqrt{\frac{f_T}{8\pi r_{bb'} C_{b'c}}} \tag{5-5-18}$$

This result points to the very important role played by the base spreading resistance. In order to decrease this resistance, the base should be doped higher (but this leads to a smaller common-emitter current gain) or the base should be made wider (which, on the other hand, introduces an additional delay and, hence, reduces $f_T$ and $f_{max}$.) Probably the best way to address this dilemma is to use a Heterojunction Bipolar Transistor (HBT), where a high doping in the base can be combined with a high gain (see Section 5.6).

### Example 5-5-1.

Simulate the base and collector current waveforms (i. e., dependencies of these currents on time) and input and output voltage waveforms, and calculate the dependence of the voltage gain on frequency for the one stage BJT amplifier shown in Fig. 5.5.1 using PSpice$^{tm}$. Use the following values of parameters: bias voltage $V_{cc} = 15V$, input signal frequency, $f = 2$ kHz (for the waveform simulation), circuit parameters $C_1 = 20$ $\mu$F, $R_{B1} = 120$ k$\Omega$, $R_{B2} = 20$ k$\Omega$, $R_C = 10$ k$\Omega$, $R_L = 2$ M$\Omega$, $R_E = 2$ k$\Omega$, $C_E = 15$ $\mu$F, $C_2 = 20$ $\mu$F, transistor parameters $\beta_f = 60$, $r_b = 100$ $\Omega$, $C_{jc} = 10$ pF. **Hint:** For the calculation of the voltage gain, assume the voltage of the input ac source to be equal to one Volt. Then the output voltage calculated using an ac analysis option will be numerically equal to the voltage gain.

**Solution:**

We use the following SPICE file for this simulation:

```
Small Signal BJT Amplifier
Vcc 6 0 DC 15V
Vs 1 0 ac 1V sin(0 0.01V 2KHz)
C1 1 2 20u
RB1 6 2 120k
RB2 2 0 20k
RC 6 4 10k
RL 5 0 2MEG
RE 3 0 2K
CE 3 0 15u
C2 4 5 20u
Q1 4 2 3 BJT
.model BJT NPN (bf=60 rb=100 cjc=10p)
.ac dec 10 0.1 100Meg
.tran 0.1ms 1ms .005ms
.probe
.end
```

(The nodes used in this file are shown in Fig. 5.5.1.) Notice that the BJT model parameters *cjc* and *rb* are different from the default values, which do not account for internal transistor capacitances and, hence, do not allow us to reproduce the decrease of the current and voltage gains at high frequencies.

Fig. 5.5.5 shows the results of PSpice$^{tm}$ simulation obtained using this input file. The base dc current is approximately 10.5 μA. The common-emitter current gain is large and positive (60). The voltage gain is large and negative (since an increase in the collector current increases the voltage drop across $R_c$ and, hence, decreases the voltage drop across $R_L$).

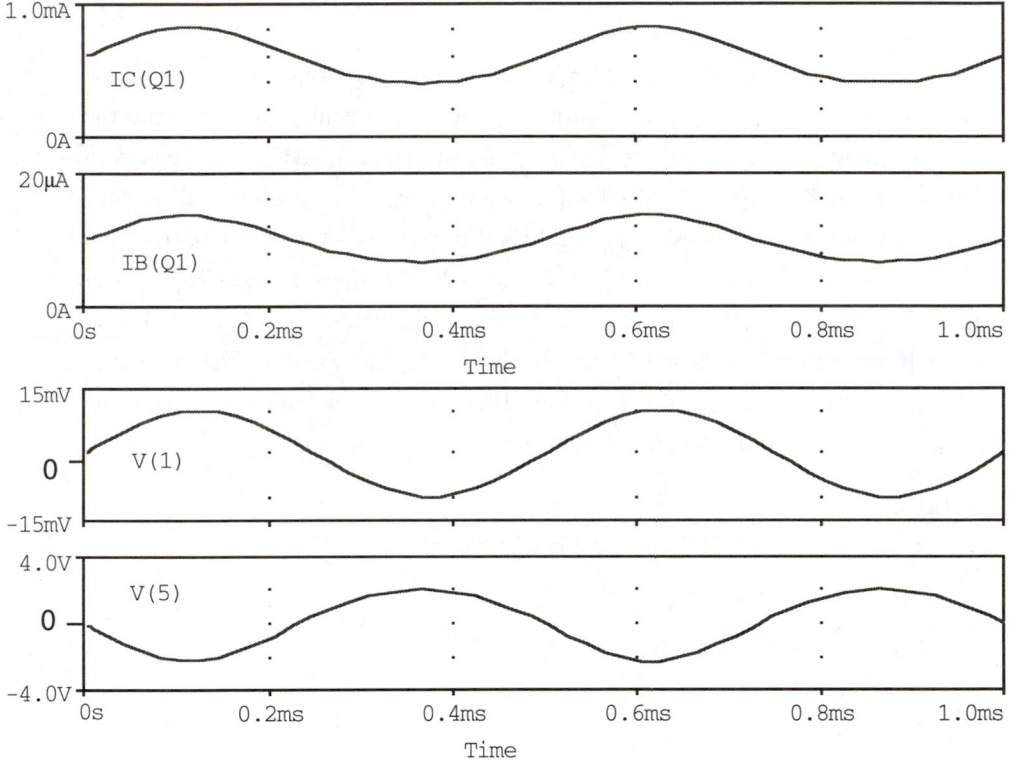

**Fig. 5.5.5.** Input and output current and voltages for one-stage common emitter BJT amplifier.

The voltage gain of the one-stage common emitter BJT amplifier simulated by PSpice$^{tm}$ is shown in Fig. 5.5.6. The fall-off of the gain at low frequencies is caused by the coupling and emitter shunt capacitances, $C_1$ and $C_E$. The decrease of the gain at high frequencies is related to the transistor cutoff frequency. The slope of the this decrease is 20 dB/decade. This is consistent with eq. (5-5-12), which predicts that, at high frequencies, the increase in frequency by one decade decreases the gain by a factor of 10 (i. e., 20 dB/decade). This example shows both the capabilities and the relative ease of using SPICE for a circuit analysis. At the same, we see that we indeed require a basic understanding of device physics so that we can choose SPICE parameters in an intelligent way, since probably in most practical cases using default SPICE parameters may lead to wrong and misleading results.

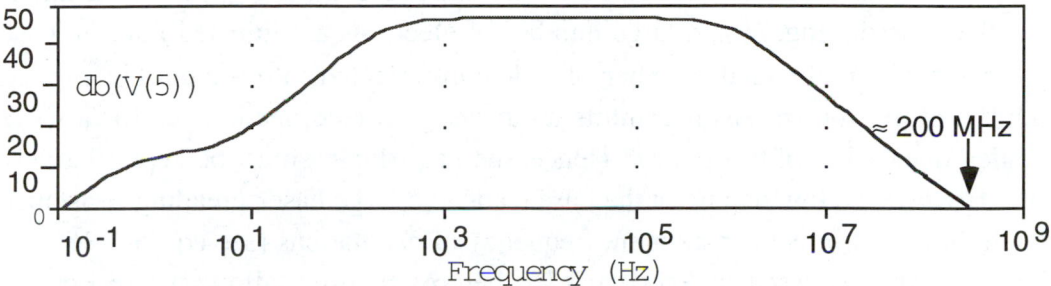

**Fig. 5.5.6.** Voltage gain for one-stage common emitter BJT amplifier.

The summary of important equations related to BJT frequency response is given in Table 5.5.3.

| Common-base current gain | $\alpha_\omega = \alpha/(1 + j\,\omega/\omega_\alpha)$ where $\omega_\alpha = 2\pi f_\alpha = 1/(C_e r_e)$ |
|---|---|
| Common-emitter current gain | $\beta_\omega = \beta/(1 + j\,\omega/\omega_\beta)$ where $\omega_\beta = 2\pi f_\beta = g_{b'e}/(C_{b'e} + C_{b'c})$ |
| Cutoff frequency | Crude estimate: $f_T \approx \dfrac{g_m}{2\pi(C_e + C_{b'c})} \approx \dfrac{g_m}{2\pi C_e}$  More accurate equation: $f_T \approx \dfrac{1}{2\pi\tau_{eff}}$ where $\tau_{eff} = \tau_e + \tau_c + \tau_{cT}$, $\tau_e = (C_e + C_{b'c} + C_p)/g_m$, $\tau_{cT} \approx x_{dcb}/v_{sn}$, $\tau_c = r_{cs} C_{b'c}$ |
| Maximum oscillation frequency, $f_{max}$ | $f_{max} = \sqrt{\dfrac{f_T}{8\pi r_{bb'} C_{b'c}}}$ |

**Table 5.5.3.** Important equations related to BJT frequency response.

## *5-6.  HETEROJUNCTION BIPOLAR TRANSISTORS

In an $n$-$p$-$n$ BJT, electrons are injected into the base, and holes are injected into the emitter region in response to the same forward bias voltage, $V_{be}$, across the emitter-base junction. However, since the emitter doping, $N_{de}$, is much larger than the base doping, $N_{ab}$, a large number of electrons are injected into the base whereas a relatively small number of holes are injected into the emitter region, and, therefore, the transistor exhibits a current gain (see Section 5.1 for a more detailed discussion of this point). Hence, the base doping must be kept relatively low. However, a low doping in the base leads to a large base spreading resistance which, in turn, limits the maximum frequency of oscillations [see eq. (5-5-18)].

As we discussed in Section 4.10, heterostructures allows us to control electron and hole flows separately by creating different barriers for electrons and holes caused by the conduction and valence band discontinuities at the heterointerface. In 1951, William Shockley proposed the idea of the **Heterojunction Bipolar Transistor** (**HBT**) with a wide band gap emitter and a more narrow band gap base for such purpose. Herbert Kroemer developed this idea further in 1957 and in the early 1980s. For many years, this idea was not practical but, with an advent of the Molecular Beam Epitaxy (MBE) technology discussed in Section 4.10, HBTs emerged as prime contenders for devices operating in microwave and millimeter wave range.

Their principle of operation becomes clear if we analyze the dependence of the common-emitter current gain, $\beta$, on doping levels and energy band discontinuity at the emitter-base interface:

$$\beta = \frac{I_c}{I_b} < \beta_{max} = \frac{D_n N_{de} X_e}{D_p N_{ab} W} \exp\left(-\frac{\Delta E_g}{k_B T}\right) \qquad (5\text{-}6\text{-}1)$$

where $I_c$ and $I_b$ are the collector and base currents, respectively; $\beta_{max}$ is the maximum value of the common-emitter current gain; $D_n$ and $D_p$ are hole and electron diffusion coefficients, respectively, $N_{de}$ and $N_{ab}$ are the donor concentration in the emitter region and acceptor concentration in the base region, respectively; $X_e$ is the width of the emitter region (from the base boundary to the emitter contact); $W$ is the base width (from the emitter boundary to the collector boundary); and $\Delta E_g = E_{gb} - E_{ge}$, $E_{gb}$ and $E_{ge}$ are the energy band gaps in the base and emitter regions. [Compare eq. (5-6-1) with eq. (5-1-17), where the exponential factor, $\exp(-\Delta E_g/k_BT)$ was not included, since we assumed that $E_{gb} + E_{ge}$.]   This factor appears because the emitter current is proportional to

$$n_{ib}^2 = \sqrt{N_{cb}N_{vb}}\ \exp\left(-\frac{E_{gb}}{k_BT}\right)$$   and   the   base   current   is   proportional   to

$$n_{ie}^2 = \sqrt{N_{ce}N_{ve}}\ \exp\left(-\frac{E_{ge}}{k_BT}\right)$$   [see Example 5-1-1 and eq. (5-1-6)].   Even in a

conventional transistor, $\Delta E_g$ is not equal to zero because of the energy band narrowing [see eq.(5-1-28)], but there it has a positive sign and reduces the maximum achievable gain.  As can be seen from the energy diagram of an HBT shown in Fig. 5.6.1, an HBT, $\Delta E_g < 0$, and we can have it both: a high gain [since the factor $\exp(-\Delta E_g/k_BT)$ can be very large] and a high base doping and, hence, a low base spreading resistance, since this factor will more than compensate for an unfavorable $N_{de}/N_{be}$ ratio.

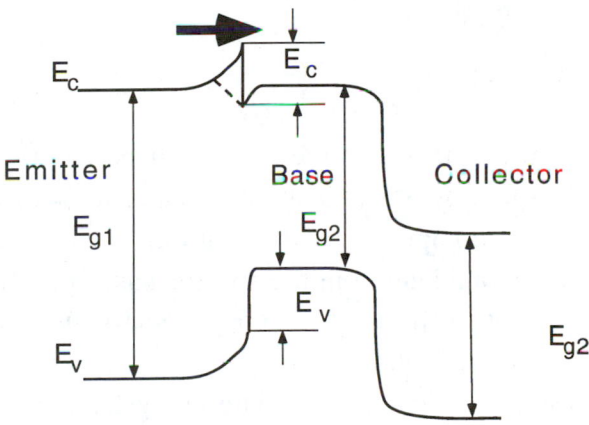

**Fig. 5.6.1.**  Schematic HBT band structure.

The conduction band discontinuity may present an impediment to the electron current flow, creating a potential barrier for the electrons entering the base from the emitter region.  On the other hand, as was pointed out by Kroemer in 1982, electrons in the "spike-notch" region of the conduction band shown in Fig. 5.6.1 enter the base from the emitter with very large kinetic energies (close to $\Delta E_c$) and, as a consequence, may have very high velocities, up to several times $10^5$ m/s.  (A large arrow in Fig. 5.6.1 represents such hot electrons injected into the base.)  The high electron velocity may reduce the base transit time and may increase an HBT cutoff frequency.

$Al_xGa_{1-x}As$/GaAs is a typical heterostructure material system used in HBTs and other heterostructure devices. $Al_xGa_{1-x}As$ has an energy band gap that varies linearly with molar fraction of Al, $x$, and is wider than the energy band gap for GaAs ($x = 0$); see Fig. 4.10.2. Hence, the magnitude of the conduction band spike in Fig. 5.6.1 can be reduced by grading the composition of the wide band gap emitter near the heterointerface (i. e., by smoothly reducing x with distance near the emitter-base interface). The resulting band diagram is shown in Fig. 5.6.1 by the dashed line.

Since the emitter injection efficiency of an HBT is very high, the transistor gain in HBTs, β, is limited primarily by the recombination current. In addition to the recombination in the base and recombination in the emitter base depletion region, the recombination current in HBTs includes the surface recombination at the emitter edges. (In GaAs, the surface recombination velocity may be quite high, on the order of $10^6$ cm/s.) Hence, the HBT common emitter current gain, β, can be estimated as follows:

$$\beta = \frac{j_{ne}}{j_{re} + j_{rb} + J_{rs} / Z} \tag{5-6-2}$$

where $S = LZ$ is the emitter area, $L$ is the emitter stripe length (long dimension), and $Z$ is the emitter stripe width (narrow dimension); $j_{ne}, j_{re}$, and $j_{rb}$ are the densities of the electron component of the emitter current, the recombination current caused by the recombination in the emitter-base space charge region and the recombination current in the base region, respectively; and $J_{rs}$ is the linear density of the surface recombination current.

Many HBTs use the AlGaAs/GaAs material system. AlGaAs/GaAs HBTs demonstrated output powers up to 12 W in X band (6.1 to 8.9 GHz) and more than 200 mW at 40 GHz. More advanced HBTs use heterostructure layers grown on InP. InP based HBTs are expected to reach cutoff frequencies above 500 GHz. Recently, Si/SiGe HBTs have been developed because their fabrication is based on mature silicon bipolar technology. Cutoff frequencies of SiGe HBTs exceed 100 GHz.

HBT modeling is more difficult than modeling conventional BJTs. AIM-Spice has an HBT model with default parameters for AlGaAs/GaAs HBTs [see Lee et al. (1993)]. Appendix A8 contains the instructions on downloading the student version of the AIM-Spice program via the Internet computer network.

# REFERENCES

I. GETREU, *Modeling the Bipolar Transistor*, Tektronix, Inc., part no. 062-2841-00 (1970)

K. LEE, M. SHUR, T. A. FJELDLY, AND T. YTTERDAL, *Semiconductor Device Modeling for VLSI*, Prentice Hall, Englewood Cliffs, NJ (1993)

M. SHUR, *Physics of Semiconductor Devices*, Series in Solid State Physical Electronics, Prentice Hall, Englewood Cliffs, NJ (1990)

S. M. SZE, *Physics of Semiconductor Devices*, Second Edition, John Wiley & Sons, New York (1981)

S. M. SZE, *Semiconductor Devices. Physics and Technology*, John Wiley & Sons, New York (1985)

R. M. WARNER AND B. L. GRUNG, *Transistors Fundamentals for the Integrated-Circuit Engineer*, John Wiley & Sons, New York (1983)

# BIBLIOGRAPHY

B. G. STREETMAN, Solid State Electronic Devices, Fifth Edition, Prentice Hall, Englewood Cliffs, NJ (1995)

*Concise and clear undergraduate text on semiconductor devices.*

G. W. NEUDECK, *The Bipolar Junction Transistors*, Addison Wesley Modular Series on Solid State Devices, Vol. III, Reading, MA (1983)

*Undergraduate text on FETs with many examples and problems.*

S. K. GHANDHI, *VLSI Fabrication Principles*, John Wiley & Sons, New York (1983)

*Detailed description of transistor fabrication technology.*

P. M. ASBECK, M. F. CHANG, K. C. WANG AND D. L. MILLER, "Heterojunction Bipolar Transistor Technology," in *Introduction to Semiconductor Technology. GaAs and Related Compounds*, Cheng T. Wang, Editor, John Wiley & Sons, New York (1990)

*A good description of HBT technology and device characteristics.*

# REVIEW QUESTIONS

**1.** a. Sketch a qualitative band diagram of a $p$-$n$-$p$ BJT for the active forward mode.

☐ 1 point

**2.** Sketch qualitative distributions of holes in the base of a $p$-$n$-$p$ BJT for (a) active forward mode, (b) saturation, (c) cutoff, and (d) reverse active mode.

☐ 4 points

**3.** Figure 5.1.3 is drawn for a typical regime of a BJT operation when the emitter-base junction is forward biased, and the collector-base junction is reverse-biased. How do we know this from the figure?

☐ 1 point

**4.** Figure 5.1.3 corresponds to higher emitter doping than base doping, and higher base doping than collector doping. How do we know this from the figure?

☐ 1 point

**5.** The base spreading resistance of a BJT is 200 $\Omega$. The common-emitter current gain is 100. At what values of the emitter current will the emitter current crowding become important for room temperature operation? Why?

☐ 2 points

**6.** What is the magnitude of the forward bias that increases the concentration of the minority carriers in the base at the emitter-base junction by a factor of a hundred?

☐ 1 point

**7.** Explain the meaning of all the terms in the right-hand side of eq. (5-2-19).

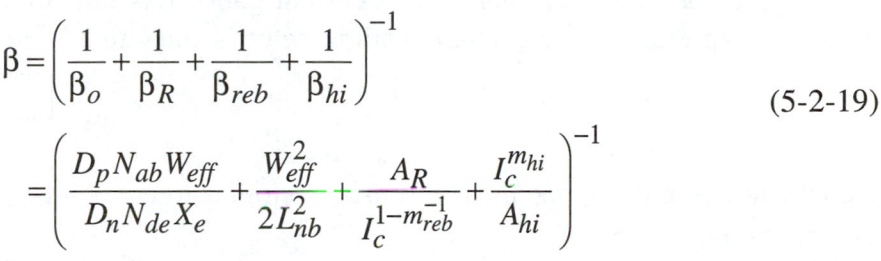

$$\beta = \left( \frac{1}{\beta_o} + \frac{1}{\beta_R} + \frac{1}{\beta_{reb}} + \frac{1}{\beta_{hi}} \right)^{-1}$$

(5-2-19)

$$= \left( \frac{D_p N_{ab} W_{eff}}{D_n N_{de} X_e} + \frac{W_{eff}^2}{2L_{nb}^2} + \frac{A_R}{I_c^{1-m_{reb}^{-1}}} + \frac{I_c^{m_{hi}}}{A_{hi}} \right)^{-1}$$

☐ 4 points

**8.** The collector current in an $n$-$p$-$n$ BJT operating in the common-emitter configuration is equal to 5 mA for the collector-emitter bias of 5 V. The output conductance $g_{ce} = 0.5$ mS. What is the value of the Early voltage?

☐ 1 point

**9.** The common-emitter short-circuit current gain of a BJT is equal to 50. What is the common-base short-circuit current gain?

☐ 1 point

**10.** The common-base short-circuit current gain $\alpha = 0.8$. What is the common-emitter short circuit current gain, $\beta$?

☐ 1 point

**11.** The BJT common-base current gain is $\alpha = 0.96$. What is the ratio of the leakage current, $I_{ceo}$, for the common-emitter configuration with an open base to the leakage current, $I_{cbo}$, for the common-base configuration with an open emitter for this BJT?

☐ 1 point

**12.** The breakdown voltage for the common-emitter circuit configuration is 5 V. The breakdown voltage for the common-base circuit configuration is 15 V. What is the common-base short-circuit current gain? (Assume that the parameter $m_b$ in the expression for the multiplication factor is equal to 3.)

☐ 2 points

**13.** Write down equations for $g$-parameters similar to eqs. (5-4-3) to (5-4-6) for $h$-parameters.

☐ 2 points

**14.** The BJT collector current is maintained at 2 mA in a temperature range from $0^o$ to $150^oC$. Sketch the variation of the BJT transconductance with temperature. Show the scales.

☐ 2 points

**15.** Which is the largest and which is the smallest out of the frequencies $f_\alpha, f_\beta$, and $f_T$? Give the definitions of these frequencies.

☐ 1 point

**16.** The BJT collector current at room temperature is 1 mA. The common-emitter current gain is 20. The emitter capacitance, $C_e = 100$ pF. Estimate the cutoff frequency, $f_T$ (for $C_e \gg C_{b'c}$).

☐ 2 points

**17.** Estimate the factor decreasing the common-emitter short-circuit current gain, $\beta$, caused by an energy gap narrowing of 0.05 eV in a silicon BJT at 300 K and 350 K.

☐ 2 points

**18.** Estimate the factor increasing the common-emitter short-circuit current gain, $\beta$, in an HBT with a wide band gap emitter with an emitter-base valence gap discontinuity of 0.2 eV at 300 K and 350 K.

☐ 1 point

# PROBLEMS

**5-1-1.** Show that for a uniformly doped base of an $n$-$p$-$n$ BJT, the base transport factor is given by $\alpha_T = 1 - \dfrac{W^2}{2L_{nb}^2}$. (Assume that the electron concentration at the emitter-base junction, $n_{be}$, is much greater than the equilibrium concentration of electrons in the base, $n_{bo}$.)

**5-1-2.** Assuming the exponential variation of the acceptor density, $N_a = N_{ao}\exp(-x/\lambda_b)$ in the base of an $n$-$p$-$n$ BJT, show that the base transport factor is given by $\alpha_T = 1 - \dfrac{W^2}{2fL_{nb}^2}$ where $f = W/(2\lambda_b)$ for $\lambda_b \ll W$ and $L_{nb}^2 \gg \lambda_b^2 W^2$.

**\*5-1-3.** Discuss how emitter energy gap narrowing affects the emitter injection efficiency.

**5-1-4.** How may an increase in the emitter doping affect the gain of a bipolar junction transistor? Why? (Describe both beneficial and detrimental effects of such an increase.)

**\*5-2-1.** Design a Si bipolar junction transistor with a common-emitter current gain of more than 100 and a collector-base breakdown voltage of more than 20 V. (Here "design" means choosing reasonable doping levels in the emitter, base, and collector regions and choosing reasonable widths of the emitter, base, and collector regions.)

**5-3-1.** Sketch a qualitative band diagram of a bipolar junction transistor for a reverse collector-base voltage equal to the critical voltage of the punch-through breakdown.

**5-4-1.** Derive the expressions given in Table 5.4.3.

**5-5-1.** Derive the following expression for the ac emitter current, $i_e$:

$$i_e = \frac{v_{b'e}\left(1 + j\omega C_e r_e\right)}{r_e} = \frac{i_{eo}}{1 + j\omega C_e r_e}$$

**5-5-2.** A Si bipolar junction transistor operates at room temperature (300 K). The collector current is 10 mA. The short-circuit common-emitter current gain $\beta = 100$. Estimate the parameters of the simplified transistor small signal equivalent circuit shown in the figure and the $h$-parameters, $h_{ie}$ and $h_{fe}$.

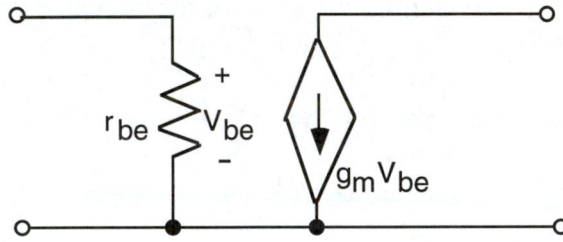

**5-6-1.** List heterostructure material systems suitable for HBTs and discuss advantages and disadvantages of these material systems.

**\*5-6-2.** Compare Si BJTs and GaAs HBTs. What are advantages and disadvantages of these materials for bipolar technology? What are the two most important advantages of HBTs compared to regular BJTs?

# 6

# MOSFETs

## 6-1. PRINCIPLE OF OPERATION

The concept of a **Field Effect Transistor** (FET) was first proposed by Lilienfeld (1930). However, the device became practical only after pioneering work done by William Shockley in the early 1950s. Figure 6.1.1 shows the basic FET structure. An output current in a FET flows between the **drain** and the **source** contacts. This current is controlled by the **gate** bias. Depending on the gate bias, a FET can be either in the **off-state,** with very few free carriers in the transistor **channel** between the source and the drain, or in the **on-state,** when free carriers enter the transistor channel from the source and may flow to the drain. In the on-state, the free carriers in the transistor channel are capacitively coupled with the gate electrode, which effectively forms a parallel plate capacitor with the channel. The gate-to-channel voltage controls the free carrier charge induced into the channel and, hence, the drain current. Since the gate contact is isolated from the channel by the gate insulator, the input impedance is very high, a major advantage compared to Bipolar Junction Transistors.

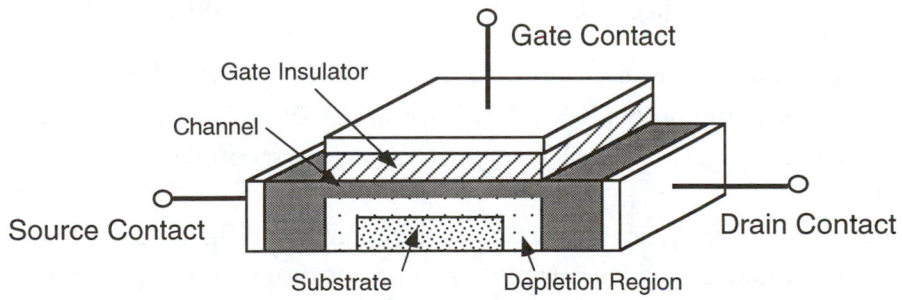

**Fig. 6.1.1.** Schematic illustration of a Field Effect Transistor (FET).

Field Effect Transistors use different semiconductors and different ways of isolating the gate from the channel.  The most important device by far is the silicon **Metal Oxide Semiconductor Field Effect Transistor – MOSFET** (see Fig. 6.1.2).  In a MOSFET, a native silicon oxide ($SiO_2$) is used as a gate dielectric.  MOSFETs can have an $n$-type channel (NMOS) or a $p$-type channel (PMOS).  Figure 6.1.2 shows an NMOS where the channel is created by a free electron gas induced by the gate voltage in the semiconductor at the silicon-silicon dioxide interface and the electrons come from an $n^+$ source and flow into an $n^+$ drain.  These contacts are implanted or diffused into a $p$-type substrate.  As can be seen from Fig. 6.1.2, a MOSFET is actually a four-terminal device with source, gate, drain, and bulk (substrate) contact.  The $n^+$ source and drain contacts and the $n$-type channel are isolated from the $p$-type substrate by the depletion regions formed in the same way as in a $p$-$n$ junction (see Chap. 4).  The substrate bias can control the width of these depletion regions.

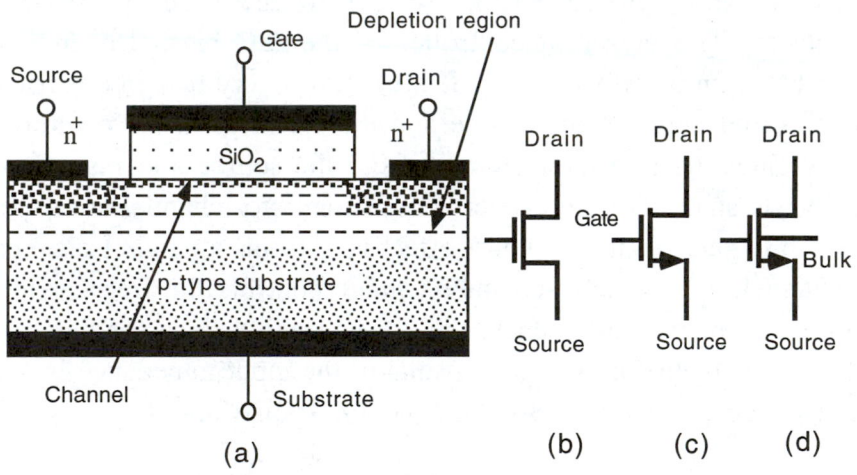

**Fig. 6.1.2.** (a) Schematic structure of $n$-channel MOSFET (NMOS) and circuit symbols for (b) MOSFET, (c) $n$-channel MOSFET, and (d) $n$-channel MOSFET when the bulk (substrate) potential has to be specified in a circuit.

In a PMOS, a free hole gas creates the channel, and the $p^+$ source and drain contacts are implanted or diffused into a $n$-type substrate.  Both $n$-channel and $p$-channel devices can be fabricated on the same **integrated circuit (IC)** chip (see Fig. 6.1.3).  This technology [called **Complementary MOS (CMOS)** technology] provides very low power consumption along with a fairly high speed. It is the most important semiconductor technology. The integration scale for

CMOS and MOSFETs has reached $10^8$ devices on a single chip and continues to increase rapidly.

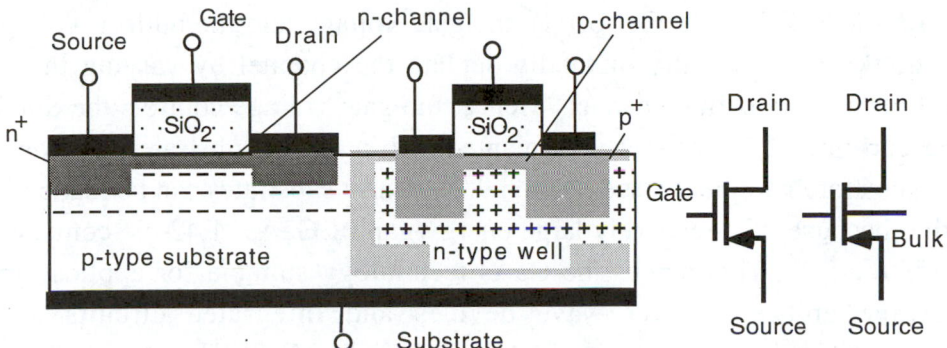

**Fig. 6.1.3.** Schematic structure of Complementary MOSFET (CMOS) and circuit symbols for *p*-channel MOSFET (PMOS). Minuses and pluses show the depletion regions.

CMOS transistors can also be fabricated on the same chip with Bipolar Junction Transistors. This technology is called **BiCMOS**. BiCMOS technology is primarily used in high-speed and high-performance **Very Large Scale Integrated (VLSI)** circuits.

Amorphous silicon and polysilicon **Thin Film Transistors (TFTs)** represent another important and rapidly growing silicon based technology. Amorphous silicon (a-Si) TFTs use a hydrogenated amorphous silicon. This material has a very low electron mobility (hundreds of times smaller than crystalline silicon). Typical gate sizes for a-Si TFTs vary between 3 and 15 μm – an order of magnitude larger or more than for other FETs. However, a-Si can be inexpensively and reliably deposited over very large areas, making this technology ideal for applications in Active Matrix Liquid Crystal Displays (AMLCD) where each dot on the screen (called **pixel**) is driven by its own transistor. TFTs also find applications in imagers, printers, copiers, and consumer electronics. The electron and hole mobilities in polysilicon TFTs are much higher than in a-Si, but their fabrication requires higher temperatures and is more expensive. The TFT technologies will be considered in Chapter 7.

Some other semiconductor materials, such as GaAs, have lighter electrons in the conduction band and, hence, a higher electron mobility and drift velocity. However, they lack a good stable native oxide. In these materials, the depletion region formed by a Schottky contact is often used as a means to isolate the gate from the FET channel. A FET with Schottky contact gate is called a **MEtal Semiconductor Field Effect Transistor – MESFET**.

In a MESFET (see Fig. 6.1.4), the gate voltage and the built-in voltage of the Schottky gate partially or totally deplete the channel by varying the cross section of the conducting channel. Hence, the gate voltage controls the drain-to-source current. The channel is implanted into a GaAs substrate (or grown on such a substrate by epitaxy; see Section 9.2). GaAs substrates can be made highly resistive because of a relatively large energy gap of GaAs (1.42 eV compared to 1.12 eV for Si). This makes the GaAs technology suitable for applications in microwave and millimeter wave devices and integrated circuits and in optoelectronic circuits (see Chap. 8). Digital GaAs MESFET integrated circuits exhibit higher speed and/or lower power consumption than comparable Si integrated circuits. The integration scale for digital GaAs MESFET integrated circuits has exceeded $10^5$ transistors on a chip. (Even though this integration scale is impressive, it is still three orders of magnitude smaller than the integration scale achieved for Si MOSFETs.)

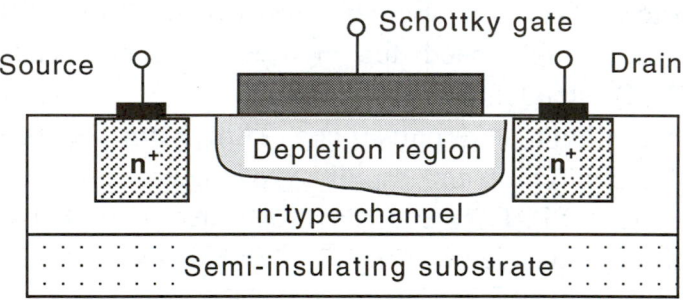

**Fig. 6.1.4.** Schematic MESFET structure.

Since GaAs substrates are usually highly resistive (semi-insulating), GaAs MESFETs often have three terminals (with no contact to the substrate).

Depending on the applied gate-to-source bias, $V_{gs}$, any Field Effect Transistor can be either in the on-state, with a conducting channel between the source and drain, or in the off-state, with practically no conduction between the source and the drain. The gate voltage, $V_T$, separating these two regimes is called the **threshold voltage**. Below the threshold, the conducting channel disappears (it is **pinchedoff**) as illustrated for a MESFET in Fig. 6.1.5. All FETs belong to one of the two types – **enhancement mode** FETs, which are turned off at zero gate bias (they are also called **normally–off** transistors) and **depletion mode** FETs, which are turned on at zero gate bias (they are also called **normally–on** transistors).

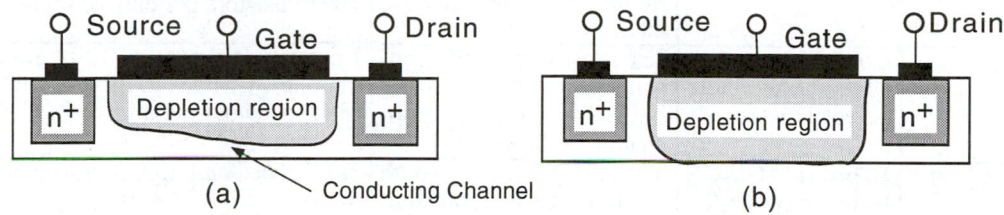

(a)    Conducting Channel                           (b)

**Fig. 6.1.5.** Metal Semiconductor Field Effect Transistor (MESFET). (a) Above threshold ($V_{gs} > V_T$). (b) Below threshold when the conducting channel is pinched off ($V_{gs} < V_T$).

Other important emerging FET technologies include fast compound semiconductor **Heterostructure Field Effect Transistors** uniquely suited for ultra-high-speed and ultra-high-frequency applications. Heterostructure transistors also allow us to develop CMOS-like technology that combines *n*-channel and *p*-channel devices on the same chip. Since this technology may combine a relatively high speed with a low power consumption, it may find applications in wireless communication systems and laptop computers.

FETs made of wide band gap semiconductors, such as SiC and GaN, may be well suited for applications in electronic systems operating in harsh or high-temperature environments and, possibly, in nonvolatile fast electronic memories. However, this potentially important technology is still in its infancy.

In this chapter, we will consider silicon MOSFETs, which are used both as discrete devices and as components of monolithic Integrated Circuits (ICs). The minimum device size in such circuits has dropped below 0.5 micron in commercial IC chips. It is expected that MOSFETs with 0.1 micron gates will become standard by year 2000. Soon Integrated Circuits (ICs) will contain up to 100 million devices or more per chip with the cost per bit of memory below $10^{-4}$ cent. It is estimated that already there are more than 4 million MOSFETs produced annually for every person in the industrialized world. New device structures and the combination of bipolar and field effect technologies may lead to further advances, yet unforeseen.

Table 6.1.1 contains a brief comparison of important FET technologies with emphasis on their features and applications.

| | Description | Comments | Applications |
|---|---|---|---|
| MOSFET | Metal Oxide Field Effect Transistor | The most important solid-state device. Almost all MOSFETs are made from Si and utilize the Si-SiO$_2$ interface | All electronics circuits, Very Large Scale Integrated Circuits (VLSI). Integration scale is up to $10^8$ transistors per chip or higher |
| CMOS | Complementary Metal Oxide Field Effect Transistor | The most important solid-state technology utilizing both $n$-channel and p-channel Si MOSFETs | Primarily low power VLSI. Short channel CMOS exhibits fairly high speed of operation |
| BiCMOS | Bipolar CMOS technology | Technology utilizing both CMOS and Bipolar Junction transistors integrated on the same semiconductor chip | High speed and high performance VLSI |
| a-Si TFTs | Amorphous Silicon Thin Film Transistors | Cheap and reliable technology for very large area integrated circuits (up to a square foot or larger), which uses thin a-Si films deposited on inexpensive glass substrates | Typical gate sizes 3 to 15 μm – an order of magnitude larger than for other FETs. Very slow technology. Applications include flat panel displays, imagers, printers, copiers, consumer products. Probably the second most important semiconductor technology (after CMOS and MOS technology) |
| poly-Si TFTs | Polysilicon Thin Film Transistors | Much better performance and higher speed than for a-Si TFTs but a much more expensive technology requiring higher processing temperatures | Applications include high-resolution projection displays. In the future, this technology may challenge a-Si TFTs |
| MESFETs | MEtal Semiconductor Field Effect Transistors | Primarily made from GaAs, which is a higher speed material than Si. Also, GaAs substrates are highly resistive, which makes this technology suitable for microwave and millimeter wave applications | Applications include microwave and millimeter wave devices and integrated circuits and, more recently, optoelectronic circuits. Digital GaAs MESFET integrated circuits exhibit higher speed and/or lower power consumption than comparable Si integrated circuits. Integration scale up to $10^5$ transistors per chip or higher |
| HFETs | Heterostructure Field Effect Transistors | New important compound semiconductor technology utilizing heterostructures based on material systems, such AlGaAs/GaAs, AlGaAs/InGaAs, GaAs, AlInAs/InGaAs/InP, etc. | Applications include microwave and millimeter wave devices (both power and low noise) and integrated circuits and optoelectronic circuits. Complementary HFETs hold promise of high-speed low-power operation |

**Table 6.1.1.** Important Field Effect Transistor technologies.

## 6-2.  CHARGE CONTROL MODELS

In a Field Effect Transistor, a gate voltage controls the charge of free carriers in the device channel and, hence, the channel current.  This **charge control** concept is very important for understanding how this device works.  In this section, we consider **charge control models** for the silicon MOSFET.  However, similar concepts also apply to other types of FETs.

As was discussed in Section 6.1, in the on-state, the MOSFET gate forms a capacitor with the FET channel at the semiconductor-insulator interface.  In an *n*-channel MOSFET, the substrate is *p*-type silicon.  The depletion regions of the *p-n* junctions formed between the *n*-type channel and the $n^+$ contacts with the *p*-type substrate provide isolation between different devices.  The substrate contact is the fourth contact of a MOSFET (see Fig. 6.2.1), and the substrate potential plays an important role in adjusting the device threshold voltage.

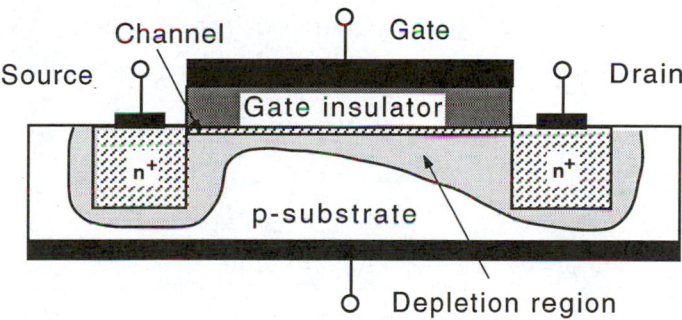

**Fig. 6.2.1.**  Schematic cross section of an *n*-channel MOSFET.  The asymmetry in the shape of the depletion region is caused by the applied drain bias.

A positive gate bias draws the electrons into the channel.  A negative gate bias repels electrons and prevents them from entering the channel from the contacts by establishing a potential barrier between the source and drain contacts.

**Example  6-2-1.**

The electron concentration in the drain and source contact regions of a Si MOSFET is $10^{18}$ cm$^{-3}$.  The electron concentration in the MOSFET channel at zero drain-to-source bias in the subthreshold regime of operation is $10^9$ cm$^{-3}$.  Estimate the barrier height between the source and the drain.

**Solution:**

Figure 6.2.2 shows a qualitative band diagram of the device.

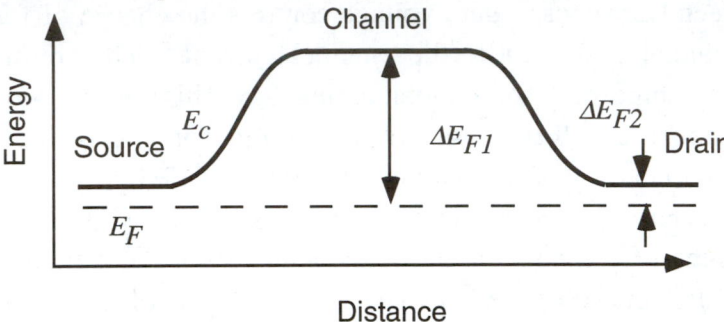

**Fig. 6.2.2.** MOSFET band diagram in subthreshold regime for the direction between source and drain.

As can be seen from the figure, the barrier height, $\Phi_b = \Delta E_{F1} - \Delta E_{F2}$. The relationship between the carrier concentration and the Fermi level is given by

$$n = N_c \exp\left(\frac{E_F - E_c}{k_B T}\right) = N_c \exp\left(-\frac{\Delta E_F}{k_B T}\right)$$

Therefore

$$\Delta E_{F1} = k_B T \ln\left(\frac{N_c}{n_1}\right) \quad \Delta E_{F2} = k_B T \ln\left(\frac{N_c}{n_2}\right)$$

$$\Phi_b = \Delta E_{F1} - \Delta E_{F2} = k_B T \ln\left(\frac{n_2}{n_1}\right) \approx 0.02584 \times \ln\left(10^9\right) = 0.535(eV)$$

In the subthreshold regime, the drain current is a small leakage current since only a small number of electrons can pass over the potential barrier separating the drain and the source. This regime is called the **subthreshold regime** of operation. An increase in a gate voltage in this regime leads to a reduction of the barrier height and, hence, to an exponential increase of this small subthreshold current (see Fig. 6.2.3).

At larger gate voltages, the barrier is reduced so much that electrons can enter the channel from the source and effectively form the second conducting "plate" of the MIS capacitor. At that point, a further increase in the gate voltage leads to a proportional increase of the electronic charge, $qn_s$, in the channel. This

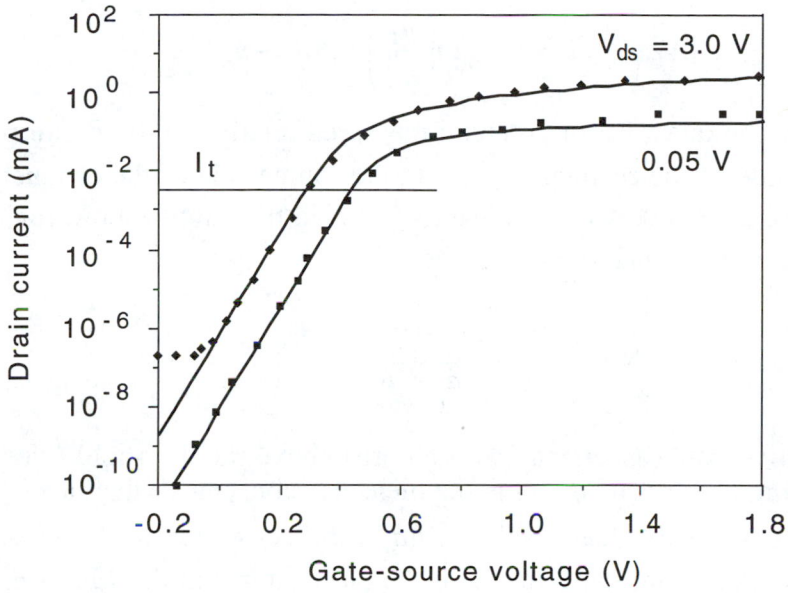

**Fig. 6.2.3**. Measured (dots) and calculated (solid lines) *I-V* characteristics of a submicron *n*-channel MOSFET. $I_t$ is the threshold current. (from M. Shur, T. Fjeldly, T. Ytterdal and K. Lee, "Unified MOSFET Model," *Solid-State Electronics*, Vol. 35, p. 1795, Dec. 1992).

increase can be approximately described by the following equation

$$qn_s = c_i(V_{GS} - V_T) = c_i V_{GT} \qquad (6\text{-}2\text{-}1)$$

(called the equation of the charge control model). Here $q$ is the electronic charge, $n_s$ is the surface carrier concentration in the channel,

$$c_i = \varepsilon_i / d_i \qquad (6\text{-}2\text{-}2)$$

is the gate capacitance per unit area, $V_{gs}$ is the gate-to-source voltage, $\varepsilon_i$ and $d_i$ are the dielectric permittivity and gate oxide thickness, respectively, and $V_T$ is the transistor **threshold voltage**. [Equation (6-2-1) is valid if the drain-to-source voltage is small compared to $V_{GS} - V_T$, so that $n_s$ can still be considered to be nearly independent of the coordinate along the channel.] Below threshold, when $V_{gs} < V_T$, the surface carrier concentration, $n_s$, is an exponential function of $V_{GT}$, as was discussed above and as can be seen from Fig. 6.2.2. Both these regimes can be described by an empirical equation called the **Unified Charge Control Model (UCCM)** equation:

$$V_{GT} - V_F = \eta V_{th} \ln\left(\frac{n_s}{n_o}\right) + a(n_s - n_o) \qquad (6\text{-}2\text{-}3)$$

where $V_F$ is the quasi-Fermi potential measured relative to the Fermi potential at the source side of the channel ($V_F = 0$ at the source side of the channel and $V_F = V_{DS}$ at the drain side), $\eta$ is the ideality factor in the subthreshold regime, $V_{th}$ is the thermal voltage, and $a \approx q/c_a$, where

$$c_a = \frac{\varepsilon_i}{d_i + \Delta d} \qquad (6\text{-}2\text{-}4)$$

is the effective gate capacitance per unit area above the threshold (when the gate voltage swing $V_{GT} \gg \eta V_{th}$), $\varepsilon_i$ is the dielectric constant of the gate insulator, $d_i$ is the thickness of the gate insulator, $\Delta d$ is the correction to the insulator layer thickness related to the capacitance of the electron gas in the MOSFET channel, and

$$n_o = \frac{c_a \eta V_{th}}{2q} \qquad (6\text{-}2\text{-}5)$$

is the surface electron concentration in the channel at $V_{GT} = 0$. The derivation of eq. (6-2-5) is beyond the scope of this book [see Lee et al. (1993)], but the physics behind it is fairly transparent: the threshold concentration is not zero because the finite thermal energy supplies some electrons into the channel even when the source and drain are separated by a potential barrier. For Si MOSFETs, $\Delta d$ is only about 4 Å and can be disregarded for all but the most advanced devices with a very thin insulator layer so that usually $c_a \approx c_i = \varepsilon_i/d_i$.

Equation (6-2-3) is the most important equation used in the FET models in AIM-Spice [see Lee et al. (1993)]. This equation reduces to the charge control model equation at large values of $V_{gs} - V_T \gg V_{th}$ when the term with the logarithm in the right-hand side of this equation becomes small so that

$$V_{GT} - V_F \approx a(n_s - n_o) \qquad (6\text{-}2\text{-}6)$$

When $-V_{GT} \gg V_{th}$, the term with the logarithm becomes dominant:

$$V_{GT} - V_F \approx \eta V_{th} \ln\left(\frac{n_s}{n_o}\right) \tag{6-2-7}$$

so that eq. (6-2-3) describes the subthreshold regime. For zero drain bias, eq. (6-2-3) becomes

$$V_{GT} = \eta V_{th} \ln\left(\frac{n_s}{n_o}\right) + a(n_s - n_o) \tag{6-2-8}$$

Differentiating both sides of this equation with respect to $V_{GT}$, we find the gate-to-channel capacitance, $c_{ch}$:

$$c_{ch} = \frac{c_a}{1 + \dfrac{c_a \eta V_{th}}{q n_s}} \tag{6-2-9}$$

The calculated dependencies of $c_{ch}$ on $V_{GT}$ are shown in Fig. 6.2.4. Above threshold, $c_{ch}$ approaches $c_a$. Below threshold, it decreases exponentially, dropping down to a certain parasitic capacitance, which, unfortunately, is always present in practical devices.

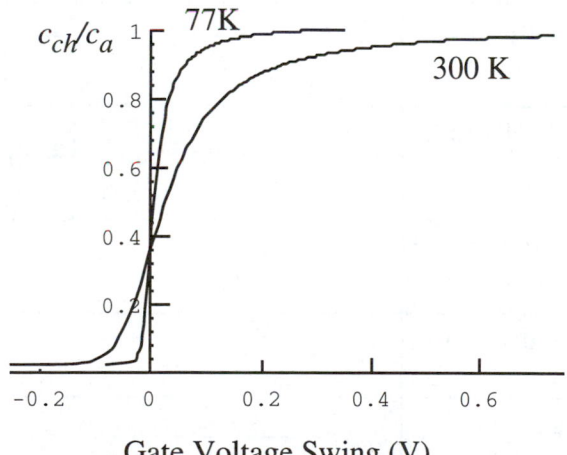

Gate Voltage Swing (V)

**Fig. 6.2.4.** Dimensionless channel capacitance, $c_{ch}/c_a$, versus gate voltage swing, $V_{GT}$ at 300 K and 77 K. Parameters used in the calculation: $\eta = 1$, $d = 250$ Å, $\varepsilon_i = 3.45 \times 10^{-11}$ F/m. A parasitic capacitance of $0.02c_a$ was added to the channel capacitance.

In order to understand MOSFET operation and to be able to simulate MOSFET characteristics, we now have to answer several important questions:

- What determines the threshold voltage?
- How to describe the effects of the drain potential?
- What is the electron velocity in the device channel?
- What is the MOSFET equivalent circuit and how to describe capacitive elements of such an equivalent circuit?
- What is the role of the substrate contact?
- How to calculate MOSFET *I-V* and *C-V* characteristics?

Once these questions are answered, we will be ready to describe the features of different models used for MOSFET simulation in different versions of SPICE.

The summary of the most important equations related to charge control models is given in Table 6.2.1.

| | |
|---|---|
| Classic Charge Control Model (CCM) | $qn_s = c_i\left(V_{gs} - V_T\right)$ |
| Gate capacitance per unit area | $c_i = \dfrac{\varepsilon_i}{d_i}$ |
| Unified Charge Control Model (UCCM) | $V_{GT} - V_F = \eta V_{th} \ln\left(\dfrac{n_s}{n_o}\right) + a\left(n_s - n_o\right)$ |
| Effective gate capacitance per unit area above threshold | $c_a = \dfrac{\varepsilon_i}{d_i + \Delta d}$ |
| Sheet electron concentration in the channel at $V_{GT} = 0$ | $n_o = \dfrac{c_a \eta V_{th}}{2q}$ |
| MOSFET channel capacitance | $c_{ch} = \dfrac{c_a}{1 + \dfrac{c_a \eta V_{th}}{qn_s}}$ |

**Table 6.2.1.** The summary of the equations of charge control models.

## 6-3.  MOSFET THRESHOLD VOLTAGE

Silicon MOSFETs usually have gate contacts made from a highly conductive $n+$ **polysilicon** material.  A metal ohmic contact to this **polysilicon gate** is used for external connections.  Figure 6.3.1 shows the band diagrams of silicon $n$-channel and $p$-channel MOSFETs.  These diagrams show the dependence of the band edges on the distance in the direction perpendicular to the semiconductor-dielectric interface for a particular case when the gate voltage is such that the bands are "flat," that is, the energies of the band edges in the semiconductor do not depend on the distance.  Such a condition is called the **flat band condition**, and the corresponding gate voltage is called the **flat band voltage**.

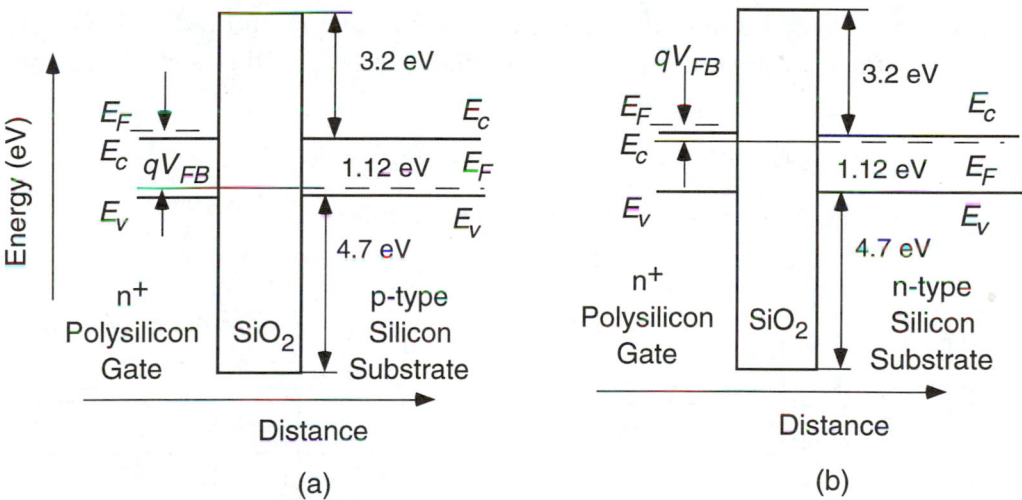

**Fig. 6.3.1.**  Flat band diagrams for (a) $n$-channel and (b) $p$-channel Si MOSFETs (after Lee et al., 1993, copyright © Prentice Hall, 1993, reproduced by permission of Prentice Hall, Inc., Englewood Cliffs, NJ).

(Please notice that since the MOSFET is a three-dimensional device, we may be interested in band diagrams drawn in different directions.  For example, in Section 6.2, we discussed a band diagram in the direction from the source to the drain.)

The band diagrams shown in Fig. 6.3.1 can be readily understood, since $n+$-polysilicon and $n$-type silicon (which serves as the substrate for a $p$-channel MOSFET) have similar differences between their Fermi levels and the bottom of the conduction band in $SiO_2$.  For the $p$-type substrate (which serves as the substrate for an $n$-channel MOSFET), the difference is close to the energy gap of Si. Thus, for an $n$-channel MOSFET, the flat band voltage is approximately $-1$ V. This voltage corresponds to the **subthreshold** regime, since there are almost no electrons in the channel to carry a current between the $n^+$ source and drain contacts (as an exercise, please estimate the electron concentration in the silicon layer at the flat band). Hence, $V_T > V_{FB}$.

Figure 6.3.2 compares the band diagrams of the $n$-channel MOSFET at the flat band and with the gate bias $V_{GS} > V_{FB}$. At the flat band, the difference between the quasi-Fermi levels in the polysilicon gate and in the silicon substrate is equal to $q|V_{FB}|$. When the gate bias $V_{GS}$ is applied, this difference becomes equal to $qV_{GS} + q|V_{FB}|$. On the other hand, as can be seen from Fig. 6.3.2b, this difference between the quasi-Fermi levels is equal to $V_i + V_s$, since the electron affinity $X \approx 3.2$ eV is approximately the same for silicon and polysilicon. Here $V_i$ is the voltage drop across the gate insulator and $V_s$ is the surface potential. Hence,

$$V_{GS} = -|V_{FB}| + V_i + V_s = V_{FB} + V_i + V_s \qquad (6\text{-}3\text{-}1)$$

(We assume that the drain-to-source bias is negligible.)

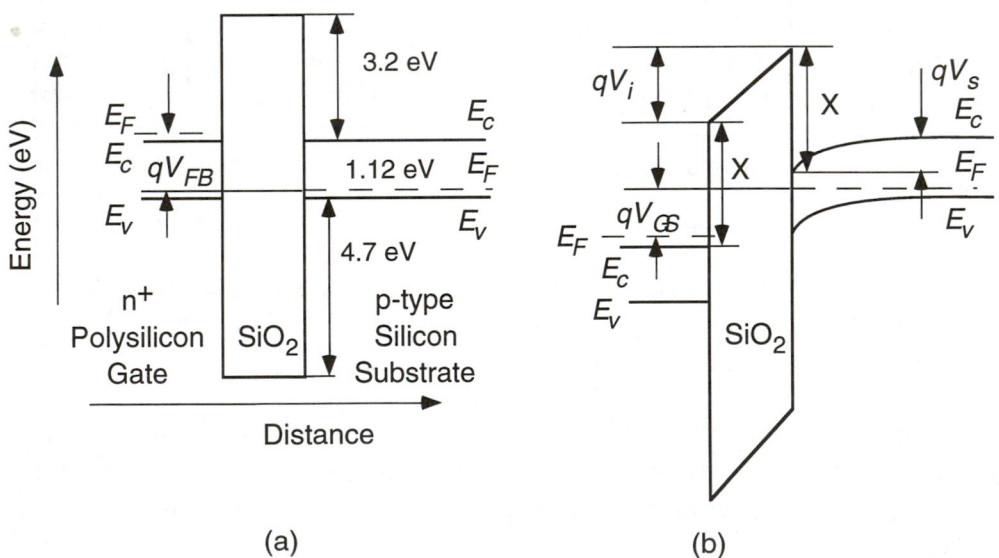

(a)                                                 (b)

**Fig. 6.3.2.** Band diagram of $n$-channel MOSFET for $V_T > V_{FB}$.
$V_{GS} = -|V_{FB}| + V_s + V_i$ where $V_{FB}$ is negative (approximately $-1$ V for NMOS) and $V_s$ and $V_i$ are positive.

Using the condition of continuity of the electric flux density at the semiconductor-gate dielectric interface

$$\varepsilon_s F_s = \varepsilon_i F_i \qquad (6\text{-}3\text{-}2)$$

we find the voltage drop $V_i = F_i d_i$ across the gate insulator. Here $\varepsilon_i$ is the dielectric permittivity of the insulator layer, $\varepsilon_s$ is the dielectric permittivity of the semiconductor, $F_s$ is the surface electric field in the semiconductor, $F_i$ is the electric field in the insulator layer, and $d_i$ is the insulator thickness. Hence, the gate voltage becomes

$$V_{GS} = V_{FB} + V_s + \varepsilon_s F_s / c_i \qquad (6\text{-}3\text{-}3)$$

where

$$c_i = \varepsilon_i / d_i \qquad (6\text{-}3\text{-}4)$$

is the gate insulator capacitance per unit area. Since according to Gauss' law, the surface field can be related to the total surface charge in the semiconductor (see Problem 6-3-1)

$$\varepsilon_s F_s = Q_s \qquad (6\text{-}3\text{-}5)$$

eq. (6-3-3) for the gate voltage can be also written as

$$V_{GS} = V_{FB} + V_s + Q_s / c_i \qquad (6\text{-}3\text{-}6)$$

When the gate voltage is still relatively close to the flat band voltage, the electron carrier concentration in the channel remains small. In this case, $Q_s = Q_d$ where the negative depletion charge density in the $p$-type substrate is

$$Q_d = \sqrt{2\varepsilon_s q N_a V_s} \qquad (6\text{-}3\text{-}7)$$

With an increase of the surface band bending potential, $V_s$, the electron concentration increases exponentially, as we already discussed in Section 6.2. At a certain value of $V_s$, the free electron charge is no longer negligible compared to the depletion charge. This value of $V_s$ corresponds to the threshold condition. (Figure 6.3.3, which shows the conduction band bending below and above threshold.)

The **threshold voltage** is usually defined as the gate voltage at which the total band bending at the surface is equal to double the difference, $\varphi_b = V_{th} \ln(N_a / n_i)$, between the intrinsic Fermi level, $E_{Fi}$, and the Fermi level in the substrate, $E_F$. At this band bending, the electron concentration at the insulator-semiconductor interface becomes equal to the hole concentration in the bulk. This situation is called **inversion** or as **strong inversion**, and the layer of free electrons induced at the surface is called an **inversion layer**. The threshold voltage corresponding to the onset of the strong inversion is given by

$$V_T = V_{FB} + 2\varphi_b + \sqrt{4\varepsilon_s q N_a \varphi_b} / c_i \qquad (6\text{-}3\text{-}8)$$

or

$$V_T = V_{FB} + 2\varphi_b + \gamma_N \sqrt{2\varphi_b} \qquad (6\text{-}3\text{-}9)$$

where

$$\gamma_N = \sqrt{2\varepsilon_s q N_a} / c_i \qquad (6\text{-}3\text{-}10)$$

is called the **body effect constant**.

### Example 6-3-1.

Plot the conduction band edge versus distance perpendicular to the insulator-semiconductor interface for an $n$-channel MOSFET (with the threshold voltage $V_T = 0.34$ V) at the flat band ($V_{GS} = -1$ V), below threshold ($V_{GS} = 0$ V), and above threshold ($V_{GS} = 1$ V). Gate oxide thickness $d_i = 150$ Å, gate oxide permittivity $\varepsilon_i = 3.45 \times 10^{-11}$ F/m, silicon permittivity $\varepsilon_s = 1.05 \times 10^{-10}$ F/m, flat band voltage $V_{FB} = -1$ V, substrate doping density $N_a = 5 \times 10^{16}$ cm$^{-3}$, substrate bias $V_{BS} = 0$ V, and room temperature $T = 300$ K.

### Solution:

We choose $E_c = 0$ far from the interface. Then we find the values of the surface potential, $V_s$, as a function of the gate bias, $V_{GS}$. Substituting eq. (6-3-7) into eq. (6-3-6), we obtain

$$V_{GS} = V_{FB} + V_s + \sqrt{2\varepsilon_s q N_a V_s} / c_i \qquad (6\text{-}3\text{-}11)$$

Equation (6-3-11) is valid for $V_{GS} \leq V_T$. At more positive $V_{GS}$, the surface potential varies relatively little, since all additional charge is induced into the conducting channel at the interface.  Equation (6-3-11) is a quadratic equation with respect to $(V_S)^{1/2}$.  Solving it, we find

$$V_s = \left( \sqrt{\frac{\varepsilon_s q N_a}{2c_i^2} - V_{FB} + V_{GS}} - \sqrt{\frac{\varepsilon_s q N_a}{2c_i^2}} \right)^2 \qquad (6\text{-}3\text{-}12)$$

At the flat band when, $V_{GS} = V_{FB}$, $V_s = 0$ as expected.  In this case, the band diagram is simply a straight line.  For the given values of parameters, $c_i = 0.0023$ F/m, and $\frac{\varepsilon_s q N_a}{2c_i^2} \approx 0.079$ (V).  From eq. (6-3-12), we find that for $V_{GS} = 0$, $V_s \approx$ 0.572 V, and for $V_{GS} = V_T = 0.34$ V, $V_s \approx 0.826$ V.  (At $V_{GS} = 1$ V, $V_s$ is also $\approx$ 0.826 V.)  The conduction band edge varies from $E_c = 0$ far from the interface to $E_c = -qV_s$ at the interface.  The **band bending** profile is found using the depletion approximation, which is the same approximation we used for a $p$-$n$ junction or for a Schottky barrier junction in Chapter 2.  The resulting plots are shown in Fig. 6.3.3.

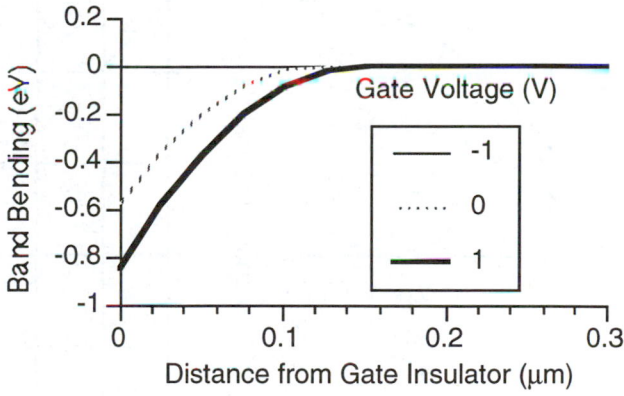

**Fig. 6.3.3.**  Conduction band edge versus distance for an $n$-channel MOSFET at flat band ($V_{GS} = -1$ V), below threshold ($V_{GS} = 0$ V), and above threshold ($V_{GS} = 1$ V).

When we apply a substrate bias, $V_{BS}$, this voltage changes the depletion width and depletion charge as illustrated in Fig. 6.3.4.  Hence, the threshold voltage becomes

$$V_T = V_{FB} + 2\varphi_b + \gamma_N \sqrt{(2\varphi_b - V_{BS})} \tag{6-3-13}$$

If the surface states are important, this equation has to be modified as follows to account for the surface state charge density, $q_{ss}$:

$$\begin{aligned} V_T &= V_{FB} + q_{ss}/c_i + 2\varphi_b + \gamma_N \sqrt{(2\varphi_b - V_{BS})} \\ &= V_{To} + \gamma_N \left[ \sqrt{(2\varphi_b - V_{BS})} - \sqrt{2\varphi_b} \right] \end{aligned} \tag{6-3-14}$$

where

$$V_{To} = V_{FB} + q_{ss}/c_i + 2\varphi_b + \gamma_N \sqrt{2\varphi_b} \tag{6-3-15}$$

is the threshold voltage at zero substrate bias.  In modern silicon MOSFETs, the effect of the surface states is usually negligible.  However, the surface states may be very important for new emerging MOSFET technologies, such as SiC MOSFET technology.

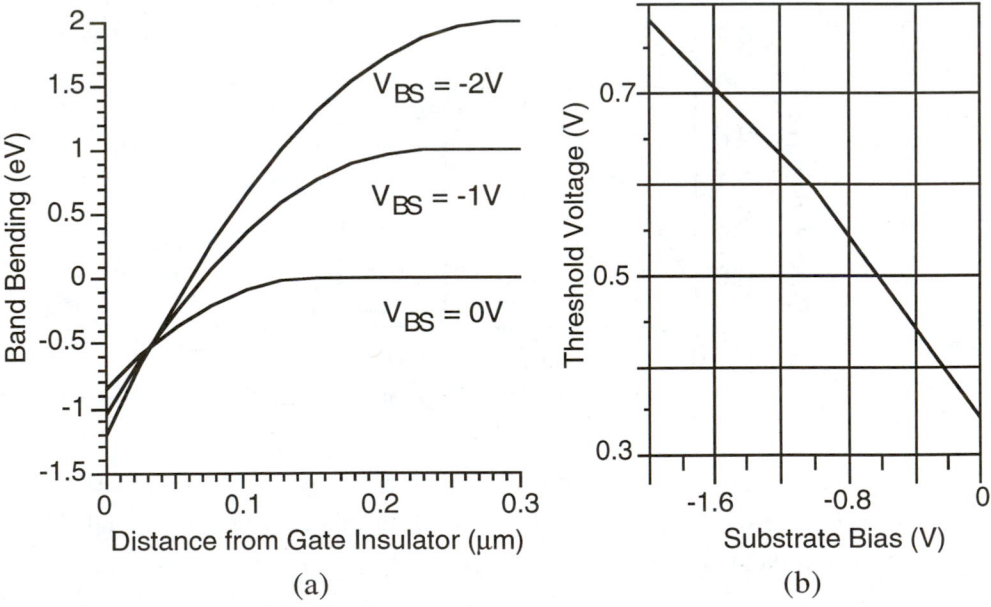

(a)                                                      (b)

**Fig. 6.3.4.**  (a) Conduction band edge versus distance for an *n*-channel Si MOSFET and (b) dependence of the threshold voltage on substrate bias. The gate oxide thickness and permittivity $d_i = 150$ Å, $\varepsilon_i = 3.45 \times 10^{-11}$ F/m, flat band voltage $V_{FB} = -1$ V, substrate doping density $N_a = 5 \times 10^{16}$ cm$^{-3}$, gate voltage $V_{GS} = 1$ V, and room temperature $T = 300$ K.

The above equations for the threshold voltage have been derived for ideal long and wide channel MOSFETs and for very small drain-to-source voltages. In practical devices, the gate length may be quite small so that the threshold voltage is affected by the depletion regions between the ohmic contacts and the substrate, by edge effects, and by the drain bias. To account for short channel effects and narrow channel effects (see the discussion below), eq. (6-3-14) has to be modified as follows:

$$V_{To} = V_{FB} + 2\varphi_b + \gamma_N \sqrt{2\varphi_b} + \Delta V_{TL} + \Delta V_{TW} \qquad (6\text{-}3\text{-}16)$$

where $\Delta V_{TL}$ is a short channel correction and $\Delta V_{TW}$ is a narrow channel correction. Typically, $\Delta V_{TL}$ is negative (see Fig. 6.3.5), and $\Delta V_{TW}$ is positive [see Lee et al. (1993) for a detailed discussion of short channel and narrow channel effects].

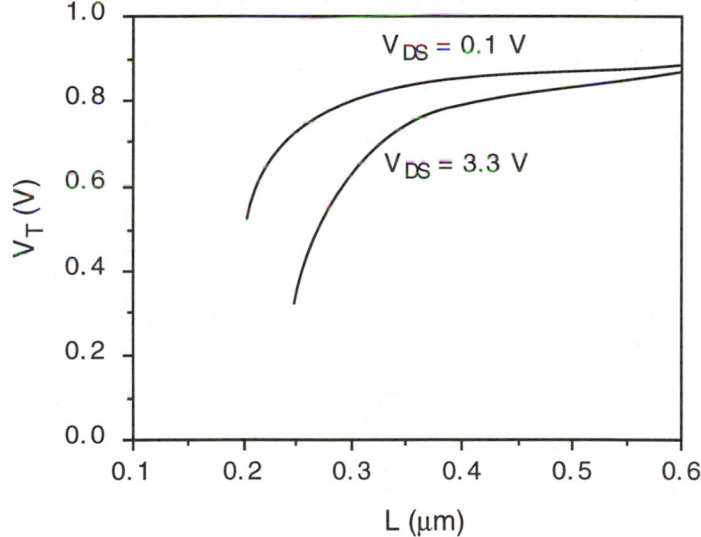

**Fig. 6.3.5.** Threshold voltage versus channel length (from T. A. Fjeldly and M. Shur, "Threshold Voltage Modeling and the Subthreshold Regime of Operation of Short-Channel MOSFETs," *IEEE Trans. Electron. Devices*, Vol. 40, pp. 137-145, Jan. 1993, copyright © IEEE, 1993).

These corrections depend on the bias voltages, temperature, and on device parameters such as a channel length, a channel width, and an oxide thickness.

Equations for *p*-channel MOSFETs (**PMOS**) are similar to those for *n*-channel devices (**NMOS**):

$$V_{TP} = V_{TPo} - \gamma_P \left[ \sqrt{(2\varphi_b + V_{BS})} - \sqrt{2\varphi_b} \right] \qquad (6\text{-}3\text{-}17)$$

$$V_{TPo} = V_{FBP} - 2\varphi_b - \gamma_P \sqrt{2\varphi_b} + \Delta V_{TL} + \Delta V_{TW} \qquad (6\text{-}3\text{-}18)$$

where now, $\varphi_b = V_{th} \ln(N_d/n_i)$, $V_{FBP}$ is the PMOS flat-band voltage, and

$$\gamma_P = \sqrt{2\varepsilon_s q N_d} / c_i \qquad (6\text{-}3\text{-}19)$$

(Note that the corrections $\Delta V_{TL}$ and $\Delta V_{TN}$ in PMOS have the opposite signs of those in NMOS.)

A popular SPICE MOSFET model called **BSIM** (meaning Berkeley Short channel Insulated gate fet Model) uses the following empirical expressions, which account for short and narrow channel effects, for the threshold voltage

$$V_{TS} = V_{TSo} + \gamma_{Neff} \sqrt{2\varphi_b - V_{BS}} - \Gamma_{Neff} (2\varphi_b - V_{BS}) \qquad (6\text{-}3\text{-}20)$$

where $V_{TSo} = V_{FB} + 2\varphi_b$, $\gamma_{Neff} = \gamma_{NL} + \gamma_{NW}$,

$$\gamma_{NL} = \gamma_{NL1} + \Delta\gamma_{NL} \left( 1 - \frac{L_o}{L} \right) \qquad (6\text{-}3\text{-}21)$$

(Expressions for $\gamma_{NW}$ are obtained from eq. (6-3-21) by replacing $L$ with $W$. Expressions for the $\Gamma$ coefficients are obtained from the expressions for the $\gamma$ coefficients by replacing $\gamma$ with $\Gamma$.) Parameters in these equations, such as $\gamma_{NL1}$, $\Delta\gamma_{NL}$, and $L_o$, are empirical parameters extracted from the measured values of the threshold voltage. This empirical model is used in designing CMOS devices and circuits. This model is also in excellent agreement with two-dimensional device simulations.

Strictly speaking, the simple threshold voltage model considered in the beginning of this section is suitable for long and wide channel devices with ideal characteristics. Nevertheless, this model is truly indispensable for a device engineer and designer. It shows how the threshold voltage can be changed by changing the substrate doping and/or substrate bias. (Both of these techniques are used for threshold voltage adjustment in practical devices.) This model can also be generalized to describe the addition of donors into an NMOS channel (or acceptors into a PMOS channel). Such donors are usually introduced by ion implantation – by bombarding the semiconductor wafer with ion beams of impurities. In an NMOS, implanted donors with surface concentration per unit area, $N_S$, shift the threshold voltage by approximately

$$\Delta V_{TN} \approx -\frac{qN_s}{c_i} \tag{6-3-22}$$

In a similar fashion, in a PMOS, implanted acceptors with surface concentration per unit area, $P_S$, shift the threshold voltage by approximately

$$\Delta V_{TP} \approx \frac{qP_s}{c_i} \tag{6-3-23}$$

Alternatively, in an NMOS, implanted acceptors increase the threshold voltage, and in a PMOS, implanted donors decrease the threshold voltage. In Section 6.5, we will use this simplified threshold voltage model to understand the design considerations for the most popular and the most important semiconductor technology – called **Complementary MOS (CMOS)**.

The summary of the most important equations related to MOS threshold voltage is given in Table 6.3.1.

| | |
|---|---|
| Threshold voltage for long and wide NMOS | $V_T = V_{FB} + 2\varphi_b + \gamma_N \sqrt{(2\varphi_b - V_{BS})}$ <br> where $\varphi_b = V_{th} \ln \dfrac{N_a}{n_i}$ |
| Body effect constant for NMOS | $\gamma_N = \sqrt{2\varepsilon_s q N_a} / c_i$ |
| Threshold voltage for long and wide PMOS | $V_{TP} = V_{TPo} - \gamma_P \left[ \sqrt{(2\varphi_b + V_{BS})} - \sqrt{2\varphi_b} \right]$ |
| Body effect constant for PMOS | $\gamma_P = \sqrt{2\varepsilon_s q N_d} / c_i$ |
| Threshold voltage for short NMOS (BSIM model) | $V_{TS} = V_{FB} + 2\varphi_b$ <br> $\quad + \gamma_{Neff} \sqrt{2\varphi_b - V_{BS}} - \Gamma_{Neff}(2\varphi_b - V_{BS})$ |
| Threshold voltage shift by implantation | $\Delta V_{TN} = -\dfrac{qN_s}{c_i} \qquad \Delta V_{TP} = \dfrac{qP_s}{c_i}$ <br> $N_s$ and $P_s$ are surface concentrations of implanted donors and acceptors |

**Table 6.3.1.** Important equations related to MOS threshold voltage.

## 6-4.  MOSFET MODELING.

### 6.4.1.   Gradual Channel Approximation.

As we discussed in Section 6.3, a gate-source voltage, $V_{GS}$, larger than the threshold voltage, $V_T$, induces an inversion charge into the MOSFET channel. Electrons (in an NMOS) come from the $n^+$-doped source, move across the channel inversion layer from the source to the drain, and leave the channel by moving into the $n^+$-doped drain.

If the drain-to-source voltage is very small, the sheet electron concentration in the channel, $n_s \approx c_i(V_{GS} - V_T)/q$, is constant along the channel. Otherwise, the potential changes from the source potential on the source side of the channel to the drain potential at the drain side of the channel.  Hence, the potential varies in two dimensions: along the channel and perpendicular to the channel (in the direction from the channel toward the gate and from the channel into the substrate).  Near the gate edges, on the MOSFET sides, the potential varies in all three dimensions!  Hence, in principle, to describe properly the current-voltage and capacitance-voltage characteristics of a field effect transistor, we may have to solve a three-dimensional problem.  Device simulators which handle such complicated problems do exist, and three-dimensional or two-dimensional simulations are indeed very helpful for short and/or narrow channel devices.  However, simple analytical models that rely on solving one dimensional equations are still useful for longer channel devices or for an approximate analysis of short channel devices.

**Gradual Channel Approximation (GCA)** provides the basis for the analytical modeling of MOSFET characteristics.  This approximation is valid when

$$|\partial F_x/\partial x| << |\partial F_y/\partial y| \approx |\rho|/\varepsilon_s \tag{6-4-1}$$

where $F_x$ and $F_y$ are the components of the electric field in the channel in the direction $x$ from the source to the drain and in the perpendicular direction $y$, respectively, $\varepsilon_s$ is the dielectric permittivity of the semiconductor, and $\rho$ is the space charge density (see Fig. 6.4.1).

GCA reduces the two-dimensional Poisson's equation

$$\frac{\partial F_x}{\partial x} + \frac{\partial F_y}{\partial y} = \frac{\rho}{\varepsilon_s} \tag{6-4-2}$$

to the one-dimensional Poisson's equation

$$\frac{\partial F_y}{\partial y} = \frac{\rho}{\varepsilon_s}$$    (6-4-3)

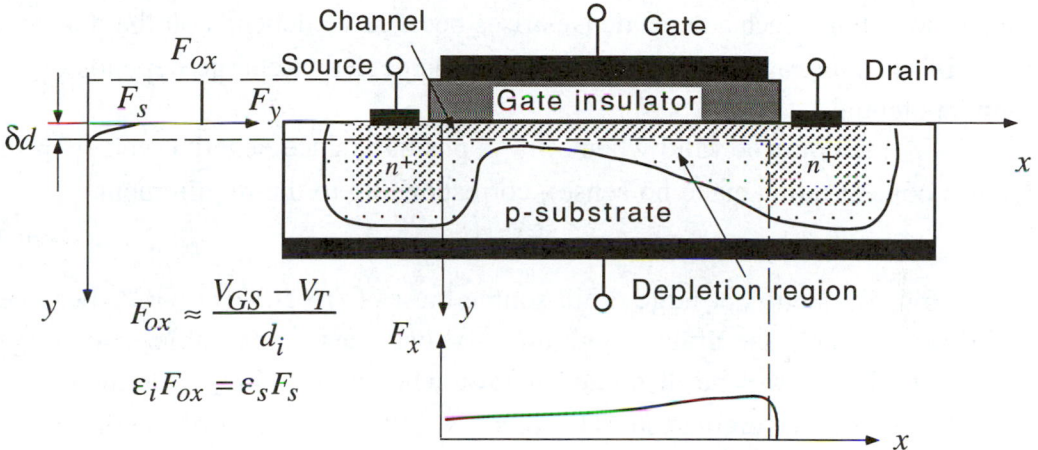

**Fig. 6.4.1.** Schematic distributions of the electric field perpendicular to the insulator-semiconductor interface ($F_y$) and of the electric field in the semiconductor at the insulator-semiconductor interface in the direction parallel to the insulator-semiconductor interface ($F_x$).

As discussed by Lee et al. (1993), GCA is valid when

$$\left(\frac{\varepsilon_i L}{\varepsilon_s d_i}\right)^2 \frac{(V_{GS} - V_T)^2}{V_{th} V_{DS}} \gg 1$$    (6-4-4)

where $L$ is the gate length, $\varepsilon_i$ and $\varepsilon_s$ are the dielectric permittivities of Si and SiO$_2$, respectively, and $V_{th}$ is the thermal voltage ($V_{th} \approx 26$ mV at room temperature). For $L = 1$ µm, $\varepsilon_i/\varepsilon_s \approx 0.33$, $V_{th} \approx 25.8$ mV (room temperature), $V_{GS} - V_T = 0.5$ V, $V_{DS} = 0.5$ V, and $d_i = 300$ Å, the left-hand side of inequality (6-4-4) is about 2,300. Hence, GCA seems to be a very good approximation for a 1 µm gate MOSFET.

When GCA is valid, the electron surface charge density, $qn_s$, can be calculated in every point of the channel exactly in the same way as for $V_{DS} = 0$ except that $V_{GS}$ should be replaced with $V_{GS} - V(x)$ where $V(x)$ is the channel potential induced by the drain-source bias. Hence, the equation of the simple

**charge control model** becomes

$$qn_s \approx c_i \left( V_{GT} - V \right) \tag{6-4-5}$$

where $c_i$ is the gate insulator capacitance per unit area and $V_{GT} = V_{GS} - V_T$. [In this equation, we also neglected the dependence of the threshold voltage on the channel potential. Such a dependence arises because $V_T$ depends on the depletion charge in the substrate (see Section 6.3) and the depletion charge depends on the channel potential.]

GCA may only be valid when $qn_s$ is positive (since negative values of the electron concentration make no sense), corresponding to the requirement

$$V_{GT} \geq V \tag{6-4-6}$$

[see eq. 6-4-5]. Hence, at large drain-source biases ($V_{DS} > V_{GT}$), GCA becomes invalid at least near the drain where $|\partial F_x/\partial x|$ becomes comparable to or larger than $|\partial F_y/\partial x|$. As will be discussed below, when $V_{DS} > V_{GT}$, and inequality (6-4-4) is invalid, the drain current becomes nearly independent of the drain bias. This regime of operation is called the saturation regime.

### 6.4.2.   Constant mobility model.

We will first consider a simple MOSFET model, assuming a constant electron mobility, $\mu_n$, neglecting diffusion effects, and using the simple charge control model given by eq. (6-4-5). In this case, the absolute value of the electron velocity

$$v_n = \mu_n F = \mu_n \frac{dV}{dx} \tag{6-4-7}$$

and, in the above-threshold regime, the absolute value of the drain current, $I_d$, is given by

$$I_d = W q \mu_n \frac{dV}{dx} n_s \tag{6-4-8}$$

where $W$ is the device width. Using eq. (6-4-5), we can rewrite Eq. (6-4-8) as

$$dx = \frac{W \mu_n c_i}{I_d} \left( V_{GT} - V \right) dV \tag{6-4-9}$$

The drain current is constant along the channel (i. e., $I_d$ is independent of $x$). Hence, the integration of eq. (6-4-9) from zero to the channel length $L$ (so that $V$ changes from zero at the source side of the gate to $V_{DS}$ at the drain side of the gate) yields the following expression for the current-voltage characteristics:

$$I_d = \frac{W\mu_n c_i}{L}\left(V_{GT} - \frac{V_{DS}}{2}\right)V_{DS} \qquad (6\text{-}4\text{-}10)$$

For very small drain-source voltages, $V_{DS} << V_{GT}$, eq. (6-4-10) may be simplified to

$$I_d = \frac{W\,\mu_n\,c_i}{L}\,V_{GT}\,V_{DS} \qquad (6\text{-}4\text{-}11)$$

Equation (6-4-11) describes the current-voltage characteristics in the linear region. This equation can be interpreted as follows. At very small drain-source bias, $V_{DS} << V_{GT}$, the charge induced into the channel does not depend on the channel potential and $qn_s = c_iV_{GT}$ everywhere in the channel. In this case, the electric field in the channel is nearly constant and given by

$$F = V_{DS}/L \qquad (6\text{-}4\text{-}12)$$

Multiplying the channel sheet charge density, $qn_s$, by the electron velocity, $v_n = \mu_n F$, and by the gate width, $W$, we obtain eq. (6-4-11).

Equation (6-4-10) is valid only for such values of $V_{DS}$ that the inversion layer still exists even at the drain side of the gate, that is,

$$n_s\left(V = V_{DS}\right) \geq 0 \qquad (6\text{-}4\text{-}13)$$

The condition $n_s = 0$ at the drain is called the **pinch-off condition**. As can be seen from eqs. (6-4-5), the pinch-off occurs at the drain side of the gate when

$$V_{DS} = V_{SAT} = V_{GS} - V_T \qquad (6\text{-}4\text{-}14)$$

where $V_{SAT}$ is the saturation voltage. Substituting $V_{DS} = V_{SAT}$ into eq. (6-4-10), we find that at the pinch-off

$$I_d = I_{sat} = \frac{W\mu_n c_i}{2L} V_{GT}^2 \qquad (6\text{-}4\text{-}15)$$

When $V_{DS}$ approaches $V_{SAT}$, the assumptions used to derive eq. (6-4-10) fail. Indeed, the electric field in the channel in the direction parallel to the semiconductor-insulator interface, $F = |F_x| = dV/dx$, is given by

$$F = \frac{I_d}{q\mu_n n_s(V)W} \qquad (6\text{-}4\text{-}16)$$

[see eq. (6-4-8)]. Thus as $V_{DS}$ approaches $V_{SAT}$ and $n_s(L)$ approaches zero, $F$

tends to infinity (see Fig. 6.4.2). This means that GCA approximation is no longer applicable. Also, since the electron velocity saturates in high electric fields, eq. (6-4-5) is no longer valid as well. When $V_{DS}$ approaches $V_{SAT}$, the channel resistivity near the drain becomes very high, since $n_s$ becomes small. Hence, most of the drain-to-source voltage drops in the region near the drain. If the drain bias is further increased, nearly all the additional voltage drop occurs near the drain, and voltage, field, and electron concentration near the source remain almost the same. Thus, the current near the source region must also remain nearly the same. Since the current must be constant along the channel, this means that the current near the pinch-off voltage should saturate, and we can expect that the current remains nearly constant at drain voltages higher than $V_{GT}$. This conclusion is confirmed by both experimental data and by more elaborate analytical and numerical modeling.

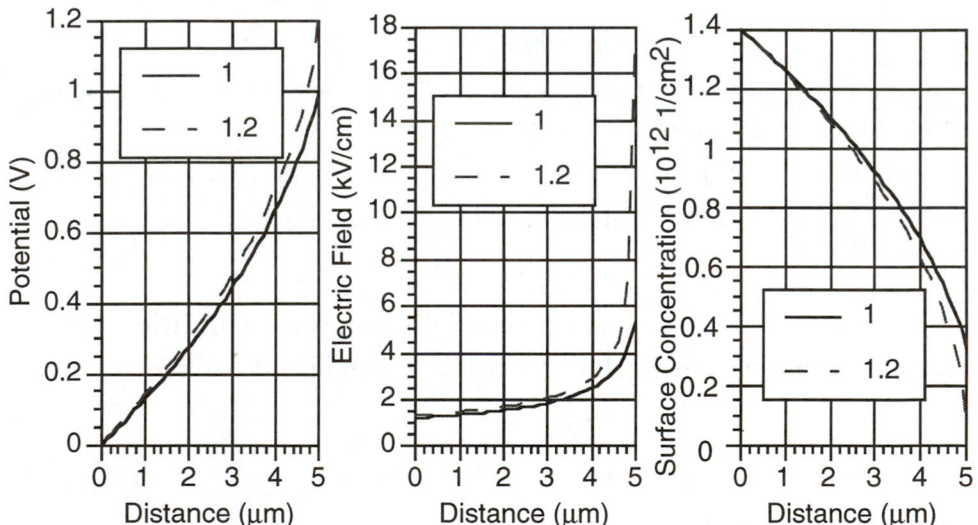

**Fig. 6.4.2.** Potential, electric field, and surface electron concentration in the channel of a Si MOSFET for $V_{DS} = 1$ and 1.2 V. $L = 5$ μm, $d_i = 200$ Å, $\mu_n = 800$ cm$^2$/Vs, $V_{GS} = 2$ V, $V_T = 1$ V.

An important device characteristic is the **transconductance**, defined as

$$g_m = \frac{dI_d}{dV_{GS}}\bigg|_{V_{DS}} \tag{6-4-17}$$

From the equations for the drain current, $I_d$, derived above, we find that

$$g_m = \begin{cases} \beta V_{DS}, & for\ V_{DS} << V_{SAT} \\ \beta V_{GT}, & for\ V_{DS} > V_{SAT} \end{cases} \tag{6-4-18}$$

where

$$\beta = \mu_n c_i \frac{W}{L} \tag{6-4-19}$$

is the transconductance parameter. As can be seen from eqs. (6-4-18) and (6-4-19), a high transconductance is obtained with high values of the low field electron mobility, thin gate insulator layers (i.e., larger gate insulator capacitance $c_i = \varepsilon_i/d_i$), and large $W/L$ ratios. The dependence of the transconductance on the low field mobility and on the gate length, $L$, is, however, strongly affected by the velocity saturation effects in short channel devices as discussed below.

The simple model described above is adequate for long channel devices (with channel lengths longer than approximately 4 µm for Si NMOS). This model is implemented in different versions of SPICE where it is called the **Level 1 MOSFET** model.

### 6.4.3.   Velocity  saturation  model.

In modern short channel devices with channel length of the order of 1 µm or less, the electric field in the channel can easily exceed the characteristic electric field of the velocity saturation

$$F_s = \frac{v_s}{\mu_n} \tag{6-4-20}$$

where $v_s$ is the electron saturation velocity ($F_s \approx 8$ kV/cm for Si NMOS). Once the electric field at the drain side of the channel (where the electric field is the highest, see Fig. 6.4.2) exceeds $F_s$, the electron velocity saturates, leading to the current saturation at the drain bias smaller than the pinch-off voltage $V_{DS} = V_{GT}$. The electric field at the drain side of the channel can be found from eq. (6-4-16)

$$F(L) = \frac{I_d}{q\mu_n n_s(V_{DS})W} \tag{6-4-21}$$

Equating $F(L)$ and $F_s$, we obtain from eq. (6-4-21), the relationship between the drain saturation voltage $V_{DS} = V_{SAT}$ and the drain saturation current $I_d = I_{SAT}$

$$F_s = \frac{I_{SAT}}{\mu_n c_i (V_{GT} - V_{SAT}) W} \tag{6-4-22}$$

A second equation linking $V_{SAT}$ and $I_{sat}$, we obtain from eq. (6-4-10)

$$I_{sat} = \frac{W \mu_n c_i}{L} \left( V_{GT} - \frac{V_{SAT}}{2} \right) V_{SAT} \tag{6-4-23}$$

Solving eqs. (6-4-22) and (6-4-23) with respect to $I_{sat}$, we obtain

$$I_{sat} = \frac{g_{ch} V_{GT}}{1 + \sqrt{1 + \left( \dfrac{V_{GT}}{V_L} \right)^2}} \tag{6-4-24}$$

where $V_L = F_s L$ and the channel conductance $g_{ch} = q \mu_n n_s W / L$.

**Example 6-4-1.**

Calculate the dependencies of the drain saturation current on the gate voltage for $n$-channel MOSFETs with gate lengths of 0.5 μm and 5 μm using first the constant mobility model and then the velocity saturation model. Use the following values of parameters: substrate doping $N_a = 5 \times 10^{16}$ cm$^{-3}$, temperature $T = 300$ K, energy gap $E_g = 1.12$ V, effective densities of states in the conduction and valence bands at 300 K, $N_c = 3.22 \times 10^{19}$ cm$^{-3}$, $N_v = 1.83 \times 10^{19}$ cm$^{-3}$, semiconductor permittivity $\varepsilon_s = 1.054 \times 10^{-10}$ F/m, flat band voltage $V_{FB} = -1$ V, oxide thickness $d_i = 200$ Å, oxide dielectric permittivity, $\varepsilon_i = 3.45 \times 10^{-11}$ F/m, electron mobility $\mu_n = 800$ cm$^2$/Vs, electron saturation velocity $v_s = 10^5$ m/s. Device width is 1 mm.

**Solution:**

We first find $n_i = (N_c N_v)^{1/2} \exp\left( -\dfrac{E_g}{2 k_B T} \right) \approx 10^{10}$ cm$^{-3}$. Then we determine

$\varphi_b = V_{th} \ln(N_a / n_i) = 0.02584 \ln(10^{16}/10^{10}) \approx 0.357$ V. Now we calculate $c_i = \varepsilon_i / d = 0.001725$ F/m and the threshold voltage $V_T = V_{FB} + 2\varphi_b + \sqrt{4\varepsilon_s q N_a \varphi_b} / c_i$ $= 0.35$ V. In the frame of the constant mobility model, the drain saturation current is given by eq. (6-4-15).   In the frame of the velocity saturation model, the saturation current is given by eq. (6-4-24).  The resulting plots are shown in Fig. 6.4.3.  As can be seen from the figure [and as can be shown using eq. (6-4-24)], velocity saturation is very important in the short channel device and makes little difference for the long channel device.  Also notice that the current is no longer inversely proportional to the gate length when the velocity saturation is accounted for.

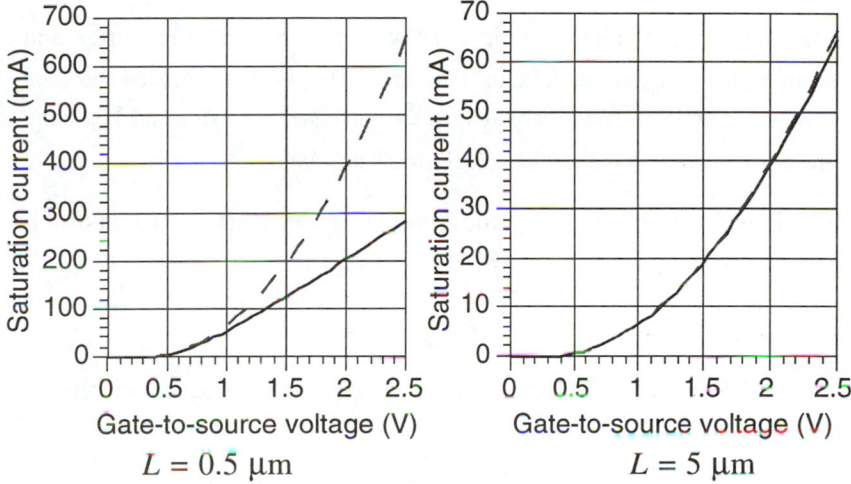

$$L = 0.5 \ \mu\text{m} \qquad\qquad L = 5 \ \mu\text{m}$$

**Fig. 6.4.3.**  Saturation current versus gate-to-source voltage for 0.5 μm gate and 5 μm gate MOSFETs.  Dashed lines: constant mobility model, solid lines: velocity saturation model.

### 6.4.4.  Source and drain series resistances.

Parasitic source and drain and source parasitic series resistances, $R_s$ and $R_d$, play an important role, especially in short channel devices where the channel resistance is smaller.  These resistances may be accounted for by using the following expressions relating the "intrinsic" gate-source and drain-source voltages, $V_{GS}$ and $V_{DS}$, to the "extrinsic" (measured) gate-source and drain-source voltages, $V_{gs}$ and $V_{ds}$, which include voltage drops across the series resistances (see Fig. 6.4.4):

$$V_{GS} = V_{gs} - I_d R_s \tag{6-4-25}$$

$$V_{DS} = V_{ds} - I_d (R_s + R_d) \tag{6-4-26}$$

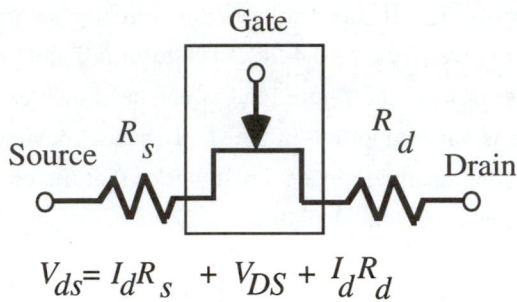

$$V_{ds} = I_d R_s + V_{DS} + I_d R_d$$

**Fig. 6.4.4.**  Equivalent circuit of MOSFET with parasitic source and drain series resistances.  The intrinsic transistor is shown inside the box. $V_{DS}$ is the drain-to-source voltage for the intrinsic transistor and $V_{ds}$ is the drain-to-source voltage for the actual (extrinsic) device.

The measured (**extrinsic**) transconductance of a field effect transistor is

$$g_m = \frac{dI_d}{dV_{gs}}\bigg|_{V_{ds}=\text{const}} \tag{6-4-27}$$

Using eqs. (6-4-25) and (6-4-26), $g_m$ can be related to the **intrinsic transconductance**

$$g_{mo} = \frac{dI_d}{dV_{GS}}\bigg|_{V_{DS}=\text{const}} \tag{6-4-28}$$

and to the **intrinsic drain conductance**

$$g_{do} = \frac{dI_d}{dV_{DS}}\bigg|_{V_{GS}=\text{const}} \tag{6-4-29}$$

of the same device as follows:

$$g_m = \frac{g_{mo}}{1 + g_{mo} R_s + g_{do}\left(R_s + R_d\right)} \tag{6-4-30}$$

In a similar fashion, the measured (extrinsic) drain conductance

$$g_d = \frac{dI_d}{dV_{ds}}\bigg|_{V_{gs}=\text{const}} \tag{6-4-31}$$

can be related to the intrinsic transconductance and intrinsic output conductance as follows (see Problem 6-4-2):

$$g_d = \frac{g_{do}}{1 + g_{mo} R_s + g_{do}(R_s + R_d)} \tag{6-4-32}$$

At very small drain-source voltages, both $I_d$ and $g_{mo}$ are small and proportional to the drain voltage, $V_{DS}$ [see eq. (6-4-11)]. Hence, the term $g_{mo}R_s$ can be neglected in eqs. (6-4-30) and (6-4-32). In the saturation regime, when $g_{do}$ is very small, the term $g_{do}(R_s + R_d)$ can be neglected in these equations.

Substituting eq. (6-4-25) into eq. (6-4-24), we find

$$I_{sat} = \frac{g_{ch}V_{gt}}{1 + g_{ch}R_s + \sqrt{1 + 2g_{ch}R_s + \left(V_{gt}/V_L\right)^2}} \tag{6-4-33}$$

where $V_L = F_sL$, $g_{ch} = g_{cho}/(1 + g_{cho}(R_s + R_d))$, $g_{cho} = c_iV_{gt}\mu_nW/L$. Figure 6.4.5 clearly shows the effects of the parasitic resistances on the output characteristics of a MOSFET. The parasitic resistances decrease the drain saturation current, decrease the drain conductance in the linear regime (at small values of $V_{ds}$), and increase the drain current saturation voltage. [A more detailed analysis of the effects related to source and drain resistances is given, for example, by Lee et al. (1993).]

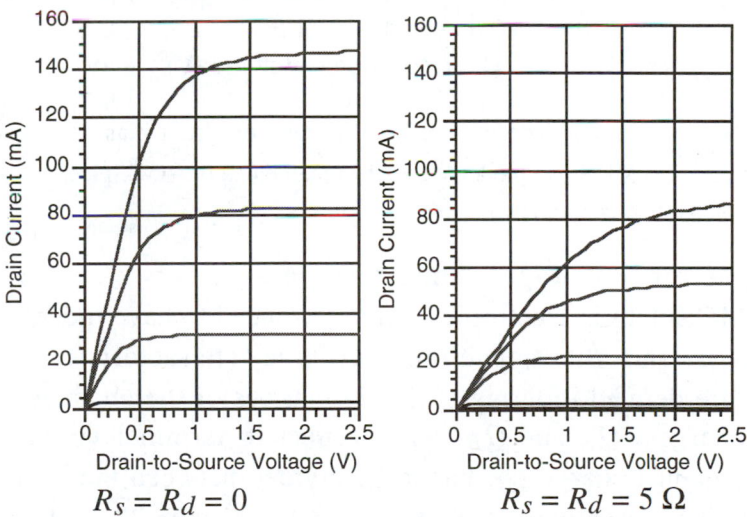

$$R_s = R_d = 0 \qquad\qquad R_s = R_d = 5\ \Omega$$

**Fig. 6.4.5.** MOSFET output characteristics calculated for zero parasitic resistances and parasitic resistances of 5 $\Omega$. Gate length is 1 $\mu$m. Other parameters are the same as those listed in Example 6-4-1.

[As an exercise, estimate the value of the threshold voltage, $V_T$, and the values of gate-to-source voltages corresponding to the curves shown in Fig. 6.4.5. **Hint:** first estimate the threshold voltage as was done in Example 6-4-1. Then extract the values of $I_{SAT}$ from the graph for $R_s = R_d = 0$ and use eq. (6-4-24) to solve for $V_{GT}$.]

Another considerable improvement of this model is to account for the dependence of the threshold voltage on the drain bias. This dependence is related to the dependence of the depletion charges in the substrate on the channel potential. An accurate modeling of this effect is difficult; however, the following empirical linear dependence of the threshold voltage is adequate (see Fig. 6.4.6):

$$V_T = V_{To} - \sigma V_{DS} \qquad (6\text{-}4\text{-}34)$$

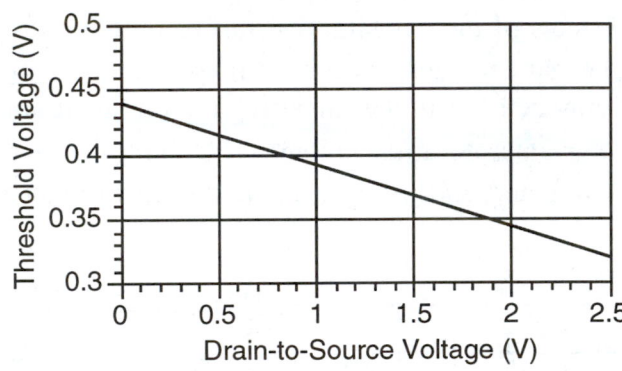

**Fig. 6.4.6.** Dependence of threshold voltage on drain bias for default parameters of MOSFET model MOSA1 implemented in AIM-Spice. ($V_{To} = 0.44$ V, $\sigma = 0.048$.)

### 6.4.5. Capacitance-voltage characteristics.

To simulate MOSFETs in electronic circuits, we need to have models for both the current-voltage and the capacitance-voltage characteristics. MOSFET capacitances are defined in terms of the derivatives of the channel and substrate (depletion) charges, $Q_N$ and $Q_B$, with respect to terminal voltages. For this purpose, the channel charge $Q_N$ has to be divided between the source and drain charges (MOSFET designers use the word "partitioned"). How should we partition the channel charge $Q_N$ between the source charge, $Q_s$, and the drain charge, $Q_d$? It is clear from the device symmetry that at zero drain bias $Q_s = Q_d$. However, in the saturation regime, the charge distribution is no longer symmetrical, with a larger channel charge near the source than near the drain. In this case, we let $Q_s = F_p Q_N$ and $Q_d = (1 - F_p)Q_N$, where $F_p$ (which is called the partitioning factor) should be larger than 0.5.

Based on a theoretical analysis and empirical data, Tsividis (1987) proposed the following empirical expression for the partitioning factor $F_p$

$$F_p = \frac{3 + 6\beta + 4\beta^2 + 2\beta^3}{5(1 + 2\beta + 2\beta^2 + \beta^3)} \qquad (6\text{-}4\text{-}35)$$

where

$$\beta = \begin{cases} 1 - V_{DS}/V_{SAT}, & \text{for } V_{DS} \le V_{SAT} \\ 0, & \text{for } V_{DS} > V_{SAT} \end{cases} \qquad (6\text{-}4\text{-}36)$$

Thus, in the saturation regime, when ß = 0, $F_p$ = 0.6. This works most of the time but, for certain simulations, choosing $F_p$ = 0 seems to give better results. Hence, one of the most popular programs for MOSFET simulation (BSIM) implemented in SPICE has a parameter XPART, which allows users to choose a preferred partitioning. This parameter can have only two values. When XPART = 0, $F_p$ = 0.6 in the saturation regime. When XPART = 1, $F_p$ = 0 in the saturation regime.

Once the charges are partitioned, all terminal capacitances, such as the gate-source capacitance, $C_{gs}$, and gate-drain capacitance, $C_{gd}$, can be defined as the derivatives of the terminal charges, $Q_s$, $Q_d$, and $Q_B$, with respect to the terminal voltages. A simpler (but a less accurate) capacitance model relates $C_{gs}$ and $C_{gd}$ to the channel capacitance per unit area, $c_g$ calculated at zero drain bias (see Section 6.2). One of the first such capacitance models was proposed by Meyer in 1970. His model is implemented in many SPICE simulators, including PSpice[tm]. One of the capacitance models implemented in AIM-Spice is the modified Meyer model given by the following equations:

$$C_{gs} = C_f + \frac{2}{3}C_{gc}\left[1 - \left(\frac{V_{sate} - V_{dse}}{2V_{sate} - V_{dse}}\right)^2\right] \qquad (6\text{-}4\text{-}37)$$

$$C_{gd} = C_f + \frac{2}{3}C_{gc}\left[1 - \left(\frac{V_{sate}}{2V_{sate} - V_{dse}}\right)^2\right] \qquad (6\text{-}4\text{-}38)$$

Here, $V_{sate} = I_{sat}/g_{ch}$ is the effective extrinsic saturation voltage, $I_{sat}$ is the saturation drain current, $g_{ch}$ is the extrinsic channel conductance at low drain-source bias, and $V_{dse}$ is an effective extrinsic drain-source voltage. $V_{dse}$ is equal to $V_{ds}$ for $V_{ds} < V_{sat}$ and is equal to $V_{sate}$ for $V_{ds} > V_{sat}$. In order to obtain a smooth transition between the two regimes, $V_{dse}$ is interpolated by the following equation:

$$V_{dse} = V_{ds}\left[1 + \left(\frac{V_{ds}}{V_{sate}}\right)^{m_c}\right]^{-1/m_c} \tag{6-4-39}$$

where $m_c$ is a constant determining the width of the transition region between the linear and the saturation regime. The capacitance $C_f$ in eqs. (6-4-37) and (6-4-38) is the side wall and fringing capacitance which can be estimated as $C_f \approx \beta_c \varepsilon_s W$ where $\beta_c$ is on the order of 0.5 or, better still, taken from experimental data.

The source-substrate, drain-substrate, and gate-substrate capacitances are calculated in terms of the depletion capacitances of the substrate-to-source and substrate-to-drain junctions [see, for example, Shur (1990), p. 393].

More accurate models should take into account depletion charges and their dependence on both gate-to-source and drain-to-source voltages. One such model that links MOSFET *C-V* characteristics to the parameters of the MOSFET body plot, $\gamma_N$ and $\Gamma_N$ (see Section 6.3), is described by Lee et al. (1993).

### 6.4.6.  MOSFET models in SPICE.

With the development of more advanced technologies, MOSFET dimensions become smaller and smaller. The characteristics of very short channel devices may be substantially different from those predicted by the simple models described above. Model improvements may be introduced (and have been introduced) in many different ways. This (and, perhaps, also the extreme importance of MOSFETs for modern electronics) explains the large number of MOSFET models incorporated into different versions of SPICE: four models in PSpice[tm] and ten models in AIM-Spice. Table 6.4.1 summarizes the features, advantages, and disadvantages of different SPICE models. Appendix A5 lists default parameters of SPICE MOSFET models. (Level 2 model is used by program MOSIS. MOSIS is the National Science Foundation sponsored program, which allows educational institutions to fabricate MOSFET integrated circuits at a relatively low cost using services of major semiconductor manufacturers.) Table 6.4.2 lists important equations related to MOSFET modeling.

| LEVEL=1 | Long channel model (such as described in Section 6.3), developed in 1968 by Shichman-Hodges, applicable for $L \geq 4$ µm |
|---|---|
| LEVEL=2 | Short channel model, developed in 1980, applicable for $L \geq 2$ µm.  Used by MOSIS program |
| LEVEL=3 | Semiempirical short channel model, developed in 1980, applicable for $L \geq 2$ µm (more sophisticated than Level 2) |
| LEVEL=4 | BSIM (Berkeley Short channel IGFET Model), developed in 1985, applicable for $L \geq 1$ µm (one of the most popular MOSFET SPICE models) |
| LEVEL=5 | New BSIM (BSIM2) model |
| LEVEL=6 | Simple empirical short channel model (developed by Sakurai and Newton in 1990) |
| LEVEL=7 | Universal extrinsic short channel MOS model [described by Lee et al. (1993)], applicable for $L \geq 0.1$ µm. Allows the description of both subthreshold and above threshold regimes. |
| LEVEL=8 | Unified long channel MOS model [described by Lee et al. (1993)], applicable for $L \geq 4$ µm.  Allows the description of both subthreshold and above threshold regimes. |
| LEVEL=9 | Short channel MOS model [described by Lee et al. (1993)], applicable for $L \geq 0.09$ µm. |
| LEVEL=10 | Unified intrinsic short channel model [described by Lee et al. (1993)], applicable for $L \geq 0.5$ µm.  Allows the description of both subthreshold and above threshold regimes. |
| LEVEL=11 | Unified extrinsic amorphous silicon thin film transistor model [described by Lee et al. (1993)]. Allows the description of both subthreshold and above threshold regimes. |
| LEVEL=12 | Polysilicon thin film transistor model [described by Lee et al. (1993)]. Allows the description of both subthreshold and above threshold regimes. |

**Table 6.4.1.**  MOSFET SPICE models.  Levels 1, 2, 3, and 4 are implemented in both PSpice[tm] and AIM-Spice.  Levels 5, 6, 7, 8, 9, and 10 implemented in AIM-Spice.  Levels 11 and 12 (implemented in AIM-Spice) are Thin Film Transistor models.

| Charge control model | $qn_s \approx c_i(V_{GT} - V)$    where $c_i = \dfrac{\varepsilon_i}{d_i}$ |
|---|---|
| Drain current for long channel MOSFET below saturation | $I_d = \dfrac{W\mu_n c_i}{L}\left(V_{GT} - \dfrac{V_{DS}}{2}\right)V_{DS}$ |
| Drain saturation current for long channel MOSFET | $I_{sat} = \dfrac{W\mu_n c_i}{2L} V_{GT}^2$ |
| Transconductance for long channel MOSFET | $g_m = \begin{cases} \beta V_{DS}, & \text{for } V_{DS} \le V_{SAT} \\ \beta V_{GT}, & \text{for } V_{DS} > V_{SAT} \end{cases}$ |
| Drain saturation current for short channel MOSFET | $I_{sat} = \dfrac{g_{ch}V_{GT}}{1 + \sqrt{1 + \left(\dfrac{V_{GT}}{V_L}\right)^2}}$ |
| Effect of series resistances on transconductance and output conductance | $g_m = \dfrac{g_{mo}}{1 + g_{mo}R_s + g_{do}(R_s + R_d)}$ <br><br> $g_d = \dfrac{g_{do}}{1 + g_{mo}R_s + g_{do}(R_s + R_d)}$ |
| Effect of series resistances on drain saturation current for short channel MOSFET | $I_{sat} = \dfrac{g_{ch}V_{gt}}{1 + g_{ch}R_s + \sqrt{1 + 2g_{ch}R_s + (V_{gt}/V_L)^2}}$ <br> where <br> $V_L = F_s L$, $g_{ch} = g_{cho}/(1 + g_{cho}(R_s + R_d))$, $g_{cho} = c_i V_{gt}\mu_n W/L$. |
| Dependence of threshold voltage on drain bias | $V_T = V_{To} - \sigma V_{DS}$ |
| Gate-to-channel capacitance | $c_{gc} = \dfrac{c_a c_b}{c_a + c_b}$   where <br><br> $c_a = \dfrac{\varepsilon_i}{d_i}$   $c_b = \dfrac{qn_o}{\eta V_h}\exp\left(\dfrac{V_{GT}}{\eta V_h}\right)$   $n_o \approx \dfrac{\varepsilon_i \eta V_{th}}{2qd_i}$ |

**Table 6.4.2.** The summary of the most important equations related to MOSFET modeling.

## 6-5.  FET SMALL SIGNAL EQUIVALENT CIRCUIT

A simplified FET small signal equivalent circuit is shown in Fig. 6.5.1a.  (For a more accurate description of the small signal response of a particular FET, we may need additional circuit elements specific for that device.  For example, substrate-to-channel capacitances may play an important role in a MOSFET and have to be included.)

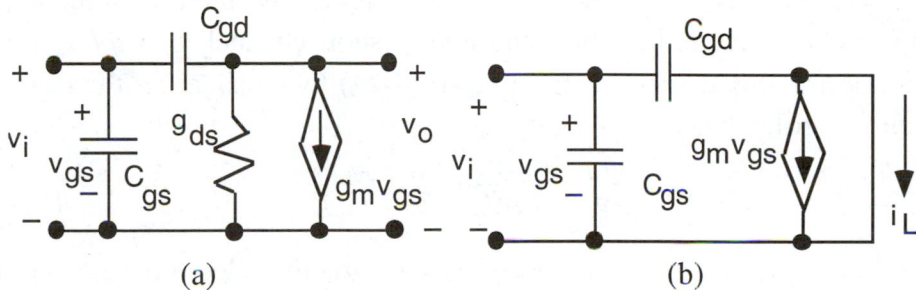

(a)                                    (b)

**Fig. 6.5.1.**  (a) Simplified MOSFET equivalent circuit and (b) equivalent circuit for the calculation of short circuit current gain.

Using the circuit shown in Fig. 6.5.1a, we can analyze the FET performance as a small signal common-source amplifier.  A standard circuit analysis leads to the following expression for the voltage gain:

$$A_v = \frac{v_o}{v_i} = \frac{-g_m + j\omega C_{gd}}{g_{ds} + j\omega C_{gd}} \tag{6-5-1}$$

Hence, the absolute value of the voltage gain is always smaller than $g_m/g_{ds}$ where $g_{ds}$ is the drain-to-source differential conductance.  Using the equivalent circuit shown in Fig. 6.5.1b, we find the short circuit current gain

$$A_i = \frac{i_L}{i_i} = \frac{-g_m}{j\omega\left(C_{gs} + C_{gd}\right)} \tag{6-5-2}$$

From this equation we can find the cutoff frequency, $f_T$, at which the absolute value of the short circuit current gain is equal to unity:

$$f_T = \frac{g_m}{2\pi\left(C_{gs} + C_{gd}\right)} \tag{6-5-3}$$

(see Problem 6-5-1).  We can assume that $C_{gs} + C_{gd} \sim C_i$ where $C_i = \varepsilon_i WL/d_i$ is the gate oxide capacitance.  The MOSFET transconductance, $g_m$, can be estimated as follows. At low drain bias, $V_{DS}$, the drain current $I_d = q n_s \mu \dfrac{V_{DS}}{L} W$ where $n_s$ is the electron concentration in the channel per unit area. Hence, $I_d \approx (Q_G / WL) v_{eff} W$ and

$$g_m = \frac{\partial I_d}{\partial V_g} = q \frac{\partial n_s}{\partial V_g} \mu \frac{V_{DS}}{L} W \approx C_i \mu \frac{V_{DS}}{L^2}$$

where $C_i = c_i WL = G_{gs} + C_{gd}$ is the gate-to-channel capacitance. In the opposite limiting case of a large drain bias and a very short channel, $I_d = q n_s v_s W$ [as an exercise, derive this expression from eq. (6-4-24) by considering the case when $V_L$ tends to zero]. Hence

$$g_m = \frac{\partial I_d}{\partial V_g} = q \frac{\partial n_s}{\partial V_g} v_s W \approx C_i \frac{v_s}{L}$$

In both cases, $g_m = C_i v_{eff}/L$ where $v_{eff}$ is the effective electron velocity in the channel, and we obtain from eq. (6-5-3)

$$f_T = 1/2\pi t_{tr} \qquad (6\text{-}5\text{-}4)$$

where $t_{tr} = L/v_{eff}$ is the transit time of electrons in the channel. Assuming that $v_{eff}$ to be of the order of $5 \times 10^4$ m/s (which is about one half of the electron saturation velocity in Si), we obtain a characteristic transit time for a MOSFET, on the order of $t_{tr}$ (ps) $\approx 20 \times L$ (μm) and $f_T$ (GHz) $\sim 8/L$ (μm). In fact, the measured switching times may be quite a bit larger because the transistor response is slowed down by the parasitic and fringing capacitances, $C_p$, which add to the gate capacitance:

$$f_T = \frac{g_m}{2\pi\left(C_{gs} + C_{gd} + C_p\right)} \qquad (6\text{-}5\text{-}5)$$

It is instructive to compare eqs. (6-5-3) to (6-5-5) with similar equations for a Bipolar Junction Transistor (see Section 5.5):

$$f_T \approx \frac{g_m}{2\pi\left(C_e + C_{b'c}\right)} = \frac{1}{2\pi\tau_{eff}} \qquad (6\text{-}5\text{-}6)$$

where $\tau_{eff}$ is the effective delay time.

    The cutoff frequency is a characteristic of the intrinsic transistor (without parasitic series resistances). The maximum oscillation frequency, $f_{max}$, is a characteristic of the extrinsic device (which includes series source, drain, input, and gate resistances, $R_s$, $R_d$, $R_i$, and $R_g$; see Fig. 6.5.2). ($f_{max}$ is defined as the frequency at which the power gain of the transistor is equal to unity under optimum matching conditions for the input and output impedances; see Section 5.5.)  The analysis of the FET equivalent circuit yields

$$f_{max} \approx \frac{f_T}{\sqrt{r + f_T \tau}} \qquad (6\text{-}5\text{-}7)$$

where  $r = g_{ds}(R_s + R_i + R_d)$  and  $\tau = 2\pi R_g C_{gd}$.    When the output drain conductance, $g_{ds}$, is small, eq. (6-5-7) becomes

$$f_{max} \approx \sqrt{\frac{f_T}{2\pi R_g C_{gd}}} \qquad (6\text{-}5\text{-}8)$$

Once, again, we can compare eq. (6-5-8) with a similar equation for a BJT (see Section 5.5)

$$f_{max} = \sqrt{\frac{f_T}{8\pi r_{bb'} C_{b'c}}} \qquad (6\text{-}5\text{-}9)$$

which shows that the base spreading resistance plays a similar role for a BJT as the gate series resistance in a FET.

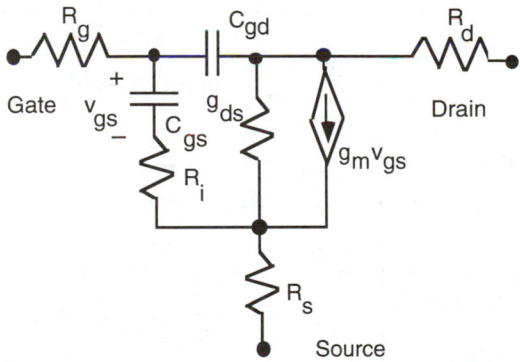

**Fig. 6.5.2.**  FET equivalent circuit including parasitic elements.  Parasitic inductances $L_s$ and $L_g$ may be added in series with $R_s$ and $R_g$, respectively, for a more realistic modeling.

**Example 6-5-1.**

Assuming that the effective electron velocity in a Si MOSFET equal to $5 \times 10^4$ m/s and the parasitic capacitance $C_{pw} = C_p/W = 0.3$ fF/micron, estimate the cutoff frequency of a Si MOSFET with the gate length $L = 0.1$ $\mu$m and the $SiO_2$ thickness $d_i = 100$ Å.  (The dielectric permittivity of $SiO_2$ is $3.45 \times 10^{-11}$ F/m.)

**Solution:**

Since the MOSFET transconductance $g_m = C_i v_{eff}/L$, and $C_{gs} + C_{gd} = C_i$, we find from eq. (6-5-5),

$$f_T = \frac{v_{eff}}{2\pi L \left(1 + \dfrac{C_p}{C_i}\right)} = \frac{v_{eff}}{2\pi L \left(1 + \dfrac{C_{pw} d_i}{\varepsilon_i L}\right)}$$

Substituting the numerical values of the MOSFET parameters, we obtain $f_T = 42.6$ GHz.  Hence, deep submicron Si MOSFET should be able to operate at microwave frequencies (provided that our estimate of the parasitic capacitance is not too optimistic)!

Table 6.5.1 lists the equations related to FET small signal response.

| FET voltage gain | $A_v = \dfrac{v_o}{v_i} = \dfrac{-g_m + j\omega C_{gd}}{g_{ds} + j\omega C_{gd}}$ |
|---|---|
| FET short circuit current gain | $A_i = \dfrac{i_L}{i_i} = \dfrac{-g_m}{j\omega\left(C_{gs} + C_{gd}\right)}$ |
| FET cutoff frequency | $f_T = \dfrac{g_m}{2\pi\left(C_{gs} + C_{gd}\right)} \qquad f_T = \dfrac{1}{2\pi t_{tr}}$ <br> where $t_{tr} = L/v_{eff}$ |
| FET maximum frequency of oscillations | $f_{max} = \dfrac{f_T}{\sqrt{r + f_T \tau}}$ where $r = g_{ds}\left(R_s + R_i + R_d\right)$ <br> and $\tau = 2\pi R_g C_{gd}$ |

**Table 6.5.1.**  The summary of the most important equations related to FET small signal response.

## 6-6.   CMOS

**Complementary Metal Oxide Semiconductor (CMOS)** technology dominates the electronics market.   In this technology, the basic element of electronic circuits – the inverter – is formed by connecting two MOSFETs – $n$-channel and $p$-channel MOSFETs fabricated on the same wafer.   In order to achieve such an integration, the $n$-channel and $p$-channel devices have to be isolated.   Just as in a regular MOSFET, depletion regions between $n$-type and $p$-type doped regions can be used for the isolation.   Figure 6.6.1 shows typical CMOS structures.   In $n$-well CMOS,   $p$-channel devices are fabricated in the ion-implanted $n$-type well.   In $p$-well CMOS, $n$-channel devices are fabricated in the ion-implanted $p$-type well.   The twin-well CMOS technology utilizes both $n$-type and $p$-type wells.

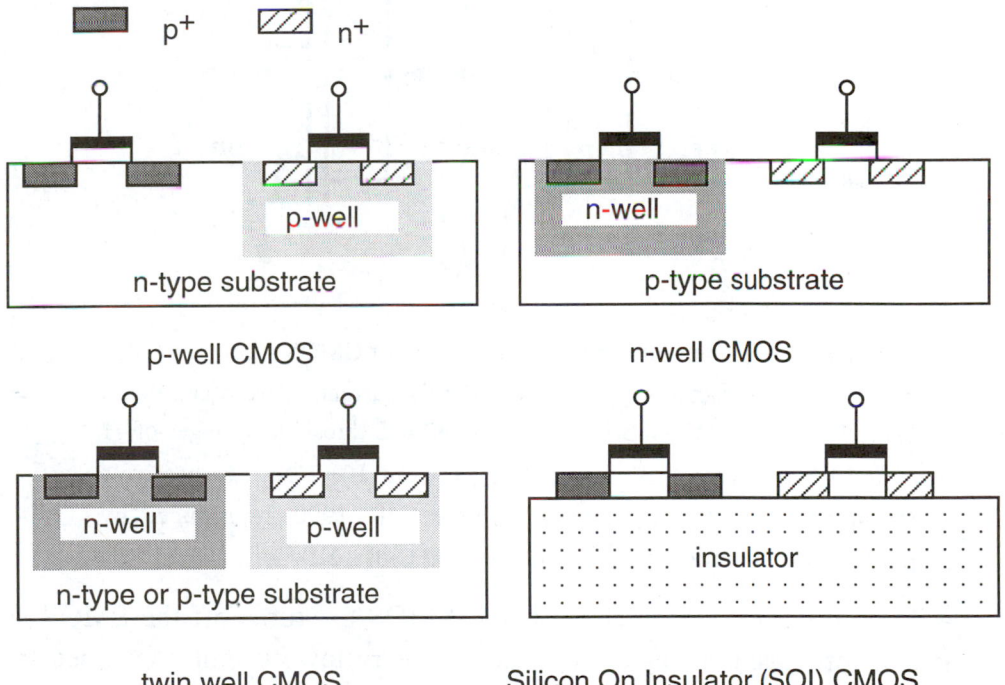

**Fig. 6.6.1.**   CMOS technologies.

The basic element of digital electronic circuits is the CMOS inverter (see Fig. 6.6.2).   In a CMOS inverter, the gates of $n$- and $p$-channel MOSFETs are tied together as shown in the figure.

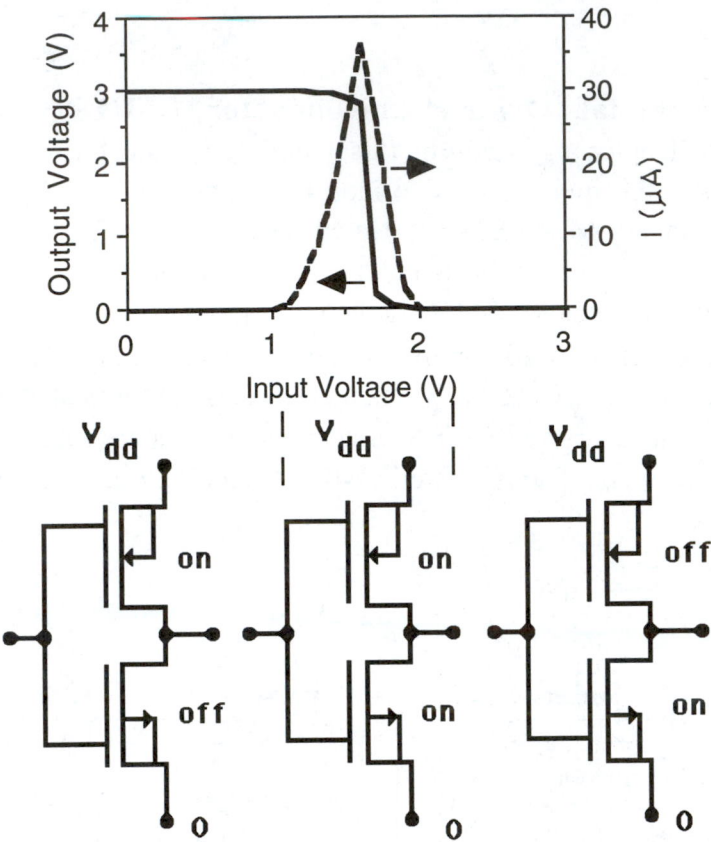

**Fig. 6.6.2.** Schematic illustration of operation of CMOS gate.

The transfer characteristic and power supply current, $I$, were computed using PSpice[tm] for $V_{dd} = 3$ V, PMOS and NMOS threshold voltages of $-1$ V and 1 V, and gate dimensions $30 \times 1$ $\mu$m$^2$ and $10 \times 1$ $\mu$m$^2$, respectively (from Shur, 1990, copyright © Prentice Hall, 1990, reproduced by permission of Prentice Hall, Inc., Englewood Cliffs, NJ).

A positive gate bias, which turns the NMOS on, turns off the PMOS, and vice versa. Both stable states correspond to a very low current consumed from the power supply. Figure 6.6.2 shows the transfer characteristic of such an inverter and the dependence of the current supplied by the power supply, $V_{dd}$, on the input voltage.

**Example 6-6-1.**

Using PSpice[tm] calculate dependencies shown in Fig. 6.6.2.

**Solution:**

This dependencies are computed using the following input PSpice[tm] file:

```
CMOS TRANSFER CURVE
* Options for the run.
.OPT ACCT NOMOD NOPAGE RELTOL=.001
.TEMP 25
VIN 2 0 DC 3
VDD 1 0 DC 3
* Sweep the voltage source VIN from*   0 volts to 3 volts in steps of 0.1 volts.
.DC VIN 0 3 0.1
M1 3 2 1 1  MP L =1u W = 30u
M2 0 2 3 0  MN L =1u W = 10u
.MODEL MP PMOS (VTO = -1)
.MODEL MN NMOS (VTO = 1)
.PRINT DC V(3)  I(VDD)
.PLOT DC V(3)  I(VDD)
.END
```

Figure 6.6.2 confirms that a CMOS inverter has a very low power consumption in either of the two stable states corresponding to output high and output low. Another advantage of CMOS technology is that it can operate in a wide range of supply voltages (from 15 V to 3 V or less) with a large output voltage swing (from zero to the supply voltage for the CMOS inverter shown in Fig. 6.6.2). There are also certain disadvantages. Since the hole mobility in Si is smaller than the electron mobility (300 $cm^2$/Vs or so compared to 800 to 1200 $cm^2$/Vs), PMOS is slower than NMOS. Hence, generally speaking, CMOS circuits are slower than NMOS circuits. To a certain extent, this can be alleviated by using clever circuits, which use fewer $p$-channel devices and more $n$-channel devices. CMOS may also require a somewhat larger number of fabrication steps than NMOS. Finally as we can see from Fig. 6.6.1, CMOS devices have $p$-$n$-$p$ and $n$-$p$-$n$ regions formed by contacts, substrates, and implanted wells. These sequences of regions form parasitic Bipolar Junction Transistors. Under certain bias conditions, these parasitic BJTs can lead to a very large currents, which may even destroy the device. This effect is called **latch-up**, and special design features have to be introduced to control it. However, modern CMOS technologies are very successful in controlling latch-up.

Of course, the fourth structure in Fig. 6.6.1, representing **Silicon-On-Insulator technology (SOI)**, is totally devoid of the latch-up problem. Except for a higher cost, SOI technology is superior in many ways. (The cost, however, is a very important factor in the modern world.) All drawbacks of the more conventional CMOS technologies can be brought under control, with the possible exception of the nonideal effects in CMOS with submicron or/and deep submicron feature sizes.  SOI technology also eliminates any substrate current and reduces parasitic capacitances related to the depletion regions formed in the semiconducting substrate.  In one popular version of this technology, silicon islands are grown on sapphire and used for $n$- and $p$-channel transistors.  The high cost of sapphire substrates and the difficulties related to the epitaxial growth of silicon on sapphire have slowed down the applications of this technology.  Still, this is a well developed technology, and Silicon-On-Sapphire chips are produced in significant quantities by a number of leading semiconductor manufacturers.  Further refinement of SOI technologies will also open up the possibility of vertical integration, leading to the development of three-dimensional integrated circuits (see Fig. 6.6.3).

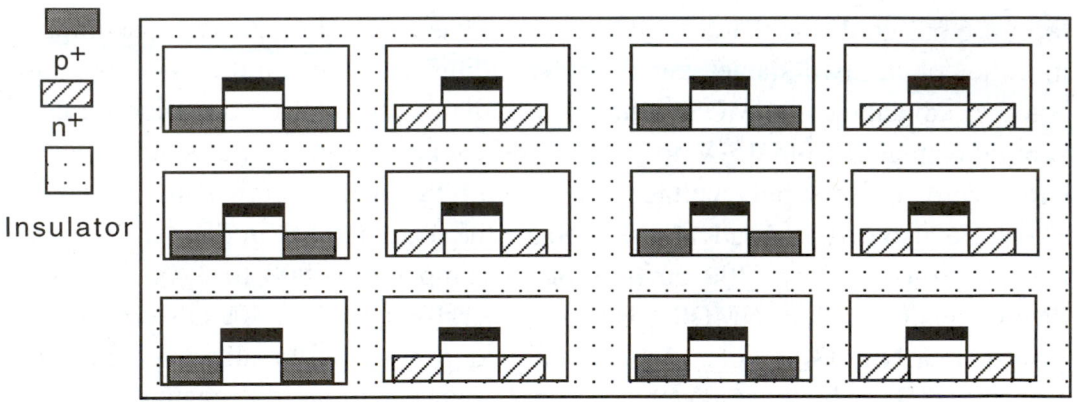

**Fig. 6.6.3.**  Schematic diagram of three-dimensional integrated CMOS circuit based on SOI technology.

As can be seen from Fig. 6.6.2, PMOS and NMOS occupy symmetrical positions in the CMOS inverter circuit.  Therefore, it is not surprising that the best performance is achieved when the PMOS and NMOS characteristics are symmetrical so that $V_{TN} = - V_{TP}$ where $V_{TN}$ and $V_{TP}$ are the threshold voltages for NMOS and PMOS, respectively.  The MOSFET gates are usually made from $n^+$ doped polysilicon.  As was discussed in Section 6.3, for such gates, the flat band voltage is approximately $-1$ V for NMOS and nearly 0 V for PMOS.  Typically, the preferred threshold voltages for CMOS are $V_{TN} \approx 0.7$ V and $V_{TP}$

$\approx - 0.7$ V. These values may be obtained by adjusting the doping level in the substrate and in the ion-implanted well. These adjustments are easier to achieve in the twin-well technology (where the doping levels in both wells can be adjusted separately). However, this technology also involves an additional cost related to a larger number of fabrication steps. Alternatively, an additional implant near the surface can be used to control the threshold voltage. The theory behind the threshold voltage adjustment was considered in Section 6.3.

For CMOS operating at low supply voltages, smaller values of threshold voltages are required. These values can be achieved using a "dual-poly" technology, that is, by using $n^+$ polysilicon gates for NMOS and $p^+$ polysilicon gates for PMOS (see Problems 6-6-2 to 6-6-4). (Unfortunately, this technology is more expensive.)

CMOS structure has both $n$-type and $p$-type regions. These regions can be used to fabricate integrated BJTs on the same wafer. Such a technology is called **BiCMOS** (Bipolar CMOS). BiCMOS technology combines the best features of both CMOS and BJTs and may be superior to CMOS for high-speed applications with a limited power consumption. Still, it is very difficult to compete with regular CMOS because of their low cost and very low power consumption. CMOS technology is becoming more and more competitive every year. The last decade of the twentieth century has truly become a CMOS decade! To emulate this success, researchers are also trying to develop complementary field effect technology using compound semiconductors (such as GaAs; see Chap. 7). This new technology promises a higher speed and a low power consumption. However, the potential impact of this technology is not yet clear.

Table 6.6.1 summarizes the basic information related to CMOS.

| | |
|---|---|
| Basic CMOS technologies | $n$-well, $p$-well, twin-well, and SOI (Silicon-On-Insulator) |
| Preferred relation between NMOS and PMOS threshold voltages in CMOS inverter | $V_{TN} = - V_{TP}$ |
| Latch-up | A large current surge in CMOS related to parasitic BJT structures |
| BiCMOS | Technology using CMOS and BJTs integrated on the same chip |

**Table 6.6.1.** Summary of basic information related to CMOS.

## 6-7.  PHYSICAL CONSTRAINTS ON MOSFET PERFORMANCE

The achievements of MOSFET technology have been nothing short of spectacular.  Modern MOSFETs operate up to gigahertz frequencies, and the power supply voltages for MOSFET integrated circuits have shrunk from 9 V to 3.3 V with a commensurate decrease in power consumption.  Even faster, less power-hungry devices will undoubtedly be developed in the future.  So what are the physical constraints that may impede this unrelentless progress?

As a rule of thumb, smaller devices operate faster (since it takes less time for electrons to travel from one contact to another) and require less power. Therefore, it is important for us to understand how small we can make devices and how their design should be changed when the dimensions are scaled down. MOSFETs with gate length as short as 0.09 μm have been demonstrated in laboratories.  We can probably decrease the gate length a bit more, perhaps down to 300 to 500 Å, before quantum effects will become so important that they will dominate the device performance.  A MOSFET with such a short gate will become a different device, and we do not really understand that well how it will work.

**Example  6-7-1.**
Calculate the de Broglie wavelength for a light hole in Si with the effective mass $m_p$ = 0.153 $m_e$ at room temperature and comment on how it relates to the 0.09 μm gate length of a $p$-channel Si MOSFET.

**Solution:**
Light holes move randomly with a typical thermal velocity, $v_{thp}$, given by

$$v_{thp} = \sqrt{\frac{3k_BT}{m_p}}$$

Hence, the typical light hole wave vector is given by (see Section 1.2)

$$k = \frac{m_p v_{thp}}{\hbar} = \frac{1}{\hbar}\sqrt{3k_BTm_p}$$

and the de Broglie wavelength is found from

$$\lambda = \frac{2\pi}{k} = \frac{2\pi\hbar}{\sqrt{3k_BTm_p}} = \frac{2\pi\times1.055\times10^{-34}}{\sqrt{3\times1.38\times10^{-23}\times300\times0.153\times9.11\times10^{-31}}} = 400\ (\text{Å})$$

This is only about half of the 0.09 Å gate length mentioned above.

The gate length, $L$, determines other geometric dimensions of a MOSFET as well as the substrate doping.  Typically, the device width, $W$, and the gate insulator thickness, $d$, scale proportionally to the gate length.  Since $n$-type and $p$-type regions in a MOSFET are isolated by depletion regions, the width of these regions should also be scaled proportionally to $L$.  This means that the substrate doping level, $N_a$, should be scaled as $1/L^2$.  (Please explain why, using the equations for the width of the depletion region for an $n^+$-$p$ junction at zero bias derived in Section 4.2.)  These scaling rules are summarized in Table 6.7.1.

| Dimension of doping level | Scaling with gate length, $L$ |
|---|---|
| Insulator thickness, $d$ | Proportional to $L$ |
| Device width, $W$ | Proportional to $L$ |
| Substrate doping, $N_a$ | Proportional to $1/L^2$ |

**Table 6.7.1.**  Scaling for MOSFETs.

Assuming that the saturation velocity in Si is close to $1\times10^5$ m/s, we obtain that the transit time in a 0.1 μm MOSFET is on the order of 1 ps.  If modern computers using MOSFETs with gate lengths on the order of 0.8 μm operate with a 100 MHz clock, it is safe to predict that a 0.1 μm technology will be at least 10 times faster.  In addition, the power consumption will decrease.  Indeed, the power-delay product of a MOSFET, $P\tau$, is equal to the energy stored in the MOSFET channel capacitance, $C$

$$P\tau = C\Delta V^2 \qquad\qquad (6\text{-}7\text{-}1)$$

Here $\Delta V$ is the gate voltage swing.  $\Delta V$ should be much greater than the thermal voltage to provide a clear transition from the below- to the above-threshold regime (i. e., from off-state to on-state) during device switching (see Problem 6-7-1).  The value of $C$ can be estimated as

$$C = \frac{\varepsilon_i WL}{d} \propto L \tag{6-7-2}$$

which means that the power-delay product decreases with the decrease in $L$. Assuming $L = 0.1$ μm, $W/L = 3$, $L/d = 15$, $\varepsilon_i = 3.45 \times 10^{-11}$ F/m (the dielectric permittivity of $SiO_2$), and $\Delta V = 1$ V, we obtain $P\tau \approx 0.16$ fJ. Such a low-power operation will truly transform our lives, making wireless communications – the area of technology that literally exploded recently – much more practical and affordable. And we are talking about an electronic device technology that is relatively mainstream and where many developments are evolutionary rather than revolutionary. Even better performance may come from novel semiconductor materials and novel semiconductor devices considered in Chapters 7 and 9 of this book. May or may not. The old, reliable MOSFET technology is really hard to beat when it comes to large-scale applications.

But the road to Very Large Scale Integrated (VLSI) circuits using the 0.1 μm MOSFET technology is not covered by roses either. In such small devices, non-ideal, **short channel effects** become dominant. Primarily, these effects arise because the electric fields in small devices become enormous (on the order of $10^7$ V/m in a 0.1 μm gate device). In such high fields, electrons in the MOSFET channel gain a large energy from the electric field, causing all kinds of troubles, such as impact ionization (see Sections 3.8 and 4.5), and the electron transfer into the traps in the gate dielectric (such a transfer leads to a change in the device characteristics with time caused by applied bias voltages). These **hot electron effects** present a major obstacle to scaling down MOSFET dimensions. Also, contact resistances become comparable with or even larger than the channel resistance. Still another limitation of the MOSFET performance is linked to the very high lateral electric field (perpendicular to the MOSFET interface). Since the gate dielectric thickness is proportional to $L$ and the gate-voltage swing $\Delta V$ is nearly independent of $L$, the surface electric field is proportional to $1/L$. Very high surface electric fields squeeze electrons in the channel closer to the gate dielectric-semiconductor interface. This increases the electron scattering by the interface roughness and decreases the field effect mobility. Nevertheless, many of these problems can be solved, and the current-voltage characteristics of short channel MOSFETs (with a 0.1 μm feature size) may look nearly ideal. Hence, the 0.1 μm VLSI may become a reality before the start of the twenty-first century.

# REFERENCES

K. LEE, M. SHUR, T. A. FJELDLY, AND T. YTTERDAL, *Semiconductor Device Modeling for VLSI*, Prentice Hall, Englewood Cliffs, NJ (1993)

M. SHUR, *Physics of Semiconductor Devices*, Series in Solid State Physical Electronics, Prentice Hall, Englewood Cliffs, NJ (1990)

Y. P. TSIVIDIS, *Operation and Modeling of the MOS Transistor*, McGraw-Hill, New York (1987)

# BIBLIOGRAPHY

S. M. SZE, *Physics of Semiconductor Devices*, Second Edition, John Wiley & Sons, New York (1981)

S. M. SZE, *Semiconductor Devices. Physics and Technology*, John Wiley & Sons, New York (1985)

S. M. SZE, Editor, *VLSI Technology*, Second Edition, McGraw-Hill, New York (1988)

R. F. PIERRET, *Field Effect Devices*, Addison Wesley Modular Series on Solid State Devices, Vol. 4, Reading, MA (1983)

*Undergraduate text on FETs with many examples and problems.*

D. K. SCHRODER, *Advanced MOS Devices*, Addison Wesley Modular Series on Solid State Devices, Reading, MA (1987)

*Textbook on FETs with many examples and problems.*

H. C. de GRAAFF AND F. M. KLAASSEN, *Compact Transistor Modeling for Circuit Design*, Computational Microelectronics Series, S. Selberherr, Editor, Springer-Verlag, New York (1990)

*Emphasis on circuit models suitable for implementation in circuit simulators, such as SPICE.*

A. D. MILNES, Editor, *MOS Devices, Design and Manufacture*, Edinburgh University Press, Edinburgh (1983)

*A review of earlier work on MOSFETs.*

E. H. NICOLLIAN AND J. R. BREWS, *MOS Physics and Technology*, John Wiley & Sons, New York (1982)

*One of the classic books on MOSFETs.*

D. G. ONG, *Modern MOS Technology, Processes, Devices, & Design*, McGraw-Hill, New York (1984)

*A clear and simple book on MOSFETs written by an INTEL designer.*

B. G. STREETMAN, *Solid State Electronic Devices*, Fifth Edition, Prentice Hall, Englewood Cliffs, NJ (1995)

*A popular undergraduate text on semiconductor devices.*

M. SHOJI, *CMOS Digital Circuit Technology*, Prentice Hall, Englewood Cliffs, NJ (1988)

*Emphasis on CMOS design for digital applications.*

A. R. ALVAREZ, *BiCMOS Technology and Applications*, Second Edition, Kluwer Academic Publishers, Boston (1993)

*Book dealing with MOS, CMOS, and BiCMOS technologies.*

P. E. ALLEN AND D. R. HOLBERG, *CMOS Analog Circuit Design*, Holt, Rinehart and Winston, New York (1987)

*Design of analog MOS circuits.*

M. SZE, Editor, *High-Speed Semiconductor Devices,* John Wiley & Sons, New York (1990)

*A very good book with a chapter on submicron MOSFETs by John Brews.*

# REVIEW QUESTIONS

**1.** List different field effect technologies and their applications.

☐ 1 point

**2.** A Si MOSFET has the subthreshold current of $10^{-7}$ A at the gate-to-source voltage $V_{gs} = 0$ V and $10^{-8}$ A at $V_{gs} = -0.1$ V (at room temperature, $T = 300$ K). What is the subthreshold ideality factor?

☐ 2 points

**3.** A silicon MOSFET has a threshold voltage of $-1$ V. The oxide thickness is 200 Å. The oxide dielectric permittivity is $3.45 \times 10^{-11}$ F/m. What gate-to-source voltage will induce a surface electron concentration $n_s = 2 \times 10^{12}$ cm$^{-2}$ into the MOSFET channel ?

☐ 1 point

**4.** A silicon MOSFET has a threshold voltage of $-1$ V. The oxide thickness is 200 Å. The oxide dielectric permittivity is $3.45 \times 10^{-11}$ F/m. The subthreshold slope factor $\eta = 2$. Estimate the surface electron concentration, $n_s$, at $T = 300$ K in the MOSFET channel at the gate bias of $-2$ V.

☐ 1 point

**5.** Define the flat band voltage.

☐ 1 point

**6.** The silicon dielectric permittivity is $1 \times 10^{-10}$ F/m. The acceptor concentration in a $p$-type MOSFET substrate is $N_a = 10^{16}$ cm$^{-3}$. Calculate the depletion charge per unit area for a 0.1 V surface potential at the semiconductor-insulator interface (with respect to the bulk.)

☐ 1 point

**7.** From Fig. 6.3.1, estimate the signs and magnitudes of the flat-band voltages for $n$-channel and $p$-channel MOSFETs.

☐ 2 points

**8.** Will the threshold voltage shift toward positive or negative values if the substrate of an $n$-channel MOSFET is implanted with acceptors?

$\square$  1 point

**9.** How much will the threshold voltage shift if the substrate of an $n$-channel MOSFET is implanted with acceptors with a total surface concentration $p_s = 5 \times 10^{11}$ cm$^{-2}$? The oxide thickness is 200 Å. The oxide dielectric permittivity is $3.45 \times 10^{-11}$ F/m.

$\square$  2 points

**10.** Explain the significance and criterion of the validity of the gradual channel approximation.

$\square$  1 point

**11.** What is the definition of the pinch-off condition?

1 point

$\square$

**12.** Define transconductance and drain-to-source conductance.

$\square$  2 points

**13.** A silicon MOSFET has the drain current of 1 mA at the gate-to-source voltage $V_{GS} = 1$ V and 2 mA at $V_{GS} = 2$ V (temperature, $T = 300$ K). The device operates in a linear regime (at small drain bias $V_{DS} = 100$ mV, much smaller than $V_{GT} = V_{GS} - V_T$ where $V_T$ is the threshold voltage and $V_{GS}$ is the gate-to-source bias). The gate length is 1 μm. The oxide thickness is 200 Å. The oxide dielectric constant is $3.45 \times 10^{-11}$ F/m. The electron mobility in the device channel is 700 cm$^2$/Vs. Parasitic source and drain resistances are negligible.

(a) What is the threshold voltage?

$\square$  2 points

(b) What is the device width?

$\square$  2 points

**14.** The MOSFET transconductance and drain-to-source conductance in the saturation region are 20 mS and 2 mS, respectively. What is the maximum achievable voltage gain for this device?

$\qquad$ 1 point

**15.** What are the signs of the threshold voltages for $n$-channel and $p$-channel devices in CMOS technology?

$\qquad$ 1 point

**16.** What are the advantages and disadvantages of SOI technology?

$\qquad$ 1 point

**17.** What is latch-up?

$\qquad$ 1 point

## PROBLEMS

**6-2-1.** For gate voltages from 0 V to 3 V, plot and compare the dependencies of the surface carrier concentrations in an FET channel at $T = 300$ K using a Charge Control Model (CCM) and the Universal Charge Control Model (UCCM). The threshold voltage is 0.7 V. The oxide thickness is 150 Å. The dielectric permittivity of $SiO_2$ is $3.45 \times 10^{-11}$ F/m. Subthreshold ideality factor, $\eta = 2$.

**6-3-1.** Sketch the band diagram of a MOS capacitor under bias and explain the relationship between the surface field and the total surface charge in the semiconductor:

$$\varepsilon_s F_s = Q_s$$

**Hint:** Use Gauss' law.

**6-3-2.** The dependence of the threshold voltage, $V_{TN}$, on $\sqrt{(2\varphi_b - V_{BS})} - \sqrt{2\varphi_b}$ for a long and wide channel Si NMOS transistor shown in the figure (this dependence is called a **body plot**). From this body plot, deduce the substrate doping and the flat-band voltage. (The oxide thickness is 150 Å.) The dielectric permittivities of Si and $SiO_2$ are $1.05\times10^{-10}$ F/m and $3.45\times10^{-11}$ F/m, respectively.

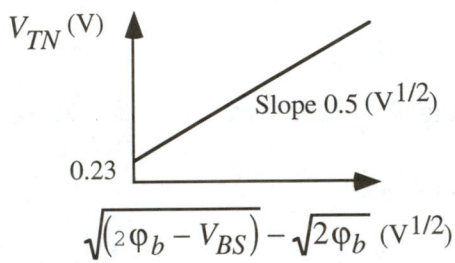

**\*6-3-3.** Calculate the depletion region width, $d_{dep}$, for a $p$-type MOSFET versus doping for substrate doping from $5\times10^{13}$ cm$^{-3}$ to $5\times10^{16}$ cm$^{-3}$ for the surface potential $V_s = 0.6$ V (measured with respect to the bulk potential). The dielectric permittivity of the semiconductor $\varepsilon_s = 1.05\times10^{-10}$ F/m. Based on this calculation, choose the doping level appropriate for a 1 µm gate device (requiring that the depletion regions between the $n^+$ source and drain contacts do not overlap).

**Hint:** Assume that the built-in potential between the $n^+$ contact and the substrate is also close to 0.6 V.

**6-3-4.** The measured threshold voltage of a fabricated $p$-channel MOSFET is equal to $V_T = -1.5$ V. The design target was $V_T = -0.8$ V. Describe the possible changes in the fabrication process and/or in the bias voltages needed in order to increase $V_T$ to $-0.8$ V. The gate oxide thickness and dielectric permittivity are 250 Å and $3.45\times10^{-11}$ F/m.

**6-3-5.** Sketch the field and voltage distribution in a MOS structure at the threshold gate voltage. Oxide thickness $d_i = 300$ Å, doping level $N_A = 10^{15}$ cm$^{-3}$. The dielectric permittivity of the semiconductor $\varepsilon_s = 1.05\times10^{-10}$ F/m, the oxide dielectric permittivity is $\varepsilon_i = 3.45\times10^{-11}$ F/m, the substrate bias $V_{sub} = 0$ V, the effective densities of states in the conduction and valence band $N_c = 3.22\times10^{19}$ cm$^{-3}$ and $N_v = $

$1.83\times10^{19}$ cm$^{-3}$, respectively, the energy gap $E_g = 1.12$ V, and $T = 300$ K. Assume a flat-band voltage of $-1$ V.

**6-3-6.** Plot the surface potential, $V_s$, versus gate bias, $V_{GS}$, for $-1$ V $< V_{GS} <$ 2 V. Use the following parameter values: gate oxide thickness $d_i = 150$ Å, gate oxide dielectric permittivity $\varepsilon_i = 3.45\times10^{-11}$ F/m, silicon dielectric permittivity $\varepsilon_s = 1.05\times10^{-10}$ F/m, flat-band voltage $V_{FB} = -1$ V, threshold voltage $V_T = 0.3$ V, substrate doping $N_a = 5\times10^{16}$ cm$^{-3}$, and substrate bias $V_{BS} = 0$ V.

**6-4-1.** Design NMOS and PMOS (i. e., choose reasonable and appropriate values of the gate oxide thickness and other parameters) with gate lengths of 2 µm and threshold voltages of approximately 0.7 V and $-0.7$ V, respectively. Using both the constant mobility model and the model accounting for the velocity saturation in the channel (and choosing reasonable and appropriate values of the electron and hole mobilities and saturation velocities), calculate the dependencies of the saturation current on the gate voltage for these devices. Choose the gate width of the NMOS to be 20 µm; choose the gate width of the PMOS in such a way that the saturation currents in both devices are close. Choose the same oxide thickness for both NMOS and PMOS. Assume an $n^+$ polysilicon gate for both devices (which results in flat-band voltages of $V_{FBN} \approx -1$ V and $V_{FBP} \approx 0$ V, respectively). Do not choose the dielectric thickness smaller than 150 Å.

**6-4-2.** Prove that

$$g_m = \frac{g_{mo}}{1 + g_{mo}R_s + g_{do}\left(R_s + R_d\right)}$$

where $g_m = \dfrac{\partial I_{ds}}{\partial V_{gs}}$ is the measured (extrinsic) transconductance of a field effect transistor, $g_{mo} = \dfrac{\partial I_{ds}}{\partial V_{GS}}$ is the intrinsic transconductance of the same device without the series resistances, $R_s$ and $R_d$, and $g_{do} = \dfrac{\partial I_{ds}}{\partial V_{DS}}$ is the intrinsic drain conductance of the same device without the series resistances, $R_s$ and $R_d$. Here the extrinsic and intrinsic gate-to-source and drain-to-source voltages are related as follows:

$$V_{gs} = V_{GS} + I_{ds} R_s \text{ and } V_{ds} = V_{DS} + I_{ds}(R_s + R_d)$$

where the gate voltage $V_{GS}$ is the voltage difference between the gate potential and the channel potential at the source side of the channel (intrinsic gate-to-source voltage) and $V_{DS}$ is the voltage difference between the channel potentials at the drain and source sides of the channel (intrinsic drain-to-source voltage).

**6-4-3.**   Calculate dependencies of the drain-to-source current on the drain-to-source voltage for a gate voltage $V_{gs} = 5$ V for a silicon MOSFET for the series resistances $R_s = R_d = 0\ \Omega$ and $R_s = R_d = 100\ \Omega$. The device gate length, $L = 4\ \mu m$, gate width $W = 100\ \mu m$, electron mobility in the device channel $\mu_n = 1{,}000$ cm$^2$/Vs, the dielectric permittivity of the gate insulator $\varepsilon_i = 3.45 \times 10^{-11}$ F/m, the oxide thickness is 350 Å, the dielectric permittivity of silicon $\varepsilon_s = 1.05 \times 10^{-10}$ F/m, flat-band voltage $V_{FB} = -1$ V, substrate bias $V_{sub} = 0$, temperature $T = 300$ K, and substrate doping $N_a = 10^{15}$ cm$^{-3}$. Neglect the velocity saturation effects.

**6-4-4.**   For gate voltages from 0 V to 2 V, plot the FET gate-to-channel capacitance for an FET with a threshold voltage of 0.7 V and an oxide thickness is 150 Å. The dielectric permittivity of SiO$_2$ is $3.45 \times 10^{-11}$ F/m. Temperature $T = 300$ K. The subthreshold ideality factor, $\eta = 2$.

**6-5-1.**   Using the circuit analysis of the FET equivalent circuit shown in Fig. 6.5.1b prove that $f_T = \dfrac{g_m}{2\pi(C_{gs} + C_{gd})}$.

**6-5-2.**   Using PSpice[tm], simulate the frequency response of an FET amplifier shown in the figure from 100 Hz to 100 MHz. Use a 2 μm gate length, a threshold voltage of 0.5 V, and a power supply of 5 V. Choose the transistor widths to obtain at least a voltage gain of 3 at low frequencies.

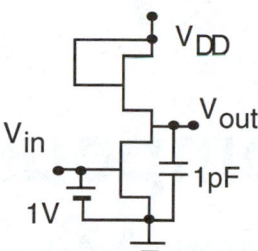

**6-6-1.** Using SPICE, simulate the circuit shown in Fig. 6.6.2. Compare your results with the output shown in the figure. Repeat these calculations for power supply voltage of 5 V and comment on the power consumption dependence on the power supply voltage.

**6-6-2.** Rewrite eq. (6-3-13) for PMOS.

**6-6-3.** Using Fig. 6.3.1, estimate the values of flat-band voltages for an NMOS with an $n^+$ polysilicon gate, for a PMOS with an $n^+$ polysilicon gate, and for a PMOS with an $p^+$ polysilicon gate.

**6-6-4.** Using the values of flat-band voltages for an NMOS with an $n^+$ polysilicon gate, for a PMOS with an $n^+$ polysilicon gate, and for a PMOS with an $p^+$ polysilicon gate obtained solving Problem 6-6-3, plot the threshold voltage for these three devices versus the substrate doping ($N_a$ for NMOS and $N_d$ for PMOS) varying $N_a$ and $N_d$ from $10^{15}$ cm$^{-3}$ to $10^{17}$ cm$^{-3}$. Use the following values of the device parameters: gate oxide thickness $d_i = 200$ Å, gate oxide permittivity $\varepsilon_i = 3.45 \times 10^{-11}$ F/m, silicon permittivity $\varepsilon_s = 1.05 \times 10^{-10}$ F/m, intrinsic carrier concentration, $n_i = 10^{10}$ cm$^{-3}$, substrate bias $V_{BS} = 0$ V, and room temperature ($T = 300$ K).

**6-7-1.** Assuming that the ratio of MOSFET currents in on-state and off-state should be at least $10^6$, find the minimum gate voltage swing at room temperature ($T = 300$ K). **Hint:** Assume that in the below-threshold regime, $I_{DS} \approx I_T \exp\left(\dfrac{V_{GT}}{\eta V_{th}}\right)$ where $I_T$ is the value of the drain current at the threshold and $\eta$ is the ideality factor (use $\eta = 1.2$).

# 7

# Compound Semiconductor FETs and Thin Film Transistors (TFTs)

## *7-1. INTRODUCTION

MOSFETs (especially CMOS) dominate the electronic device market. However, in many niche markets, this dominance has been successfully challenged by compound semiconductor FETs, especially at microwave and millimeter wave frequencies and in ultra-high-speed digital integrated circuits.

More recently relatively low performance devices – Thin Film Transistors – have found important applications in flat panel displays (see Section 8.7), where they are used in large area integrated circuits (with sizes that can be larger than 21 inches in diagonal).

This chapter describes these two families of FETs, which occupy two opposite sides of the electronic device market – very high- and very low-speed electronics.

## *7-2. COMPOUND SEMICONDUCTOR FETs

### *7.2.1. Applications of compound semiconductor technology.

Compound semiconductor field-effect transistors occupy an important niche in the electronics industry. Their applications include GaAs FET millimeter wave and microwave amplifiers, oscillators, mixers, switches, attenuators, modulators, and current limiters. High-speed integrated digital circuits based on GaAs FETs have been developed. The basic advantages of GaAs devices include a higher electron velocity and mobility, which lead to a smaller transit time and faster response, while semi-insulating GaAs substrates reduce parasitic capacitances and simplify the fabrication process. Other material systems such as AlGaAs/InGaAs, InGaAs/InP, and so on, have also exhibited superior device properties for applications in very high speed circuits.

Recently, the new approach of oxidizing a thin silicon layer grown by Molecular Beam Epitaxy on GaAs surface offered hope for the development of a viable GaAs MOSFET technology. (Still another approach to the development of a GaAs MOSFET involves using a more exotic compound for the gate insulator – $BaF_2$.) At the present time, Schottky barrier Metal Semiconductor Field Effect Transistors (MESFETs), Junction Field Effect Transistors (JFETs), and Heterostructure Field Effect Transistors (HFETs) are the most commonly used compound semiconductor FETs.

In MESFETs, the Schottky barrier gate contact is used in order to modulate the channel conductivity (see Fig. 6.1.4). This allows one to avoid problems related to traps in the gate insulator in MOSFETs, such as hot electron trapping in the gate insulator, threshold voltage shift due to charge trapped in the gate insulator, and so on. Silicon has a very stable natural oxide that can be grown with a very low density of traps, and the problems related to the gate insulators have been minimized for silicon MOSFETs. However, compound semiconductors, such as GaAs, do not have such a stable oxide, and most of the compound semiconductor devices use Schottky gates. The drawback of normally-off MESFET technology is a limited gate voltage swing due to the low turn-on voltage of the Schottky gate. However, this limitation is not important in depletion mode FETs with a negative threshold voltage (which are also called normally-on FETs) and is less important in low power digital circuits operating with a low supply voltage. Also, in many cases, GaAs MESFETs and JFETs are fabricated by direct ion implantation into a GaAs semi-insulating substrate, making GaAs IC fabrication less complicated than CMOS fabrication.

Compound semiconductor Heterostructure Field Effect Transistors use a depleted or an undoped layer of a wide band gap semiconductor separating the gate electrode from the channel. This leads to a larger gate voltage swing. Also, these transistors may have a higher field effect mobility since, in most HFET designs, the device channel is devoid of ionized impurities. This minimizes the carrier scattering and leads to a higher field effect mobility.

Compound semiconductor technology has both advantages and disadvantages when compared to silicon technology. The advantages over silicon include higher speed and/or smaller power dissipation (at a comparable speed), and higher radiation hardness. However, compound semiconductor technology is much less developed. In an elemental semiconductor, such as silicon, the device quality depends on the material purity. In a compound semiconductor, such as GaAs, the material composition is of utmost importance. For example, defects that degrade device performance may be caused by a deficiency of arsenic atoms. Another serious problem associated with GaAs MESFETs, JFETs, and heterostructure field effect transistors is the **gate leakage current**. This current limits the allowed gate voltage swing, thus reducing the noise margin in digital circuits utilizing normally-off FETs (i. e., FETs with a positive threshold voltage). Finally, the fabrication processes for compound semiconductor devices are not as well developed or understood as for silicon.

Compound semiconductor technology dominates microwave, millimeter wave, and optoelectronic applications. More recently, this technology emerged as a leading contender for applications in ultra-high-speed low-power integrated circuits primarily because of the higher electron velocity in GaAs and related compounds. The higher velocity is a consequence of a lighter electron effective mass. This advantage is especially pronounced in short-channel devices where ballistic and overshoot effects play a dominant role.

In this section, we consider GaAs MESFETs and compound semiconductor Heterostructure Field Effect Transistors (HFETs).

## *7.2.2.   MESFETs.

In $n$-channel MESFETs, $n^+$ drain and source regions are connected by an $n$-type channel. This channel is partially depleted by voltage applied to the gate. In the **normally-off (enhancement mode)** MESFETs, the channel is totally depleted by the gate built-in potential even at zero gate voltage. The threshold voltage of normally-off devices is positive. In **normally-on (depletion mode)** MESFETs, the conducting channel has a finite cross section at zero gate voltage.

A schematic diagram of the depletion region under the gate of a MESFET for a finite drain-to-source voltage is shown in Fig. 6.1.4. The depletion region is wider closer to the drain because the positive drain voltage provides an additional reverse bias across the channel-to-gate junction. The shape of the depletion region and the device current-voltage characteristics may be found using the gradual channel approximation (see Section 6.4).

The model that related the MESFET current-voltage characteristics to the electron mobility, the electron saturation velocity, the device dimensions, and applied voltages can be developed using an approach similar to that described for MOSFETs in Section 6.3. Such an approach leads to the following equation for the drain saturation current:

$$I_{sat} = \beta(V_{GS} - V_T)^2 \tag{7-2-1}$$

where

$$\beta = \frac{2\varepsilon_s \mu_n v_s W}{A(\mu_n V_{po} + 3v_s L)} \tag{7-2-2}$$

is the transconductance parameter,

$$V_T = V_{bi} - V_{po} \tag{7-2-3}$$

is the threshold voltage, $V_{GS}$ is the intrinsic gate-to-source voltage, and

$$V_{po} = \frac{qN_dA^2}{2\varepsilon_s} \tag{7-2-4}$$

is the pinch-off voltage. Equation (7-2-1) does not account for the gate leakage current which is discussed in Subsection 7.2.4. [For the derivation of eq. (7-2-1) see, for example, Shur (1990)].

In practice, this "square law" model [i. e., eq. (7-2-1)] is fairly accurate for devices with relatively low pinch-off voltages ($V_{po} = V_{bi} - V_T \leq 1.5 \sim 2$ V). For devices with higher pinch-off voltages, the model called the **Raytheon model** (which is implemented in many versions of SPICE) yields better agreement with experimental data:

$$I_{sat} = \frac{\beta(V_{GS} - V_T)^2}{1 + t_c(V_{GS} - V_T)} \tag{7-2-5}$$

Here $t_c$ is an empirical parameter that depends on the doping profile in the MESFET channel. Another empirical model (called the **Sakurai-Newton model**) is also quite useful for MESFET modeling:

$$I_{sat} = \beta_{sn}(V_{GS} - V_T)^{m_{sn}} \qquad (7\text{-}2\text{-}6)$$

The advantage of this model is simplicity. The disadvantage is that the empirical parameters $\beta_{sn}$ and $m_{sn}$ cannot be directly related to the device and material parameters. (The Sakurai-Newton model is implemented in several versions of SPICE. In AIM-Spice, this model is implemented as Level 6 MOSFET model.)

The source and drain series resistances, $R_s$ and $R_d$, may play an important role in determining the current-voltage characteristics of GaAs MESFETs. These resistances can be taken into account in the same way as it was done for MOSFETs in Section 6.4. The intrinsic gate-to-source voltage, $V_{GS}$, is given by

$$V_{GS} = V_{gs} - I_{ds}R_s \qquad (7\text{-}2\text{-}7)$$

where $V_{gs}$ is the applied (extrinsic) gate-to-source voltage. Substituting eq. (7-2-7) into eq. (7-2-1) and solving for $I_{sat}$ we obtain

$$I_{sat} = \frac{2\beta V_{gt}^2}{1 + 2\beta V_{gt}R_s + \sqrt{1 + 4\beta V_{gt}R_s}} \qquad (7\text{-}2\text{-}8)$$

In device modeling suitable for Computer-Aided Design, one has to model the current-voltage characteristics in the entire range of drain-to-source voltages, not only in the saturation regime. In 1980, Curtice proposed the use of a hyperbolic tangent function for the interpolation of MESFET current-voltage characteristics

$$I_d = I_{sat}(1+\lambda V_{ds})\tanh\left(\frac{g_{ch}}{I_{sat}}\right) \qquad (7\text{-}2\text{-}9)$$

where

$$g_{ch} = \frac{g_i}{1 + g_i(R_s + R_d)} \qquad (7\text{-}2\text{-}10)$$

is the MESFET conductance at low drain-to-source voltages, and

$$g_i = g_{cho}\left(1 - \sqrt{\frac{V_{bi} - V_{GS}}{V_{po}}}\right) \qquad (7\text{-}2\text{-}11)$$

is the intrinsic channel conductance at low drain-to-source voltages predicted by the Shockley model.

The constant $\lambda$ in eq. (7-2-9) is an empirical constant that accounts for the output conductance in the saturation regime. This output conductance may be related to short channel effects and also to parasitic leakage currents in the substrate. Hence, output conductance may be reduced by using a heterojunction buffer layer between the device channel and the substrate or by using a $p$-type buffer layer. Such a layer creates an additional barrier which prevents carrier injection into the substrate.

The **Curtice model** is implemented in PSpice[tm]. The Curtice model and the Raytheon model [see eq. (7-2-3)] have become the most popular models used for MESFET circuit modeling. A more sophisticated model, which describes both subthreshold and above-threshold regimes of MESFET operation, is implemented in AIM-Spice [see Lee et al. (1993)]. This model accurately reproduces current-voltage characteristics over several decades of currents (see Fig. 7.2.1).

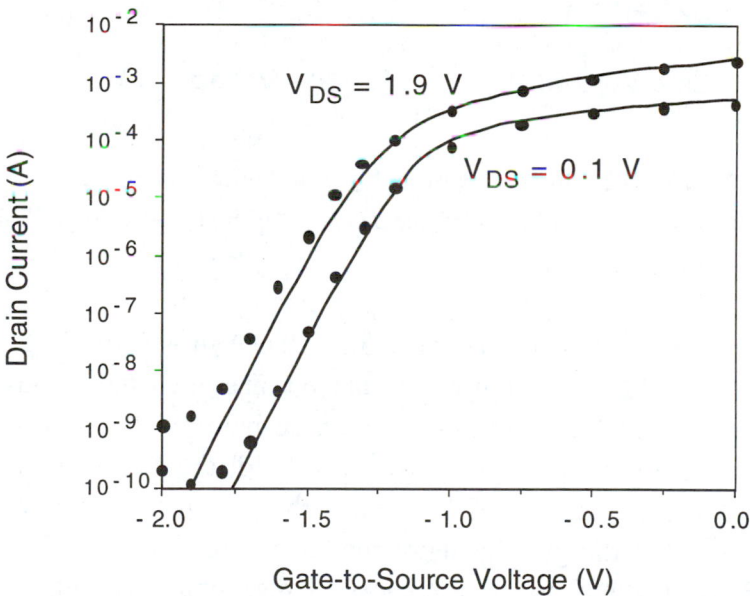

**Fig. 7.2.1.** Subthreshold experimental (symbols) and calculated (solid lines) *I-V* characteristics for ion-implanted MESFET with nominal gate length $L = 1\ \mu m$ (from M. Shur, T. Fjeldly, Y. Ytterdal, and K. Lee, "Unified GaAs MESFET Model for Circuit Simulations," *International Journal of High Speed Electronics*, Vol. 3, No. 2, pp. 201-233, June 1992).

Figure 7-2-2 shows the comparison between the dependencies of the device saturation current, $I_{sat}$, and the device transconductance, $g_m = dI_{sat}/dV_{gs}$, for a Si MOSFET and a GaAs MESFET. (Both devices have the same gate length $L = 0.5$ μm.) These dependencies were calculated using the models implemented in AIM-Spice. This comparison clearly shows the advantages and disadvantages of the GaAs MESFET technology. The advantage is a higher transconductance and a larger saturation current.

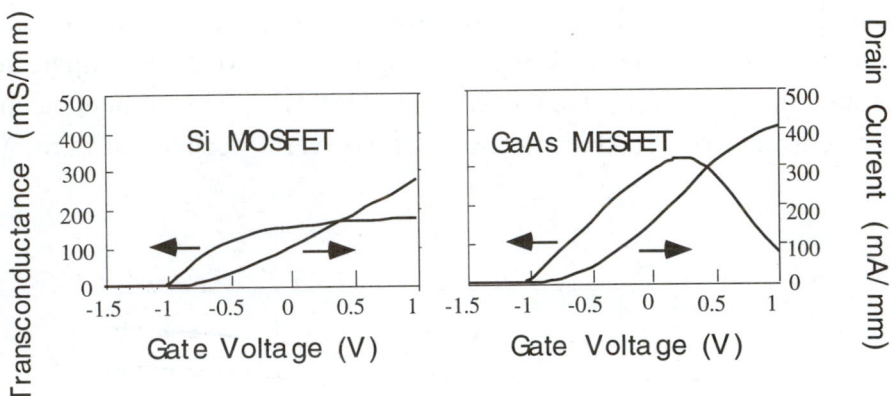

**Fig. 7.2.2.** Device saturation current, $I_{sat}$, and device transconductance, $g_m = dI_{sat}/dV_{gs}$, for a Si MOSFET and a GaAs MESFET. Gate length $L = 0.5$ μm for both devices.

The disadvantage is the drop in the transconductance in MESFETs at high gate bias, whereas in MOSFETs the transconductance remains virtually constant. This difference can be understood from the comparison of the gate-to-channel capacitances in these two devices. Above threshold, the MOSFET capacitance remains fairly constant (see Fig. 6.2.4), since the gate-to-channel separation is approximately equal to the gate insulator thickness. In a MESFET, the gate-to-channel separation is equal to the thickness of the depletion region, $d_d$. For a uniform doping profile

$$d_d = \left[\frac{2(V_{bi} - V_G)}{qN_d}\right]^{1/2} \tag{7-2-12}$$

where $V_{bi}$ is the built-in voltage. As a consequence, the channel capacitance per unit area

$$c_{ch} = \varepsilon/d_d \tag{7-2-13}$$

varies with the gate bias.

The electron charge induced into the channel by the gate-to-channel voltage swing, $V_G - V_T$, where $V_T$ is the MESFET threshold voltage, is proportional to $c_{ch}$. Hence, the intrinsic device transconductance $g_m = dI_d/dV_G$ is also proportional to $c_{ch}$ and increases with increasing gate bias. However, the Schottky contact gate becomes leaky (conductive) once the gate-to-channel voltage approaches the built-in voltage, $V_{bi}$ (which is approximately 0.7 V for a typical Schottky contact on GaAs). As a consequence of this leakage, the MESFET transconductance decreases once the gate-to-source voltage, $V_{gs}$, approaches $V_{bi}$.

A higher MESFET transconductance is related to a higher electron mobility and higher electron velocity in GaAs. These advantages also allow us to obtain the same current in a GaAs MESFET as in a comparable MOSFET but at a smaller voltage. This translates into the same or higher speed of operation at a lower power consumption. Are these advantages worth the trouble of using GaAs technology, which is less developed and more expensive than silicon technology? For many applications, the answer to this question has been a raging debate in the electronics community. As was mentioned in Section 7.1, for millimeter wave applications (30 GHz $< f <$ 300 GHz), the answer is definitely in favor of GaAs and related materials. As a proof, watch for small (a foot and a half) TV satellite dishes, which are replacing old huge satellite dishes. Both systems use GaAs devices, but smaller dishes use devices operating at higher frequencies. Just a short time ago, a superiority of GaAs-based devices was just as apparent for the upper part of the microwave range of frequencies (0.1 GHz $<$ $f <$ 30 GHz). However, nowadays innovative Si MOSFETs seem to approach this frequency range. In digital applications, the Si MOSFETs (CMOS, mostly) give MESFETs a much tougher competition. In his lecture at the University of Virginia in 1994, former chief IBM scientist, Dr. John Armstrong, called the development of digital GaAs technology at IBM one of the costliest IBM mistakes! However, I believe that for applications demanding top speed (or smaller power at the same speed as the best Si technology), GaAs MESFET technology (or its sister HFET technology) may be superior. Many niche applications definitely exist and will grow in size and importance. However, in order to compete with the mainstream silicon technology, we may need to develop new device concepts that do not just clone FETs but are specifically geared to take a full advantage of higher speed inherent in compound semiconductor materials.

Figure 7.2.2 shows the device characteristics representing an average GaAs MESFET. Advanced GaAs MESFETs with shorter channels have achieved a much more impressive performance. In addition to making the gate length shorter, better devices can be made by using thinner and higher doped device channels. Thinner channels lead to a higher effective gate capacitance

$$C_g = \frac{\varepsilon_s LW}{A}$$

where $A$ is the channel thickness, and to a commensurate increase of the transconductance parameter, $\beta$ (and, hence, the transconductance for a given voltage swing); see eq. (7.2.1). This should lead to a better device performance since the parasitic capacitances (which are independent of $A$) should play a smaller role since their ratio to $C_g$ decreases. [If the device threshold voltage is kept constant, the decrease in $A$ requires an increase in doping proportional to $A^2$, see eq. (7-2-4).] Another important advantage of thinner channels is a smaller short channel effect since the ratio $L/A$ increases (see Section 6.3 where we discussed the gradual channel approximation). Hence, thin and highly doped active layers should lead to a higher speed of operation and more ideal device characteristics. However, the impurity scattering in highly doped channels decreases the electron mobility. In Heterostructure Field Effect Transistors (HFETs) , we can achieve a smaller gate-to-channel separation, a higher mobility, and a smaller leakage current, as discussed below.

### *7.2.3.  Heterostructure Field Effect Transistors (HFETs)

In Heterostructure Field Effect Transistors (HFETs) a wide band gap semiconductor layer separates the gate electrode from the channel. The drain-to-source current is carried by the two-dimensional (2D) electron gas at the heterointerface between the wide band gap semiconductor layer and the narrow gap semiconductor channel. This 2D electron gas is similar to that forming at Si-SiO$_2$ interface in silicon MOSFETs (see Section 6.1). However, in compound semiconductors, electrons typically have a much smaller effective mass than in Si, and, hence, quantum effects in the 2D electron gas in HFETs are much more pronounced. In most HFET structures, the dopants in the wide gap semiconductor layer control the device threshold voltage. Usually, these donors are separated from the 2D electron gas by a thin spacer layer. Such a layer reduces the impurity scattering of channel electrons by the remote donors.

In the structure shown in Fig. 7.2.3a, the threshold voltage is controlled by the dopants in the doped AlGaAs layer. A Molecular Beam Epitaxy (MBE) technology, which was briefly discussed in Section 4.10, also allows us to put all the donors into just one atomic plane. Such doping is called **δ-doping**. In the δ-doped structure shown in Fig. 7.2.3b, the threshold voltage is controlled by the dopants in the doped plane in the AlGaAs layer.

Both $n$-channel and $p$-channel complementary Heterostructure Insulated Gate Field Effect Transistors (HIGFETs) can be fabricated on the same wafer (see Fig. 7.2.4). This technology, which is similar to Si CMOS, has promise of low-power high-speed operation.

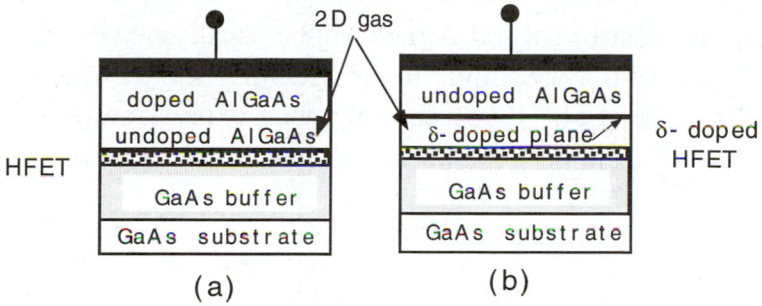

**Fig. 7.2.3.** Examples of modulation doped layers for HFETs: (a) conventional modulation doped layer, (b) delta-doped layer (from M. Shur and T. A. Fjeldly, "HEMT Modeling," in *Compound Semiconductor Device Modeling*, C. Snowden and R. Miles, Editors, Springer-Verlag, London, 1993, pp. 56-73).

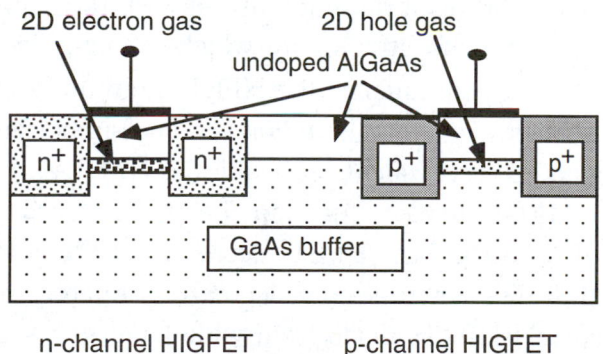

**Fig. 7.2.4.** Complementary $n$-channel and $p$-channel HIGFETs (from M. Shur and T. A. Fjeldly, "HEMT Modeling," in *Compound Semiconductor Device Modeling*, C. Snowden and R. Miles, Editors, Springer-Verlag, London, 1993, pp. 56-73).

HFET structures can be implemented in a large variety of heterostructure systems, such as AlGaAs/GaAs, AlGaAs/InGaAs/GaAs, and AlInAs/InGaAs/InP.

The equation of the charge control model for an HFET can be written similar to that for a MOSFET [compare with eq. (6-2-1)]:

$$n_s = \frac{\varepsilon_i V_{GT}}{q(d_i + \Delta d)} \qquad (7\text{-}2\text{-}14)$$

where $\Delta d$ can be interpreted as the effective thickness of the 2D electron gas. Typically, $\Delta d \approx 40$ to $80$ Å in AlGaAs/GaAs HFETs. [In silicon MOSFETs, $\Delta d$ is very small (on the order of several Å) and can be usually neglected.] Therefore, the basic HFET model is very similar to the MOSFET model of Section 6.4. As was discussed by Lee et al. (1993), the following expression describes the $I$-$V$ characteristics for FETs in both the linear and saturation regions:

$$I_d \approx \frac{g_{ch} V_{ds}}{\left[1 + \left(g_{ch} V_{ds} / I_{sat}\right)^m\right]^{1/m}} \qquad (7\text{-}2\text{-}15)$$

Here, $m$ is a parameter that determines the shape of the characteristics in the knee region, $g_{ch} = g_{chi}/(1 + g_{chi} R_t)$ is the extrinsic channel conductance, $R_t = R_s + R_d$ is the sum of the source and the drain series resistances, and $g_{chi} = q\mu n_s W/L$ is the intrinsic channel conductance. The drain saturation current can be calculated using eq. (6-4-33).

Since the conduction band discontinuity in HFETs is limited (see Section 4.10), the surface electron concentration, $n_s$, which can be induced into the HFET channel is limited. As a consequence the HFET transconductance decreases at large gate voltage swings when this limitation becomes important (see Fig. 7.2.5). (The curves in Fig. 7.2.5 were calculated using the HFET and MESFET models implemented in AIM-Spice.) As can be seen from the figure, the transconductance of AlGaAs/GaAs HFETs decreases at the voltages smaller than those for GaAs MESFETs where such a decrease is related to the gate leakage current. In AlGaN/GaN HFETs made from wide band gap semiconductors (see Section 9.7), the conduction band discontinuity is large, and such decrease may be avoided.

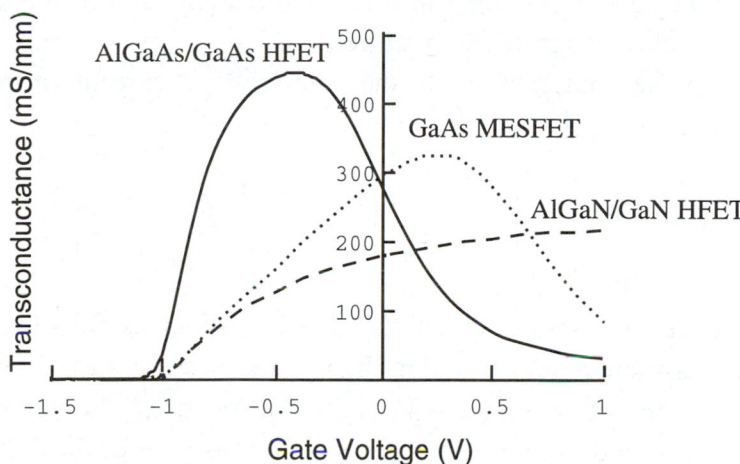

**Fig. 7.2.5.** Computed device transconductance in the saturation region versus gate bias for three devices: 0.5 μm gate GaAs MESFET, 0.5 μm gate AlGaAs/GaAs HFET, and 0.25 μm AlGaN/GaN HFET. The source series resistance per unit width for the MESFET and the AlGaAs/GaAs HFET is 0.3 Ωmm. The source resistance per unit width for the AlGaN/GaN HFET is 2 Ωmm (from M. Shur et al., "GaN/AlGaN Field Effect Transistors for High Temperature Applications," in *Proceedings of the 21st International Symposium on Compound Semiconductors*, San Diego, CA, Aug. 19-22, 1994).

A more realistic HFET model implemented in AIM-Spice [see Lee et al. (1993)] accounts for the subthreshold regime of operation, for the limitation of the maximum concentration of the 2D gas, and for the gate leakage current (see Section 7.2.4).

**\*7.2.4.   Gate leakage current.**

The gate contact in MESFETs is isolated from the device channel only by the depletion layer. (In HFETs, a wide band gap semiconductor can provide an additional barrier separating the gate from the channel.) At forward gate bias, the gate contact becomes conductive just like a regular Schottky diode (see Chapter 4). Figure 7.2.6 shows an equivalent circuit that accounts for the gate current in MESFETs and HFETs by introducing two equivalent Schottky diodes connected between the gate and the source and drain contacts. The current-controlled current source included into this equivalent circuit accounts for the effect of the gate current on the drain current.

Adding up the contributions to the gate leakage current from both these diodes, we obtain

$$I_g = J_{ss} LW \left[ \exp\left( \frac{V_{GS}}{m_{gs} V_{th}} \right) + \exp\left( \frac{V_{GD}}{m_{gd} V_{th}} \right) - 2 \right] \tag{7-2-16}$$

where $J_{ss}$ is the reverse saturation current density, which is calculated using the thermionic or thermionic-field emission theory (see Section 4.8), $L$ and $W$ are the gate length and gate width, $V_{GS}$ and $V_{GD}$ are the intrinsic gate-source and gate-drain voltages, $m_{gs}$ and $m_{gd}$ are the gate-source and gate-drain Schottky diode ideality factors, and $V_{th}$ is the thermal voltage.

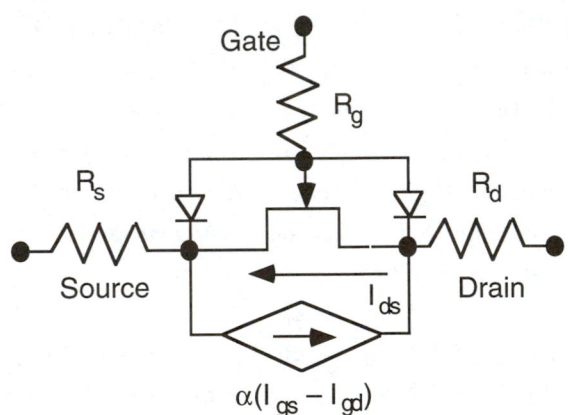

**Fig. 7.2.6.** MESFET or HFET equivalent circuit (from P. P. Ruden, M. Shur, A. I. Akinwande, and P. Jenkins, "Distributive Nature of Gate Current and Negative Transconductance in Heterostructure Field Effect Transistors," *IEEE Transactions on Electron Devices*, ED-36, No. 2, pp. 453-456, Feb. 1989, © IEEE 1989).

A more accurate model has to account for the electron heating in the channel. This can be done by introducing effective electron temperatures, $T_s$ and $T_d$, at the source side and drain sides of the channel. $T_s$ is close to the lattice temperature, $T$, and $T_d$ increases with $V_{ds}$. Hence, eq. (7-2-16) becomes

$$I_g = J_{gs}LW\left[\exp\left(\frac{V_{GS}}{m_{gs}V_{ths}}\right)-1\right] + J_{gd}LW\left[\exp\left(\frac{V_{GD}}{m_{gd}V_{thd}}\right)-1\right] \quad (7\text{-}2\text{-}17)$$

where $J_{gs}$ and $J_{gd}$ are the reverse saturation current densities for the gate-source and the gate-drain diodes, respectively, and $V_{ths} = k_B T_s/q$ and $V_{thd} = k_B T_d/q$.

Such a model for the leakage current can be easily implemented in SPICE. The effect of the leakage current is illustrated in Fig. 7.2.2, where it is responsible for the decrease of the MESFET transconductance at large gate biases.

A more detailed description of the gate leakage models is given by Lee et al. (1993). Tables 7.2.1 and 7.2.2 summarize important equations related to MESFET and HFET modeling.

| | |
|---|---|
| Depletion width in a MESFET | $A_d(x) = \sqrt{\dfrac{2\varepsilon_s\left[V(x)+V_{bi}-V_{GS}\right]}{qN_d}}$ |
| MESFET pinch-off voltage | $V_{po} = \dfrac{qN_D A^2}{2\varepsilon_s}$ |
| MESFET drain saturation current (square law model) | $I_{sat} = \beta(V_G - V_T)^2$ <br> where $\beta = \dfrac{2\varepsilon_s\mu_n v_s W}{A\left(\mu_n V_{po}+3v_s L\right)}, \quad V_T = V_{bi}-V_{po}$ |
| MESFET drain saturation current (extrinsic square law model) | $I_{sat} = \dfrac{2\beta V_{gt}^2}{1+2\beta V_{gt}R_s + \sqrt{1+4\beta V_{gt}R_s}}$ |
| MESFET drain saturation current (Raytheon model) | $I_{sat} = \dfrac{\beta(V_G - V_T)^2}{1+t_c(V_G - V_T)}$ |
| MESFET drain saturation current (Sakurai-Newton model) | $I_{sat} = \beta_s(V_G - V_T)^{n_m}$ |
| Curtice MESFET model | $I_d = I_{sat}(1+\lambda V_{ds})\tanh\left(\dfrac{g_{ch}}{I_{sat}}\right)$ <br> $g_{ch} = \dfrac{g_i}{1+g_i(R_s+R_d)}, \quad g_i = g_{cho}\left(1-\sqrt{\dfrac{V_{bi}-V_{GS}}{V_{po}}}\right)$ |

**Table 7.2.1.** Important equations related to MESFET modeling.

| Charge control model | $n_s = \dfrac{\varepsilon_i V_{GT}}{q(d_i + \Delta d)}$ |
|---|---|
| Current-voltage characteristics | $I_{ds} \approx \dfrac{g_{ch} V_{ds}}{\left[ 1 + \left( g_{ch} V_{ds} / I_{sat} \right)^m \right]^{1/m}}$ <br><br> where <br><br> $g_{ch} = g_{chi}/(1 + g_{cho} R_t)$ <br><br> $R_t = R_s + R_d$ <br><br> $g_{cho} = c_i V_{gt} \mu_n W/L$ |
| Saturation current (see Section 6.4) | $I_{sat} = \dfrac{g_{ch} V_{gt}}{1 + g_{ch} R_s + \sqrt{1 + 2 g_{ch} R_s + \left( V_{gt} / V_L \right)^2}}$ <br><br> where $V_L = F_s L$ |
| Gate leakage current (applies to MESFETs also) | $I_g = J_{gs} L W \left[ \exp\left( \dfrac{V_{GS}}{m_{gs} V_{ths}} \right) - 1 \right]$ <br><br> $+ J_{gd} L W \left[ \exp\left( \dfrac{V_{GD}}{m_{gd} V_{thd}} \right) - 1 \right]$ |

**Table 7.2.2.**  Important equations related to HFET modeling.

# *7-3.   THIN FILM TRANSISTORS

### *7.3.1.   Amorphous silicon and polysilicon.

Most electronic devices are fabricated on crystalline wafers of silicon – thin silicon disks cut from a crystal boule grown from a melt (see Section 9.2). However, in the last ten years or so, amorphous and polycrystalline devices have occupied an important market niche, primarily in applications involving Liquid Crystal Displays (LCDs); see Section 8.7.

As was discussed in Section 2.2, the interatomic distances between the nearest neighbors in an amorphous solid are more or less the same as in a crystalline material of the same substance (see Fig. 2.2.1). The bond angles are also similar but a long-range order is absent. An amorphous material is said to have a "short-range" order. Since the long-range order is absent, an amorphous material has a large number of dangling bonds which are not attached to the nearest neighbors. These dangling bonds trap the conduction band electrons. The dangling bonds correspond to levels in the energy gap of the amorphous material. These levels are also called **localized states**. In pure amorphous silicon (a-Si), the number of the localized states is so high that the material is unsuitable for most device applications.

In 1972, Spear and LeComber demonstrated that amorphous silicon films prepared by glow discharge decomposition of silane gas ($SiH_4$) have a relatively low density of defect states in the energy gap. The amorphous silicon obtained by this process has a very large concentration of hydrogen. The hydrogen atoms tie up these dangling bonds and decrease the density of localized states in the energy gap. Compared to pure a-Si, in amorphous Si:H (a-Si:H), the number of localized states is reduced by many orders of magnitude. However, the remaining localized states still determine the transport properties of a-Si:H. Usually, a larger fraction of the electronic charge in a-Si:H resides in the localized states and not in the conduction band. The density of localized states depends on material growth conditions (such as the substrate temperature, etc.). Also, new localized states may be created by doping, by illumination, and even by a prolonged application of bias voltages. This is why the characteristics of a-Si:H devices are quite different from those of more conventional crystalline devices.

Electron and hole mobilities in a-Si:H films are orders of magnitude smaller than in crystalline silicon. (Electron and hole band mobilities are on the order of 5 to 10 $cm^2$/Vs. in a-Si:H. The effective electron mobility in a-Si thin film transistors is smaller still, on the order of 1 $cm^2$/Vs.) However, very large-area (2×4 feet and larger) high-quality a-Si:H films may be inexpensively deposited in a continuous process on a large variety of different substrates, from glass to mylar film, making this material very attractive for applications requiring large-area devices and circuits, for example, in solar cells or in large-area displays. Other amorphous materials, similar to a-Si:H, include amorphous Ge:H, amorphous C:H, amorphous SiC:H, and amorphous Si:F. Growing successive layer of different hydrogenated amorphous films, one can produce heterostructure devices. As will be discussed in Section 8.2, such heterostructure films are especially important for solar cell applications.

In this section, we consider a-Si Thin Film Transistors (TFTs), which have become a viable and important technology for large-area, low-cost integrated circuits. As was mentioned above, these circuits are currently being used to drive large-area Liquid Crystal Displays (LCDs). Basic Integrated Circuits (ICs) and addressable image sensing arrays have also been implemented.

Another material used in Thin Film Transistors is polycrystalline silicon (poly-Si). This material consists of small Si crystallites (grains) of somewhat irregular sizes separated by crystal boundaries (see Fig. 2.2.1). These boundaries contain localized states that may trap conduction band electrons. The grain boundaries present an impediment to the electric current, and the material quality improves with an increase in the grain size. In poly-Si, the electron and hole mobilities are much larger than in amorphous Si (typically 30 to 300 $cm^2$/Vs for electrons and 10 to 100 $cm^2$/Vs for holes). Poly-Si can be produced by recrystallizing amorphous Si films. However, poly-Si technology involves higher processing temperatures and therefore requires much more expensive substrates than a-Si technology, since these substrates have to withstand a much higher processing temperature. A promising new technology involves laser recrystallization of amorphous Si films. In this process, an intensive laser radiation is concentrated on the recrystallized film, and the substrate temperatures do not increase as much as in the more conventional recrystallization process. Hence, a lower cost glass may be acceptable as a substrate material.

## *7.3.2. Amorphous silicon Thin Film Transistors.

Figure 7.3.1 shows an approximate distribution of localized states for intrinsic (undoped) a-Si:H. The localized states in the upper half of the energy gap (closer to the bottom of the conduction band) are neutral when empty and negatively charged when filled. Hence, these states behave as acceptorlike states (see Section 3.4). The states in the bottom half of the energy gap are positively charged when empty and neutral when filled. Hence, these states behave as donorlike states. As shown in the figure, the energy gap of a-Si:H is close to 1.7 eV. (a-Si:H behaves like a direct gap semiconductor.) The localized states can be roughly divided into tail acceptorlike states, deep acceptorlike states, deep donorlike localized states, and tail donorlike states. A detailed account of material properties of a-Si:H and related devices is given by Kanicki (1991).

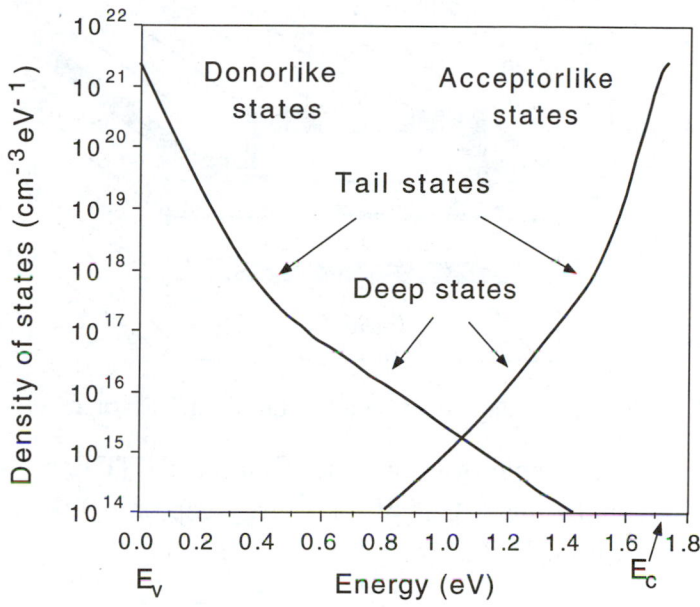

**Fig. 7.3.1** Simplified density of localized states in amorphous silicon. In a-Si TFTs, the effective density of localized states is affected by surface states, by temperature treatment, and by long-term application of drain and gate bias voltages (called temperature and bias stress, respectively) (after M. Shur, M. Hack, and J. G. Shaw, "A New Analytic Model for Amorphous Silicon Thin Film Transistors," *J. Appl. Phys.*, 66(7), 3371-3380, Oct. 1989.

The position of the Fermi level in an undoped, uniform a-Si:H sample in the dark, $E_{Fo}$, should be found from the neutrality condition. In different samples, $E_c - E_{Fo}$ varies from 600 meV to 800 meV. a-Si:H can be doped $n$-type (typically by phosphorus) or $p$-type (by boron). In 1976, Spear and LeComber demonstrated the first a-Si:H $p$-$n$ junction.

a-Si TFTs (see Fig. 7.3.2) are usually fabricated by first depositing metal gates on glass, then a layer of a gate dielectric (typically silicon nitride), a layer of intrinsic a-Si:H, and a layer of a passivating dielectric (once again, silicon nitride). The gate insulator thickness is on the order of 1,000 to 2,000 Å. The intrinsic a-Si:H layer is 500 Å to 1000 Å. The passivating dielectric thickness is of the order of 1,000 Å. A thin ($\approx$ 100 Å) layer of highly doped $n^+$ a-Si is required for source and drain contacts. Typical gate lengths are 10 to 15 μm.

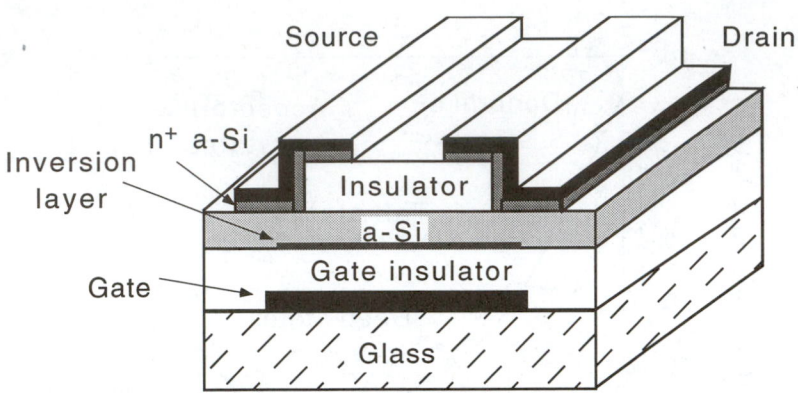

**Fig. 7.3.2.** Schematic diagram of an a-Si Thin Film Transistor.

To explain the principle of operation of an a-Si:H TFT, we shall consider the localized states in the upper half of the energy gap of a-Si:H at the a-Si-gate insulator interface (see Fig. 7.3.3).

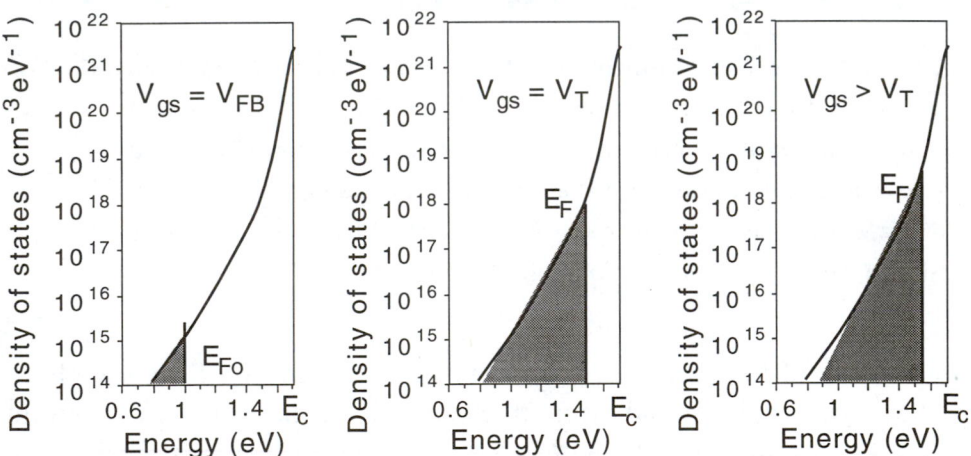

**Fig. 7.3.3.** Localized states in the upper half of the energy gap of a-Si:H at the a-Si-insulator interface at different gate voltages. Dashed regions indicate filled localized states.

The localized states below the Fermi level are nearly completely filled with electrons, and the localized states above the Fermi level are practically empty. A gate bias induces electrons into the a-Si:H channel. These electrons fill the localized states and shift the Fermi level toward the conduction band. When the Fermi level is in the deep localized states, a relatively modest electron concentration is enough for a considerable shift of the Fermi level, since the concentration of deep localized states is relatively small. A large shift of the Fermi level corresponds to a large relative increase of the electron concentration in the conduction band, $n$, since $n = N_c \exp[(E_F - E_c)/k_B T]$ (see Section 3.4). This corresponds to the below-threshold regime of operation. In this regime, the free carrier density is very small and increases rapidly with increasing gate-source voltage, which shifts the position of the Fermi level. However, once the Fermi level enters the tail states, its position changes much less with a further increase of the gate bias, since the concentration of the tail states rapidly increases with energy (see Fig. 7.3.3). The gate bias at which the Fermi level enters the tail states corresponds to the threshold voltage. In the above-threshold regime, the free carrier concentration is a sizable fraction of the total induced charge. In crystalline MOSFETs, the field effect mobility is close to the drift mobility of free electrons since, above threshold, practically all the charge induced into the channel by the gate voltage is the charge of free electrons. However, in a-Si TFTs (and, to a certain extent, even in polysilicon TFTs) a large fraction of the induced charge resides in the localized states even in the above-threshold regime. These electrons do not contribute to the channel conductance. Hence, we can define the field effect mobility in the following way:

$$\mu_{FET} = \mu \frac{n_s}{n_{ss}} \qquad (7\text{-}3\text{-}1)$$

where $n_s$ is the sheet electron concentration in the conduction band,

$$n_{ss} = c_i(V_G - V_T)/q \qquad (7\text{-}3\text{-}2)$$

is the total surface electron concentration in the channel ($n_{ss} = n_s + n_t$ where $n_t$ is the surface concentration of the electrons trapped in the localized states), $q$ is the electronic charge, $\mu_{FET}$ is the field-effect mobility, $\mu$ is the mobility of the electrons in the conduction band, $c_i$ is the gate-to-channel capacitance per unit area, $V_G$ is the gate-to-channel voltage, and $V_T$ is the threshold voltage. The field-effect mobility is a function of the surface electron charge, $n_{ss}$, induced into the a-Si channel since at larger charge densities, more localized states are filled and, hence, a larger fraction of charge goes into conduction band states. According to Lee et al. (1993), this dependence can be approximated by

$$\mu_{FET} = \mu_{th} \left( \frac{n_{ss}}{n_{th}} \right)^m \qquad (7\text{-}3\text{-}3)$$

where $q n_{th}$ is the surface charge in the channel at the threshold voltage, $\mu_{th}$ is the field-effect mobility at the threshold voltage, and $m$ is a constant (typically $\approx 0.5$) (see Fig. 7.3.4).

Using eqs. (7-3-1) to (7-3-3), one can derive expressions for a-Si TFT characteristics using an approach similar to that of Section 6.4 (see Problem 7-3-1). Such a model, which accounts for space-charge injection effects and describes the TFT characteristics in both below- and above-threshold regimes, is implemented in AIM-Spice [see Lee et al. (1993)]. The model is in good agreement with measured current-voltage characteristics (see Fig. 7.3.5).

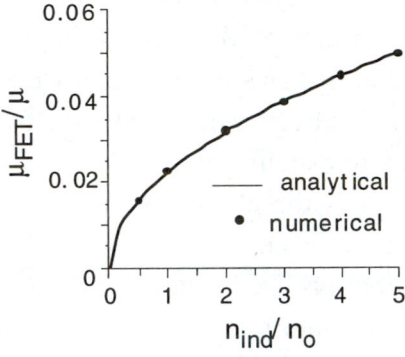

**Fig. 7.3.4.** Field-effect mobility versus the electron sheet density induced into a-Si channel (from Lee et al. 1993, copyright © Prentice Hall, 1993, reproduced by permission of Prentice Hall, Inc., Englewood Cliffs, NJ).

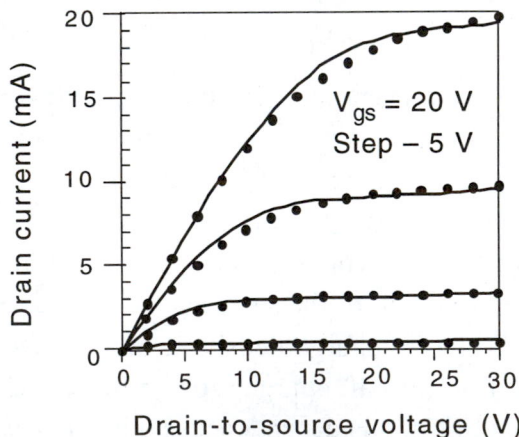

**Fig. 7.3.5.** Current-voltage characteristics of an a-Si TFT (from M. Shur, M. Hack , and J. G. Shaw, *J. Appl. Phys.*, 66(7), 3371-3380, Oct. 1989). Symbols: experimental data; solid lines: calculated curves.

## *7-3-3. Polysilicon Thin Film Transistors.

Polysilicon consists of small crystallites (called grains) separated by grain boundaries. The electron and hole mobilities in a polysilicon sample increase with an increase in the grain size. The carrier transport across the grain boundaries plays a dominant role in polysilicon. Nevertheless, many features of polysilicon TFTs can be understood using a model that treats the nonuniform polycrystalline sample with grain boundaries as some uniform effective medium with effective electron and hole mobilities. These mobilities (50 to 300 $cm^2/Vs$ for electrons and 30 to 70 $cm^2/Vs$ for holes) are much larger than typical field effect mobilities in a-Si:H (0.2 to 1 $cm^2/Vs$ for electrons and $10^{-3}$ to $10^{-2}$ $cm^2/Vs$ for holes) but much smaller than the electron and hole mobilities in crystalline silicon (600 to 1200 $cm^2/Vs$ for electrons and 150 to 300 $cm^2/Vs$ for holes).

Figure 7.3.6 shows a typical poly-Si TFT structure.

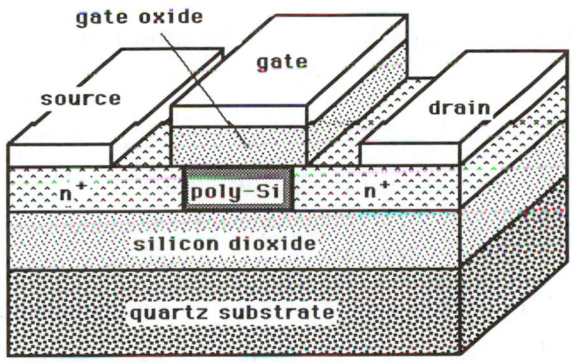

**Fig. 7.3.6.** Schematic diagram of poly-Si TFT (from Lee et al., 1993, copyright © Prentice Hall, 1993, reproduced by permission of Prentice Hall, Inc., Englewood Cliffs, NJ.)

The characteristics of poly-Si TFTs can be described using a charge control model similar to that used for long channel MOSFETs. However, the effective field effect mobility in both *p*-channel and *n*-channel depends on the carrier concentration in the channel (similar to such a dependence for a-Si TFTs).

Just as for crystalline MOSFETs, we distinguish between two regions of operation: subthreshold and above-threshold regimes. Above the threshold, the current is given by

$$I_{ds} = \frac{q\mu_{FET}n_{ss}WV_{DS}}{L}\left(1 - \frac{V_{DS}}{2V_{SAT}}\right)$$

(7-3-4)

where

$$n_{ss} = \frac{c_i V_{GT}}{q}$$

(7-3-5)

$V_{DS}$ is the intrinsic drain-source voltage, $L$ is the gate length, $W$ is the gate width, and $\mu_{FET}$ is the effective electron field effect mobility, which can be approximated as

$$\frac{1}{\mu_{FET}} = \frac{1}{\mu_o} + \frac{1}{\mu_1(n_{ss}/n_o)^m}$$

(7-3-6)

[compare with eq. (7-3-3) for an a-Si TFT]. The saturation voltage

$$V_{SAT} = \alpha_{sat}V_{GT}$$

(7-3-7)

where parameter $\alpha_{sat}$ accounts for the deviation of $V_{SAT}$ from the value predicted by the constant mobility model given by $V_{SAT} = V_{GT}$. Typically, $\alpha_{sat}$ varies from 0.8 to 1. (With trivial changes in signs, the above equations apply also to $p$-channel polysilicon TFTs.)

In a poly-Si TFT, the electrons in the channel have a small effective mobility at low electron concentrations [see eq. (7-3-6)]. As a consequence the current at the threshold voltage is very small, and the apparent threshold voltage extracted from the current-voltage characteristics is substantially higher than the threshold voltage extracted from the capacitance-voltage characteristics. This difference can be as large as 1 to 2 V for an $n$-channel poly-Si TFT and 2 to 4 V for a $p$-channel poly-Si TFT. (In a crystalline MOSFET, these two threshold voltages coincide.)

Figure 7.3.7 shows a comparison between experimental data for a $p$-channel polysilicon TFT and the model described above. As can be seen from the figure, the model is in adequate agreement with experimental data, except for the rising region of the current-voltage characteristics at large $V_{DS}$ (called the kink region). AIM-Spice includes a more complicated empirical model that allows us to reproduce the kink region as well.

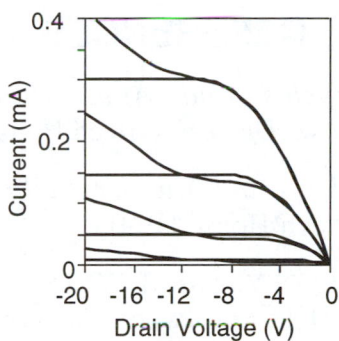

**Fig. 7.3.7.** $I_{ds}$ versus $V_{DS}$ for $p$-channel polysilicon TFT for $V_{GS} = -8$, $-11$, $-14$, and $-17$ V. Saturated curves are calculated; unsaturated curves with the kink are measured (from M. Shur, M. Hack, and Y. H. Byun, "Circuit Model and Parameter Extraction Technique for Polysilicon Thin Film Transistors," in *Proceedings of 2d International Semiconductor Device Research Symposium*, Charlottesville, pp. 165-168, Dec. 1993).

Tables 7.3.1 and 7.3.2 summarize equations related to TFTs.

| Definition of field effect mobility in a-Si TFTs | $\mu_{FET} = \mu \dfrac{n_s}{n_{ss}}$ |
|---|---|
| Field-effect mobility versus induced surface electron concentration in a-Si TFTs | $\mu_{FET} = \mu_1 \left( \dfrac{n_{ss}}{n_o} \right)^m$ |

**Table 7.3.1.**  Important equations related to a-Si TFTs.

| Field-effect mobility versus induced surface electron concentration in a-Si TFTs | $\dfrac{1}{\mu_{FET}} = \dfrac{1}{\mu_o} + \dfrac{1}{\mu_1 \left( n_{ss} / n_o \right)^m}$ |
|---|---|
| Drain current in poly-Si TFTs below saturation | $I_{ds} = \dfrac{q \mu_{FET} n_{ss} W V_{DS}}{L} \left( 1 - \dfrac{V_{DS}}{2 V_{SAT}} \right); \; n_{ss} = \dfrac{c_i V_{GT}}{q}$ |
| Drain-to-source saturation voltage in poly-Si TFTs | $V_{SAT} = \alpha_{sat} V_{GT}$ |

**Table 7.3.2.**  Important equations related to poly-Si TFTs.

# REFERENCES

J. KANICKI, Editor, *Physics and Applications of Amorphous and Microcrystalline Semiconductor Devices*, Artech House (1991)

J. KANICKI, Editor, *Amorphous and Microcrystalline Semiconductor Devices*: *Optoelectronics Devices*, Artech House (1991)

K. LEE, M. SHUR, T. A. FJELDLY, AND T. YTTERDAL, *Semiconductor Device Modeling for VLSI*, Prentice Hall, Englewood Cliffs, NJ (1993)

M. SHUR, *Physics of Semiconductor Devices*, Series in Solid State Physical Electronics, Prentice Hall, Englewood Cliffs, NJ (1990)

S. M. SZE, *Physics of Semiconductor Devices*, Second Edition, John Wiley & Sons, New York (1981)

# BIBLIOGRAPHY

S. I. LONG AND S. E. BUTNER, *Gallium Arsenide Digital Integrated Circuit Design*, McGraw-Hill, New York (1990)

H. MORKOÇ, H. UNLU, AND G. LI, *Principles and Technology of MODFETs*, Vols. 1 and 2, John Wiley & Sons, New York (1991)

M. SHUR, *GaAs Devices and Circuits*, Plenum, New York (1987)

J. SINGH, *Physics of Semiconductors and Their Heterostructures*, McGraw-Hill, New York (1993)

C. SNOWDEN AND R. MILNES, Editors, *Semiconductor Device Modeling*, Springer-Verlag, London (1993)

S. M. SZE, *Semiconductor Devices. Physics and Technology*, John Wiley & Sons, New York (1985)

M. SZE, Editor, *High-Speed Semiconductor Devices,* John Wiley & Sons, New York (1990)

S. TIWARI, *Compound Semiconductor Device Physics*, Academic Press, Boston (1992)

C. T. WANG, Editor, *Introduction to GaAs Technology*, John Wiley & Sons, New York, (1990)

# REVIEW QUESTIONS

**1.**   Comment  on  the  advantages  and  the  disadvantages  of  compound semiconductor technology.

☐   2 points

**2.**  Describe  the  Raytheon,  the  square  law,  and  the  Sakurai-Newton  models  for MESFETs.

☐   2 points

**3.**  Sketch  an  equivalent  circuit  of  an  HFET  that  accounts  for  the  gate  leakage current.

☐   1 point

**4.**  What are the advantages and the disadvantages of HFET technology compared to silicon MOSFETs?

☐   2 points

**5.**   An  HFET  has  a  threshold  voltage  of  0.2  V.   The  wide  band  gap semiconductor separating  the  channel  from  the  gate  is  200  Å  thick  and  has dielectric permittivity of $10^{-10}$ F/m.  What gate-to-source voltage will induce the surface electron concentration $n_s = 2 \times 10^{12}$ cm$^{-2}$ in the HFET channel?  Compare this  problem  with  a  similar  problem  for  a  silicon  MOSFET.   What  are  the differences? (In a MOSFET, the gate dielectric permittivity is $3.45 \times 10^{-11}$ F/m.)

☐   3 points

**6.**  What  is  the  order  of  magnitude  of  the  mobility  of  the  electrons  in  the conduction band of a-Si:H?

☐   1 point

**7.** Compare typical values of the field effect mobilities in $n$-channel and $p$-channel poly-Si TFTs, crystalline MOSFETs, and a-Si TFTs.

2 points

**8.** Design an a-Si TFT that has a saturation current of approximately 10 µA at 20 V of the gate bias. Assume that the threshold voltage is 3 V, and choose the gate insulator thickness of 3,000 Å. **Hint:** For this design use the simplest MOSFET model and assume the that the effective field effect mobility is 1 $cm^2/Vs$. The gate dielectric ($Si_3N_4$) permittivity is equal to $6.64\times10^{-11}$ F/m.

4 points

# PROBLEMS

**7-2-1.** Sketch proposed dimensions, material composition, and doping profile for a MESFET with a threshold voltage, $V_T = -1$ V, and a device transconductance at zero gate bias in the saturation region of at least 200 mS/mm. Justify your choices.

**7-2-2.** Sketch proposed dimensions, material composition, and doping profiles for an HFET with a threshold voltage, $V_T = -1$ V, and a device transconductance at zero gate bias in the saturation region of at least 300 mS/mm. Justify your choices.

**7-3-1.** Using eqs. (7-3-1) to (7-3-2), derive expressions for a-Si TFT characteristics using an approach similar to eqs. (6-4-10) and (6-4-15) for crystalline MOSFETs.

**7-3-2.** (a) The field distribution in a poly-Si $n$-channel TFT structure is shown in the figure. Sketch the potential distribution.
(b) Why is the electric field not continuous at the $Si-SiO_2$ interface?
(c) What is the surface potential, $V_s$?
(d) What is the threshold voltage, $V_T$, if the flat-band voltage is $-1$ V?

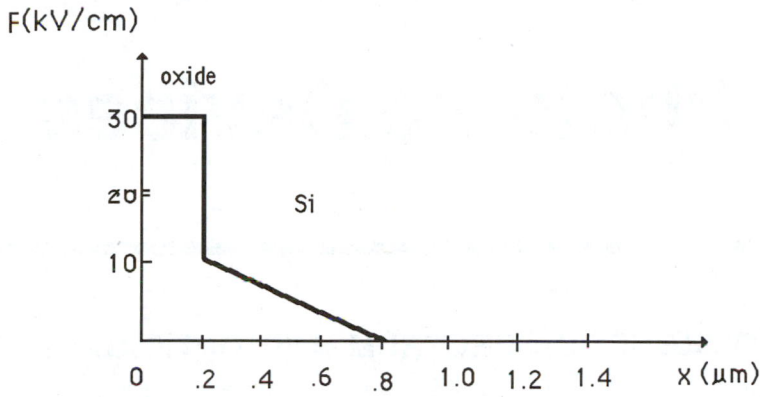

**7-3-3.** (a) Propose a set of parameters for PSpice[tm] MOSFET model that will approximately correspond to an a-Si TFT.

(b) Propose a set of parameters for PSpice[tm] MOSFET model that will approximately correspond to a poly-Si *n*-channel TFT.

(c) Propose a set of parameters for PSpice[tm] MOSFET model that will approximately correspond to a poly-Si *p*-channel TFT.

(d) Simulate the *I-V* characteristics for these three devices for a reasonable range of bias voltages, compare, and comment.

# 8

# Photonic Devices

## 8-1. ELECTROMAGNETIC SPECTRUM AND OPTICAL PROPERTIES OF SOLIDS

The electromagnetic spectrum covers a huge range of frequencies – from dc to cosmic rays – which correspond to frequencies higher than $10^{22}$ Hz or to wavelengths shorter than $3 \times 10^{-14}$ m. Since transmission of information requires a certain frequency band, more communication channels can be used at higher frequencies. This explains why optical fibers carrying light signals are rapidly replacing telephone wires. Light signals are also immune to electrical noise, leading to an improved sound quality of phone systems. Even more importantly, most information we perceive is visual, and most energy we use comes from the sun, either directly or indirectly.

Solid-state photonic devices using quanta of light – photons – include
- Light sources, such as **Light Emitting Diodes** (**LEDs**) and **semiconductor lasers.**
- Devices that detect optical signals (**photodetectors** or **optical sensors**).
- **Photovoltaic devices** (**solar cells**).

This chapter deals with all these photonic devices and optical fibers used to transmit light.

The optical part of the electromagnetic spectrum corresponds to the wavelengths from extreme infrared radiation ($\lambda = 1,000$ µm) to extreme ultraviolet radiation ($\lambda = 0.01$ µm); see Tables 8.1.1 and 8.1.2. Longer wavelengths ($\lambda > 1,000$ µm) correspond to millimeter waves (1 mm $< \lambda <$ 10 mm) and microwave radiation (1 cm $< \lambda <$ 30 cm); shorter wavelength correspond to X-rays ($10^{-6}$ µm $< \lambda <$ 0.01 µm) and gamma rays ($10^{-8}$ µm $< \lambda <$ $10^{-4}$ µm). (These ranges may overlap, and we may refer to the electromagnetic radiation with a given wavelength, for example, 0.5 mm, as either submillimeter waves or extreme infrared.)

| | Wavelength, $\lambda$ ($\mu$m) | Frequency, $f$ (Hz) | Photon Energy, $E$ (eV) |
|---|---|---|---|
| Cosmic Rays | $\lambda < 3\times 10^{-7}$ | $f > 10^{21}$ | $E > 4.1\times 10^6$ |
| Gamma Rays | $10^{-8} < \lambda < 8\times 10^{-3}$ | $4\times 10^{16} < f < 3\times 10^{22}$ | $155 < E < 1.24\times 10^8$ |
| X-rays | $2\times 10^{-6} < \lambda < 0.2$ | $1.5\times 10^{15} < f < 1.5\times 10^{20}$ | $6.2 < E < 6.2\times 10^5$ |
| Ultraviolet | $10^{-2} < \lambda < 0.39$ | $7.7\times 10^{14} < f < 3\times 10^{16}$ | $3.18 < E < 124$ |
| Visible light | $0.39 < \lambda < 0.77$ | $3.9\times 10^{14} < f < 7.7\times 10^{14}$ | $1.61 < E < 3.18$ |
| Infrared | $0.77 < \lambda < 10^3$ | $3\times 10^{11} < f < 3.9\times 10^{14}$ | $1.24\times 10^{-3} < E < 1.61$ |
| Millimeter waves | $10^3 < \lambda < 10^4$ | $3\times 10^{10} < f < 3\times 10^{11}$ | $1.24\times 10^{-4} < E < 1.24\times 10^{-3}$ |
| Microwave | $10^4 < \lambda < 3\times 10^6$ | $10^8 < f < 3\times 10^{10}$ | $4.13\times 10^{-7} < E < 1.24\times 10^{-4}$ |
| Radio waves | $3\times 10^6 < \lambda < 2\times 10^{11}$ | $1.5\times 10^3 < f < 10^8$ | $6.2\times 10^{-12} < E < 4.13\times 10^{-7}$ |
| Long waves | $\lambda > 2\times 10^{11}$ | $f < 1.5\times 10^3$ | $E < 6.2\times 10^{-12}$ |

**Table 8.1.1.** Electromagnetic spectrum.

| | Wavelength, $\lambda$ ($\mu$m) | Frequency, $f$ (Hz) | Photon Energy, $E$ (eV) |
|---|---|---|---|
| Extreme Ultraviolet | $10^{-2} < \lambda < 0.2$ | $7.69\times 10^{14} < f < 3\times 10^{16}$ | $6.2 < E < 124$ |
| Far Ultraviolet | $0.2 < \lambda < 0.30$ | $10^{15} < f < 1.5\times 10^{15}$ | $4.13 < E < 6.2$ |
| Near Ultraviolet | $0.30 < \lambda < 0.390$ | $7.69\times 10^{14} < f < 10^{15}$ | $3.18 < E < 4.13$ |
| Violet | $0.390 < \lambda < 0.455$ | $6.59\times 10^{14} < f < 7.69\times 10^{14}$ | $2.73 < E < 3.18$ |
| Blue | $0.455 < \lambda < 0.492$ | $6.09\times 10^{14} < f < 6.59\times 10^{14}$ | $2.52 < E < 2.73$ |
| Green | $0.492 < \lambda < 0.577$ | $5.20\times 10^{14} < f < 6.09\times 10^{14}$ | $2.15 < E < 2.52$ |
| Yellow | $0.577 < \lambda < 0.597$ | $5.03\times 10^{14} < f < 5.20\times 10^{14}$ | $2.08 < E < 2.15$ |
| Orange | $0.597 < \lambda < 0.622$ | $4.82\times 10^{14} < f < 5.03\times 10^{14}$ | $1.99 < E < 2.08$ |
| Red | $0.622 < \lambda < 0.770$ | $3.90\times 10^{14} < f < 4.82\times 10^{14}$ | $1.61 < E < 1.99$ |
| Near Infrared | $0.77 < \lambda < 1.5$ | $2\times 10^{14} < f < 3.90\times 10^{14}$ | $0.827 < E < 1.61$ |
| Medium Infrared | $1.5 < \lambda < 6$ | $5\times 10^{13} < f < 2\times 10^{14}$ | $0.207 < E < 0.827$ |
| Far Infrared | $6 < \lambda < 40$ | $7.5\times 10^{12} < f < 5\times 10^{13}$ | $0.031 < E < 0.207$ |
| Extreme Infrared | $40 < \lambda < 1000$ | $3\times 10^{11} < f < 7.5\times 10^{12}$ | $0.031 < E < 0.00124$ |

**Table 8.1.2.** Optical part of electromagnetic spectrum.

Figure 8.1.1 shows a semiconductor film and reflected and transmitted beams of light beams. The reflection, absorption, and transmission coefficients are related to the optical properties of the semiconductor material and the sample geometry.

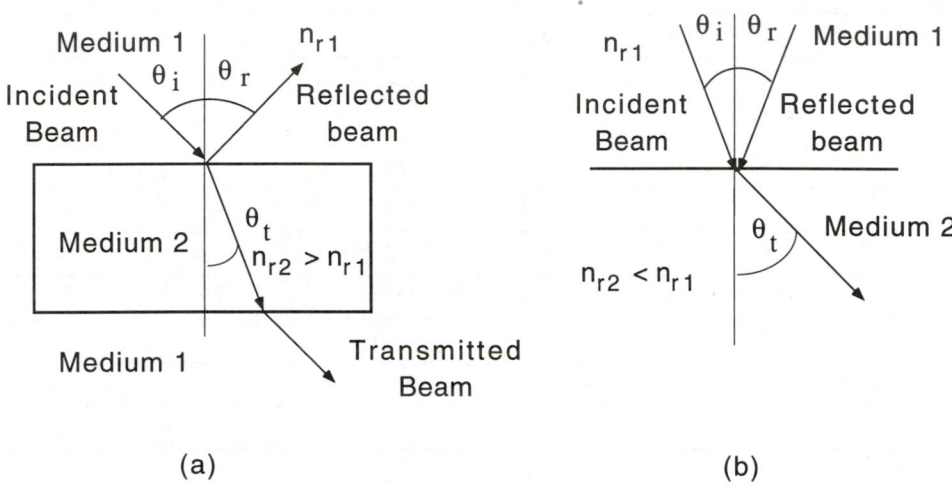

(a)　　　　　　　　　　　　　　　　　　　(b)

**Fig. 8.1.1.** (a) Incident, reflected, absorbed, and transmitted light ($n_{r2} > n_{r1}$) and (b) reflection for $n_{r2} < n_{r1}$.

In vacuum, the speed of light, $c$, is equal to $2.99792458 \times 10^8$ m/s $\approx 3 \times 10^8$ m/s; see Appendix A2). In any medium, the speed of light, $c_m$, is smaller

$$c_m = c/n_r \tag{8-1-1}$$

where $n_r$ is called the refractive index. In order to account for the absorption of light, we introduce a complex refractive index, $n_r^*$:

$$n_r^* = n_r(1 - i\chi) \tag{8-1-2}$$

where $\chi$ is called the absorption index related to the absorption coefficient

$$\alpha = \frac{4\pi\chi n_r}{\lambda} \tag{8-1-3}$$

Here $\lambda = 2\pi c/\omega$ is the wavelength of the electromagnetic radiation in vacuum. The equation describing the propagation of the electromagnetic wave of a particular **polarization** (i. e., a particular direction of the wave electric field) can be written as follows:

$$F = F_m \exp\left(-\frac{\alpha x}{2}\right)\exp\left[-i\omega(t - n_r x / c)\right] \qquad (8\text{-}1\text{-}4)$$

where $F$ is the electric field of the electromagnetic wave, $F_m$ is the amplitude, $\omega$ is frequency, and $x$ is the direction of propagation. The complex refractive index can be related to the complex dielectric permittivity, $\varepsilon^*$:

$$n_r^* = \left(\frac{\varepsilon^*}{\varepsilon_o}\right)^{1/2} \qquad (8\text{-}1\text{-}5)$$

where $\varepsilon_o = 8.85418 \times 10^{-12}$ F/m is the dielectric permittivity in vacuum (see Appendix A2),

$$\varepsilon^* = \varepsilon - i\sigma/\omega \qquad (8\text{-}1\text{-}6)$$

and $\sigma$ is the conductivity of the semiconductor. Equating $\varepsilon_o^2\left(n_r^*\right)^2$ to $\varepsilon^*$ and using eq. (8-1-2), we find

$$\varepsilon = \varepsilon_o n_r^2 \left(1 - \chi^2\right) \qquad (8\text{-}1\text{-}7)$$

$$\sigma = 2n_r^2 \chi \omega \varepsilon_o \qquad (8\text{-}1\text{-}8)$$

Substituting $\chi$ from eq. (8-1-8) into eq. (8-1-3) yields

$$\alpha = \frac{\sigma}{n_r c \varepsilon_o} \qquad (8\text{-}1\text{-}9)$$

The **reflection coefficient**, $R$, is defined as the ratio of the intensity of the reflected wave to the intensity of the incident wave. When the directions of the incident and reflected electromagnetic waves are perpendicular to the sample surface and the sample is in vacuum, the reflection coefficient is related to $n_r^*$ as follows [see, for example, Yariv (1985)]:

$$R = \left|\frac{n_r^* - 1}{n_r^* + 1}\right|^2 = \frac{(n_r - 1)^2 + n_r^2 \chi^2}{(n_r + 1)^2 + n_r^2 \chi^2} \qquad (8\text{-}1\text{-}10)$$

If we neglect losses, then the reflection coefficient at the interface between two materials with refractive indexes $n_{r1}$ and $n_{r2}$ is given by

$$R = \frac{\left(\dfrac{n_{r2}}{n_{r1}} - 1\right)^2}{\left(\dfrac{n_{r2}}{n_{r1}} + 1\right)^2} \qquad (8\text{-}1\text{-}11)$$

This equation applies to the normal incidence of the incoming wave. For an arbitrary incidence angle, the reflection and refraction are governed by **Snell's law** (see Fig. 8.1.1):

$$\theta_i = \theta_r$$
$$n_{r1}\sin(\theta_i) = n_{r2}\sin(\theta_t) \qquad (8\text{-}1\text{-}12)$$

An important consequence of Snell's law is the effect of total reflection that may take place when the incoming wave is reflected from the material with a smaller refractive index $n_{r2} < n_{r1}$. According to eq. (8-1-12), if

$$\sin(\theta_i) > \sin(\theta_{cr}) = \frac{n_{r2}}{n_{r1}} \qquad (8\text{-}1\text{-}13)$$

the wave undergoes a **total internal reflection** since there is no solution for $\sin(\theta_t) = n_{r1}\sin(\theta_i)/n_{r2}$ if $n_{r1}\sin(\theta_i)/n_{r2} > 1$ (see Fig. 8.1.2). This effect is used in **optical fibers** – silica fibers used in optical communications. The light beam is contained within such a fiber because it undergoes total internal reflections.

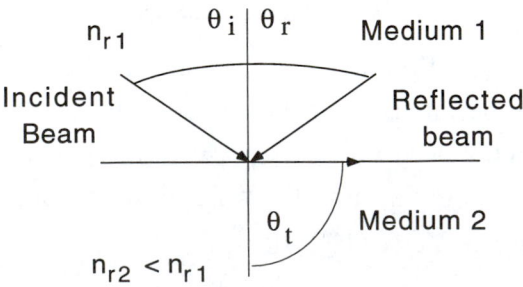

**Fig. 8.1.2.** Total internal reflection.

As we discussed in Section 3.8 [see eqs. (3.8.9) and (3.8.10)], the intensity,

$P_l$, of an electromagnetic wave propagating in a semiconductor in direction $x$ is governed by the following equation:

$$dP_l / dx = -\alpha P_l \tag{8-1-14}$$

$1/\alpha$ is called the **penetration depth**; see Fig. 8.1.3.

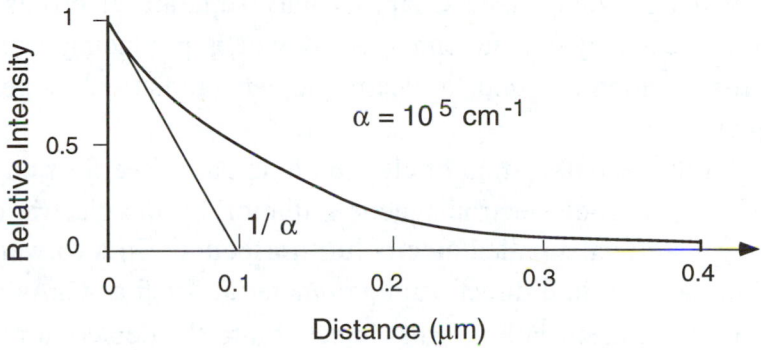

**Fig. 8.1.3.** Light intensity versus distance from surface.

Figure 8.1.4 shows the dependencies of absorption coefficients, $\alpha$, on the photon wavelength, $\lambda$ ($\mu$m), and energy, $E$ (eV) = 1.24/$\lambda$ ($\mu$m).

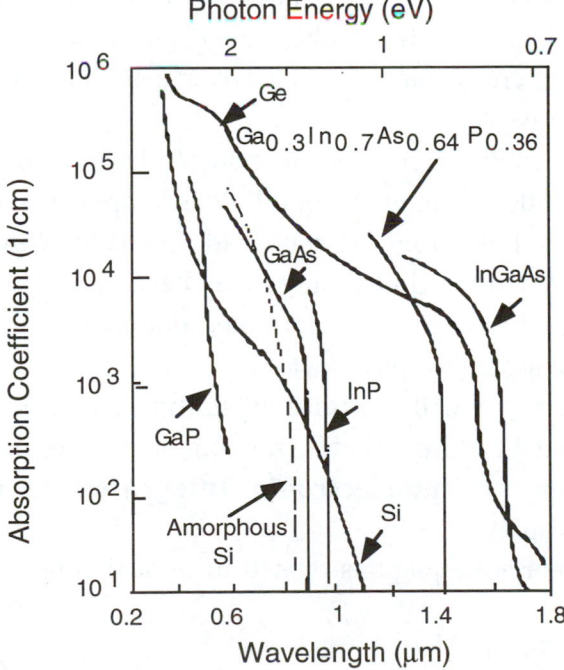

**Fig. 8.1.4.** Absorption coefficients for different semiconductors.

The electromagnetic radiation with frequency $\omega < E_g / \hbar$ passes through a semiconductor with relatively little absorption. The radiation with frequency $\omega > E_g / \hbar$ is absorbed much more stronger. Such a radiation may create electron-hole pairs. A built-in electric field that exists in a *p-n* junction diode or in a Schottky barrier diode (see Chap. 4) may separate electrons and holes. Hence, these devices may operate either as **solar cells** producing electricity or as **photodetectors**, which respond to electromagnetic radiation by generating an electrical signal.

The radiative recombination of electron-hole pairs (see Section 3.8) can be used for the generation of electromagnetic radiation by the electric current in a *p-n* junction. This effect is called **electroluminescence**. In a forward-biased *p-n* junction fabricated from a direct band gap material, such as GaAs or GaN, the recombination of the electron-hole pairs injected into the depletion region causes the emission of electromagnetic radiation. Such a device is called a Light Emitting Diode (LED). If mirrors are provided (usually by cleaved crystallographic surfaces of the semiconductor) and the concentration of the electron hole pairs (called the injection level) exceeds some critical value, this device may function as a semiconductor laser that emits a coherent electromagnetic radiation with all photons in phase with each other. LEDs fabricated from different semiconductors cover a broad range of wavelengths, from infrared to ultraviolet.

An important application of semiconductor lasers and LEDs is in optical communications. Optical communication links use optical fibers: thin strands of silica glass that carry light from the source to a detector with little attenuation. Soon, most households in the United States may have optical communication links that allow us to receive and send enormous amounts of information. These optical links will use novel photonic devices.

In this chapter, we will consider crystalline and amorphous solar cells, photodetectors, Light Emitting Diodes, semiconductor lasers, integrated optical and electronic devices (**Optoelectronic Integrated Circuits – OEICs**), optical fibers, and displays.

The most important equations related to optical constants are summarized in Table 8.1.3.

| | |
|---|---|
| Complex refractive index | $n_r^* = n_r(1 - i\chi)$ |
| Absorption coefficient | $\alpha = \dfrac{4\pi\chi n_r}{\lambda}$ |
| Link between $n_r^*$ and complex dielectric permittivity, $\varepsilon^*$ | $n_r^* = \left(\dfrac{\varepsilon^*}{\varepsilon_o}\right)^{1/2}$ <br><br> where $\varepsilon^* = \varepsilon - i\sigma/\omega$ |
| Dielectric permittivity and refractive index | $\varepsilon = \varepsilon_o n_r^2\left(1 - \chi^2\right)$ |
| Conductivity | $\sigma = 2n_r^2\chi\omega\varepsilon_o$ |
| Absorption coefficient and conductivity | $\alpha = \dfrac{\sigma}{n_r c\varepsilon_o}$ |
| Reflection coefficient for directions of incident and reflected waves perpendicular to sample surface (sample is in vacuum) | $R = \left\|\dfrac{n_r^* - 1}{n_r^* + 1}\right\|^2 = \dfrac{(n_r - 1)^2 + n_r^2\chi^2}{(n_r + 1)^2 + n_r^2\chi^2}$ |
| Reflection coefficient for directions of incident and reflected waves perpendicular to interface of two materials (with losses neglected) | $R = \dfrac{\left(\dfrac{n_{r2}}{n_{r1}} - 1\right)^2}{\left(\dfrac{n_{r2}}{n_{r1}} + 1\right)^2}$ |
| Snell's law | $\theta_i = \theta_r$ <br><br> $n_{r1}\sin(\theta_i) = n_{r2}\sin(\theta_t)$ |
| Condition of total reflection | $\sin(\theta_i) > \sin(\theta_{cr}) = \dfrac{n_{r2}}{n_{r1}}$ |
| Equation describing distribution of light intensity | $\dfrac{dP_l}{dx} = -\alpha P_l$ |
| Relationship between photon energy and wavelength | $E \text{ (eV)} = 1.24/\lambda \text{ (}\mu\text{m)}$ |

**Table 8.1.3.** The summary of the most important equations related to optical constants.

## 8-2.  SOLAR CELLS

It is not an exaggeration to say that the U.S. industry runs on oil.  However, in the future, solar cell technology (called **photovoltaic technology**), which converts sunlight into electricity, should be able to provide a sizable portion of our overall energy needs.  Relatively cheap and efficient large-area solar cells made from amorphous silicon (a-Si) hold promise for such large-scale applications.  Smaller, but more efficient, crystalline solar cells have found numerous applications in remote locations or in space.  Solar cells of all kinds are used in different consumer products – from watches and calculators to power supplies for laptop computers.

It is probably easier to understand the principle of solar cell operation using a hydrogenated a-Si $p$-$i$-$n$ solar cell as an example (see Fig. 8.2.1).

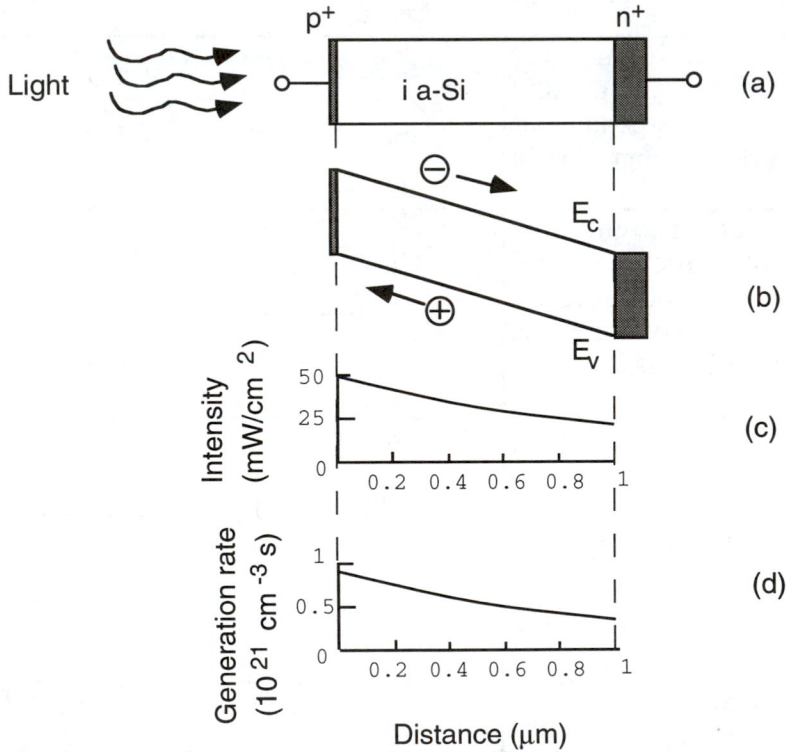

**Fig. 8.2.1.**  (a) Schematic structure, (b) band diagram, (c) intensity of absorbed light, and (d) generation rate of electron-hole pairs for a-Si $p$-$i$-$n$ solar cell.

(The properties of a-Si:H material were briefly discussed in Section 7.3, where we considered a-Si TFTs.)

As shown in Fig. 8.2.1, the light absorbed in the $i$-region creates electron-hole pairs whose concentration is proportional to number of absorbed photons with energies larger that the energy gap of amorphous silicon. These electrons and holes are separated by the built-in electric field. These separated charges produce an electric current, $I_L$, called the **light-generated current**:

$$I_L = qS \int_0^L G(x)dx \qquad (8\text{-}2\text{-}1)$$

Here $G(x)$ is the generation rate of the electron-hole pairs in the $i$-region (see Section 3.8), and $L$ is the length of the $i$-region. For the values shown in Fig. 8.2.1, the light-generated current per unit area is approximately 9.3 mA/cm$^2$. Later on in this section, we will discus how this value can be increased.

The direction (and, hence, the sign) of the light-generated current is opposite to the direction of the current in the dark (called the **dark current**), which can be calculated using the diode equation (see Section 4.3):

$$|I_{dark}| = I_s \left[ \exp\left( \frac{V}{\eta V_{th}} \right) - 1 \right] \qquad (8\text{-}2\text{-}2)$$

If the light-generated current is larger than the dark current, the device acts as a current source, that is, as a source of electricity.

The total diode current under illumination is given by

$$I = I_L - I_s \left[ \exp\left( \frac{V}{\eta V_{th}} \right) - 1 \right] \qquad (8\text{-}2\text{-}3)$$

where, as customary for solar cells, we chose the sign of the light-generated current as positive. If we account for an unavoidable parasitic series and shunt resistances, $R_s$ and $R_{sh}$, eq. (8-2-3) becomes

$$I = I_L - I_s \left[ \exp\left( \frac{V + IR_s}{\eta V_{th}} \right) - 1 \right] - \frac{V}{R_{sh}} \qquad (8\text{-}2\text{-}4)$$

It is instructive to compare this equation with the nonideal diode equation introduced in Section 4.3. [Please notice the different (positive) sign in front of the $IR_s$ term and explain why this sign is positive.]

Figure 8.2.2 shows the equivalent circuit of a solar cell.

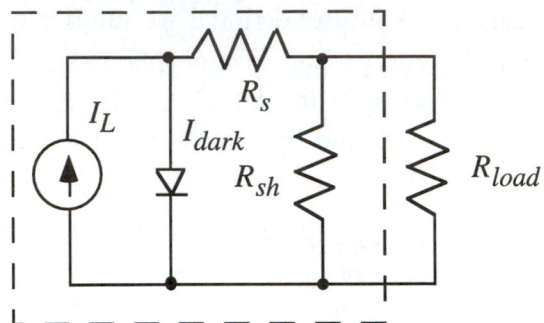

**Fig. 8.2.2.**   Equivalent circuit of a solar cell connected to a load resistance.

In a crystalline *p-n* junction solar cell, the electrons and holes generated by light in the neutral regions within a few diffusion lengths from the junction, contribute to the light-generated current. This can be demonstrated by analyzing the distribution of the minority carriers (holes in the *n*-type region and electrons in the *p*-region); see Problem 8-2-1. (In *p*- and *n*-regions of a *p-i-n* a-Si solar cell, where the diffusion lengths of minority carriers are quite small, the largest contribution to the light-generated current comes from the electron-hole pairs generated in the *i*-region.)

In a solar cell, the generation rate is not uniform and depends on the incident light spectrum and reflection coefficients of the cell surfaces. Also, for a practical theory of a solar cell, we have to relate the electron-hole pair generation rate to the solar spectrum and intensity. However, eq. (8-2-4) is very useful for a phenomenological modeling of solar cells. Equation (8-2-4) can be represented by the equivalent circuit shown in Fig. 8.2.2. Using this equivalent circuit, it is easy to simulate a solar cell using SPICE. Such a simulation can be useful, for example, in designing a power supply that includes many solar cells connected in parallel and/or in series.

Figure 8.2.3 shows two solar spectra: the spectra for the solar radiation at sea level when the sun is at zenith [called for AM1 (Air Mass One) illumination] and the spectra for the solar radiation just outside the earth's atmosphere [called AM0 (Air Mass Zero) illumination]. The intensity of the AM1 radiation is $P_{in} = 92.5$ mW/cm$^2$. The intensity of the  AM0 illumination is approximately 135 mW/cm$^2$. (The losses in the atmosphere are primarily caused by the ultraviolet absorption in the ozone and the infrared absorption in the water vapor.) We also showed in Fig. 8.2.3 wavelengths corresponding to the energy gaps of three semiconductor materials used in photovoltaic applications. Since the light-generated current is caused by the valence-to-conduction band transitions, only photons with energies larger than the band gap $\left(\hbar\omega \geq E_g\right)$, that is, with wavelengths $\lambda \leq \lambda_g \, (\mu m) \approx 1.24 / E_g$ (eV) contribute to the light-generated current.

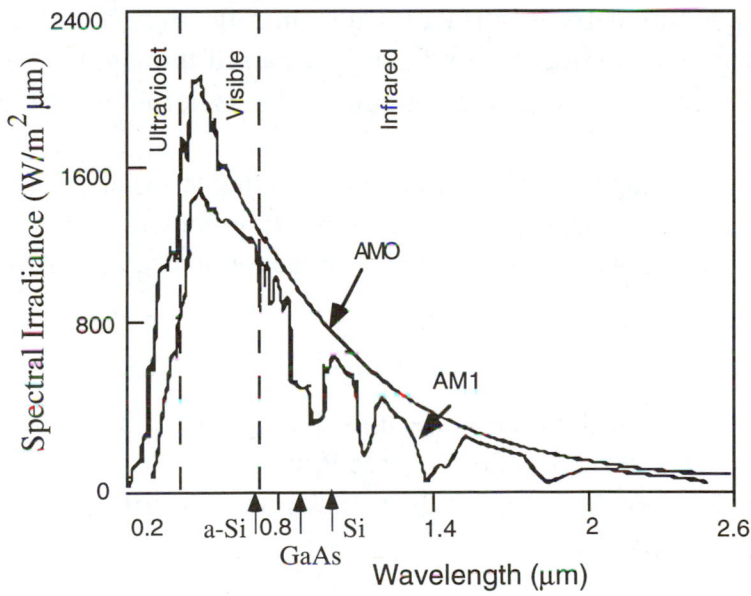

**Fig. 8.2.3.** AM1 and AM0 solar spectra (after Shur, 1990, copyright © Prentice Hall, 1990, reproduced by permission of Prentice Hall, Inc., Englewood Cliffs, NJ).

The light-generated current can be related to the solar spectrum as follows:

$$I_L = qS \int_0^{\lambda_g} \alpha_c(\lambda) N_{ph}(\lambda) Q_c(\lambda) d\lambda \tag{8-2-5}$$

Here $\alpha_c(\lambda)$ is the fraction of incident photons absorbed in the cell, $N_{ph}(\lambda)d\lambda$ is the number of the incident photons with wavelengths between $\lambda$ and $\lambda + d\lambda$ per second per unit area, and $Q_c(\lambda)$ is the fraction of these absorbed photons that generate electron-hole pairs ($Q_c$ is called the collection efficiency). Assuming that all photons in the solar spectrum with energies higher than the silicon energy band gap are absorbed in a solar cell and produce electricity, we can obtain an estimate for the upper bound of the light generated current. Such an approach yields 54 mA/cm$^2$ for the AM0 illumination and 44 mA/cm$^2$ under the standard test conditions (when AM1.5 radiation with the solar radiation intensity $P_{in}$ = 100 mW/cm$^2$ is used). Efficient silicon solar cells exhibit light-generated current close to 42 mA/cm$^2$ under these test conditions.

Such high values of the light-generated current have been achieved by using special textured front surfaces with antireflection coatings and back-reflecting surfaces that effectively reflect light passed through the cell back into the active region, increasing the length of the material where the light is absorbed.

The idea of using a special antireflection coating is based on the dependence of the reflection coefficient from the boundary between two media with refraction indexes $n_1$ and $n_2$ on the relative refraction index, $n_r = n_2/n_1$:

$$R = \left(\frac{n_r - 1}{n_r + 1}\right)^2 \qquad (8\text{-}2\text{-}6)$$

Covering the surface with an antireflection coating with a refraction index $n_3 = (n_1 n_2)^{1/2}$ minimizes the total reflection (see Problem 8-2-4).

As can be seen from eq. (8-2-4), the **open circuit voltage**, $V_{oc}$, of a solar cell is determined by

$$0 = I_L - I_s\left[\exp\left(\frac{V_{oc}}{\eta V_{th}}\right) - 1\right] \qquad (8\text{-}2\text{-}7)$$

(Here we neglected parasitic resistances $R_s$ and $R_{sh}$.) Hence,

$$V_{oc} = \eta V_{th} \ln\left(\frac{I_L}{I_s} + 1\right) \qquad (8\text{-}2\text{-}8)$$

A typical silicon *p-n* junction solar cell is made using either a *p+-n* junction or an *n+-p* junction because *p+-n* and *n+-p* junctions have higher built-in voltages than *p-n* junctions. The built-in electric field separating the electrons and holes generated by light can at most provide the built-in potential, $V_{bi}$. Hence, the built-in voltage gives the upper bound of the open circuit voltage, and *p+-n* and *n+-p* have higher open-circuit voltages. The *p+* layer in *p+-n* cells and the *n+*

layer in $n^+$-$p$ cells  are made very thin because the diffusion length of minority carriers in highly doped layers is very small.

As was discussed in Chapter 4, $V_{bi}$ is roughly proportional to the energy gap, $E_g$.  On the other hand, only photons with energies larger than the band gap are absorbed in a semiconductor and, hence, the light-generated current decreases with an increase in $E_g$.  As a consequence, a solar cell efficiency is the largest for the energy gap about 1.4 eV at room temperature, such as GaAs, and GaAs-based solar cells exhibit record breaking efficiencies.  However, crystalline silicon solar cells are nor far behind.

The measured current-voltage characteristics of a high-efficiency silicon solar cell are shown in Fig. 8.2.4.  Open circle in the figure shows the operating point at which the cell delivers the maximum power

$$P_{\max} = I_{pm}V_{pm} \qquad (8\text{-}2\text{-}9)$$

that can be obtained from this solar cell.  The load resistance corresponding to this operating point is given by

$$R_L = V_{pm} / I_{pm} \qquad (8\text{-}2\text{-}10)$$

The ratio

$$FF = \frac{P_{\max}}{I_{sc}V_{oc}} \qquad (8\text{-}2\text{-}11)$$

is called the **fill factor**.  Here $I_{sc}$ is the cell current for $V = 0$ (called the **short-circuit current**).  The maximum solar cell efficiency is

$$\eta_s = \frac{I_{sc}V_{oc}FF}{P_{in}} \qquad (8\text{-}2\text{-}12)$$

where $P_{in}$ is the power of the incident solar radiation.

As can be seen from eq. (8-2-8) $V_{oc}$ is roughly proportional to $\ln(I_L/I_s)$. The short-circuit current is close to the light-generated current, $I_L$, which, in turn, is proportional to the input radiation power, $P_{in}$.  Hence, at large intensities, the efficiency of a solar cell, $\eta_s = I_L V_{oc}FF/P_{in}$ where $P_{in}$ is the power of the incident radiation given by

$$\eta_s \sim \text{const} \times FF \ln(P_{in}) \qquad (8\text{-}2\text{-}13)$$

that is, the solar cell efficiency should increase with intensity until thermal effects at very large intensities cause the efficiency to decrease.

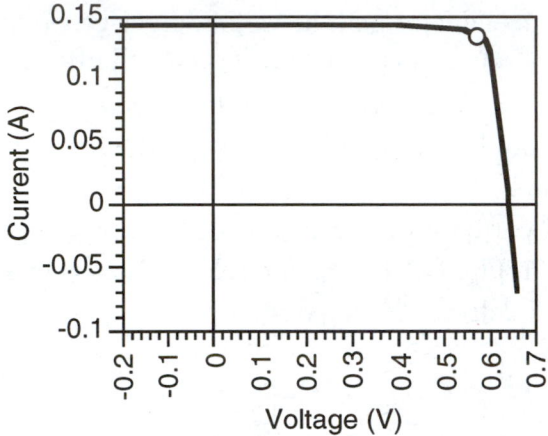

**Fig. 8.2.4.** Measured current-voltage characteristics of a high efficiency silicon solar cell. Open circle voltage, $V_{oc} = 0.6411$ V, short-circuit current density, $J_{sc} = 35.48$ mA/cm$^2$, fill factor 0.822, efficiency 18.70%. Open circle denotes the maximum power point (after M. A. Green, A. W. Blakers, J. Shi, E. M. Keller, and S. R. Wenham, *IEEE Trans. Electron. Dev.*, ED-31, No. 5, p. 679, 1984, © IEEE, 1984).

The solar cell operation is based on the separation of the photogenerated carriers caused by the built-in potential. Hence, *p-i-n* diodes, Schottky barrier diodes, and heterostructure diodes can also operate as solar cells.

In addition to crystalline and hydrogenated amorphous silicon, many other materials such as GaAs, CdTe, CuInSe$_2$, and polycrystalline silicon have been used for photovoltaic cells. Probably the most important photovoltaic technology suitable for generating very large amounts of electricity is the a-Si:H technology. Even though typical efficiencies of a-Si solar cells are under 10%, these cells with large area can be produced fairly inexpensively. The first a-Si solar cell was described by Carlson and Wronski in 1976. Since that time remarkable progress has been achieved in this technology, leading to continuous mass production of large-area a-Si solar cells at reasonable cost.

Amorphous germanium-hydrogen and amorphous silicon carbide-hydrogen alloys have also been obtained by the same process, with the energy gap changing from 1.1 eV for amorphous germanium alloys up to 2.5 eV for amorphous silicon carbide alloys.

Amorphous silicon is highly photoconductive. Typically the dark conductivity of a-Si is on the order of $10^{-10}$ to $10^{-9}$ $(\Omega cm)^{-1}$ and its conductivity under AM1 illumination is on the order of $10^{-5}$ to $10^{-4}$ $(\Omega cm)^{-1}$. The possibility to dope amorphous silicon in a controlled fashion, high photoconductivity, and high absorption coefficient of amorphous silicon (see Fig. 8.1.3) make this material very attractive for applications in solar cells, sensors, imagers, and other photonic devices as well as in temperature and pressure sensors.

A typical a-Si solar cell is shown in Fig. 8.2.5. This particular cell uses an a-SiC/a-Si heterostructure since a-SiC has a larger band gap, which allows light to penetrate into the active *i*-region with smaller losses. The top contact is made of indium tin oxide – a transparent and highly conductive material.

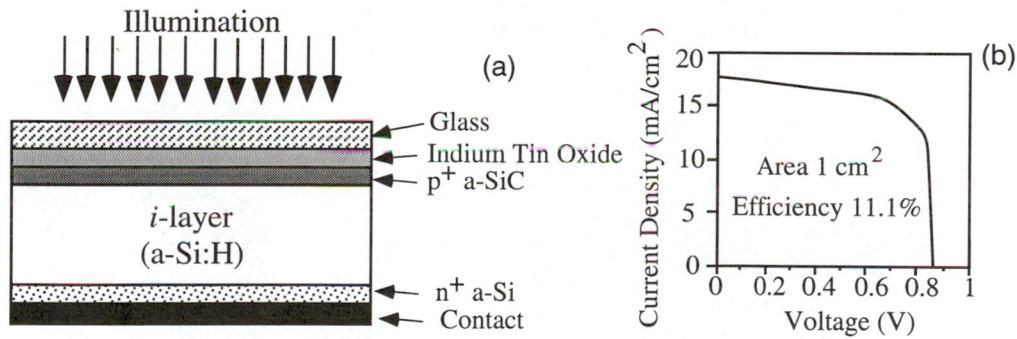

**Fig. 8-2-5.** (a) Schematic diagram of an a-Si:H solar cell with amorphous hydrogenated silicon carbide window and (b) measured current-voltage characteristics of amorphous silicon carbide-amorphous silicon cell (after H. Sakai, K. Maruyama, T. Yoshida, Y. Ichikawa, T. Hama, M. Ueno, M. Kamiyama, and Y. Uchida, *Technical Digest of the International PVSEC-I*, Kobe, Japan, p. 591, 1984, © IEEE 1984).

Large-area mass-produced a-Si solar cells have exhibited high efficiencies (up to 10% or so for 1 square foot cells; maximum efficiencies in small area a-Si solar cells reach 13 to 14 %). The obtained values of the short-circuit current for these cells are on the order of 15 mA/cm², and typical open-circuit voltages are close to 900 to 950 mV with fill factors on the order of 0.7 to 0.75. Such high efficiencies have been obtained by using multiple layer cells (called tandem cells), such as shown in Fig. 8.2.6. A tandem cell consists of several (two or even three) solar cells connected in series by tunneling $p^+$-$n^+$ junctions, which, ideally, have a very small resistance. The top cell is made of a wider band gap semiconductor, the bottom cell (or cells) from a semiconductor with a more narrow gap. The high-energy photons are absorbed in the top wide band gap cell. In the narrow gap semiconductor, the absorption edge is shifted toward smaller energies. Hence, this part of the tandem cell can use an additional part of the solar spectrum, increasing the overall efficiency. The thicknesses of the $i$-layers should be chosen in such a way that the light-generated current produced by the wide band gap and narrow band gap sections of the cell match (see Problem 8-2-3). In amorphous solar cells, the wide band gap material (compared to a-Si:H) may be produced using hydrogenated a-SiC. A more narrow band gap material may be an a-SiGe:H alloy. Crystalline tandem solar cells use heterostructure material systems, such as AlGaAs/GaAs. An antireflection coating and a reflecting bottom contact (to reflect light back into the cell active region) also lead to noticeable improvements in efficiency.

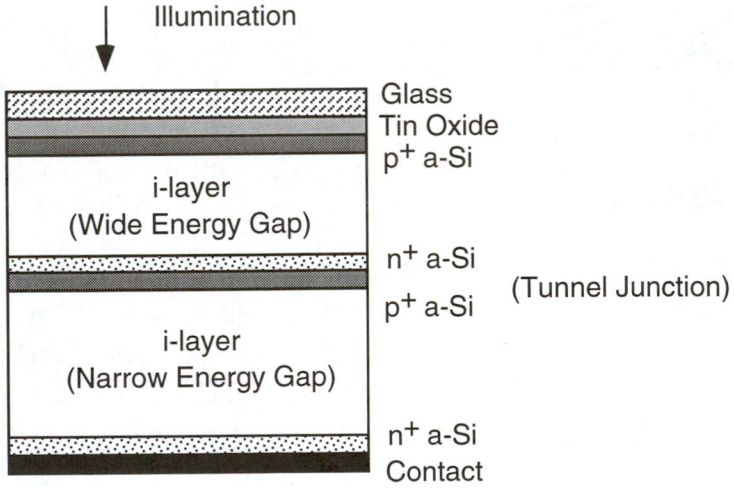

**Fig. 8.2.6.** Schematic diagram of multiple layer (tandem) a-Si solar cell where wide gap and narrow gap $p$-$i$-$n$ cells are connected in series.

Such a connection is achieved by doping the connecting $p^+$ and $n^+$ layers very high so that they form a tunnel junction (which has a low resistance) in the middle of the cell.

Efficiencies of a-Si solar cells strongly depend on the density of the localized states in the intrinsic amorphous silicon (see Section 7.3).  As was first shown by Stabler and Wronski in 1977, the illumination by light with photon energies larger than the energy gap may produce new light-induced defect localized states.  Additional electron-hole pair recombination caused by these states degrades the a-Si solar cell performance with time.  An a-Si cell may lose nearly half of its efficiency after illumination (light soaking) for 24 hours.  However, such a degradation can be drastically reduced in carefully designed a-Si solar cells.  For example, a-Si cells with a thinner *i*-layer are more stable since such cells have a higher built-in electric field (please explain why).  Hence, the electron-hole pairs generated in the *i*-layer by illumination are separated more effectively by the built-in electric field.  As a consequence, the solar cells with a thinner *i*-layer are less sensitive to the increase in the density of the localized states caused by light.

Tandem cells (where each of the two *i* layers is thinner than in a conventional cell) are more stable.

The most important equations related to solar cells are summarized in Table 8.2.1.

| Solar cell dark current | $$\left| I_{dark} \right| = I_s \left[ \exp\left( \frac{V}{\eta V_{th}} \right) - 1 \right]$$ |
| --- | --- |
| Current-voltage characteristic of a solar cell | $$I = I_L - I_s \left[ \exp\left( \frac{V + IR_s}{\eta V_{th}} \right) - 1 \right] - \frac{V}{R_{sh}}$$ |
| Light-generated current | $$I_L = qS\int \alpha_c(\lambda) N_{ph}(\lambda) Q_c(\lambda) d\lambda$$ |
| Open circuit voltage | $$V_{oc} = \eta V_{th} \ln\left( \frac{I_L}{I_s} + 1 \right)$$ |
| Fill factor | $$FF = \frac{P_{max}}{I_{sc} V_{oc}}$$ |
| Solar cell efficiency | $$\eta_s = \frac{I_{sc} V_{oc} FF}{P_{in}}$$ |
| Solar cell efficiency versus illumination intensity | $$\eta_s \sim \text{const} \times FF \ln(P_{in})$$ |

**Table 8.2.1.**  Important equations related to solar cells.

## *8-3.  PHOTODETECTORS

Photodetectors respond to light by producing an electrical signal.  As was mentioned in the Introduction to the book, the history of photodetectors dates back to 1873 when the British engineer W. Smith discovered that the resistivity of selenium – a semiconductor material – is very sensitive to light.  In 1875, Werner von Siemens invented a selenium photometer and in 1878 Alexander Graham Bell used this device for a wireless telephone communication system.  This started practical applications of semiconductor devices.  Nowadays, photodetectors are used in  fiber optics communication systems, in image processing, in establishing optical links between electrical circuits, and so on.

In this section, we consider different types of photodetectors: a photoconductor, a reverse-biased $p$-$i$-$n$ diode (which is similar to a reverse biased solar cell), an Avalanche Photo Diode (APD), a phototransistor, and a Metal-Semiconductor-Metal (MSM) photodetector.

### *8.3.1.  Photoconductor.

Light quanta – photons – may cause different types of transitions in a semiconductor that increase the concentrations of electrons in the conduction band and/or holes in the valence band and, hence, the semiconductor conductivity (see Fig. 8.3.1).  This effect is called photoconductivity.

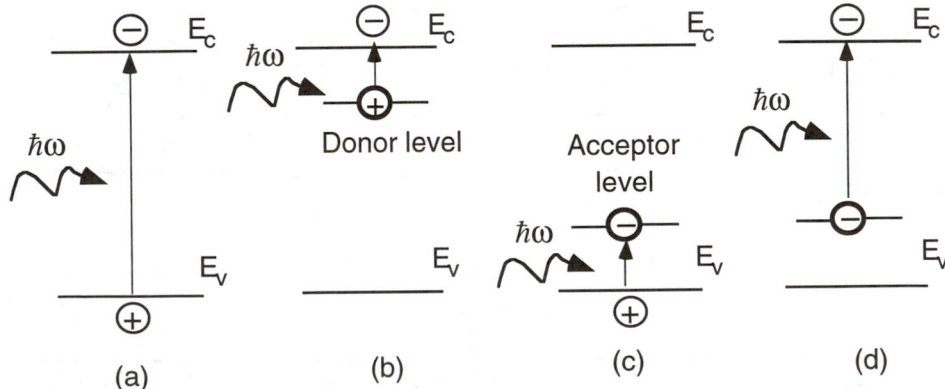

**Fig. 8.3.1.**  Transitions caused by photons in a semiconductors:
(a) intrinsic (valence band to conduction band) transition, (b) extrinsic (deep donor to conduction band) transition, (c) extrinsic (deep acceptor to valence band) transition, (d) extrinsic (deep acceptor to conduction band) transition.

An **intrinsic photoconductor** uses band-to-band transitions. It responds to the radiation with the wavelength in vacuum $\lambda$ ($\mu$m) < $1.24/E_g$ (eV) where $E_g$ is the energy gap. (Figure 8.1.3 shows the wavelength dependence of the absorption coefficient is shown for several important semiconductors.)

**Extrinsic photoconductors** (relying on extrinsic transitions, see Figs. 8.3.1b, c, and d) often operate at cryogenic temperatures, since at higher temperatures the conductivity may be dominated by carriers thermally excited from impurity levels.

As was discussed in Section 3.8, for a uniform illumination, the generation rate of electron-hole pairs is given by

$$G_L(\omega) = \frac{Q_c P_l}{\hbar \omega L} \tag{8-3-1}$$

where $P_l$ is the absorbed optical power per unit area (in W/m$^2$), $L$ is the active thickness of the photoconductive film, and $Q_c$ is the quantum efficiency, that is, the number of electron-hole pairs created per absorbed photon.

If the concentration of the light-generated electron-hole pairs, $n$, is much greater than the concentration of carriers in the dark, the recombination rate is given by

$$R_L = \frac{n}{\tau_l} \tag{8-3-2}$$

where $\tau_l$ is the effective lifetime. In the steady state, $R_L = G_L(\omega)$. Hence $n = G_L(\omega)\tau_l$. If electron photoconductivity is dominant, then the photocurrent is given by

$$I_{ph} = \frac{q\mu_n nVS}{L} = \frac{q\mu_n G_L \tau_l VS}{L} = \frac{q\mu_n Q_c P_l \tau_l VS}{\hbar \omega L^2} \tag{8-3-3}$$

where $\mu_n$ is the electron mobility, $V$ is the applied voltage, and $S$ is the device area. It is instructive to compare this current with the maximum light-generated current, $I_L$, produced in a $p^+$-$n$ solar cell with the same quantum efficiency, $Q_c$, by the same radiation. This current, $I_L$, determined by the number of photons absorbed per unit time, is called the **primary photocurrent**:

$$I_L = \frac{q Q_c P_l S}{\hbar\omega}$$ (8-3-4)

The **photoconductor gain**, $A_{ph}$, is defined as

$$A_{ph} = \frac{I_{ph}}{I_L}$$ (8-3-5)

Using eqs. (8-3-2) to (8-3-4) we find

$$A_{ph} = \frac{\tau_l}{t_{tr}}$$ (8-3-6)

where

$$t_{tr} = \frac{L^2}{\mu_n V}$$ (8-3-7)

is the carrier transit time. In short photoconductors made from materials with long lifetimes and high values of the low field mobility, the photoconductor gain may be as high as a million. More typically, gains on the order of 1,000 for silicon photoconductors and 50 to 100 for InGaAs photoconductors have been achieved.

Quantum efficiency is another important characteristic of a photoconductor. From eq. (8-3-4), the quantum efficiency can be expressed as

$$Q_c = \frac{I_L \hbar\omega}{q P_l S}$$ (8-3-8)

A related characteristic is responsivity, $R_{ph}$, which is defined as the ratio of the photocurrent to the optical power, $P_l S$:

$$R_{ph} = \frac{q Q_c A_{ph}}{\hbar\omega} = \frac{Q_c \lambda A_{ph}(\mu m)}{1.24}\left(\frac{A}{W}\right)$$ (8-3-9)

Typically, responsivities vary between $0.1 \times A_{ph}$ to $2 \times A_{ph}$ (A/W).

The transit time, $t_{tr}$, of the fastest carriers (usually electrons) determines the detection time of a photodetector. However, the device remains conductive until all photogenerated carriers recombine. Hence, the response time of a photoconductor is proportional to the lifetime of the photogenerated carriers, $\tau_l$, which, typically, varies between $10^{-3}$ s and $10^{-10}$ s. In many applications, a detector has to respond to an optical signal with intensity modulated at a certain frequency, $\omega$. The maximum frequency at which the detector can follow the variations in the intensity (**detector bandwidth**) is smaller than $1/\tau_l$. Since the gain is proportional to the carrier lifetime, there is a clear trade-off between the gain and the speed of a photoconductor. The detector gain-bandwidth product is approximately equal to $1/t_{tr}$.

Another important figure of merit for photodetectors is the detectivity, $D$, which is inversely proportional to Noise Equivalent Power (NEP), defined as the incident optical power required to produce a signal-to-noise ratio of one in a bandwidth of one hertz. (NEP can be interpreted as the minimum power that the photodetector can detect.)

### *8.3.2.  *p-i-n*  photodiode.

Reverse-biased *p-i-n* diodes are probably the most widely used photodetectors. To understand how this device operates, let us consider a diode current-voltage characteristic under illumination  (see Fig. 8.3.2). In this figure, the direction of the dark current is chosen as negative, which is also customary for a solar cell. In the first quadrant ($V > 0$, $I > 0$), the diode operates as a solar cell. In the second quadrant ($V < 0$, $I > 0$), the diode operates as a photodetector.

A typical design of a *p-i-n* photodetector is shown in Fig. 8.3.3a. In this device, electrons and holes are generated in the undoped region and separated by the built-in electric field as shown in Fig. 8.3.3b. (This is different from a typical crystalline solar cell where the carriers are collected from the several diffusion lengths in the neutral regions of a *p-n* diode.)

The maximum current that can be collected by a *p-i-n* diode is given by

$$I_L = qS \int_0^L G_L(x)dx \qquad (8\text{-}3\text{-}10)$$

where $S$ is the device area, $L$ is the length of the intrinsic region,

$$G_L = \frac{P_l}{\hbar\omega}\alpha\exp(-\alpha x) \qquad (8\text{-}3\text{-}11)$$

is the electron-hole pair generation rate (assuming that each absorbed photon produces an electron-hole pair), $P_l$ is the incident radiation intensity (in W/m$^2$), and $P_l/\hbar\omega$ is photon flux (per unit area per second).

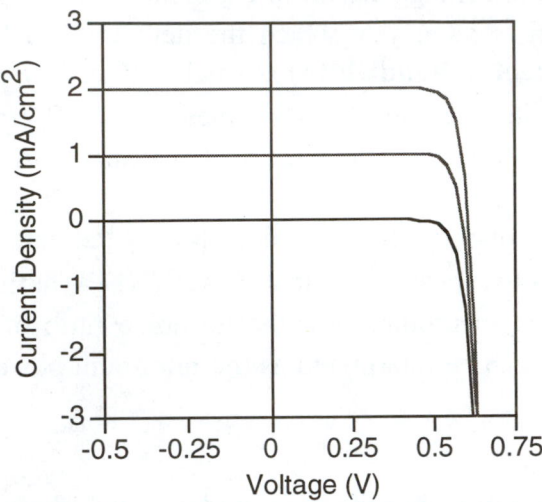

**Fig. 8.3.2.** Current-voltage characteristics of a *p-n* diode for light-generated currents, $I_L = 0$, 1, and 2 mA/cm$^2$. Parameters used: dark saturation current $10^{-9}$ mA/cm$^2$, ideality factor 1.1, series resistance, $R_s = 0$, parasitic shunt resistance, $R_{sh} \to \infty$, temperature $T = 300$ K.

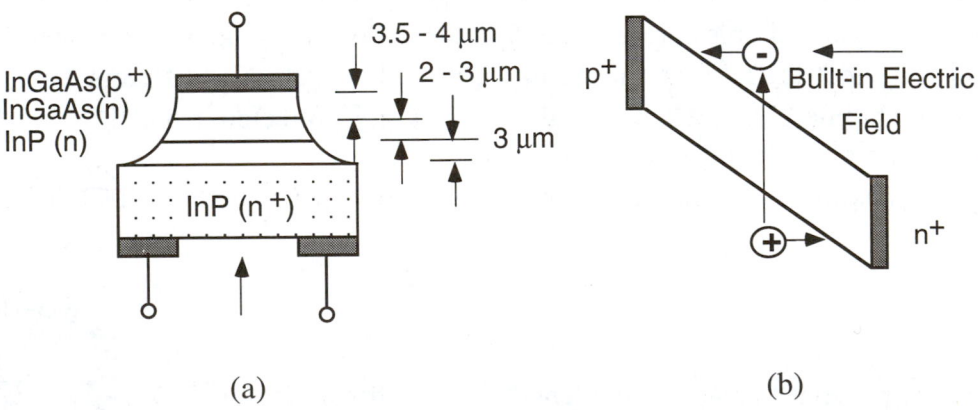

(a)                                                        (b)

**Fig. 8.3.3.** (a) *p-i-n* photodiode illuminated from the substrate (after C. A. Burrus, A. G. Dantai, and T. P. Lee, *Electron. Lett.*, 15, No. 20, pp. 655-657, 1979), and (b) band diagram of a reverse-biased *p-i-n* diode.

Substituting eq. (8-3-11) into eq.(8-3-10) and performing the integration we find

$$I_L = \frac{qSP_l[1 - \exp(-\alpha L)]}{\hbar\omega} \qquad (8\text{-}3\text{-}12)$$

Comparing eq. (8-3-12) with eq. (8-3-4), we obtain the following estimate for the maximum collection efficiency of a *p-i-n* diode:

$$Q_c = 1 - \exp(-\alpha L) \qquad (8\text{-}3\text{-}13)$$

If we take into account that some of the light is reflected from the surface of the photodetector, then this equation should be rewritten as follows:

$$Q_c = (1 - R)[1 - \exp(-\alpha L)] \qquad (8\text{-}3\text{-}14)$$

where $R$ is the reflection coefficient for incident light so that $1 - R$ is the fraction of photons entering the semiconductor.

As can be seen from eq. (8-3-14) *p-i-n* diodes with longer intrinsic regions have a higher quantum efficiency. On the other hand, the carrier transit time (which determines the frequency response of *p-i-n* diodes) increases proportionally to $L$. This establishes a clear trade-off in designing *p-i-n* diodes. The maximum frequency of operation of InGaAs *p-i-n* photodiodes is the order of 10 GHz. This is much larger than typical response frequencies of photoconductors, which are close to 100 MHz. Also, the noise in a *p-i-n* diode is usually much smaller than in a photoconductor. On the other hand, photodiodes exhibit no gain ($A_{ph} = 1$), and the photodiode responsivity, $R_{ph} \approx Q_c\lambda(\mu m)/1.24$ is much smaller than that for a photoconductor [compare with eq. (8-3-9)].

### *8.3.3. Avalanche Photo Diode (APD).

As we discussed above, *p-i-n* photodetectors have numerous advantages compared to photoconductors except one – they have no gain. This drawback is corrected in the **Avalanche Photo Diode (APD)**. An APD is a specially designed reverse-biased *p-i-n* diode where an incoming light signal initiates an avalanche breakdown. Hence, carriers generated by light create other carriers via impact ionization, providing an internal gain. In regular *p-i-n* diodes, parasitic effects (such as a breakdown along the edge of the device or the formation of an electron-hole microplasma near a local defect) usually occur at voltages lower than the critical voltage of bulk impact ionization. APDs use special design features in order to prevent edge breakdown.

Quantum efficiencies of APDs could reach nearly 100%. They exhibit high gain (up to $10^4$) and high speed, with response times as short as several tens of picoseconds. However, the stochastic (probabilistic) nature of the avalanche breakdown leads to a much higher noise than in *p-i-n* diodes (which have about the same speed of operation but a unity gain).

### *8.3.4. Phototransistor.

As we discussed in Chapter 5, in an *n-p-n* bipolar junction transistor, each hole entering the base region causes many electrons from the emitter region to enter the base. In a forward active mode of operation, these electrons diffuse through the base and enter the collector region, contributing to the collector current, $I_c$. This leads to a large current gain, $\beta$, for a common-emitter configuration:

$$I_c = \beta I_b + I_{ceo} \qquad (8\text{-}3\text{-}15)$$

Here $I_b$ is the base current and $I_{ceo}$ is the collector current with the open base (i. e., $I_b = 0$). In the bipolar junction phototransistor, incident light generates electron-hole pairs in the base-collector depletion region. In an *n-p-n* device, electrons generated in this depletion region directly contribute to the collector current, whereas holes contribute to the base current. As a consequence, the collector current under illumination is given by

$$I_c = \beta I_b^{dark} + I_{ceo}^{dark} + I_{Le}(1+\beta) \qquad (8\text{-}3\text{-}16)$$

where $I_{Le}$ is the current caused by the photogenerated electrons. Hence, a phototransistor has a large internal gain.

The Metal Oxide Semiconductor Field Effect Transistor (MOSFET) (see Chap. 6) can also be used as a photosensor. The operation of this device is based on a change in the surface potential in the MOSFET channel under illumination, leading to a modulation of the drain-to-source current.

### *8.3.5. Metal-Semiconductor-Metal Photodetector.

Metal-Semiconductor-Metal (MSM) Photodetectors operate at the same principle as *p-i-n* photodetectors. Their response is related to the current caused by the electron-hole pairs separated by the electric field in the depletion region of two Schottky diodes. These devices usually have a simple planar design, often with interdigitated fingers as shown in Fig. 8.3.4. The interdigitated design minimizes parasitic resistances.  Sometimes, these contacts are made circular to suit circular light beams.  As for *p-i-n* diode, the MSM speed is determined by the transit time, and submicron MSMs have bandwidths up to 350 GHz.

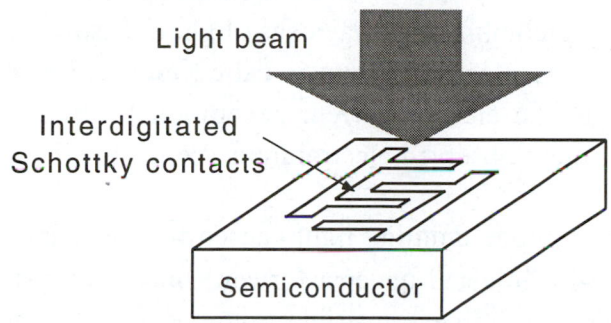

**Fig. 8.3.4.**  Schematic diagram of an MSM photodetector.

Table 8.3.1 summarizes important equations related to photodetectors.

| Generation rate for uniform illumination of photoconductor | $G_L(\omega) = \dfrac{Q_c P_l}{\hbar \omega L}$ |
|---|---|
| Photocurrent (when electron photoconductivity is dominant) | $I_{ph} = \dfrac{q\mu_n G_L \tau_l VS}{L} = \dfrac{q\mu_n Q_c P_l \tau_l VS}{\hbar \omega L^2}$ |
| Light-generated current produced in a $p^+$-$n$ solar cell | $I_L = \dfrac{qQ_c P_l S}{\hbar \omega}$ |
| Photocurrent gain | $A_{ph} = \dfrac{I_{ph}}{I_L}$   $A_{ph} = \dfrac{\tau_l}{t_{tr}}$   where   $t_{tr} = \dfrac{L^2}{\mu_n V}$ |
| Quantum (collection) efficiency | $Q_c = \dfrac{I_L \hbar \omega}{q P_l S}$ |
| Responsivity (ratio of photocurrent to the optical power) | $R_{ph} = \dfrac{qQ_c A_{ph}}{\hbar \omega} \approx \dfrac{Q_c A_{ph} \lambda(\mu m)}{1.24} \left(\dfrac{A}{W}\right)$ |
| Noise Equivalent Power (NEP) | Incident optical power producing a signal-to-noise ratio of one in a bandwidth of 1 Hz |
| Detectivity, $D$ | Inversely proportional to NEP |
| The maximum current that can be collected by a *p-i-n* diode | $I_L = qS \int\limits_0^L G_L(x)\,dx$ |
| Maximum collection efficiency of a *p-i-n* diode | $Q_c = (1-R)\left[1 - \exp(-\alpha L)\right]$ |

**Table 8.3.1.**  Important equations related to photodetectors.

## 8-4. ELECTROLUMINESCENCE AND LIGHT-EMITTING DIODES

When an electric current flows through a *p-n* junction, excess electrons and holes recombine, and this recombination may cause light emission. The effect of light emission caused by an electric current is called **electroluminescence**. (This effect is opposite to the electric current generation by light in solar cells and photodetectors.) Round observed electroluminescence in silicon carbide samples in 1907!

     A *p-n* junction diode emitting light caused by the radiative recombination of electrons and holes injected under a forward bias is called a Light Emitting Diode (LED). The radiative transitions may be band-to-impurity (see Fig. 8.4.1a) or band-to-band (see Fig. 8.4.1b).

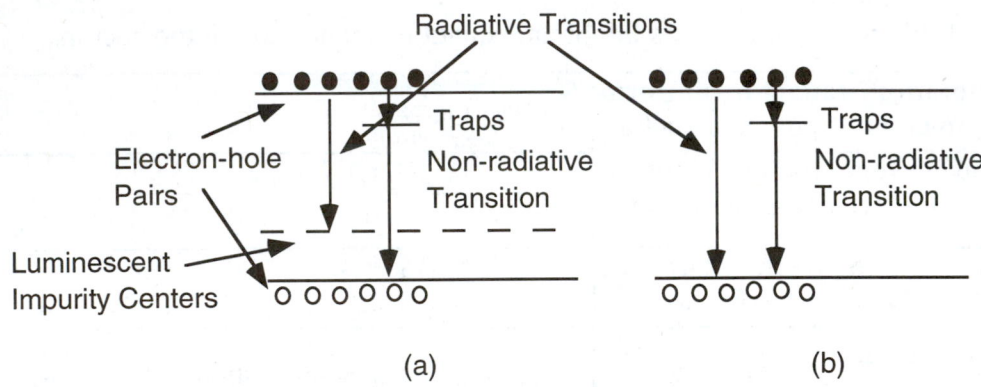

(a)             (b)

**Fig. 8.4.1.**   Radiative and nonradiative transitions in LEDs. (a) Radiative transitions via luminescent impurity centers. (b) Band-to-band radiative transitions.

     The frequency of the emitted light increases with an increasing band gap (see Fig. 8.4.2), since the energy has to be conserved during a radiative transition. The momentum must be conserved as well. As was discussed in Section 2.3, a photon momentum, $p = \hbar\omega/c$ where $\omega$ is the radian frequency and $c$ is the velocity of light (close to $3\times10^8$ m/s) is very small compared to a typical electron momentum when and $\hbar\omega \approx E_g$. The conservation of momentum requires that the electron momentum should change only due to the photon momentum. Since the wave vector $k = p/\hbar$ (see Sections 1.2 and 2.3), the electron transition from the valence into the conduction band looks practically vertical on a band diagram $E(k)$ (see Fig. 8.4.3). That is why direct gap semiconductors, which have the top of the valence band and the bottom of the conduction band aligned in the same $k$-point, are better suited to emit or absorb light and generally better suited for applications in optoelectronic and light-emitting devices. Radiative recombination is much more efficient in direct gap semiconductors such as GaAs, InP, and $Al_xGa_{1-x}As$ (for $x$ less than 0.45).

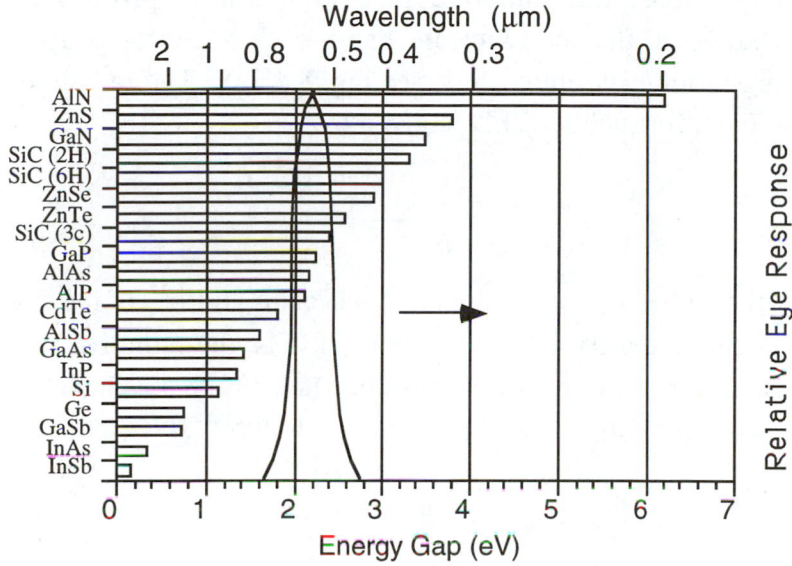

**Fig. 8.4.2.**  Energy gaps of semiconductor materials compared with spectral sensitivity of human eye.

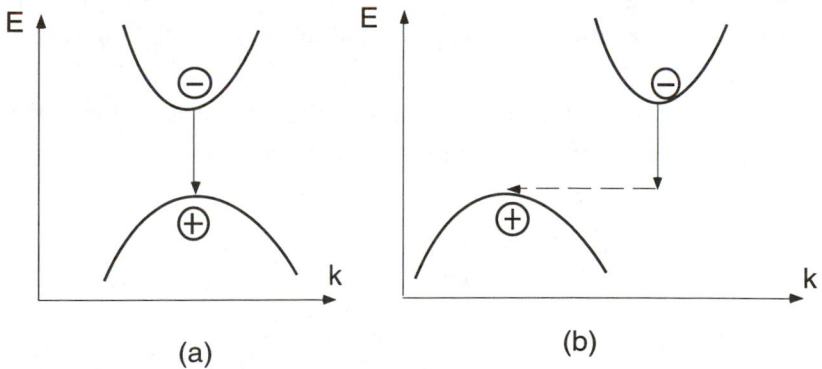

**Fig. 8.4.3.**  Radiative transitions in (a) direct and (b) indirect semiconductors. An additional wave vector, $k_i$ (shown by a dashed arrow), is required for a radiative transition in an indirect semiconductor. The additional momentum $p_i = \hbar k_i$ needed for such a transition has to come from lattice vibrations or from an impurity.

In indirect band gap semiconductors, radiative recombination may be enhanced by adding special impurities in order to help conserve momentum during indirect transitions. GaP is widely used for LED emitting visible light (see Fig. 8.4.2). Impurities (such as nitrogen or sulfur) are often incorporated into GaP to enhance radiative recombination.

In many LEDs, the radiative recombination is primarily related to transitions between the conduction band and acceptor impurity levels (corresponding to luminescent centers, see Fig. 8.4.2a). The radiative lifetime of electron-hole pairs for such an LED is given by

$$\tau_r = \frac{1}{B_R N_A} \tag{8-4-1}$$

where $B_R$ is the radiative recombination coefficient and $N_A$ is the concentration of the luminescent impurity centers. In other LEDs, the radiative recombination is related primarily to band-to-band transitions (see Fig. 8.4.2b). In this case, the radiative recombination rate, $R_r$, is proportional to the $pn$ product:

$$R_r = Bpn \tag{8-4-2}$$

and the radiative lifetime depends on the excitation level.

Nonradiative recombination processes, such as nonradiative transitions involving traps and Auger recombination, reduce the light emission efficiency. These nonradiative transitions can be characterized by a nonradiative lifetime of electron-hole pairs, $\tau_{nr}$.

The fraction of the electron-hole pairs that recombine emitting light is called the quantum efficiency (or radiative efficiency), $\eta_q$:

$$\eta_q = \frac{\tau_l}{\tau_r} \tag{8-4-3}$$

where

$$\tau_l = \frac{\tau_r \tau_{nr}}{\tau_r + \tau_{nr}} \tag{8-4-4}$$

is the overall lifetime. Another important LED characteristic is the external quantum efficiency, $\eta_{ex}$, which is defined as the fraction of electron-hole pairs producing photons actually emitted by the LED:

$$\eta_{ex} = \Gamma_l \eta_q \qquad (8\text{-}4\text{-}5)$$

Here $\Gamma_l$ is the loss factor related primarily to the reflection of emitted photons back into the semiconductor from the semiconductor surface.

For good LEDs, external quantum efficiencies may be a few percent or higher at room temperature. At cryogenic temperatures, nonradiative transitions are suppressed, and external quantum efficiencies are much higher.

The characteristic response time of an LED is determined by the overall recombination time, $\tau_l$. The LED cutoff frequency is given by

$$f_T = \frac{1}{2\pi\tau_l} \qquad (8\text{-}4\text{-}6)$$

GaAs homojunction LEDs operate at the wavelength $\lambda = 0.9$ μm, that is, they emit infrared light. Materials such as GaP with different impurities or $GaAs_{1-x}P_x$ and $Al_xGa_{1-x}As$ are used to make visible yellow, red, or green LEDs (see Fig. 8.4.4). More recently, blue SiC and violet GaN/InGaN LEDs have been fabricated.

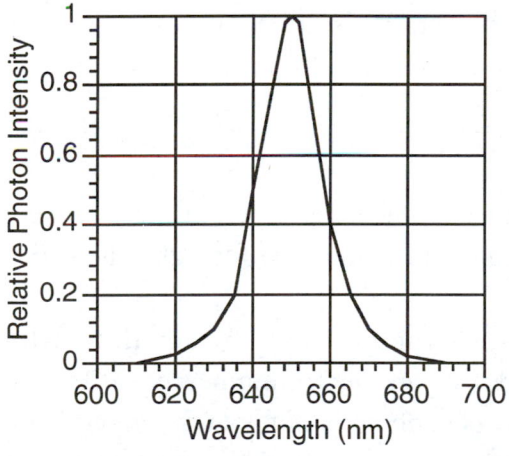 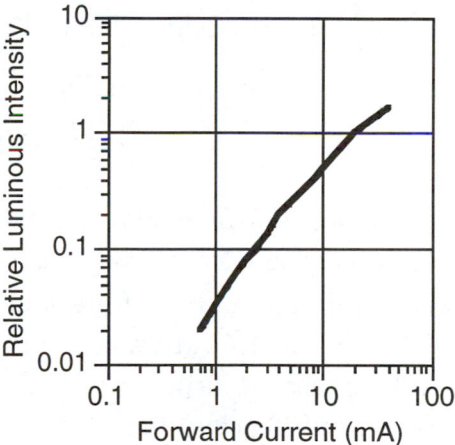

**Fig. 8.4.4.** $GaAs_{1-x}P_x$ Light Emitting Diodes. (a) Relative spectral characteristics at 25 ºC and 20 mA forward current. (b) Relative light intensity versus current level (from *Optoelectronics Data Book*, TI Instruments, 1983. Courtesy of Texas Instruments, Inc.).

LEDs are used in displays, LED printers, and optical fiber communication links. (More often, optical fiber communication links use semiconductor lasers, considered in Section 8.5.) Still another application involves opto-isolators (also called optocouplers) that electrically isolate the input and output signal and prevent noise transfer (see Fig. 8.4.5).

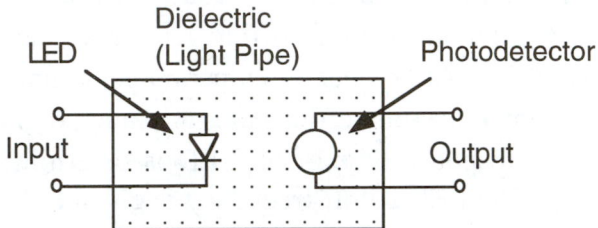

**Fig. 8.4.5.** A schematic diagram of an opto-isolator.

Another type of electroluminescence (called ac electroluminescence) occurs when an ac voltage is applied to the structure containing a semiconductor powder (typically ZnS) embedded into a transparent material with a high dielectric constant (glass or plastic), sandwiched between two conducting electrodes (see Fig. 8.4.6).

One suggested mechanism links the ac electroluminescence to the impact ionization at the grain boundaries caused by the acceleration of electrons in the nonuniform electric field. The light output power, $P_l$, is given by

$$P_l = P_{lo}(\omega)\exp\left(\sqrt{\frac{V_o}{V}}\right) \tag{8-4-7}$$

where $P_{lo}(\omega)$ is a frequency-dependent constant, $V$ is the applied bias voltage, and $V_o$ is the characteristic voltage. (This equation is satisfied for light intensities varying over eight orders of magnitude.)

Figure 8.4.6 shows a schematic structure of an ac thin film electroluminescent display pixel (i. e., a dot of an electroluminescent display). The active substance in this display (called phosphor) is manganese doped ZnS. One of the conducting contacts is made transparent (usually from the transparent conducting material indium tin oxide). Recently electroluminescent displays have been displaced by Liquid Crystal Displays (LCDs); see Section 8.7.

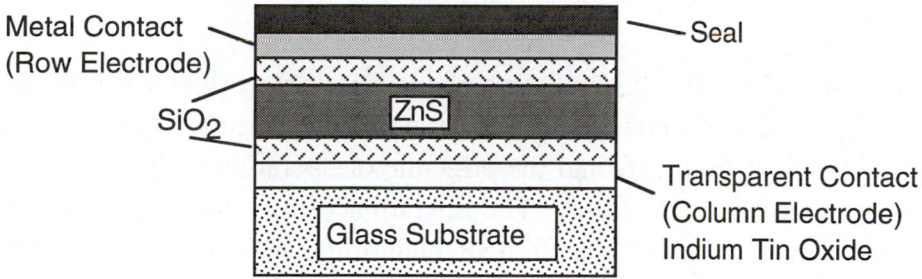

**Fig. 8.4.6.** Schematic structure of an ac thin film electroluminescent display pixel. Each pixel is addressed by one column and one row electrode.

Table 8.4.1 summarizes important equations related to LEDs and electroluminescence.

| | |
|---|---|
| Radiative lifetime for LEDs using radiative transitions between conduction band and acceptor level. | $\tau_r = \dfrac{1}{B_R N_A}$ |
| Radiative recombination rate in LEDs using band-to-band transitions | $R_r = Bpn$ |
| Radiative efficiency, overall lifetime, and radiative lifetime | $\eta_q = \dfrac{\tau_l}{\tau_r}$<br><br>where $\tau_l = \dfrac{\tau_r \tau_{nr}}{\tau_r + \tau_{nr}}$ |
| External quantum efficiency | $\eta_{ex} = \Gamma_l \eta_q$<br>where $\Gamma_l$ is the loss factor |
| Quantum (collection) efficiency | $Q_c = \dfrac{I_p \hbar \omega}{q P_L S}$ |
| LED cutoff frequency | $f_T = \dfrac{1}{2\pi \tau_l}$ |
| Light output power of ac electroluminescence | $P_l = P_{lo}(\omega) \exp\left(\sqrt{\dfrac{V_o}{V}}\right)$ |

**Table 8.4.1.** Equations related to LEDs and electroluminescence.

## 8-5.  LASERS

Albert Einstein was the first to develop a phenomenological theory describing the interaction of electromagnetic waves with systems with discrete quantum levels. His theory allows us to understand the principle of operation of **lasers** – sources of coherent monochromatic light.  The operation of any light-emitting device involves the emission and absorption of electromagnetic radiation related to **radiative transitions** between energy levels.  This processes can be understood by considering a system with only two states – ground state and excited state (see Fig. 8.5.1).  The absorption of a photon (electromagnetic wave quantum) with the energy $\hbar\omega = E_2 - E_1$ causes a transition from the ground state to the excited state.  The transitions from the excited state to the ground state lead to the emission of photons with energies $\hbar\omega = E_2 - E_1$. This emission may be **spontaneous** or **induced**.  Spontaneous emission of  photons occurs without an inducement from electromagnetic radiation.  Induced (stimulated) emission is caused by electromagnetic radiation.

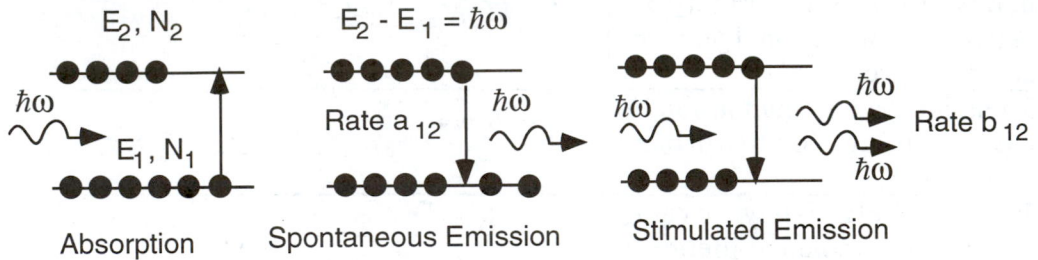

**Fig. 8.5.1.** System with only two energy states (two level system).  $N_2$ particles occupy energy level 2, and $N_1$ particles occupy energy level 1.

In practical objects, the number of energy states is very large.  However, quite often, radiative transitions involve only two states, and the model of a system with only two energy states applies.  For a system with two energy states, the spontaneous emission is described by the following equation:

$$\left(\frac{dN_2}{dt}\right)_{spont} = -a_{12}N_2 \tag{8-5-1}$$

where $N_2$ is the number of particles in state 2 and $a_{12}$ is the spontaneous emission rate.  The transition probability of spontaneous emission is independent of the electromagnetic wave intensity (i. e., independent of the number of photons available).  The spontaneous emission rate $a_{21}$ from state 1 (the ground state) with low energy to state 2 (excited state) with higher energy is zero.

The transition probabilities for induced transitions from state 2 to state 1 and from state 1 to state 2 are proportional to the radiation density, $f(\omega)$ (measured in Joules/m$^3$):

$$\left(\frac{dN_2}{dt}\right)_{induced} = -b_{12}f(\omega)N_2 \qquad (8\text{-}5\text{-}2)$$

$$\left(\frac{dN_1}{dt}\right)_{induced} = -b_{21}f(\omega)N_1 \qquad (8\text{-}5\text{-}3)$$

where $\omega$ is the radian frequency.  (As will be shown below, $b_{12}$ must be equal $b_{21}$.)

Under equilibrium conditions the total transition rates from state 2 to state 1 and from state 1 to state 2 must be equal, that is

$$\left(\frac{dN_2}{dt}\right)_{spont} + \left(\frac{dN_2}{dt}\right)_{induced} = \left(\frac{dN_1}{dt}\right)_{induced} \qquad (8\text{-}5\text{-}4)$$

and, thus,

$$N_2\left[b_{12}f(\omega) + a_{12}\right] = N_1 b_{21}f(\omega) \qquad (8\text{-}5\text{-}5)$$

In thermal equilibrium, the ratio $N_2/N_1$ is equal to the Boltzmann factor

$$\frac{N_2}{N_1} = \exp\left(-\frac{\Delta E_{21}}{k_B T}\right) \qquad (8\text{-}5\text{-}6)$$

where $\Delta E_{21} = E_2 - E_1 = \hbar\omega$.

For **black-body radiation** (see Section 1.2)

$$f(\omega) = \frac{2n_r^3 \hbar\omega^3}{\pi c^3 \left[\exp(\hbar\omega/k_B T) - 1\right]} \qquad (8\text{-}5\text{-}7)$$

where $c$ is the speed of light in vacuum and $n_r$ is the refraction index. Substituting eqs. (8-5-6) and (8-5-7) into eq. (8-5-5) we find

$$\frac{2n_r^3\hbar\omega^3}{\pi c^3\left[\exp(\hbar\omega/k_BT)-1\right]} = \frac{a_{12}}{b_{21}\exp(\hbar\omega/k_BT)-b_{12}} \tag{8-5-8}$$

Equation (8-5-7) can be satisfied at all temperatures and frequencies only if

$$b_{12}=b_{21} \quad\text{and}\quad \frac{a_{12}}{b_{12}}=\frac{2n_r^3\hbar\omega^3}{\pi c^3} \tag{8-5-9}$$

as was first shown by Einstein in 1917.

Let us now consider a medium with two energy states, $E_2$ and $E_1$, with $N_2$ particles per unit volume in state 2 and $N_1$ particles per unit volume in state 1 (see Fig. 8.5.1). The intensity of the electromagnetic wave, $I_\omega$, with frequency $\omega=(E_2-E_1)/\hbar$ propagating in this medium in direction $x$ is governed by the following differential equation [compare with eq. (8-1-15)]:

$$\frac{dI_\omega}{dx}=\gamma I_\omega \tag{8-5-10}$$

where $\gamma=K(N_2-N_1)$. This means that the intensity of the propagating wave either decays or grows exponentially in the medium depending on the sign of $\gamma$:

$$I_\omega = I_\omega(x=0)\exp(\gamma x) \tag{8-5-11}$$

The gain constant $\gamma$ is negative when $N_2<N_1$ and positive when $N_2>N_1$. Under thermal equilibrium conditions, $N_2/N_1$ is equal to the Boltzmann factor [see eq. (8-5-6)] and $N_2<N_1$. However, if the **inversion condition** $N_2>N_1$ is created, then the propagating wave will be amplified. Such a nonequilibrium condition can be formally expressed by choosing a negative value for temperature in eq. (8-5-6) and is frequently referred to as the "negative temperature" condition.

When an amplifying medium is placed between two parallel mirrors separated by an integer number of half wavelengths (the Fabry-Perot etalon, see Fig. 8.5.2 and Problem 8-1-2) and the amplification in the medium exceeds the losses caused by an imperfect reflection from the mirrors, the generation of a coherent light beam can take place in the absence of an incident wave. This light emission is coherent since photons generated as a result of induced transitions are in phase. The light emission is monochromatic since the stimulated emission occurs at the same frequency, $\omega=(E_2-E_1)/\hbar$.

In 1958, Schawlow and Townes first demonstrated such a device operating at microwave frequencies. Their device was called a **MASER** (which is an acronym for **Microwave Amplification by Stimulated Emission of Radiation**). A similar device operating in the visible range is called a **LASER** (which is an acronym for **Light Amplification by Stimulated Emission of Radiation**). The laser was first demonstrated by Thomas Maiman.

In a $p$-$n$ junction semiconductor laser, the inversion is caused by the injection of electrons and holes. The radiative recombination of the injected electron-hole pairs leads to the photon emission.

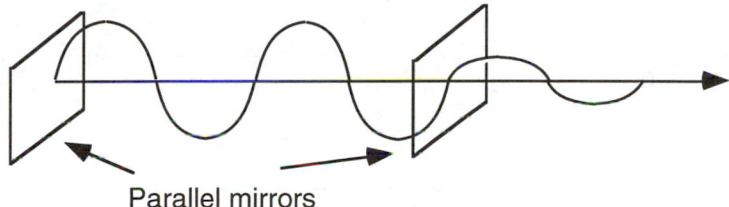

Parallel mirrors

**Fig. 8.5.2.** Wave in Fabry-Perot etalon. The separation between the mirrors is equal to an integer number of $\lambda/2$, where $\lambda$ is the wavelength of electromagnetic radiation. In a laser, one of the mirrors partially transmits light, allowing for the output of the electromagnetic radiation. Typically, mirrors are formed by two polished or cleaved parallel surfaces of the device.

Most semiconductor lasers use degenerately doped $Al_xGa_{1-x}As/GaAs$ (or $Al_xGa_{1-x}As/Al_yGa_{1-y}As$) and $In_xGa_{1-x}As_yP_{1-y}/InP$ heterojunction diodes. In a heterojunction laser, electrons are injected into the narrow active region of the material, with the smaller band gap sandwiched between the $p$ and $n$ regions of the material with the wider band gap. This provides a confinement for the injected electrons and holes, which recombine in the active layer. AlGaAs also has a smaller index of refraction than GaAs. This provides optical confinement, which enhances the stimulated emission. The energy gaps of the active region and confining regions are controlled by controlling the composition (i. e., atomic percentages of Al in $Al_xGa_{1-x}As$ and In and P in $In_xGa_{1-x}As_yP_{1-y}$). The energy gaps of GaAs and AlAs at 300 K are 1.424 eV and 2.168 eV, respectively. The energy gap of $Al_xGa_{1-x}As$ depends on the mole fraction of Al, $x$, as follows:

$$E_g = 1.424 + 1.247x \quad \text{for} \quad x < 0.45$$

$$E_g = 1.900 + 0.125x + 0.143x^2 \quad \text{for } x > 0.45$$

For $x < 0.45$, $Al_xGa_{1-x}As$ is a direct gap material; for $x \geq 0.45$, $Al_xGa_{1-x}As$ is an indirect gap material.

The energy gap, refractive index, doping, and composition profiles for a typical AlGaAs/GaAs laser are shown in Fig. 8.5.3.

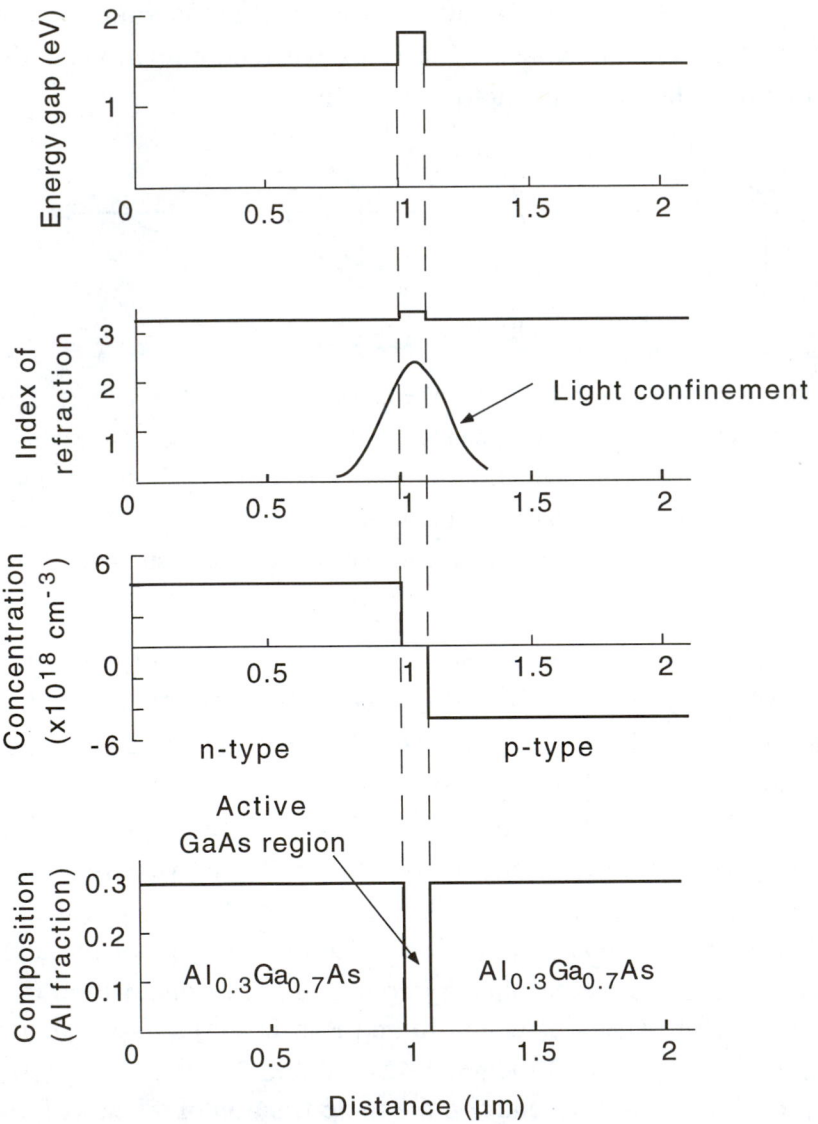

**Fig. 8.5.3.**  The energy gap, refractive index, doping, and composition profiles for a typical AlGaAs/GaAs laser.

Even a fairly small variation of the refractive index allows us to obtain a fairly good localization of light. The understanding of this fact relies on Snell's law [see eqs. (8-1-12) to (8-1-14)]. When the incidence angle $\theta_i$ is close to 90°, even a small difference in the refractive indexes is sufficient to confine light.

A schematic diagram of an AlGaAs/GaAs laser is shown in Fig. 8.5.4. AlGaAs/GaAs lasers emit light with wavelengths from 0.750 to 0.870 μm.

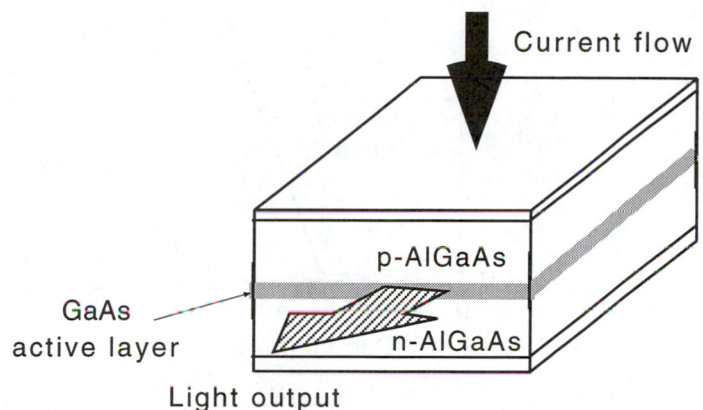

**Fig. 8.5.4.** Schematic diagram of an AlGaAs/GaAs laser. Cleaved faces (perpendicular to light output) form a Fabry-Perot etalon.

The current density, $j$, in a semiconductor laser is primarily the recombination current density

$$j = qN_{inj}d / \tau_l \qquad (8\text{-}5\text{-}12)$$

Here $N_{inj}$ is the concentration of the injected electron-hole pairs, $d$ is the thickness of the active region, and $\tau_l$ is the recombination lifetime.

The stimulated emission starts when the diode current exceeds a certain threshold value. At the threshold, the gain related to the injection of electron-hole pairs becomes larger than the losses. These losses are caused by imperfect reflections from the cleaved surfaces of the semiconductor laser and by nonradiative recombination. The power of the emitted electromagnetic radiation above the threshold is roughly proportional to the diode current (see Fig. 8.5.5).

AlGaAs/GaAs lasers have found numerous applications, including applications in CD players, optical computer disk systems, laser printers, and medical equipment. These lasers can be combined into arrays with output power exceeding 10 W.

An important application of LEDs and semiconductor lasers is in optical communications links using optical fibers. Internal reflections confine light within an optical fiber, which is a thin thread of optical silica glass. The loss coefficient of an optical silica fiber widely used in optical communications as a function of the wavelength has local minima at $\lambda = 1.55$ μm (loss of $\approx 0.2$ db/km) and at $\lambda = 1.3$ μm (loss of $\approx 0.6$ db/km, see Fig. 8.5.6a).

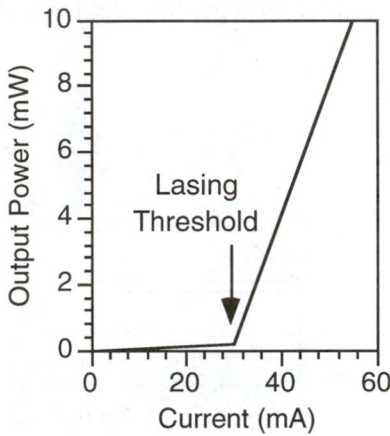

**Fig. 8.5.5.** Light output power from an AlGaAs/GaAs laser diode vs. current (from T. Sugino, A. Yoshikawa, A. Yamamoto, M. Hirose, G. Kano, and I. Teramoto, *IEDM Technical Digest*, p. 618, Los Angeles, CA, IEEE Publications, 1986. © 1986 by IEEE).

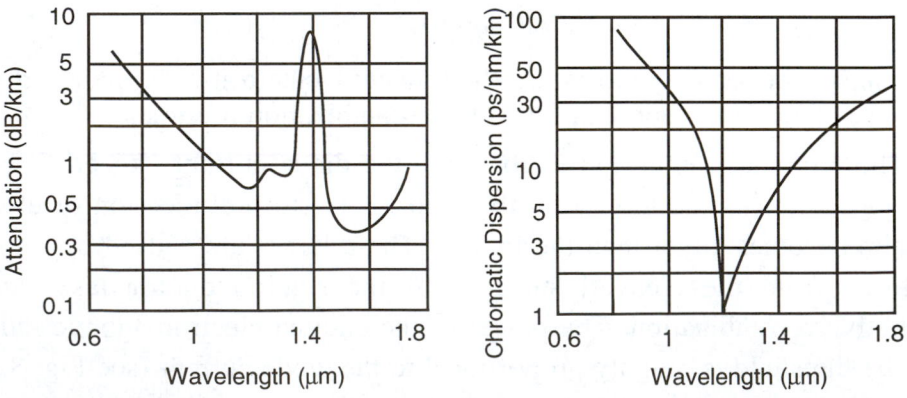

**Fig. 8.5.6.** (a) Attenuation loss and (b) chromatic dispersion versus wavelength for a silica optical fiber (from T. Bell, *Spectrum*, pp. 38-45, Dec. 1983. © 1983 by IEEE).

Another important characteristic of the optical fiber is the **chromatic**

**dispersion**, which is a measure of the dependence of the optical signal propagation velocity in a fiber on the light wavelength. Typically, at currents close to the threshold for the stimulated emission, the laser diode emits several lines corresponding to different modes of operation. The differences in the wavelengths of these modes could be on the order of a nanometer. (At higher currents, just one mode may remain. This single-mode operation is highly desirable for optical communications using optical fibers.) As can be seen from Fig. 8.5.6b, optical signals with wavelengths $\lambda = 1.55$ $\mu$m and $\lambda = 1.551$ $\mu$m will travel a 1 km long fiber with the time difference of approximately 20 ps. For $\lambda = 1.3$ $\mu$m, the chromatic dispersion is much smaller.

Since InGaAsP/InP lasers operate in the range between 1.3 and 1.6 $\mu$m, depending on the composition of the active layer, they have become the lasers of choice for long-distance fiber optics communication links.

Recently, blue semiconductor lasers using ZnSe-ZnTe heterojunctions have been developed. Other wide band semiconductors, such as AlN-GaN system (see Section 9.7) look very promising for blue, violet, and ultraviolet lasers. Another very interesting recent development is the demonstration of a far infrared semiconductor laser by an AT&T group (see Section 9.4).

Table 8.5.1 summarizes important equations related to lasers.

| | |
|---|---|
| Radiation density for black-body radiation | $$f(v) = \frac{8\pi n_r^3 h v^3}{c^3\left[\exp(hv/k_B T)-1\right]}$$ |
| Equation for electromagnetic wave intensity | $\dfrac{dI_v}{dx} = \gamma I_v$ <br><br> where $\gamma = K(N_2 - N_1)$ for a two-level system |
| Occupation of levels in two-level system under equilibrium conditions | $\dfrac{N_2}{N_1} = \exp\left(-\dfrac{hv}{k_B T}\right)$ |
| Current density in semiconductor laser | $j = q N_{inj} d / \tau_l$ |

**Table 8.5.1.** The summary of important equations related to lasers.

## *8-6.  INTEGRATED OPTOELECTRONICS

Monolithic integration of semiconductor electronic devices led to a dramatic decrease in cost, higher reliability, much smaller size and power consumption, and higher speed compared to circuits consisting of discrete components. Monolithic integration of semiconductor lasers, photodetectors, LEDs, and semiconductor electronic devices should allow us to achieve similar advantages for optoelectronics systems.

The applications of **Opto Electronic Integrated Circuits (OEICs)** include optical telecommunications, signal processing, and optical interconnects. In the longer run, OEICs may find applications in optical computing and parallel data processing.

We already discussed optical fibers in Section 8.5, but similar waveguiding properties have ridges made from a semiconductor material, such as GaAs. This allows us to make optical waveguides directly on a semiconductor wafer. Another important effect used in OEICs and other optical devices is the **electro-optic effect**, that is, the electric field dependence of the relative dielectric constant $\varepsilon_r = n_r^2$ where $n_r$ is the refractive index. This effect can be crudely described by the following equation:

$$\frac{d(1/\varepsilon_r)}{dF} = e_{opt} \tag{8-6-1}$$

where $e_{opt}$ is the electro-optic coefficient and where we completely ignore complicated tensor properties of optical materials. In other words, we do not pay attention to the direction in which the electric field, $F$, has to be applied to cause this change in $\varepsilon_r$ and the direction of the light propagation for which this change will be important. However, such a gross simplification allows us to compare the electro-optic effect in different materials (see Table 8.6.1).

| Material | Index of refraction | Electrooptic coefficient $(10^{-12} \text{ m/V})$ |
|----------|---------------------|--------------------------------------------------|
| GaAs | 3.3 (at $T = 300$ K) | 1.43 |
| BaTiO$_3$ | 2.49 | 32.6 |
| LiNbO$_3$ | 2.29 | 820 |

**Table 8.6.1.**  Electro-optical properties.

The electro-optic effect can be used in a variety of optical devices, such as a phase shifter shown schematically in Fig. 8.6.1.

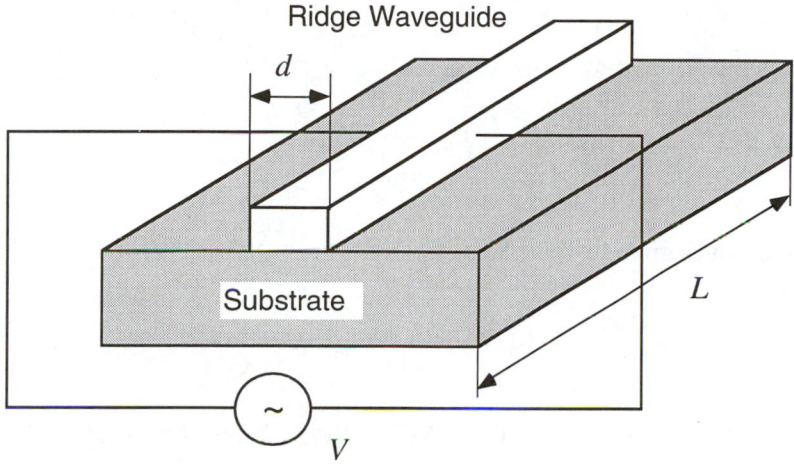

**Fig. 8.6.1.**  Schematic diagram of a phase shifter.

In this device, the electrodes are placed on the sides of a ridge waveguide fabricated on a semiconductor substrate.  The electric field, $F = V/d$, changes the refraction index by $\Delta n$, resulting in a phase shift

$$\alpha = \frac{2\pi\Delta n_r L}{\lambda} \tag{8-6-2}$$

**Example  8-6-1.**

Estimate the phase shift in a GaAs phase shifter.  The device length, $L$, is 100 $\mu$m. The wavelength of light is 0.8 $\mu$m.  The waveguide width, $d$, is 1 $\mu$m.  The applied voltage $V$ is 20 V.

**Solution:**

From eq. (8-6-1),

$$\frac{d(1/\varepsilon_r)}{dF} = \frac{d(1/n_r^2)}{dF} = e_{opt} \tag{8-6-3}$$

Hence,

$$\frac{dn_r}{dF}=-\frac{1}{2}e_{opt}n_r^3 \tag{8-6-4}$$

$$\Delta n_r=-\frac{1}{2}e_{opt}n_r^3\frac{V}{d} \tag{8-6-5}$$

$$\alpha=-\frac{\pi L}{\lambda}e_{opt}n_r^3\frac{V}{d} \tag{8-6-6}$$

Using the values given in Table 8.6.1, we find

$$\alpha=-\frac{\pi\times100}{0.8}1.43\times10^{-12}\times3.3^3\frac{20}{1\times10^{-6}}=0.403=145^\circ$$

InP based OEICs (using GaInAsP alloys) have been applied for fiber transmission systems since these systems can be used at 1.3 μm and 1.55 μm wavelengths where the attenuation and chromatic dispersion in optical fibers is minimal (see Section 8.5).  Optical fibers have a very large bandwidth (of many terabits).  Hence, a very large number of signals can be transmitted through the same optical fiber.  In fiber optics communication systems, the overall speed is often limited not by the laser response but rather by the electronics available for the modulation of the laser light.  Here again, integration of optical and electronics components may help to increase the overall system speed achieving multigigabit data links.

AlGaAs/GaAs OEICs operating at wavelengths close to 0.8 μm have been used for signal processing.  An example of an AlGaAs/GaAs OEIC is a transmitter/receiver array.  Such circuits operate with bit rates up to several gigabits per second.  A schematic diagram of optical interconnects is shown in Fig. 8.6.2.

In conventional Very Large Scale Integrated Circuits where delays caused by interconnect wires become very essential, the replacement of these connections by optical links may drastically improve performance.  Small transceiver chips, such as shown in Fig. 8.6.2, can connect silicon or GaAs VLSI chips and/or circuits boards (see also Fig. 8.6.3). Optical links have advantages of ultra-high speed, low power dissipation, and immunity to outside interference.  However, it is more difficult to localize light beams than electrical signals.  Also, light leakage from an optical waveguide on a chip may present a problem.

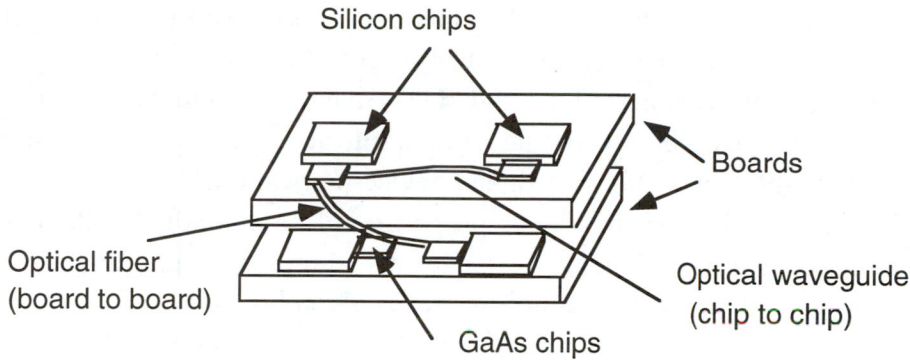

**Fig. 8.6.2.** Schematic diagram of optical interconnects between VLSI silicon chips and circuit boards using GaAs transceiver chips (after L. D. Hutcheson, P. Haugen, and A. Husain, *Spectrum*, pp. 30-35, March 1987. © 1987 by IEEE).

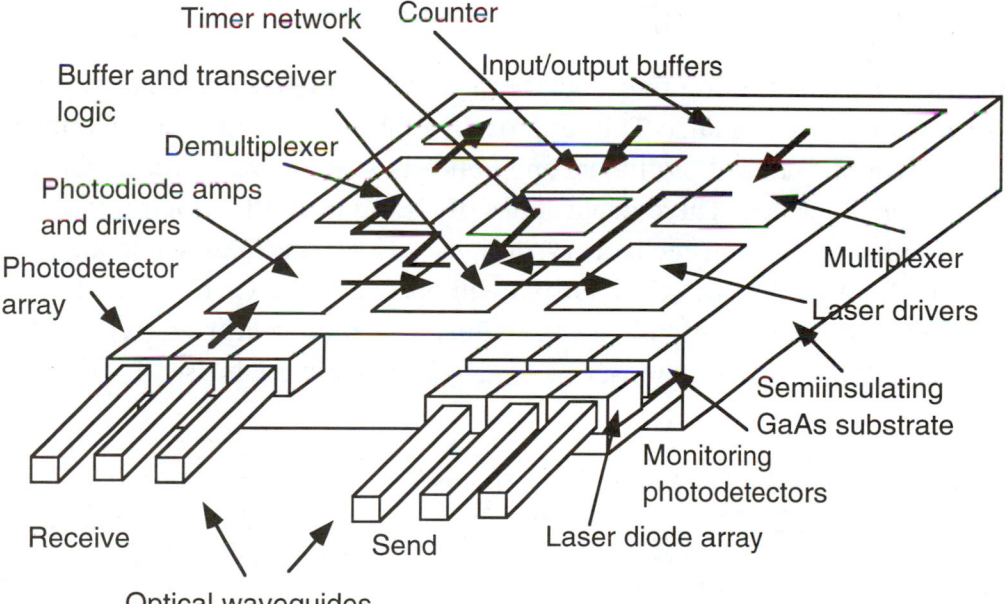

**Fig. 8.6.3.** Schematic diagram of a GaAs transceiver chip (after L. D. Hutcheson, P. Haugen, and A. Husain, *Spectrum*, pp. 30-35, March 1987. © 1987 by IEEE).

Very simple OEICs may be  implemented with 20 to 30 elements.  More complex functions may require 2,000 elements or more.  The basic difficulty in optoelectronic integration is that optical elements (lasers, detectors, and light waveguides) and transistors (FETs and BJTs) have different layer structures. These difficulties will be overcome with a further development of modern technologies of semiconductor growth and device fabrication.

OEICs may also use nonlinear properties of semiconductor devices for mixing optical signals.  When two laser beams illuminate a photodetector, the optical electric field, $\mathbf{F_0}$, in the semiconductor is given by

$$\mathbf{F_0} = \mathbf{F_1} \exp(j\omega_1 t) + \mathbf{F_2} \exp(j\omega_2 t) \tag{8-6-7}$$

where $\mathbf{F_1}$, $\mathbf{F_2}$ and $\omega_1$, $\omega_2$ are the optical electric field and frequencies for laser 1 and 2, respectively.  If the photodetector characteristic is nonlinear, the photocurrent has a component which is proportional to $|\mathbf{F_0}|^2$:

$$i_{12} = a|\mathbf{F_0}|^2 \approx a\left\{|\mathbf{F_1}|^2 + |\mathbf{F_2}|^2 + 2|\mathbf{F_1}||\mathbf{F_2}|\cos[(\omega_1 - \omega_2)]t\right\} \tag{8-6-8}$$

Using two lasers with close frequencies, it is possible to generate a microwave or a millimeter wave signal.  This approach allows one to fabricate optically controlled amplifiers and oscillators and optically steered arrays for applications in phased-array radars.  These applications of OEICs are still in their infancy.

Another promising technology may use wide band gap semiconductors, such as GaN, AlN, and SiC. Such OEICs will operate in the visible and ultraviolet ranges.  Only the future will tell how soon this technology will be developed.

Table 8.6.2 gives a summary of semiconductor materials used for OEICs

| | |
|---|---|
| Materials for 0.8 μm OEICs and typical applications | AlGaAs/GaAs, optical interconnects, optical transceivers |
| Materials for 1.3 and 1.55 μm OEICs and typical applications | InGaAsP, fiber optics communications |
| Materials for visible and ultraviolet light OEICs | GaN, AlN, SiC |

**Table 8.6.2.** Semiconductor materials used for OEICs.

## 8-7.  DISPLAYS

From a television set and a computer monitor to a digital watch, we encounter displays everywhere.  In this section, we consider **Cathode Ray Tube** (**CRT**) displays and **Liquid Crystal Displays** (**LCDs**).    [Another important technology is **Electroluminescent   Display** (**ELD**) technology based on the effect of electroluminescence was briefly considered in Section 8.4.]

### 8.7.1.  Cathode Ray Tube.

A CRT is a vacuum tube with a cathode that emits a beam of electrons accelerated by the voltage difference between the anode and the cathode.  The deflection system steers the beam into any position on the screen using either electric or magnetic fields (see Fig. 8.7.1).

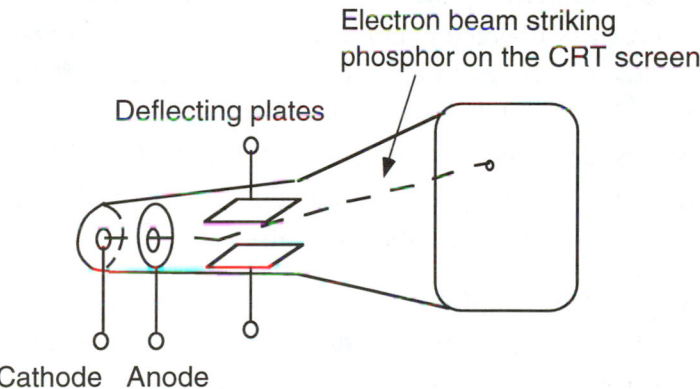

**Fig. 8.7.1.**  Schematic diagram of Cathode Ray Tube.

The electron beam striking the phosphor on the CRT screen causes this point (pixel) on the screen to emit light.  The light intensity can be controlled by controlling the energy of the electrons in the beam (i. e., by changing the anode voltage).  In a color display, the screen is coated with a pattern of phosphor dots doped to emit the three primary colors – red, green, and blue.  Depending on their relative intensity, the combination of these three primary colors can reproduce a large number of colors and hues.  A set of three very closely spaced red, green, and blue phosphor dots form one color pixel, since the human eye perceives the light coming from these three close locations as coming from one spot.  Three separate electron beams are used to excite this color pixel, and the color of the emitted light depends on their relative intensity.  Good-quality color CRTs have shadow mask tubes where a metal mask with patterns of holes is placed close to the screen.  The three electron beams come through the holes in the metal mask creating a sharper image.

The image in a CRT is created by moving the electron beam horizontally from left to right, one line after the other.  (A special voltage pulse blanks the screen when the electron beam returns from right to left.)  The number of lines varies depending on the resolution of the CRT.  A high-quality computer monitor may have, for example, 768 lines with 1,024 pixels in each line for a total number of pixels equal to $1,024 \times 768 = 786,432$.

The human eye does not notice the flicker of the image if the image changes every 30 ms or so.  (This rate is somewhat different for different people.)  Therefore, if the visual information on the screen is updated faster than 30 ms or so, the human eye does not perceive any screen flicker.  Good CRTs typically have a vertical refresh rate of 60 Hz or even more.  At 60 Hz vertical refresh rate, the screen is updated every $1/60 \approx 16.7$ ms, and nobody can notice any flicker.  The electron beam in a CRT hits a pixel during the time $t_h \approx 16.7$ ms$/N_{hor}/N_l$ where $N_{hor}$ and $N_l$ are the number of pixels in the line and the number of lines, respectively.  For $N_{hor} = 1,024$ and $N_l = 768$, $t_h \approx 21.2$ ns.  However, the light emission by a phosphor dot persist during 16.7 ms (i. e., nearly a million times longer!) until the next refresh.

### 8-7-2.  Liquid Crystal Displays.

Just a short time ago, CRTs totally dominated the display market.  However, in 1990, the market for liquid crystal displays worldwide reached 1.5 billion dollars.  This market has been growing since at the rate of nearly 40% a year.

A **liquid crystal** consists of molecules that are not spherical in shape but have an elongated or disc shape.  Their orientation can be characterized by a vector **L** directed along the molecule axis or perpendicular to the molecule plane (see Fig. 8.7.2).  This vector is called the **molecular optical axis** since the molecular interaction with light depends on its orientation.

In a liquid formed by such molecules, these vectors may be parallel to each other (at least at moderate temperatures).  In this case, the liquid is called a liquid crystal because it has anisotropic properties.  The liquid crystal with oriented molecules of the kind shown in Fig. 8.7.2 is called a **nematic liquid crystal** (from the Greek word νημα, which means thread).  Typically, a liquid crystal is divided into domains with different orientation of molecules (see Fig. 8.7.2c).  However, this can be prevented by placing a nematic liquid crystal between two glass plates polished in one direction. Such a polishing makes microscopic oriented channels on the glass surface, and the molecules of the nematic liquid crystal orient themselves in the direction of polishing. (In a typical display using liquid crystals, the distance between the glass plates is only several microns.) If the directions of polishing for both glass plates containing the liquid crystal are parallel, then the optical axes of the molecules are in the same direction. If the directions of polishing for both plates containing the liquid crystal are perpendicular, the molecule optical axes rotate between the plates (Fig. 8.7.3).

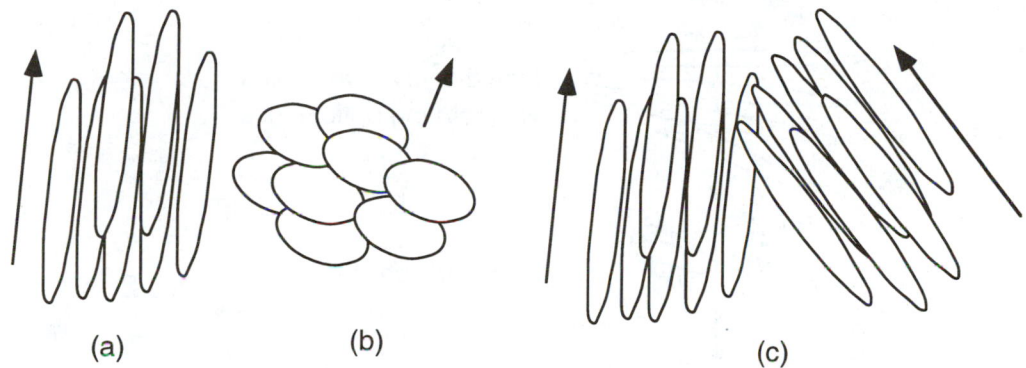

(a)                    (b)                    (c)

**Fig. 8.7.2.** Liquid crystal molecules. (a) Elongated shape molecules. (b) Disc shape molecules. (c) Illustration of how the liquid crystal is typically divided into domains with different orientation of molecules.

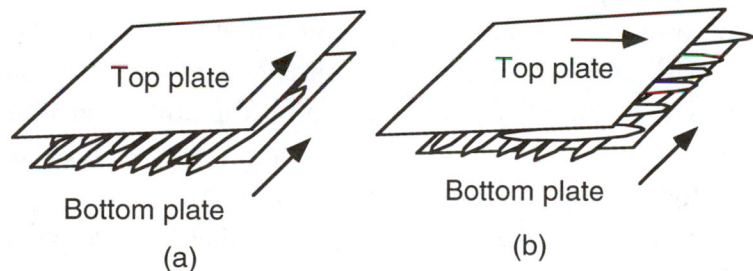

(a)                    (b)

**Fig. 8.7.3.** Schematic orientation of molecules in nematic liquid crystal between two glass plates. Arrows show the direction of polishing.
(a) Plates polished in the same direction and (b) in perpendicular directions .

When we shine light on a liquid crystal, the interaction of the electric field of the electromagnetic wave with the elongated molecules strongly depends on the orientation of the electric field. In an electric field, molecules are polarized (which means that electrons shift with respect to positive ion cores, creating a dipole moment). This polarization is drastically different for the directions along vector **L** and perpendicular to vector **L**. Hence, the optical properties of a liquid crystal are different in these two directions. The axis **L** is called the optical axis of the liquid crystal, and its rotation in space changes the polarization of the light (see Fig. 8.7.4).

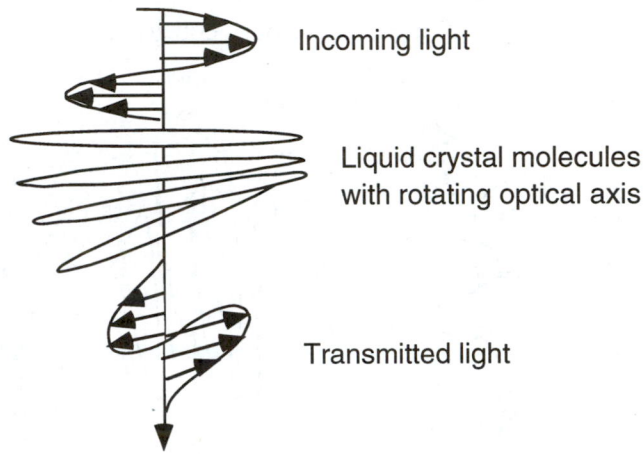

**Fig. 8.7.4.** Rotation of the polarization plane by nematic liquid crystal. Arrows show the direction of electric field of light wave.

The light can be polarized using a polarizer plate, which transmits light of one polarization only. If we place the glass plates with a liquid crystal between two polarizers oriented in the same direction and if the direction of the liquid crystal optical axis is parallel to the polarization direction, the light will come through this system. If the two polarizers have perpendicular directions of polarization, no light will come through (see Fig. 8.7.5).

The important property of a liquid crystal is that the applied electric field rotates its optical axis. Hence, if an electric field perpendicular to the glass plates is applied to the liquid crystal in Fig. 8.7.5a, no light will come through. Alternatively, the rotation of the optical axis by the electric field in the liquid crystal shown in Fig. 8.7.5b will allow light to pass. This is the principle of operation of Liquid Crystal Displays (LCD). These displays use liquid crystal pixels very similar to the structures shown in Fig. 8.7.5. The only difference is that the glass plates containing the liquid crystal are covered by thin layers of a transparent conducting material (usually indium tin oxide), which is used to apply voltage to the liquid crystal pixels.

These displays may either use the reflected light or have a fluorescent back light (backlit displays). The pixels in the simplest displays may be either black or white, depending on the applied signal. More sophisticated display have several gray levels (created by choosing different applied voltages and, hence, different degrees of rotation of the optical axis). If three closely located pixels are covered by red, green, and blue filters forming one logical pixel, the LCD becomes a color display, since a very large number of colors can be obtained by combining these three primary colors – red, green, and blue.

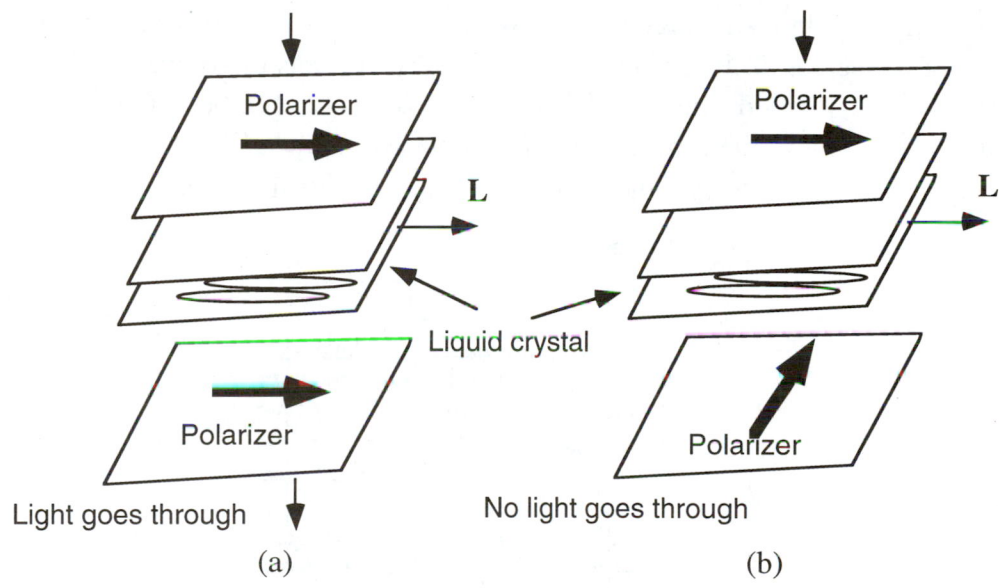

**Fig. 8.7.5.** Light propagation through the liquid crystal between polarizer plates.

This book was written on an Apple Powerbook 170. This laptop computer has a backlit liquid crystal display with 640 pixels in the horizontal direction and 400 hundred pixels in the vertical direction. How to drive all these pixels is an interesting electrical engineering problem! The simplest (and the cheapest) solution is to use a **passive LCD technology** described below. A passive LCD uses a matrix formed by metal lines as shown in Fig. 8.7.6.

The pixels are located at the intersection of the horizontal and vertical metal lines shown in the figure. First, we notice that the display shown in Fig. 8.7.6 has only $11+12 = 33$ inputs for $11 \times 12 = 131$ elements. In principle, we could have applied the input voltage between a certain column line and a certain row line in order to turn on a pixel at the intersection of this row and this column. For example, the input voltage to row 7 and column 2 would turn on pixel (7, 2). This would be similar to an electron beam in a CRT tube, which illuminates a point on a screen, making the screen phosphor emit light. However, this scheme cannot work with a liquid crystal since liquid crystals do not have as long a "memory" as screen phosphors. Therefore, the signals are applied to all columns and one row at a time. When the voltage pulse of one polarity is applied, for example, to row 7, voltage pulses of the opposite polarity are applied to all column lines corresponding to the pixels in row 7 that have to be turned on (column lines 2 and 7 in Fig. 8.7.6). Since the human eye does not notice the flicker of the screen if the screen is updated every 16.7 ms or so, we have to turn on the row of pixel during $16.7/N$ ms where $N$ is the number of rows in the display. If $N = 400$, then the write time is approximately $16.7/400 \approx 42$ μs, and the line of pixels has to store the signal for approximately 16.7 ms.

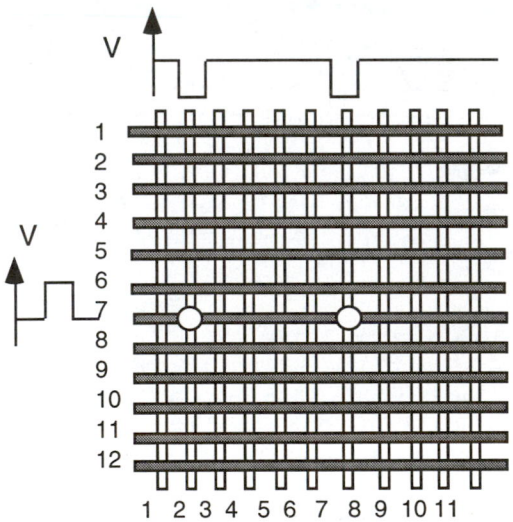

**Fig. 8.7.6.** Schematics of passive LCD. Open circles denote pixels turned on.

This technique is suitable for relatively low-quality displays. First, the number of rows may be limited, since it is difficult to obtain liquid crystals that allow one to obtain a large ratio of on to off time (which is equal to the number of rows). Second, it is difficult to eliminate cross talk or obtain a large number of levels of gray since the voltage pulse applied to the row line affects all pixels located on this line to some degree.  Therefore, passive LCDs are used in relatively cheap laptop computers.

**Active Matrix Liquid Crystal Displays (AMLCDs)** use a superior (but considerably more sophisticated and expensive) technology.  In AMLCDs, each pixel is driven by its own Thin Film Transistor (TFT) as shown in Fig. 8.7.7.

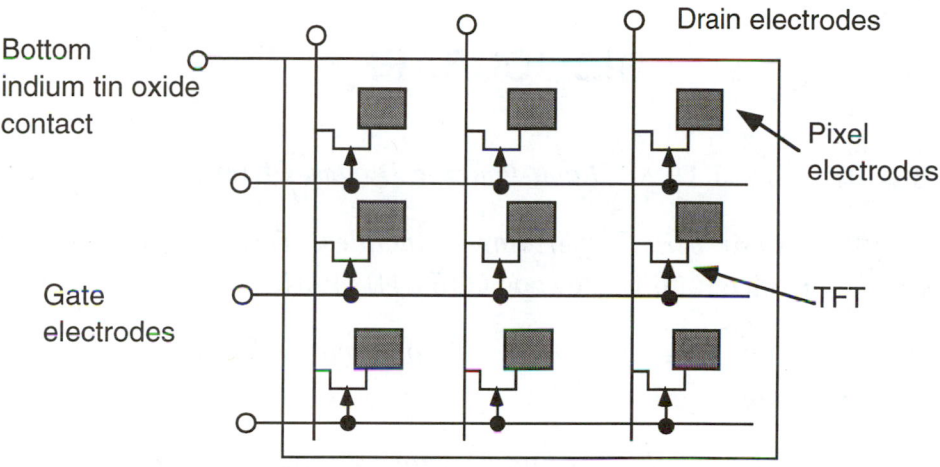

**Fig. 8.7.7.**  Schematics of AMLCD.

The bottom transparent indium tin oxide electrode is common for all pixels.  Amorphous silicon or polysilicon TFTs operate as pass transistors that connect the pixel to the drain power supply when the transistors are turned on. Just as in passive LCDs, the image information is transmitted one line at the time. By choosing different levels of the drain bias, it is possible to have many levels of gray by charging the liquid crystal pixels to different voltages.  (To the first order, an LCD pixel behaves as a capacitor that is charged to the drain potential when the TFT connected to the pixel is turned on.)  Three pixels with red, green, and blue filters in close proximity to each other form one color pixel.  The great advantages of this technology become apparent if one compares passive LCDs and AMLCDs for similar computers, such as different models of Apple Powerbook. AMLCDs have a much, much better image quality.  However, this advantage comes at a price.  Also, having millions of TFTs in high-resolution color displays requires a very high yield in  the manufacturing process.  It is not rare to find defective pixels even in new AMLCD displays.

# REFERENCES

S. M. SZE, *Physics of Semiconductor Devices*, Second Edition, John Wiley & Sons, New York (1981)

A. YARIV, *Optical Electronics*, Third Edition, New York, Holt, Rinehart, and Winston (1985)

# BIBLIOGRAPHY

A. A. BERGH AND P. J. DEAN, *Light-Emitting Diodes*, Clarendon, Oxford (1976)

M. A. GREEN, *Solar cells: Operating Principles, Technology, and Device Applications*, Prentice Hall, Englewood Cliffs, NJ (1982)

R. G. HUNSPERGER, *Integrated Optics: Theory and Technology*, Third Edition, Springer-Verlag, New York (1991)

J. KANICKI, Editor, *Physics and Applications of Amorphous and Microcrystalline Semiconductor Devices*, Artech House, MA (1991)

H. KRESSEL AND J. K. BUTLER, *Semiconductor Lasers and Heterojunction LEDs*, Academic Press, New York (1977)

J. I. PANKOVE, *Optical Processes in Semiconductors*, Dover Publications, New York (1975)

M. SHUR, *Physics of Semiconductor Devices*, Prentice Hall, Englewood Cliffs, NJ (1990)

A. VAN DER ZIEL, *Solid State Electronics*, Prentice Hall, Englewood Cliffs, NJ (1976)

C. T. WANG, Editor, *Introduction to Semiconductor Technology. GaAs and Related Compounds*, Wiley Interscience, New York (1990)

A. YARIV, *Quantum Electronics*, Second Edition, New York, John Wiley & Sons (1975)

# REVIEW QUESTIONS

**1.** The absorption coefficient of a semiconductor film is $\alpha = 10^5$ cm$^{-1}$. What thickness of the semiconductor film will absorb 90% of the light intensity shining on the film? The light rays are perpendicular to the film. Assume that there is no reflection.

☐ 1 point

**2.** What are the advantages of using GaAs for optoelectronics applications compared to silicon? Give an example of another semiconductor material suitable for optoelectronics applications.

☐ 1 point

**3.** The $I$-$V$ characteristics of two solar cells under the AM1 illumination are shown in the figure. Both cells have active areas of 1 cm$^2$ each.

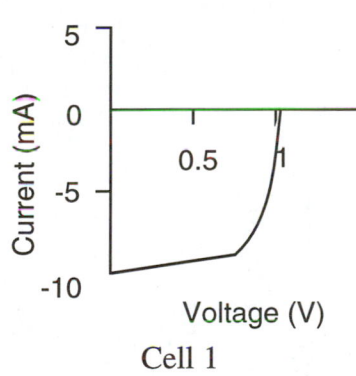
Cell 1

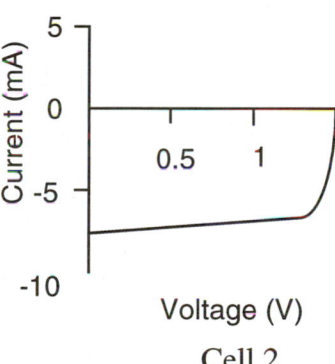
Cell 2

(a) What are the light-generated currents for cells 1 and 2?

☐ 1 point

(b) What are the open circuit voltages for cells 1 and 2?

☐ 1 point

(c) What are the approximate efficiencies for cells 1 and 2?

☐ 2 points

(d) Which cell (1 or 2) corresponds to the higher energy gap material?

☐ 1 point

# PROBLEMS

**8-1-1.**  The absorption coefficient of a semiconductor film is $\alpha = 10^5$ cm$^{-1}$. Determine the thicknesses of semiconductor films required to design filters that will absorb 90% and 99.9% of the light shining on the film. The light rays are perpendicular to the film. Assume that there is no reflection.

**8-1-2.**  Consider the system shown in the figure. (This system is called a Fabry-Perot etalon.) The film thickness is $L$. The reflection coefficient from the film boundaries is $R$. The film generates light that can be described in terms of the film material having a negative absorption coefficient, $G$. How large does $G$ have to be to exactly compensate for the losses related to the reflections from the boundaries?

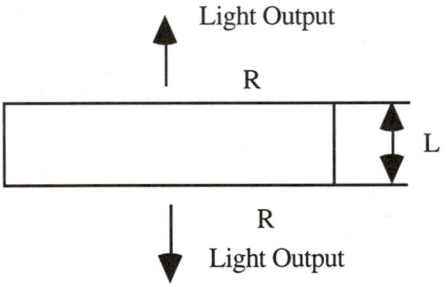

**\*8-2-1.**  Solve the steady-state continuity equation for holes in the illuminated $n$-type region. The steady-state continuity equation for holes in the illuminated $n$-type region is

$$D_p \frac{d^2 \Delta p_n}{dx^2} - \frac{\Delta p_n}{\tau_{pl}} + G_L = 0$$

where $\Delta p_n = p_n - p_{no}$, $D_p$ is the hole diffusion coefficient, and $\tau_{pl}$ is the minority carrier (hole) lifetime [compare with eq. (3-7-28)]. Use the following boundary conditions:

$$\Delta p_n(x \to \infty) = G_L \tau_{pl}$$

and

$$\Delta p_n(0) = p_{no}\left[\exp\left(\frac{V}{V_{th}}\right) - 1\right]$$

Here $V_{th} = k_B T/q$ is the thermal voltage and $x = 0$ corresponds to the p-n junction boundary. [Assume that the width of the depletion region, $x_d$, is much smaller than the hole diffusion length, $L_p = (D_p \tau_{pl})^{1/2}$, and the width of the n region, $X_n$, is much larger than $L_p$.] Use the results to calculate the light-generated current in a crystalline solar cell.

**\*8-2-2.** Design a silicon solar cell supplying 10 W of power under AM1 illumination. Assume a 12% efficiency of the solar cell, zero shunt resistance, and series resistance of 2 $\Omega\text{cm}^2$, and estimate other solar cell parameters from Fig. 8.2.4. Estimate a reasonable load resistance.

**8-2-3.** Explain why the thicknesses of the i-layers in a tandem solar cell (see Fig. 8.2.6) should be chosen in such a way that the light-generated current produced by the wide band gap and narrow band gap sections of a tandem cell should match (i. e., be close to each other).

**8-2-4.** Assume that the refraction indexes of two media are 3.3 and 1. Find the reflection coefficients with and without an antireflection coating with refraction index 1.83.

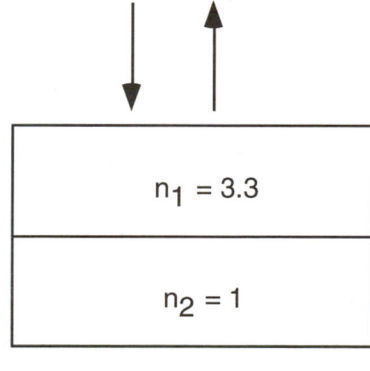
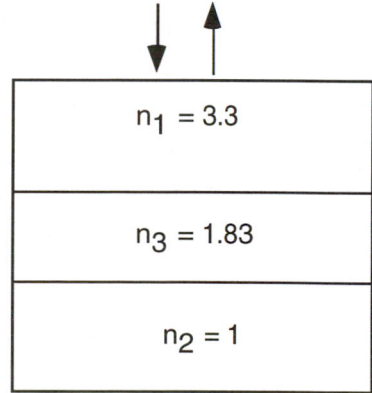

**8-2-5.** Calculate and plot the current-voltage characteristics of a solar cell with a 1 cm$^2$ area using eq. (8-2-4) for (a) $R_s = 0$, $R_s = 2$, $R_s = 4$ $\Omega$ for $R_{sh} \Rightarrow \infty$ and (b) $R_{sh} \Rightarrow \infty$, $R_{sh} = 100$ $\Omega$, $R_{sh} = 40$ $\Omega$ for $R_s = 0$. Use the following parameters: light-generated current density, $I_L = 41$ mA/cm$^2$, dark saturation current density, $I_s = 10^{-9}$ mA/cm$^2$, ideality factor $\eta = 1.1$, temperature $T = 300$ K.

**8-3-1.** Assuming a quantum efficiency of 80% and the absorption coefficient of a semiconductor material, $\alpha = 10^5$ cm$^{-1}$, design a *p-i-n* photodetector that produces a 1 $\mu$A current when illuminated with radiation with a photon energy of 1.5 eV with an intensity of 10 mW/cm$^2$. Assume that external and reflection losses are 50%.

**8-3-2.** Assuming a quantum efficiency of 80% and the absorption coefficient of a semiconductor material, $\alpha = 10^4$ cm$^{-1}$, design a photoconducting detector that produces a 1 $\mu$A current when illuminated with the radiation with a photon energy of 1.5 eV with an intensity of 10 mW/cm$^2$. The minority carrier lifetime is 1 $\mu$s. The electron and hole saturation velocities are $10^5$ m/s. Assume that external and reflection losses are 50% and choose the applied voltage high enough to saturate the electron and hole velocities (assuming the saturation electric fields for electrons and holes to be 2 kV/cm and 20 kV/cm, respectively).

**8-3-3.** Assuming zero collection efficiency for photons with energies smaller than the energy gap and a quantum efficiency of 100% for the photons with energies higher than the energy gap, choose a semiconductor material appropriate for a "solar blind" photodetector (i. e., for a detector that does not respond to visible electromagnetic radiation) and calculate the responsivity as a function of the radiation wavelength for a *p-i-n* detector made from this material.

# 9

# *Device Fabrication and Novel Devices

## *9-1. INTRODUCTION

Each year thousands of new technical articles and patents describe new electronic materials, new electronic device technologies, and novel device concepts. Some of these discoveries and inventions may lead to real breakthroughs that enable new applications, reduce cost, and improve reliability. Others may never find applications at all or become important only after a long delay when technological improvements will make these ideas feasible. Thin Film Transistors (TFTs) considered in Chapter 7 is a case in point. The first TFTs were very rudimentary, their characteristics varied quite a bit, and their performance was very poor. Nowadays, millions of a-Si TFTs are used in Active Matrix Liquid Crystal Displays (AMLCDs). AMLCDs represent one of the largest and fastest growing electronics markets, which is expected to reach more than 15 billion dollars a year by 1998. Compound semiconductor transistors discussed in Chapter 7 is another example of important technology that is taking a long time to mature.

In this chapter, I will not try to predict which novel materials, technologies, and devices will become technologies of the future (even though I may have an opinion on that). Rather, I will try to give examples that are fairly representative of approaches to innovations in electronics.

Most of these innovations rely on new materials and/or on new advanced device fabrication technologies. Some recently explored ideas have been around for decades waiting for practical realization, which became possible only recently because of breakthroughs in material and device technologies.

Many of these technologies are linked to new semiconductive and superconductive materials. As mentioned in Chapter 2, compound semiconductors, such as GaAs or InP, may have certain advantages over silicon, primarily because the electrons in these materials have a lighter effective mass. These materials have been intensively studied since the 1950s so that they probably can't fully qualify as "novel." Rather, novel devices based on these materials use very short dimensions and/or heterostructures that allow us an additional degree of control over the carrier flow in a semiconductor. In this chapter, we will give several examples of such devices.

Many compound semiconductors have a wider energy gap than silicon. For example, the energy gaps of GaAs, GaN, and SiC are 1.42 eV, 3.5 eV, and 2.9 eV, respectively, compared to 1.12 eV for Si. Hence, these materials have much smaller intrinsic carrier densities and may operate at higher temperatures. This is especially true for SiC and GaN, which have potential applications in high-temperature, harsh-environment electronics. Other compound semiconductors, such as HgCdTe, have a small energy gap (0.1 eV for HgCdTe, for example). These narrow gap semiconductors are used for infrared detectors, since the energy band gap is equal to the energy of infrared photons that lead to valence-to-conduction band transitions in these compounds.

Table 9.1.1 summarizes advantages and disadvantages of compound semiconductor technology. Table 9.1.2 lists several important compound semiconductor materials and their applications.

An example of a recent dramatic breakthrough in new materials which we will consider in this chapter is the discovery of superconducting materials, which have zero resistance to an electric current up to temperatures as high as 160 K.

For many novel devices an accurate control of the vertical  device dimensions and doping profile is needed to maintain device uniformity across the wafer, which is a necessary precondition of achieving high-yield and/or large-scale integration. The  necessity to achieve a very low contact resistance in vertical devices is another practical limitation. Finally, parasitic elements and interconnects play an important role in practical devices. All these factors make it difficult to reach and, in some cases, even to estimate the potential performance limits of novel devices. But technological difficulties have never stopped the emergence of new ideas, new materials, and new device concepts, and in Section 9.2, we discuss novel material and device fabrication technologies that make many novel devices possible.

| Advantages of compound semiconductor technology | •Higher speed materials<br>•Semi-insulating substrates available (leading to smaller parasitic capacitances)<br>•Heterostructure systems allow for "energy band engineering"<br>•Superior optoelectronic properties<br>•Wide band gap semiconductors for high temperature electronics<br>•Narrow gap semiconductors are a natural choice for infrared electronics |
|---|---|
| Disadvantages of compound semiconductor technology | •Absence of good natural oxides<br>•Defects related to composition<br>•Less developed technology<br>•Higher price |

**Table 9.1.1.** Advantages and disadvantages of compound semiconductor technology.

| GaAs | Most popular compound semiconductor |
|---|---|
| InP | GaAs competitor |
| AlGaAs | Wide band gap semiconductor forming a heterostructure with GaAs |
| AlInAs | Wide band gap semiconductor forming a heterostructure with InGaAs |
| InGaAs | High mobility semiconductor |
| SiC | High thermal conductivity wide band gap material |
| GaN | High mobility, high thermal conductivity wide band gap material |
| AlN | Wide band gap material forming a heterostructure with GaN |
| GaP | Material for visible LEDs |
| HgCdTe | Narrow gap material for infrared detectors |
| ZnSe and ZnTe | Important II-VI compounds used in blue lasers |

**Table 9.1.2.** Important compound semiconductor materials.

## *9-2.  MATERIAL GROWTH AND DEVICE FABRICATION

My first experience in device fabrication was more than 40 years ago when I was building a simple radio using a semiconductor Schottky diode as a detector.  The diode was a small selenium crystal, an eighth of an inch or so in diameter, with a metal needle attached to it.  I had to move the needle around the crystal trying to find a "good spot" where the diode would work.  I guess that particular spot may have corresponded to a relatively low trap density at the surface.  Or, perhaps, the doping was very nonuniform, and moving the needle allowed me to find a spot with an appropriate local doping level.  (This radio set, by the way, did not require a battery, and all the parts, except for the diode, were homemade.)

How can I describe that feeling of excitement and awe that I experienced when the set actually worked?  I also remember that I thought how incredibly lucky those few fortunate people were who, instead of doing all kinds of boring and routine jobs, had a wonderful opportunity to make a living by thinking how to make and improve these incredible crystals and diodes.

The progress of electronic device technology has been unrelenting since that time and nothing short of spectacular, and in this section we will give a brief overview of semiconductor technology.

### *9.2.1.  Bulk material growth.

About 98% of semiconductor devices are made from silicon.  Silicon represents 25.7% of matter by weight in the outer 10 miles of the earth's crust.  Only oxygen is more abundant in terrestrial matter.  One of the readily available silicon compounds is quartzite (which is fairly pure $SiO_2$).  This material, mixed with coal, coke, and wood chips, is processed at temperatures around 1,800 ºC to produce Metallurgical Grade polycrystalline Silicon (MGS), which is 98% pure.  MGS is used to produce a purer Electronic Grade  Silicon (EGS).  To this end, MGS silicon is mechanically pulverized and reacted with HCl, forming $SiHCl_3$ gas.  This substance is purified by fractional distillation.  Then, the reaction of this gas with hydrogen produces EGS.  The final product is a polycrystalline rod of EGS about 8 inches in diameter and several meters long.  The annual worldwide production of EGS is on the order of 5,000 tons.  Since Si density is 2.33  $g/cm^3$, all Si used worldwide in a year can be placed in a cube of approximately $13 \times 13 \times 13$ m$^3$.

Silicon used for integrated circuits is grown by the **Czochralski method**.  In this method, a silicon crystal grows around a silicon crystal seed slowly pulled out of the silicon melt with a simultaneous rotating motion (see Fig. 9.2.1). The grown crystal reproduces the crystal orientation of the seed. [A semiconductor ingot (or boule) is referred to as **bulk material**.] Typically, ingots oriented along the (100) direction are used for MOSFET integrated circuits. Material grown in the (111) direction is sometimes preferred for bipolar junction transistors.

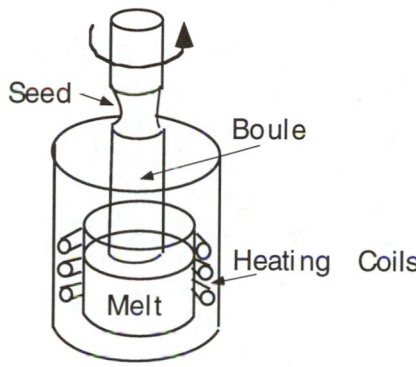

**Fig. 9.2.1.** Czochralski crystal puller.

Semiconductor material can be purified by **zone melting**. In this technique, a narrow slice of semiconductor ingot is melted (see Fig. 9.2.2).

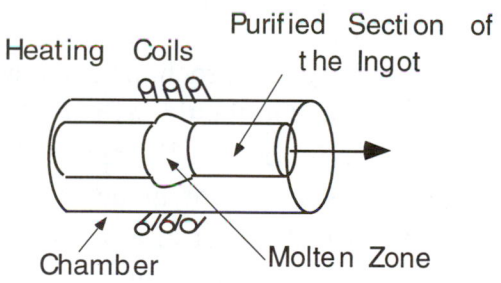

**Fig. 9.2.2.** Zone melting.

The molten zone is moved across the ingot (by moving the ingot with respect to the heating coils). Impurities have a tendency to concentrate in the liquid phase, and, therefore, are moved by the molten zone to the edge of the ingot. This property is described by the distribution coefficient

$$k_d = \frac{N_{solid}}{N_{liquid}} \tag{9-2-1}$$

Usually, $k_d < 1$ and varies between $10^{-5}$ and 0.8 for different impurities. After a large number of passes, the material becomes very pure, especially if a special technique called **floating zone purification** is used to avoid contact of the semiconductor ingot with the container walls. Silicon – the purest electronic material on earth – may have as little as $10^{12}$ cm$^{-3}$ concentration of residual impurities. Since Si has $5.02 \times 10^{22}$ atoms/cm$^3$ (see Appendix A6), this means fewer than one impurity atom per 50 billion Si atoms!

**Example 9-1-1.**

Determine how many zone melting passes are required to reduce the impurity concentration in a silicon ingot from $10^{20}$ cm$^{-3}$ to $10^{12}$ cm$^{-3}$ if the distribution coefficient, $k_d$, for this type of impurities is equal to 0.1.

**Solution:**

The impurity concentration after $m$ passes is related to the initial impurity concentration, $N_{impurity}$, as follows

$$\frac{N_m}{N_{impurity}} = k_d^m$$

Hence, the required number of passes is

$$m = \frac{\log\left(\frac{N_m}{N_{impurity}}\right)}{\log(k_d)} = \frac{\log\left(\frac{10^{12}}{10^{20}}\right)}{\log(0.1)} = 8$$

A semiconductor ingot grown by Czochralski method is cut into slices, which are mechanically lapped to obtain flat thin silicon disks wafers of uniform thickness (called **wafers**). These wafers are chemically etched and polished to obtain semiconductor substrates suitable for manufacturing of semiconductor devices and integrated circuits. The diameter of a Si wafer can be as large as 12 inches. The surface dimensions of a typical transistor can be smaller than 10 μm$^2$. Try to figure out the mind boggling number of such small devices that can fit on the surface of a 12 inch wafer!

**\*9.2.2. Epitaxial growth.**

Modern electronics technology is the technology of thin films. Millions of devices are fabricated at the surface of a semiconductor wafer using but a micron or two of the surface layer. All intricate doping and, sometimes, composition profiles needed for the fabrication of semiconductor devices are with a few microns or less from the surface of the wafer. Therefore, very often, the surface layer is grown by **epitaxial technique**. (**Epitaxy** means "arranged upon" in

Greek.) We already mentioned **Molecular Beam Epitaxy** (MBE) in Section 4.10. In an MBE system, a deep vacuum is maintained inside the MBE chamber, and a heated semiconductor substrate is bombarded by ion beams produced by ion sources. These ions form a high-quality semiconductor film on the substrate. Closing or opening shutters (see Fig. 4.10.1) can control the film composition and/or doping levels literally within one atomic distance. Probably, most of the recent progress in novel device structures has been achieved due to the development and improvements of the MBE technology. However, MBE growth is slow (with typical rates on the order of 10 nm/min), the MBE machines are very expensive (up to a million dollars apiece at the moment of writing), and there may be reproducibility problems from run to run. Still, MBE is already used for laser manufacturing on a mass scale. (If you own a CD player, you have already benefited from the MBE technology, since such a player uses an MBE grown AlGaAs/GaAs semiconductor laser.) Small satellite dishes already popular in Europe and coming to the United States use MBE grown Heterostructure Field Effect Transistors (HFETs); see Chapter 7.

**Chemical Vapor Deposition** (**CVD**) and **Liquid Phase Epitaxy** (**LPE**) are more often epitaxial technologies of choice for mass production of semiconductor devices. In a chemical vapor deposition system, $SiCl_4$ and $H_2$ gases are passed over a silicon substrate held at approximately 1,250 °C (see Fig. 9.2.3). The chemical reaction

$$SiCl_4 + 2H_2 \leftrightarrow Si + 4HCl \qquad (9\text{-}2\text{-}2)$$

results in the deposition of epitaxial silicon. The silane ($SiH_4$) reaction is also used in certain cases, especially in the deposition of amorphous silicon:

$$SiH_4 \leftrightarrow Si + 2H_2 \qquad (9\text{-}2\text{-}3)$$

**Metal Organic Chemical Vapor Deposition** (**MOCVD**) is another CVD process. MOCVD utilizes organic compounds incorporating semiconductor and dopant atoms. (In any CVD process, dopants can be introduced through gases mixed into the gas flow.) This technique has many advantages of MBE and can be even combined with MBE.

Still another epitaxial process that has recently experienced a renaissance of sorts is Liquid Phase Epitaxy (LPE). In this process, a semiconductor substrate is put in contact with a semiconductor melt for a strictly controlled period of time, resulting in a deposition of an epitaxial film.

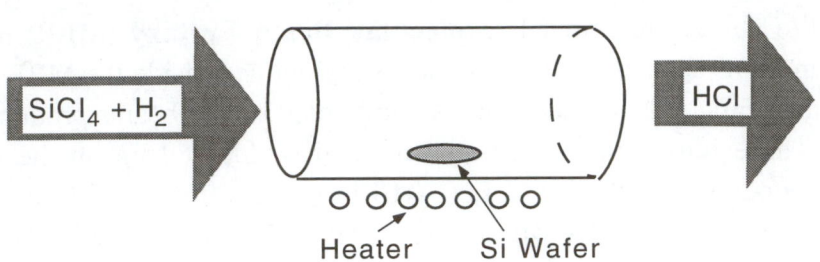

**Fig. 9.2.3.** Chemical Vapor Deposition.

Epitaxial techniques allow us to obtain tailored doping and composition profiles to suit our needs for the fabrication of particular devices. Another challenge is to define these devices on the surface of a semiconductor wafer, selectively dope and isolate different areas, and make contacts.

### *9.2.3.  Ion implantation.

In Section 4.2 , we briefly discussed ion implantation. Ion implantation is a process during which a semiconductor sample is bombarded by highly energetic impurity atoms. These ions usually penetrate into a thin surface region of the semiconductor and heavily damage the crystal structure of this surface layer. The total number of ions penetrating into the implanted sample is called the **implantation dose.** The penetration depth (called **range**) depends on the implanted species and on the ion energy and may vary from only a few hundred angstroms to a micron or so. After the implantation, the sample is heated and kept for some time at a fairly high temperature. This procedure (called **annealing**) removes the damage. As a result, we introduce donors or acceptors (depending on implanted species) into the surface region. The fraction of the implanted ions that become active as dopants is called the **activation efficiency**. A typical concentration profile is very nonuniform (see Fig. 4.2.1). However, for a simplified analysis, we can often replace this nonuniformly doped layer with a uniformly doped surface layer as shown in Fig. 4.2.1.

The penetration depth of an ion-implanted profile can be controlled by varying the energy of the incoming ion beam. Two or more different implants may be used to obtain a doping profile suited to particular device requirements. The greatest advantage of ion implantation is that it allows us to dope only certain regions of the semiconductor substrate. For example, ion implantation is used for implanting *p*-wells and/or *n*-wells into the semiconductor substrate for CMOS fabrication (see Fig. 6.6.1). Another example is ion implantation of highly doped regions to facilitate the formation of low resistive ohmic contacts (see Section 4.9).

**Example 9-2-2.**

A $p$-type Si substrate is doped by boron at $10^{16}$ cm$^{-3}$. What is the implantation dose required for implanting an $n$-type well with the average doping level of $4 \times 10^{16}$ cm$^{-3}$ if the implantation range $d_{impl} = 0.5$ μm and the activation efficiency $\eta_{impl} = 0.8$?

**Solution:**

The average donor concentration in this $n$-well must be $5 \times 10^{16}$ cm$^{-3}$ (since $4 \times 10^{16}$ cm$^{-3}$ is required by the specs and $1 \times 10^{16}$ cm$^{-3}$ is needed to compensate the acceptors). Hence, the required sheet concentration of donors is $5 \times 10^{16}$ cm$^{-3} \times 0.5 \times 10^{-4}$ cm = $2.5 \times 10^{12}$ cm$^{-2}$. The implantation dose, $Q = 2.5 \times 10^{12}$ cm$^{-2}/0.8 = 3.13 \times 10^{12}$ cm$^{-2}$.

## *9.2.4. Lithography and IC fabrication.

The growth of bulk material, epitaxy of high-quality semiconductor films with varying doping profiles (or material composition profiles for heterostructure devices), ion implantation, and many other technological steps, such as the deposition of passivating dielectric films, or chemical and plasma etching of materials are all very important for the fabrication of electronic devices and integrated circuits. But it is **lithography**, the process of transferring images onto the surface of a semiconductor wafer, that made integrated circuits possible. Lithography is somewhat similar to photography. This process allows us to fabricate many millions of monolithically integrated devices. The best way to explain this process is to consider how we can fabricate a simple MOSFET shown in Fig. 9.2.4. This process involves covering the wafer with a photosensitive polymer film (called **photoresist**), exposing the photoresist (usually to ultraviolet light), developing the photoresist (chemically dissolving the exposed photoresist areas), and using these openings in the photoresist for selective ion implantation, oxide removal, or contact deposition (see Fig. 9.2.4). The actual integrated circuit fabrication differs from this simplified process in sophistication and in a much greater number of steps but not in principle.

The masks of the kind shown schematically in Fig. 9.2.4 play a key role. Computer Aided Design (CAD) systems allow us to design such masks using a tablet or a joy stick and to provide an output on a magnetic tape that is used to drive a **pattern generator** to make the mask.

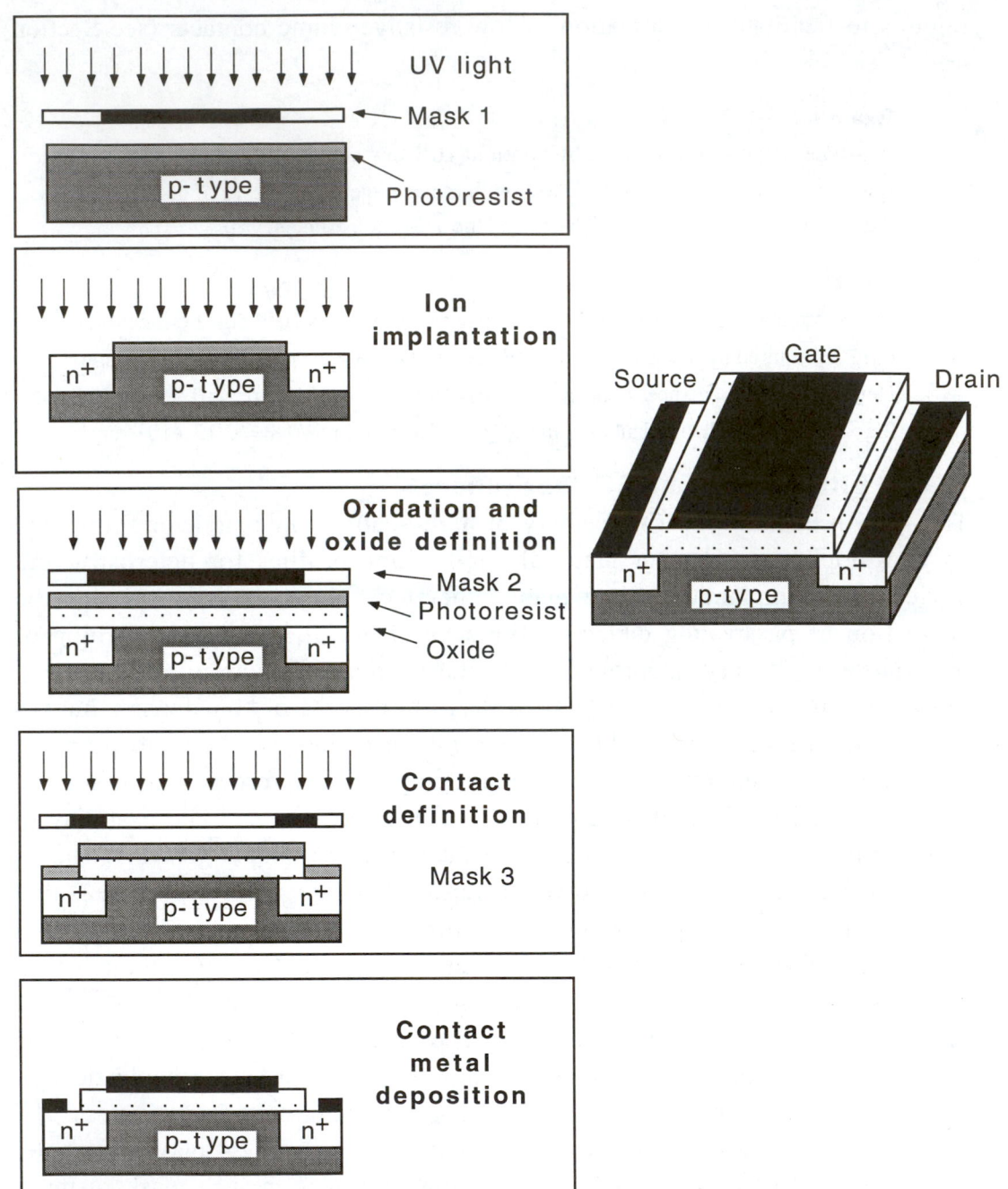

**Fig. 9.2.4.** Simplified process flow for an *n*-channel MOSFET.

(The Mentor[tm] software is used for this purpose by faculty and students at the University of Virginia.)

For custom-made integrated circuits, a much more interesting approach of the "**silicon compiler**" has emerged. In this approach (also referred to as a **Hardware Description Language**), we specify the circuit description, and a computer does the rest. In time (but not yet), the silicon compiler may become better than a human designer, just as surely as a chess computer will claim the world championship title in a few years from now.

The resolution that can be achieved by **optical lithography** is limited by the wavelength of the ultraviolet (UV) light used in the process. Using clever techniques and advanced photoresists, engineers and scientists have been able to push this limit into the submicron range (below 0.5 $\mu$m or so).

This progress slowed down the widespread use of the very powerful but very expensive **x-ray lithography**, which uses X-rays with wavelengths from 1 Å to 50 Å, with a commensurate improvement in resolution.

The **electron beam lithography**, which uses a focused electron beam (often for the **direct writing** on the electron beam resists without even using a mask), allows us to reproduce features as short as 50 Å. This technology (along with MBE) allowed electronic device engineers and scientists to fabricate most of the advanced electronic devices developed in the last decade.

After making millions of devices on a semiconductor wafer using a fabrication sequence like the one shown in Fig. 9.2.4, we are still faced with a number of less glorious but not less important tasks. Since many circuits are fabricated on the wafer, the wafer has to be diced into individual **chips** (also called **dice**), and silicon dice have to be **wire bonded** to external leads and packaged. Since integrated circuits become more and more sophisticated and complex, the number of pins increases to provide a higher input/output (I/O) count, and the cost of packaging may easily become comparable to the cost of the integrated circuit itself, if not higher. These challenges (still far from being resolved) become more difficult for high-speed, high-performance devices and circuits or for devices and circuits operating in harsh environment or at elevated temperatures. All these issues, which are beyond the scope of this book are discussed, for example, in the books on VLSI technology edited by Sze (1988) and Ghandhi (1983).

## *9-3.  HOT ELECTRON TRANSISTORS

Hot Electron Transistors are based on the effect called **Real Space Transfer**. This effect is illustrated in Fig. 9.3.1 which shows electrons moving in a GaAs layer sandwiched between AlGaAs layers. In high electric fields, electrons acquire energy from the electric field so that their effective temperature increases (electrons become **hot**). At high electron temperatures, the probability of the thermionic emission into the AlGaAs regions increases sharply. Since the electron mobility and velocity are very low in the AlGaAs regions, the electronic current decreases, and the device exhibits a negative differential resistance, similar to that caused by the intervalley transfer (see Section 4.11).

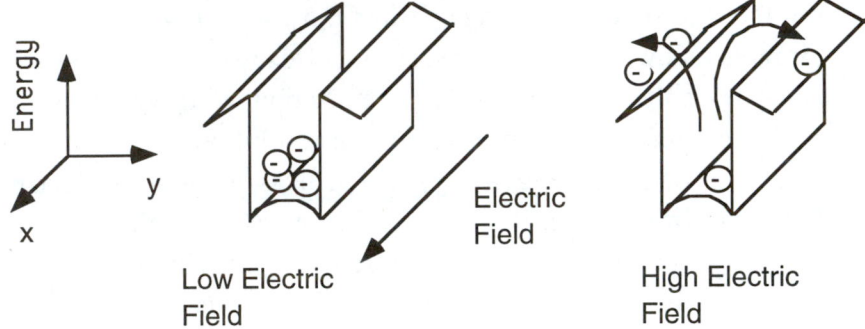

**Fig. 9.3.1.** Illustration of real space transfer (after Shur, 1990, copyright © Prentice Hall, 1990, reproduced by permission of Prentice Hall, Inc., Englewood Cliffs, NJ). In low electric field, electrons are localized in a two-dimensional quantum well; in high electric field, many electrons gain enough energy from the field to transfer over the barrier.

In 1983, Kastalski and Luryi invented the first three-terminal device (NEgative Resistance Field Effect Transistor – NERFET) using real space transfer. This device became one of several devices using **hot electrons**, which are called **Hot Electron Transistors**. Kastalski and Luryi illustrated the basic idea of their device by comparing it to a vacuum tube diode (see Fig. 9.3.2). In a vacuum tube diode, the electrons are emitted by a hot cathode filament. The cathode temperature controls the anode current, since the thermionic electron emission current, $I_{HOT}$, increases exponentially with this temperature (see Section 4.8). The cathode temperature can be varied by varying the current, $I_{cat}$, heating the cathode filament. Hence, the vacuum diode can be used as an amplifier with current gain $I_{HOT}/I_{cat}$. Such an amplifier is probably not very practical, since heating and cooling a cathode filament would take a long time, and the vacuum tube will operate in such a regime only at very low frequencies. In the Hot Electron Transistor, proposed by Kastalski and Luryi, the current, $I_{HOT}$, over the barrier separating the channel from the substrate is controlled by varying the electron temperature in the channel. The characteristic time scale of the electron temperature variation is the energy relaxation time, which is on the order of a picosecond. Hence, Hot Electron Transistors may operate up to extremely high frequencies.

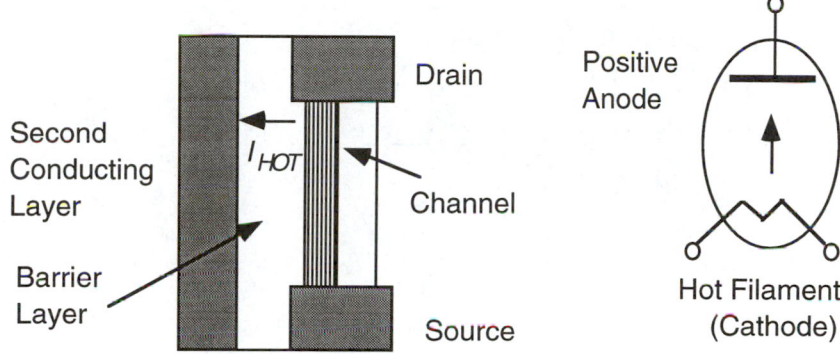

**Fig. 9.3.2.** Comparison between Hot Electron Transistor and vacuum tube (after A. Kastalski, S. Luryi, A. C. Gossard, and R. Hendel, *IEEE Electron Device Lett.*, EDL-5, p. 57, 1984. © IEEE 1984).

More accurately, the change in electron temperature is limited by the longest of the two time constants: the energy relaxation time (of the order of 1 ps or so for GaAs) and the time constant of the electric field variation in the device. The latter time is determined by the electron transit time across the high field region near the drain and may be quite short as well.

A similar effect also occurs in conventional Heterostructure Transistors at high gate and drain biases where the real space transfer leads to a sharp increase in the gate leakage current accompanied by a drop in the drain current. Possible practical applications of this new regime of operation are now under investigation.

A detailed discussion of Hot Electron Transistors is given by Luryi (1991). So far, these devices have been used for the studies of electronic properties of semiconductors in high electric fields. In the future, these devices may find applications as millimeter wave oscillators and as ultra-high-speed switching elements.

## *9-4.  SUPERLATTICE DEVICES

In 1969, Esaki and Tsu proposed to use a semiconductor superlattice based on a periodic structure of alternating layers of semiconductor materials with wide and narrow band gaps.  The first superlattices were fabricated using an AlGaAs/GaAs material system (see Fig. 9.4.1).

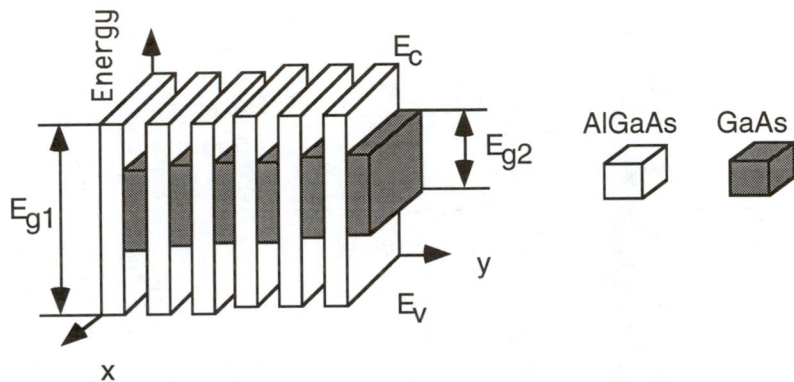

**Fig. 9.4.1.**  AlGaAs-GaAs superlattice (from Shur, 1990, copyright © Prentice Hall, 1990, reproduced by permission of Prentice Hall, Inc., Englewood Cliffs, NJ).

Let us first review what is happening in one individual quantum well formed in the GaAs  layer sandwiched between two AlGaAs barrier layers. If the GaAs layer thickness is small enough, then electronic motion in the quantum well is quantized in the direction perpendicular to the heterointerfaces (see Sections 1.2 and 2.5 and Fig. 2.5.1).  The carriers move freely in the direction parallel to the heterointerfaces so that the wave function is proportional to $\exp[i(k_x x + k_z z)f(y)$. Here $k_x$ and $k_z$ are components of the wave vector in the plane of the superlattice, $x$ and $z$ are the coordinates in the superlattice plane, and $y$ is the coordinate perpendicular to the superlattice plane.  As was discussed in Section 2.5,  each energy level found for a quantum well from the solution of the one-dimensional Schrödinger equation corresponds to a subband of states with the density of states, $D$, in each subband given by [see eq. (2-5-10) and Fig. 2.5.2]

$$D = \frac{m_n}{\pi\hbar^2} \qquad\qquad (9\text{-}4\text{-}1)$$

If the thickness of the wide-band-gap barriers layers is small enough so that electrons may tunnel through, then the situation becomes similar to what happens when individual atoms are brought together in a crystal. In this case, individual levels in the quantum wells are split into bands (called the minibands). In a crystal, the periodic atomic potential leads to band formation. In a superlattice, an artificial, human-made periodical potential causes the formation of minibands.

In 1970, Esaki and Tsu proposed to create an artificial periodic potential in a semiconductor crystal using periodic $n$-type and $p$-type doped layers. Such a superlattice is called a doping superlattice (see Problem 9-4-1).

Superlattice structures have been be used in field effect transistors where several quantum wells provide parallel conducting channels, increasing the device current carrying capabilities and, hence, the output power (see Fig. 9.4.2). Superlattices are also used for photodetectors and for novel light-emitting diodes.

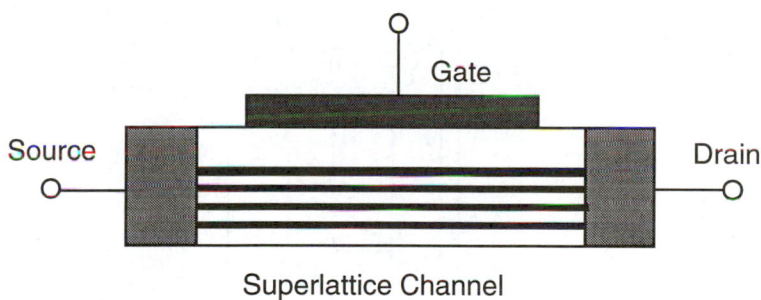

**Fig. 9.4.2.** Heterostructure Field Effect Transistor with a superlattice channel.

In heterostructure devices, superlattice buffers are used to create an intermediate layer between a substrate and an active layer. This allows us to alleviate strain caused by lattice constant mismatch and to obtain a much better quality active layer material.

One of the most interesting applications of superlattices is quite recent. In the April 1994 issue of *Science* magazine, Faist, Capasso, Sivco, Sirtori, Hutchnson, and Cho of AT&T Bell Labs reported on a new type of a far infrared laser called the Quantum Cascade Laser. [The idea of this superlattice device was proposed by Rudolf Kazarinov (who is now with AT&T Bell Labs) and Robert Suris of A. F. Ioffe Institute in 1971.] The band diagram of the active region of this device is shown in Fig. 9.4.3. The laser consists of 25 active regions connected by superlattices with a very short period (much shorter than the electron de Broglie wavelength). Electrons are injected into the first $Ga_{0.47}In_{0.53}As$ quantum well (with a thickness of only 8 Å) through a 45 Å $Al_{0.48}In_{0.52}As$ barrier. (This material system was chosen to obtain lattice match to the InP substrate.) The laser transition is between the third subband ($n = 3$, see Section 2.5) in the narrow 8 Å quantum well and the second subband ($n = 2$) in the adjacent, wider (35 Å) quantum well. The emission wavelength ($\lambda = 4.26$ μm) is in excellent agreement with the calculated energy difference between the subbands. This work represents a real breakthrough that we have been awaiting for almost a quarter of a century since the first pioneering proposal of Kazarinov and Suris.

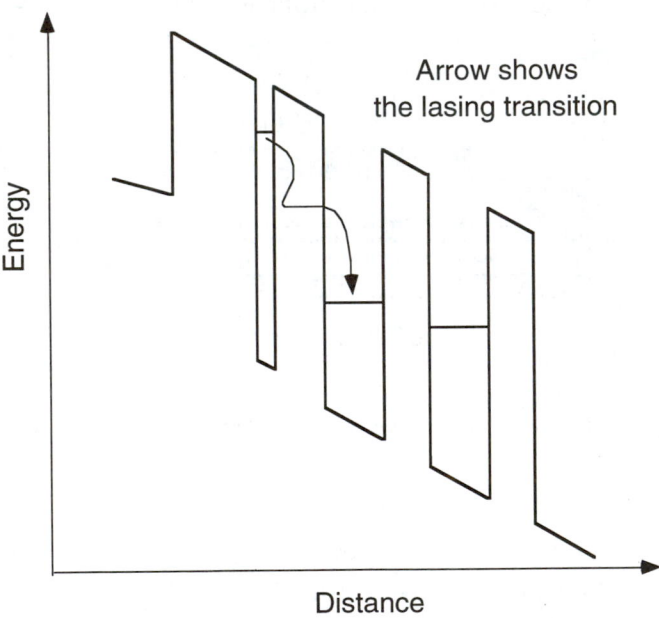

**Fig. 9.4.3.** Active region of quantum cascade laser (after J. Faist, F. Capasso, D. L. Sivco, C. Sirtori, A. L. Hutchnson, and A. Y. Cho, *Science*, Vol. 264, p. 553, April, 1994).

# *9-5.  RESONANT TUNNELING DEVICES

In a Resonant Tunneling Diode (RTD), two symmetrical barriers separate a quantum well in the middle from two $n$-type doped regions (see Fig. 9.5.1).

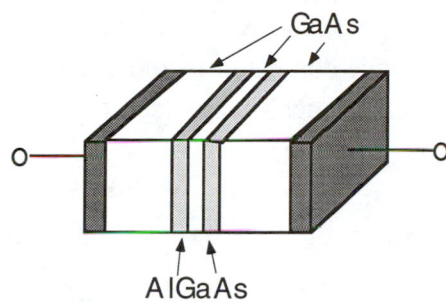

**Fig. 9.5.1.**  AlGaAs/GaAs Resonant Tunneling Diode (RTD).

A typical AlGaAs/GaAs RTD has two AlGaAs barrier layers with a thickness of 50 to 80 Å.  The principle of the RTD operation can be explained using the simplified RTD band diagram shown in Fig. 9.5.2.

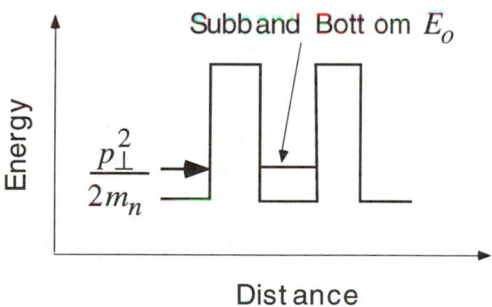

**Fig. 9.5.2.**  Simplified band diagram of Resonant Tunneling Diode.

This simplified band diagram corresponds to a double-barrier structure. The electron transmission coefficient through such a structure can be calculated using the same approach as was used in Section 1.4 for a single barrier.  (This calculation is not difficult but is fairly laborious.)  The result of such a calculation shows that the transmission probability has a sharp peak when the component of the electron momentum perpendicular to the double-barrier structure is such that

$$\frac{p_\perp^2}{2m_n} = E_o - E_c \qquad (9\text{-}5\text{-}1)$$

This corresponds to a resonance condition. (For an RTD under no bias, this problem is very similar to that of a Fabry-Perot etalon; see Problem 8-1-2. Under bias, the band diagram is no longer symmetrical; see Problem 9-5-2.)

In an RTD, the resonance condition is achieved at a certain bias voltage at which the diode current reaches a peak value (see Fig. 9.5.3).

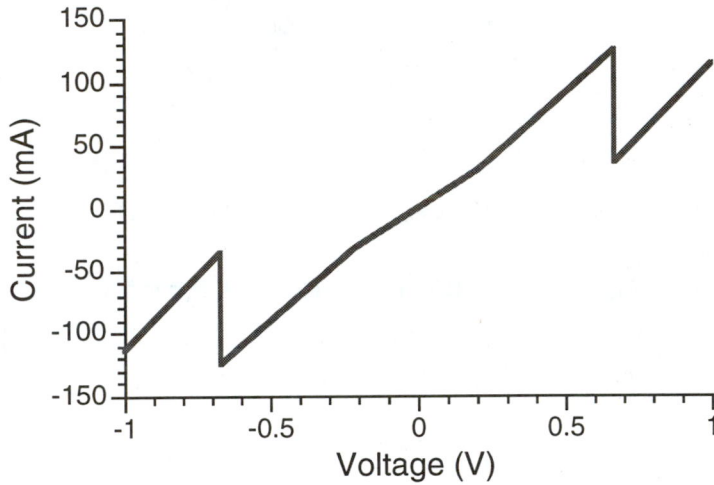

**Fig. 9.5.3.** *I-V* characteristics of $Al_{0.8}In_{0.52}As/Ga_{0.57}In_{0.53}As_{0.47}$ Resonant Tunneling Diode (RTD) (from F. Capasso, S. Sen, and F. Beltram, "Quantum-Effect Devices," in *High Speed Semiconductor Devices*, S. M. Sze, Editor, Wiley Interscience, John Wiley & Sons, Inc., New York, 1990).

Numerous novel devices using resonant tunneling structures have been proposed. These device structures include the Resonant Tunneling Bipolar Transistor (RTBT), Resonant Tunneling Transistor, the Surface Resonant Tunneling Transistor [see Capasso (1989)], and the Schottky-Gated Resonant Tunneling Transistor (see Section 9.6). These devices have the promise of becoming the fastest solid-state devices. The negative resistance associated with resonant tunneling devices can be used for switching applications in digital integrated circuits.

## *9-6.   HETERODIMENSIONAL DEVICES

All semiconductor devices use interfaces between different regions – ohmic, *p-n* junctions, Schottky barrier junctions, heterointerfaces, interfaces between a semiconductor and an insulator.  Typically, these interfaces are planes separating different regions.  However, recently a new generation of semiconductor devices has emerged.  These devices use interfaces between semiconductor regions of different dimensions and are called heterodimensional devices.  An example of such an interface is a Schottky barrier between a three-dimensional (3D) metal and a two-dimensional (2D) electron gas.  Other possible configurations include the interface between a 2D electron gas and a 2D Schottky metal and an interface between a one-dimensional (1D) electron gas and a 2D Schottky metal, and an interface between a 1D electron gas and a 3D Schottky metal (see Fig. 9.6.1).

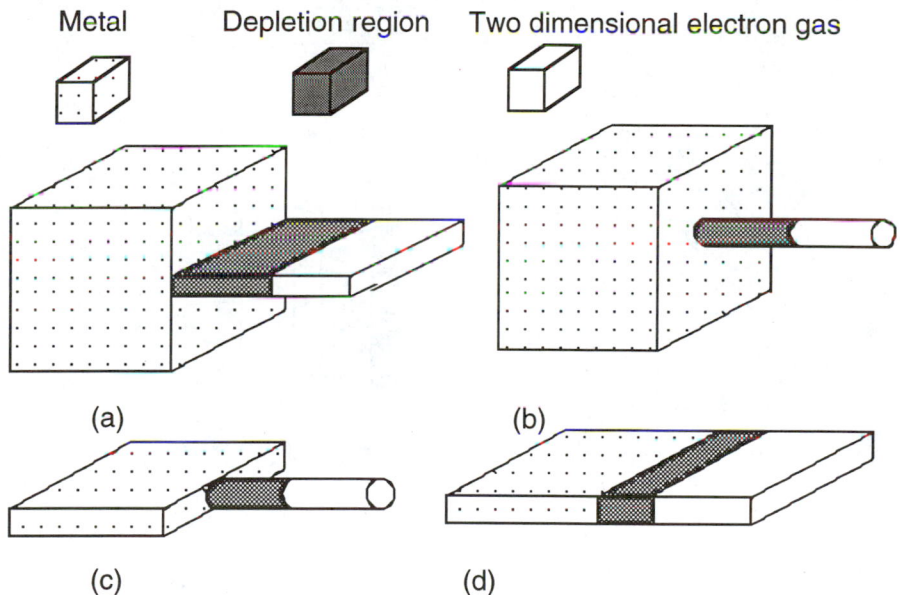

**Fig. 9.6.1.**  Heterodimensional Schottky barriers (a) 2D-3D, (b) 1D-3D, (c) 1D-2D, and (d) 2D-2D (from B. Gelmont, W. Peatman, and M. Shur, "Heterodimensional Schottky Metal-Two Dimensional Electron Gas Interfaces," *J. Vac. Sci. Technol.* B11 (4), pp. 1670-1674, July/August 1993).

Different heterodimensional Schottky contacts have several features in common: smaller capacitance because of a smaller effective cross section and a wider depletion region, a high carrier mobility related to properties of the 2D electron gas, and a smaller maximum electric field (and, hence, a higher breakdown voltage) because of the fringing electric field streamlines and a wider depletion region.   These features make these devices very promising for applications in ultra-high-frequency varactors and mixers.

In the 2D-3D Schottky contact (see Fig. 9.6.2), the streamlines of the electric field spread out from the 2D gas and terminate on the much larger surface of the Schottky metal.  This leads to a wider depletion width, smaller capacitance, and a higher breakdown voltage.  The high electron mobility of the 2D electron gas results in a small series resistance, especially at cryogenic temperatures, making this new Schottky diode very promising for applications at terahertz frequencies. This new technology is uniquely suited for applications in novel vertical transistor structures, such as described below.

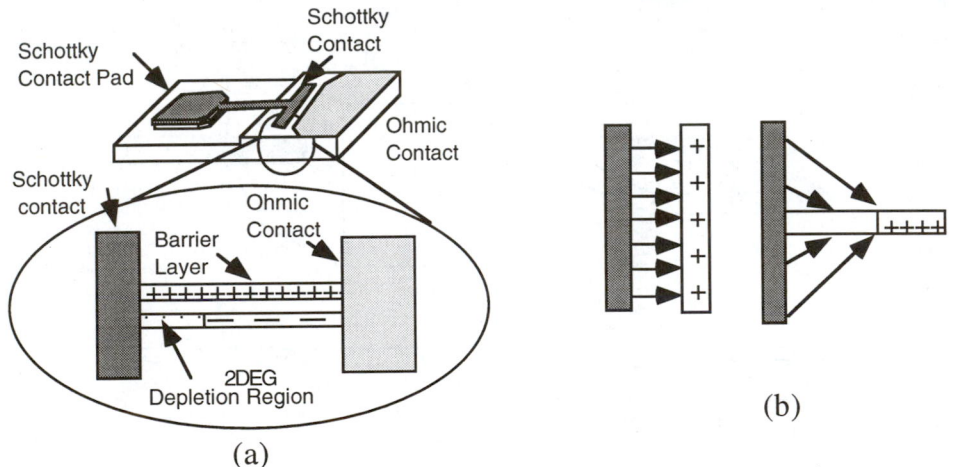

**Fig. 9.6.2.**  (a) Heterodimensional Schottky diode and (b) a comparison of the electric field streamlines for a parallel plate capacitor and a 2D-3D contact where the streamlines terminate on the much larger surface of the Schottky metal (after W. C. B. Peatman, T. W. Crowe, and M. Shur, "A Novel Schottky/2-DEG Diode for Millimeter and Submillimeter Wave Multiplier Applications," *IEEE Electron Device Letters*, Vol. 13, No. 1, pp. 11-13, Jan. 1992).

Fig. 9.6.3 shows a heterodimensional Schottky-gated Resonant Tunneling Transistor (RTT). This device uses the resonant tunneling structure discussed in Section 9.5. Two heterodimensional Schottky contacts on the sides of this structure modulate the electrons accumulated in the source region in a fashion analogous to the gate modulation of the bulk semiconductor in a conventional vertical field effect transistor. The advantages of using a double-barrier resonant tunneling structure include more efficient modulation of the 2-DEG and ballistic transport in a short structure. The RTT characteristics are shown in Fig. 9.6.3 for gate voltages 0.2 V, 0.6 V, and 1 V.

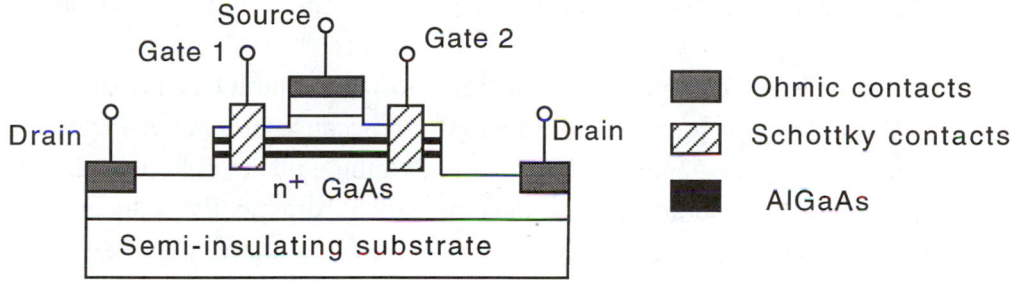

**Fig. 9.6.3.** Resonant Tunneling Transistor (RTT) (after W. Peatman, W. Grimm, H. Park, M. Shur, E. Brown, and M. Rooks, "Novel Resonant Tunneling Transistor, with High Transconductance at Room Temperature," *IEEE Electron Device Letters*, Vol. 15, No. 7, pp. 236-238, July 1994, © IEEE 1994).

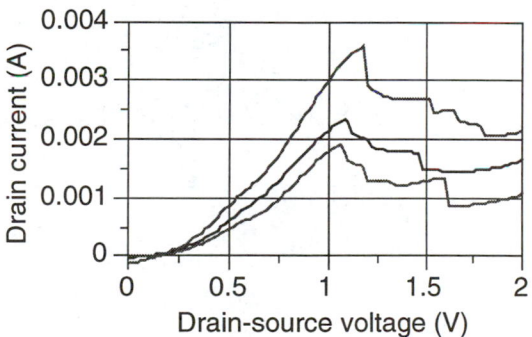

**Fig. 9.6.4.** RTT *I-V* characteristics (after W. Peatman, W. Grimm, H. Park, M. Shur, E. Brown, and M. Rooks, Novel Resonant Tunneling Transistor, with High Transconductance at Room Temperature, *IEEE Electron Device Letters,* Vol. 15, No. 7, pp. 236-238, July 1994, © IEEE 1994).

The current-voltage characteristic of this device also exhibits a negative differential resistance (see Fig. 9.6.4). This naturally leads to applications in high-speed, highly functional digital logic circuits and high-speed switching devices.

Figure 9.6.5 shows the schematic diagram of a device called a 2D MESFET. In this structure, the gate bias applied to the side Schottky gates modulates the width of the 2D electron gas. The side gate-to-channel capacitance is smaller than for a conventional HFET because of the geometry of heterodimensional junction. The geometries are as different as the geometry of a parallel plate capacitor and a geometry of capacitor formed by two perpendicular metal plates (see Fig. 9.6.1).

The subthreshold slope of the current-voltage characteristic of the 2D MESFET is smaller than for a conventional HFET because of a better localization of the electron gas in the channel. This is an advantage of the side Schottky gate device, and the subthreshold leakage current can be drastically reduced in 2D MESFETs with an improved design. This should lead to a lower current operation and to a lower power consumption in integrated circuits. 2D MESFETs should find applications in ultra-low-power electronic circuits because they have a very small device volume and a very small number of electrons in the channels.

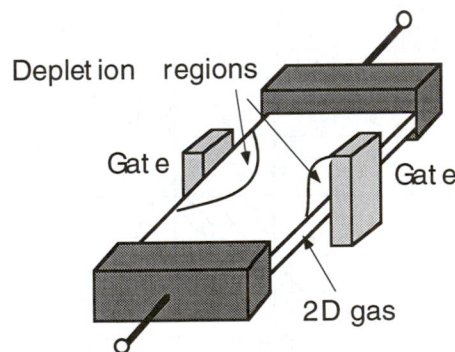

**Fig. 9.6.5.** Schematic diagram of a 2D MESFET (after W. C. B. Peatman, H. Park, M. Shur, "Novel Two-Dimensional Metal-Semiconductor Field Effect Transistor (2-D MESFET) for Ultra Low Power Circuit Applications," *IEEE Electron Device Letters*, Vol. 15, No. 7, pp. 245-247, July 1994, © IEEE 1994).

## *9-7.   WIDE BAND GAP SEMICONDUCTOR MATERIALS AND DEVICES

The typical semiconductors, silicon or gallium arsenide, have energy gaps of 1.12 and 1.42 eV, respectively.  Recent improvements in material growth and device fabrication of wide band semiconductors, such as SiC (energy gap 2.86 eV for one of the SiC modifications), GaN (energy gap 3.5 eV), and diamond (energy gap 5.5 eV) opened up new possible applications of these materials.

These and related wide band gap semiconductors (such as AlN with the energy gap of 6.2 eV) should have very low leakage currents [since the intrinsic carrier concentration, $n_i$, is proportional to $\exp(-E_g/2k_BT)$ where $E_g$ is the energy gap and $k_BT$ is the thermal energy].  They have an extremely large breakdown voltage (since in order to cause an impact ionization event, an electron or a hole has to obtain an energy in excess of $E_g$ from the electric field).  These materials are thermally and chemically stable and uniquely suited for applications in electronic circuits and systems operating in a harsh environment and/or at high temperatures.

Silicon carbide is one of the first semiconductor materials to be discovered.  As early as 1907, Round observed electroluminescence in silicon carbide, which he reported in his article published in *Electrical World*, Vol. 19, p. 309, 1907.  In 1955, Lely developed a new technique of crystal growth for SiC, and the studies of material properties and device applications of silicon carbide started to develop on a more sound basis.  Only recently, with the development of modern epitaxial techniques, have practical applications of silicon carbide devices become a reality.  Still, the applications of silicon carbide are very far from reaching its full potential.

Silicon carbide exists in more than 170 different polytypes.  The properties of the polytypes are so different that, in fact, SiC may be more accurately considered as a group of closely related materials.  Depending on the polytype crystal structure, the energy gap of silicon carbide varies from 2.2 to 3.3 eV.  It has a predicted electron saturation drift velocity, $v_s$, of $2\times10^7$ cm/s (approximately two times larger than in silicon, for which $v_s \approx 10^7$ cm/s), a breakdown field larger than 2,500 to 5,000 kV/cm (compared to 300 kV/cm for silicon), and a high thermal conductivity of 3.5 W/cm°C (compared to 1.3 W/cm°C for silicon and 0.5 W/cm°C for GaAs).  These properties make SiC important for potential applications in high-power, high-frequency devices as well as in devices operating at high temperatures and/or in a harsh environment. Applications of SiC include high-power devices, microwave devices (both avalanche diodes and microwave field effect transistors), and optoelectronic devices such as light-emitting diodes covering the visible electromagnetic spectrum and even the ultraviolet range.

Recently, new solid-state solutions of AlN/SiC/InN/GaN have been demonstrated. This exciting development opens up the possibility of a new generation of heterostructure devices based on SiC. Solid-state solutions of AlN-SiC are also expected to lead to direct gap ternary materials for UV and deep blue optoelectronics, including the development of visible light-emitting diodes and lasers.

Modern epitaxial techniques have led to a rapid improvement in the material quality of wide band gap semiconductors, such as SiC, GaN, InN, AlN, and diamond and made practical wide band gap semiconductor devices a reality. All basic device elements – from ohmic contacts to Schottky diodes and $p$-$n$ junctions – and most of the semiconductor devices – from Field Effect Transistor to a Bipolar Junction Transistor and Light-Emitting Diodes – have been demonstrated in materials ranging from different polytypes of SiC to the GaN/AlN material system.

The device potential of a semiconductor material is often estimated in terms of figures of merit. The Johnson's and Keyes' figures of merit are frequently used. Johnson's Figure of Merit ($JMF$) addresses the potential of a material for high-frequency and high-power applications:

$$JMF = \frac{E_B^2 v_s^2}{4\pi^2}$$

(9-7-1)

where $E_B$ is the breakdown electric field and $v_s$ is the electron saturation velocity. In terms of this figure of merit, SiC is 260 times better than Si and is inferior only to diamond.

Keyes' Figure of Merit ($KFM$) is relevant to integrated circuits applications:

$$KMF = \chi \sqrt{\frac{c v_s}{4\,\pi\varepsilon_o}}$$

(9-7-2)

where $c$ is the velocity of light, $\varepsilon_o$ is the static dielectric constant, and $\chi$ is the thermal conductivity.  In terms of this figure of merit, SiC is 5.1 times better than Si and, again, is inferior only to diamond.  These figures of merit illustrate SiC device potential for power and high-voltage devices, which are prime target areas for SiC applications.

SiC $p$-$n$ diodes clearly illustrate the potential advantages of wide band gap semiconductors, and SiC in particular.  The elementary theory of $p$-$n$ junctions yields the following expression for the reverse saturation current density, $j_R$, for a $p^+n$ junction (see Chap. 2):

$$j_R = q \sqrt{\frac{D_p}{\tau_p} \frac{n_i^2}{N_d}} + \frac{q n_i W}{\tau_e} \tag{9-7-3}$$

where $q$ is the electronic charge, $n_i$ is the intrinsic carrier density, $N_d$ is the doping density in the $n$-type region, $D_p$ is the hole diffusion coefficient, $W$ is the thickness of the depletion region, and $\tau_e$ is the effective lifetime:

$$W = \sqrt{\frac{2\varepsilon_s (V_{bi} - V)}{q N_d}} \tag{9-7-4}$$

where $V_{bi}$ is the built-in voltage and $\varepsilon_s$ is the dielectric permittivity of the semiconductor.

In wide band gap materials, the second term in the right-hand part of eq. (9-7-3) is dominant.  In  cubic SiC at room temperature, the theoretical value of $n_i$ is as low as $10^{-6}$ cm$^{-3}$!  For an estimate, let us assume certain values for $m_n$ and $m_p$, independent of the energy gap, since this assumption will not change the order of magnitude of the resulting number.  Let us take $m_n = 0.3\, m_e$ and $m_p = 0.6\, m_e$, to be specific.  Then we find

$$n_i\left(\text{m}^{-3}\right) = 1.34 \times 10^{21} \times T^{3/2} \exp\left(-\frac{E_g}{2k_B T}\right) (\text{K}) \tag{9-7-5}$$

Choosing the device volume to be $1 \times 0.1 \times 10$ µm$^3$ and assuming a generation time of 1 ns, we obtain

$$I_{leakage}(\text{A}) = 2.14 \times 10^{-7} \times T^{3/2}(\text{K}) \times \exp\left(-\frac{E_g}{2k_B T}\right) \tag{9-7-6}$$

(see Section 4.3).  The resulting dependence of the minimum leakage current on the energy gap for room temperature is shown in Fig. 4.3.4.  However, experimentally, such low values of the saturation current density have not yet been observed, and certainly the conventional theory of a $p$-$n$ junction is not valid when $n_i$ is very small. Nevertheless, these estimates clearly illustrate the potential of wide band gap semiconductors for applications in nonvolatile memory integrated circuits operating both at room temperature and at elevated temperatures.

The wide band gap semiconductors, AlN, GaN, InN, and the AlGaN solid-state solutions have a great potential for applications in high-power and optoelectronic devices. Depending on the molar fraction of Al, the AlGaN energy gap varies from 3.5 to 6.2 eV (see Fig. 9.7.1). The electron saturation velocity in GaN remains quite high even at elevated temperatures, which makes this material very attractive for high-temperature applications. The large difference in the energy gaps of AlGaN and GaN and the good quality of the AlGaN/GaN interface are useful for applications in heterostructure devices.

A blue light-emitting GaN and AlGaN/InGaN diodes have already been demonstrated. An AlGaN/GaN Heterostructure Field Effect Transistor and a GaN MESFET have been demonstrated as well and operated at microwave frequencies at elevated temperatures up to 350 ºC. All in all, wide band gap semiconductors represent one of the new frontiers of semiconductor device technology.

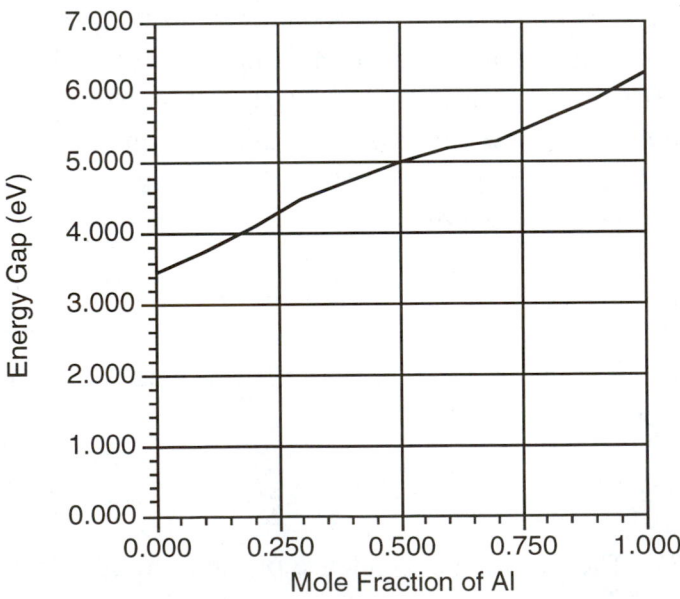

**Fig. 9.7.1.** Energy gap of AlGaN versus mole fraction of Al (after Paul Ruden, private communication, 1993).

## *9-8.  SUPERCONDUCTING DEVICES

Superconductivity – zero resistance to an electric current – is a very interesting and exciting phenomenon.  Until recently, superconductivity was observed in metals or highly doped semiconductors at very low temperatures, $T < T_c$ (where the **critical temperature** $T_c$ was less than 4 to 25 K or so).  However, in 1986, Georg Bednorz and K. Alex Müller discovered that certain copper-oxide-based ceramic materials exhibit superconductivity at much higher temperatures.  The **high temperature superconductors** have layered crystal structures.  At $T > T_c$, these materials have very anisotropic semiconducting properties.  Examples of these layered compounds are yttrium-barium-copper-oxide (code name YBCO), bismuth-strontium-calcium-copper-oxide (code name BSCCO), and thallium-barium-calcium-copper-oxide (code name TBCCO).  The superconductor with the highest $T_c \approx 160$ K (at the moment of writing) contains mercury.  Materials with the highest values of $T_c$ are usually $p$-type.  Many features of the high-temperature superconductivity are not yet understood, and here we will consider only the mechanism of low-temperature superconductivity.

To understand this phenomenon, let us consider electrons in a metal (see Fig. 9.8.1).  Since electrons are negatively charged, we may expect that they should repel each other.  However, electrons in a metal propagate among positive ions.  A moving electron attracts positive ions causing a local compression of the crystal lattice.  Since such a compression has a larger ion density, that is, a larger density of positive charges, the other moving electron may be attracted to this compression.  This attraction may cause two moving electrons with opposite spins to form a pair, called a **Cooper pair**.  (More precisely, such an interaction can be described in terms of electron interaction via phonons – the quanta of lattice vibrations.)  The electron energy of a Cooper pair is less than the energy of two separate electrons.  This provides a gap in the energy spectrum between the state with Cooper pairs and the state where electrons are separate, and unless the energy supplied to a Cooper pair from an electric field or thermal vibrations is larger than this gap, the material exhibits superconductivity.  This energy gap is quite small and decreases with temperature, vanishing at a certain critical temperature, $T_c$, at which the transition from the superconducting to the normal state occurs.  Similarly, the density of the current, $j$, flowing through a superconductor should be smaller than a certain critical value, $j_c$.  At higher current densities, the energy of the electronic motion becomes comparable to the gap in the energy spectrum, and the superconducting state is destroyed. Typical values of $j_c$ are of the order of $10^7$ A/cm$^2$. The formation of Cooper pairs is the key element of the theory of the superconductivity developed by Bardeen, Cooper, and Schrieffer in 1957.

In a superconductor, Cooper pairs are intermixed with normal electrons. As the superconductor temperature approaches $T_c$, the density of the Cooper pairs decreases, vanishing at $T = T_c$ when the superconductor experiences the transition to the normal state.

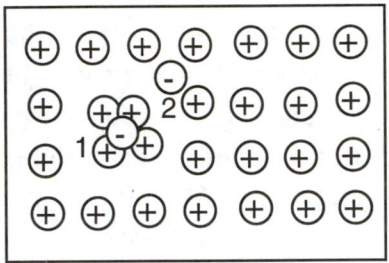

**Fig. 9.8.1.** Formation of Cooper pair. Electron 2 is attracted to the region of compressed positive ions caused by electron 1.

A very important characteristic of a superconductor is its current carrying capability, since at large current densities superconductivity is destroyed. The critical current density depends on the magnetic field. If the magnetic field is smaller than a certain critical value, $B_c$, which depends on temperature

$$B_c = B_o\left[1 - \left(\frac{T}{T_c}\right)\right]^2 \tag{9-8-1}$$

magnetic field streamlines are expelled from a superconductor. (This effect is called the **Meissner-Ochsenfeld effect**.) There are two classes of superconductors (types I and II), which behave differently when $B > B_c$. For type I superconductors, the superconductive state is destroyed when $B$ exceeds $B_c$. [The constant $B_o$ in eq. (9-8-1) is on the order of 0.1 to 1 Tesla.] Type II superconductors are characterized by two critical magnetic fields, $B_{c1}$ and $B_{c2}$. When $B < B_{c1}$, the magnetic field streamlines are pushed out of the superconductor, just as in a type I superconductor. When $B > B_{c2}$, the superconductive state is destroyed. When $B_{c1} < B < B_{c2}$, the magnetic field streamlines penetrate into the type 2 superconductor forming flux lines (magnetic flux quanta) called fluxoids or vortices.

The critical current density above which the superconducting state is destroyed depends also on temperature. High $T_C$ superconducting wires demonstrated recently are capable of carrying current with densities on the order of 10,000 A/cm$^2$ at 77 K over hundreds of meters.

A typical superconducting device is the **Josephson junction**, which consists of two superconductors or a normal metal and superconductor separated by a thin insulator barrier (see Fig. 9.8.2). The direct current related to tunneling of Cooper pairs appears when the applied constant bias exceeds the energy gap, $\Delta$, in the superconductor energy spectrum. (This energy gap is the energy per electron gained when a Cooper pair is formed. The theory of the superconductivity predicts that $\Delta \approx 1.57k_BT_c$.) However, the solution of the Schrödinger equation shows that when the Josephson junction is subjected to microwave or millimeter wave radiation with frequency $\omega$, the current flows at the voltage, $V_O$, such that

$$V_o = \frac{\hbar\omega}{q} \tag{9-8-2}$$

Therefore, a Josephson junction can be used as an extremely sensitive detector of microwave and millimeter wave radiation.

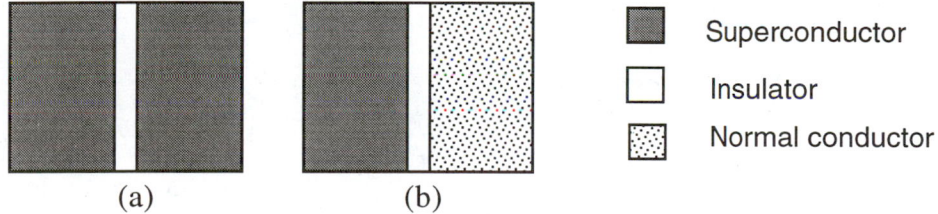

(a)                                (b)

**Fig. 9.8.2.** Josephson junctions between (a) two superconductors and (b) a superconductor and a normal conductor.

Josephson junctions are also used in extremely sensitive magnetometers, which are capable of measuring magnetic fields on the order of $10^{-10}$ Gs.

The discovery of high $T_c$ superconductivity led to the hopes of large-scale applications of superconductors in electronics and power transmission. It also stimulated dreams of discovering room-temperature superconductivity. Only the future will tell if these dreams and high expectations are fully justified.

## *9-9.  SINGLE ELECTRONICS

As semiconductor device sizes shrink, so does the number of electrons in the device active region.  For example, let us consider the number of electrons in a silicon MOSFET channel with dimensions $0.1\times1$ $\mu m^2$.  In the above-threshold regime, when the surface carrier density in the MOSFET channel is on the order of $10^{12}$ $cm^{-2}$, the total number of electrons is only $10^{12}\times0.1\times1\times10^{-8} = 1,000$. In the below-threshold regime, when the surface carrier density drops to $10^9$ $cm^{-2}$, the total number of electrons is only $10^9\times0.1\times1\times10^{-8} = 1$.  This shows us that with small devices it is possible to reach the ultimate limit of the electronic operation – single electron electronics.  Even more intriguing is that effects related to a single electron charge may appear in systems containing billions of electrons.

The electric current through a given cross section is equal to the amount of electric charge crossing this cross section per unit time.  Since this charge can be a polarization charge, that is, the charge related to the relative displacement of electrons (which are negatively charged) with respect to positively charged nuclei, the amount of the electric charge crossing a given cross section need not be larger than or equal to a single electron charge (see Fig. 9.9.1).

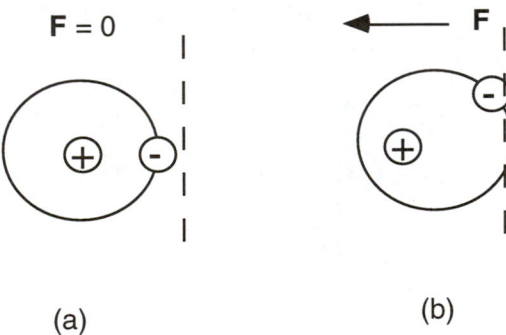

**Fig. 9.9.1.** (a) Atom in zero electric field.  (b) Electric field shifts the electronic cloud, resulting in the transfer of a fraction of a free electron charge across the surface represented by the dashed line.

On the other hand, if we consider a tunneling junction made of two semiconductor regions separated by a thin barrier layer (Fig. 9.9.2), electrons can tunnel through the barrier only in units of 1.  (A half an electron, for example, can't tunnel through since it is not a particle.)  If the area of such a junction is sufficiently small (so that the junction capacitance is also small) the voltage corresponding to one electron charge $U = q/C$ will be greater than or comparable to the thermal voltage $k_BT/q$ where $C$ is the junction capacitance. This happens when

$$C < \frac{q^2}{k_BT} \qquad (9\text{-}9\text{-}1)$$

For liquid helium temperature (4.2 K), this corresponds to $C < 0.4$ fF. If the tunneling barrier is 100 Å thick, the junction cross section should be less than

$$S < \frac{Cd}{\varepsilon_s} \qquad (9\text{-}9\text{-}2)$$

which is approximately $3.5 \times 10^{-15}$ m$^2$ for $d = 100$ Å and the dielectric permittivity $\varepsilon_s = 1.14 \times 10^{-10}$ F/m (the value for GaAs). For a device with a square cross section, this corresponds to dimensions less than $600 \times 600$ Å, which is easily achievable with modern fabrication techniques. These estimates give an idea of dimensions and temperatures required to observe single electron events.

In a small tunnel junction, the tunneling is suppressed when

$$-\frac{q}{2} < Q < \frac{q}{2} \qquad (9\text{-}9\text{-}3)$$

or

$$-\frac{q}{2C} < U < \frac{q}{2C} \qquad (9\text{-}9\text{-}4)$$

If the voltage exceeds these limits, the tunneling of one electron decreases the energy of the system. This effect (called the **Coulomb blockade**) is illustrated by Fig. 9.9.2. [The Coulomb blockade is observed when many (at least two) tunnel junctions are connected in series in order to reduce the equivalent capacitance and to increase the voltage scale. For a single junction, the capacitance of wires attached to the device may be dominant.]

The Coulomb blockade leads to a very interesting effect observed when a constant current, $I$, is flowing through a small tunnel junction. The voltage, $U$, across the device will oscillate with the frequency, $f$, given by

$$f = \frac{I}{q} \qquad (9\text{-}9\text{-}5)$$

where $q$ is the electronic charge. [These oscillations are called **Single Electron Tunneling (SET)** oscillations.]

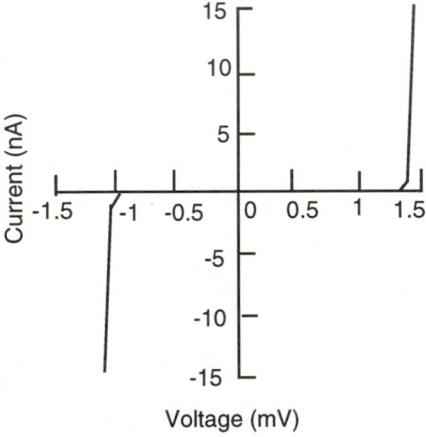

**Fig. 9.9.2.** Current-voltage characteristic of an array of 25 small tunnel junction illustrating the Coulomb blockade (from K. K. Likharev and T. Claeson, "Single Electronics," *Scientific American*, pp. 80-85, June 1992).

It is possible to make a tunneling single electron transistor by combining two tunnel junctions with the middle contact controlling the current flow (see Fig. 9.9.3). Such a control can be achieved by placing a single electron in the middle region. This is just an example of many proposed single electron structures. Such structures may allow us to achieve a much higher integration scale and a much smaller power consumption in digital integrated circuits.

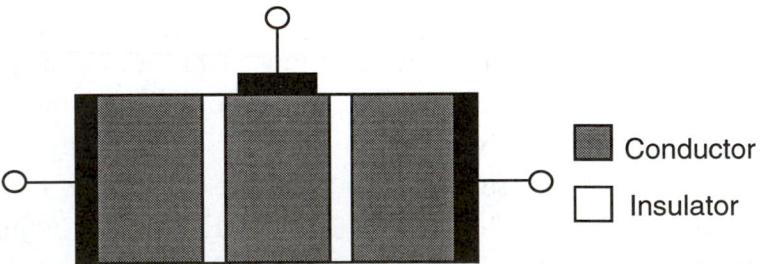

**Fig. 9.9.3.** Schematic diagram of a single electron transistor using two tunnel junctions.

# REFERENCES

F. CAPASSO, Editor, *Physics of Quantum Devices*, Springer-Verlag, New York (1989)

R. F. DAVIS, III-V Nitrides for Electronic and Optoelectronic Applications, *Proc. of IEEE*, 79, No. 5, pp. 702-712, May (1991)

S. K. GHANDHI, *VLSI Fabrication Principles*, John Wiley & Sons, New York (1983)

H. GRABERT AND D. H. DEVORET, Editors, *Single Charge Tunneling*, Plenum (1992)

K. K. LIKHAREV AND T. CLAESON, Single Electronics, *Scientific American*, pp. 80-85, June (1992)

S. LURYI, Hot-Electron Injection and Resonant-Tunneling Heterojunction Devices, Chap. 12 in *Heterojunction Band Discontinuities: Physics and Device Applications*, F. Capasso and G. Margaritondo, Editors, Elsevier Science Publishers B. V., Amsterdam (1987), pp. 489-564

S. LURYI, Chap. 7 in *High-Speed Semiconductor Devices,* S. M. SZE, Editor, John Wiley & Sons, New York (1990)

M. SHUR, *Physics of Semiconductor Devices*, Prentice Hall, Englewood Cliffs, NJ (1990)

S. M. SZE, Editor, *VLSI Technology*, Second Edition, McGraw-Hill, New York (1988)

# BIBLIOGRAPHY

B. L. ALTSHULER, P. A. LEE, AND R. A. WEBB, Editors, *Mesoscopic Phenomena in Solids*, Elsevier Science Publishers, Amsterdam (1991)

D. FERRY, Editor, *Granular Nanoelectronics*, Plenum, New York (1991)

M. SHUR AND T. A. FJELDLY, Editors, *Supercomputer Simulations of Semiconductor Devices*, Computer Physics Communications, Special Issue, August (1991)

# REVIEW QUESTIONS

**1.** List advantages and disadvantages of compound semiconductor technology.

☐ 1 point

**2.** (a) Describe the principle of operation of a Hot Electron Transistor (HET).

☐ 1 point

(b) Explain how an HFET can operate as an HET.

☐ 2 points

(c) Sketch a circuit that will allow you to use an HFET operating in a Hot Electron Transistor regime as an amplifier.

☐ 2 points

**3.** How will the peak current in the resonant tunneling diode change with the change in the barrier thickness?

☐ 1 point

**4.** You are hired as the Department Manager responsible for the development of SiC FET Integrated Circuit Technology. (a) What possible applications of such technology come to mind and why?

☐ 2 points

(b) Name a few technological issues that should be addressed in order to develop such a technology.

☐ 2 points

**5.** (a) Describe the mechanism of the formation of a Cooper pair.

☐ 1 point

(b) Why is the critical temperature of conventional superconductors low?

☐ 1 point

(c) Why is the superconducting state destroyed when the current density exceeds a certain critical value?

☐ 1 point

# PROBLEMS

**9-1-1.** Consider the band diagram shown in the figure.

(a) How can one obtain such a band diagram?
(b) Show in the figure the relevant energy levels for electrons in the quantum wells.
The electron effective mass is $6 \times 10^{-32}$ kg.
(c) Plot the densities of states for the electrons.
(d) Estimate the electron concentration in the well at which the Fermi level coincides with the lowest energy state in the quantum well in the conduction band. Temperature, $T = 300$ K.

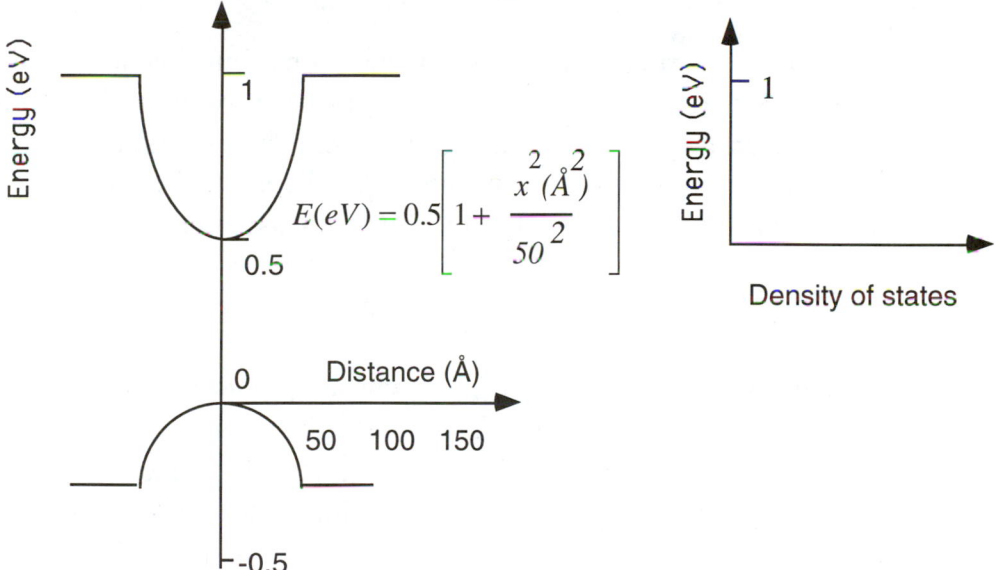

$$E(eV) = 0.5\left[1 + \frac{x^2 (\text{Å}^2)}{50^2}\right]$$

**9-1-2.** Consider a flow of electrons in a region of constant electric field, $F = 5$ kV/cm. The electron energy at $x = 0$ is equal to $E_o = 3k_BT/2$ where $T$ is the device temperature. Assume that the electron velocity, $v$, and energy, $E$, are described by the following equations:

$$mv\frac{dv}{dx} = qF - \frac{mv}{\tau_m}$$

$$v\frac{dE}{dx} = qvF - \frac{E - E_o}{\tau_E}$$

with the boundary conditions

$$v(x = 0) = v_o, \quad E(x = 0) = E_o$$

Assume that $v_o = v_{th}/4$ where $v_{th}$ is the thermal velocity. The device temperature is 77 K, the low field mobility is 50,000 cm$^2$/Vs, the energy relaxation time $\tau_E = 10^{-12}$ s. Assume that the momentum relaxation time $\tau_m = \tau_{mo}(E_o/E)$ where $\mu = q\tau_{mo}/m$. The electron effective mass is $m = 6 \times 10^{-32}$ kg. Solve these differential equations numerically and plot $v(x)$ for $0 \le x \le 0.5$ μm.

**9-1-3.** For a crude estimate, assume that electrons entering a region of constant electric field with a very low velocity travel ballistically until their energy acquired from the electric field reaches the energy of a lattice vibration quantum – a phonon, $E_{ph}$. Estimate the time duration of the ballistic flight if the electron effective mass is $m*$.

**9-3-1.** Explain the shape of the band diagram shown in Fig. 9.3.1.

**9-4-1.** Design a *p-n-p-n* superlattice (i. e., chose appropriate doping levels and dimensions of the *p*-type and *n*-type regions) and plot the resulting band diagram. (Base your design on the appropriate relationship between the periodic potential and the widths of the depletion regions.) Suggest a possible way of making contacts to the *p*-type and *n*-type regions.

**9-4-2.** Design the active layer of an AlGaAs/GaAs quantum cascade laser operating at the wavelength $\lambda = 40$ μm.

**9-5-1.** The measured characteristics of Resonant Tunneling Diodes (RTDs) may be affected by parasitic series and shunt resistances. Explain how these resistances change the measured characteristics of the tunnel diode shown in Fig. 9.5.3.

**9-5-2.** Assuming doped GaAs regions outside the double-barrier structure in the RTD shown in Fig. 9.5.1, undoped AlGaAs barriers, and an undoped GaAs quantum well between the barriers, sketch a qualitative band diagram of the RTD under bias and compare with Fig. 9.5.2.

**9-7-1.** The thermal conductivity of SiC is 5 times higher than that of GaAs. The temperature increase of a semiconductor integrated circuit is roughly proportional to the power dissipation and to the inverse thermal conductivity (the coefficient of proportionality depends on the heat sink and cooling conditions). Compare the temperature increase for two integrated circuits, made from SiC and GaAs, which operate at 5 V and 3.3 V, respectively, assuming that the current through the power supply is the same and the heat sinks are similar.

**9-8-1.** Assume that the critical temperature, $T_c = 25$ K. What is the maximum frequency of the radiation that can be detected by the Josephson junction using this superconductor?

# Appendices

## APPENDIX A1.   UNITS

| Quantity | Unit | Symbol | |
|---|---|---|---|
| **Fundamental Units:** | | | |
| Length | Meter | m | |
| Time | Second | s | |
| Mass | Kilogram | kg | |
| Temperature | Degree Kelvin | K | |
| Current | Ampere | A | |
| Light intensity | Candela | Cd | |
| **Additional Units** | | | |
| Angle | Radian | rad | |
| Solid angle | Radian | rad | |
| **Other Named Units** | | | |
| Frequency | Hertz | Hz | (1/s) |
| Force | Newton | N | (kg m/s2) |
| Energy | Joule | J | (Nm) |
| Pressure | Pascal | Pa | (N/m2) |
| Power | Watt | W | (J/s) |
| Electric charge | Coulomb | C | (A s) |
| Potential | Volt | V | (J/C) |
| Resistance | Siemens | S | (A/V) |
| Capacitance | Farad | F | (C/V) |
| Magnetic flux | Weber | Wb | (V s) |
| Inductance | Henry | H | (Wb/A) |
| Magnetic induction | Tesla | T | (Wb/m2) |
| Light flux | Lumen | Lm | (Cd rad) |

## APPENDIX A2.   PHYSICAL CONSTANTS

| Quantity | Symbol | Value |
|---|---|---|
| Avogadro number | $N_{AV}$ | $6.0221367 \times 10^{23}$ 1/mol |
| Bohr energy | $E_B$ | 13.6060 eV |
| Bohr magneton | $\mu_B$ | $5.78832 \times 10^{-5}$ eV/T |
| Bohr radius | $a_B$ | 0.52917 Å |
| Boltzmann constant | $k_B$ | $1.38066 \times 10^{-23}$ J/K |
| Boltzmann constant/$q$ | $k_B/q$ | $8.61738 \times 10^{-5}$ eV/K |
| Electronic charge | $q$ | $1.60218 \times 10^{-19}$ C |
| Electronvolt | eV | $1.60218 \times 10^{-19}$ J |
| Fine structure constant | $\alpha$ | $0.00729735308$ ($\approx 1/137$) |
| Gas constant | $R$ | $1.98719$ cal mol$^{-1}$ K$^{-1}$ |
| Gravitational constant | $\gamma$ | $6.67259 \times 10^{-11}$ m$^3$/(kg s$^2$) |
| Impedance of free space | $1/c\varepsilon_o = \mu_o c$ | $376.732 \ \Omega$ |
| Mass of electron at rest | $m_e$ | $0.91093897 \times 10^{-30}$ kg |
| Mass of proton at rest | $M_p$ | $1.6726231 \times 10^{-27}$ kg |
| Permeability in vacuum | $\mu_o$ | $1.26231 \times 10^{-8}$ H/cm ($4\pi \times 10^{-9}$) |
| Permittivity in vacuum | $\varepsilon_o$ | $8.85418 \times 10^{-12}$ F/m ($1/\mu_o c^2$) |
| Planck constant | $h$ | $6.6260755 \times 10^{-34}$ J-s |
| Reduced Planck constant | $\hbar = h/(2\pi)$ | $1.0545727 \times 10^{-34}$ J-s |
| Speed of light in vacuum | $c$ | $2.99792458 \times 10^8$ m/s |
| Standard atmosphere | | $1.01325 \times 10^5$ N/m$^2$ |
| Thermal voltage at 300 K | $k_B T/q$ | 0.025860 V |
| Wavelengths of visible light | $\lambda$ | 0.4 to 0.7 μm |

# APPENDIX A.3. PERIODIC TABLE OF THE ELEMENTS

Relative Atomic Mass Based on Carbon 12

Atomic Number → | 4.00260 / 2 **He** / Helium | ← Symbol

Name

| IA | | | | | | | | | | | | | | | | | |
|---|---|---|---|---|---|---|---|---|---|---|---|---|---|---|---|---|---|
| 1.00794 1 **H** Hydrogen | IIA | | | | | | | | | | | IIIA | IVA | VA | VIA | VIIA | 4.00260 2 **He** Helium |
| 6.941 3 **Li** Lithium | 9.01218 4 **Be** Beryllium | | | | | | | | | | | 10.81 5 **B** Boron | 12.0111 6 **C** Carbon | 14.0067 7 **N** Nitrogen | 15.9994 8 **O** Oxygen | 18.9984 9 **F** Fluorine | 20.179 10 **Ne** Neon |
| 22.9898 11 **Na** Sodium | 24.305 12 **Mg** Magnesium | IIIB | IVB | VB | VIB | VIIB | VIII | | | IB | IIB | 26.9815 13 **Al** Aluminum | 28.0855 14 **Si** Silicon | 30.9738 15 **P** Phosphorus | 32.06 16 **S** Sulfur | 35.453 17 **Cl** Chlorine | 39.948 18 **Ar** Argon |
| 39.0983 19 **K** Potassium | 40.08 20 **Ca** Calcium | 44.9559 21 **Sc** Scandium | 47.88 22 **Ti** Titanium | 50.9415 23 **V** Vanadium | 51.996 24 **Cr** Chromium | 54.9380 25 **Mn** Manganese | 55.847 26 **Fe** Iron | 58.9332 27 **Co** Cobalt | 58.69 28 **Ni** Nickel | 63.546 29 **Cu** Copper | 65.39 30 **Zn** Zinc | 69.72 31 **Ga** Gallium | 72.59 32 **Ge** Germanium | 74.9216 33 **As** Arsenic | 78.96 34 **Se** Selenium | 79.904 35 **Br** Bromine | 83.80 36 **Kr** Krypton |
| 85.4678 37 **Rb** Rubidium | 87.62 38 **Sr** Strontium | 88.9059 39 **Y** Yttrium | 91.224 40 **Zr** Zirconium | 92.9064 41 **Nb** Niobium | 95.94 42 **Mo** Molybdenum | (98) 43 **Tc** Technetium | 101.07 44 **Ru** Ruthenium | 102.906 45 **Rh** Rhodium | 106.42 46 **Pd** Palladium | 107.868 47 **Ag** Silver | 112.41 48 **Cd** Cadmium | 114.82 49 **In** Indium | 118.71 50 **Sn** Tin | 121.75 51 **Sb** Antimony | 127.60 52 **Te** Tellurium | 126.905 53 **I** Iodine | 131.29 54 **Xe** Xenon |
| 132.905 55 **Cs** Cesium | 137.33 56 **Ba** Barium | **La-Lu** 57 71 | 178.49 72 **Hf** Hafnium | 180.948 73 **Ta** Tantalum | 183.85 74 **W** Tungsten | 186.207 75 **Re** Rhenium | 190.2 76 **Os** Osmium | 192.22 77 **Ir** Iridium | 195.08 78 **Pt** Platinum | 196.967 79 **Au** Gold | 200.59 80 **Hg** Mercury | 204.383 81 **Tl** Thallium | 207.2 82 **Pb** Lead | 208.980 83 **Bi** Bismuth | (209) 84 **Po** Polonium | (210) 85 **At** Astatine | (222) 86 **Rn** Radon |
| (223) 87 **Fr** Francium | 226.025 88 **Ra** Radium | **Ac-Lr** 89 103 | (261) 104 **Unq** Unnilquadium | (262) 105 **Unp** Unnilpentium | (263) 106 **Unh** Unnilhexium | (262) 107 **Uns** Unnilseptium | **Uno** 108 Unniloctium | **Une** 109 Unnilennium | | | | | | | | | |

| | Lanthanoid series | | | | | | | | | | | | | | |
|---|---|---|---|---|---|---|---|---|---|---|---|---|---|---|---|
| 138.906 57 **La** Lanthanum | | 140.12 58 **Ce** Cerium | 140.908 59 **Pr** Praseodymium | 144.24 60 **Nd** Neodymium | (145) 61 **Pm** Promethium | 150.36 62 **Sm** Samarium | 151.96 63 **Eu** Europium | 157.25 64 **Gd** Gadolinium | 158.925 65 **Tb** Terbium | 162.50 66 **Dy** Dysprosium | 164.930 67 **Ho** Holmium | 167.26 68 **Er** Erbium | 168.934 69 **Tm** Thulium | 173.04 70 **Yb** Ytterbium | 174.967 71 **Lu** Lutetium |
| 227.028 89 **Ac** Actinium | Actinoid series | 232.038 90 **Th** Thorium | 231.036 91 **Pa** Protactinium | 238.029 92 **U** Uranium | 237.048 93 **Np** Neptunium | (244) 94 **Pu** Plutonium | (243) 95 **Am** Americium | (247) 96 **Cm** Curium | (247) 97 **Bk** Berkelium | (251) 98 **Cf** Californium | (252) 99 **Es** Einsteinium | (257) 100 **Fm** Fermium | (258) 101 **Md** Mendelevium | (259) 102 **No** Nobelium | (260) 103 **Lr** Lawrencium |

Masses in parentheses are masses of the most stable isotopes

## APPENDIX A4.  PARAMETERS OF GUMMEL-POON MODEL

The BJT parameters of the Gummel-Poon model related to the $I$-$V$ characteristics of the emitter-base, collector-base, and substrate diodes are summarized in Table A4.1.   Parameters characterizing the BJT depletion capacitances are given in Table A4.2.   Parameters characterizing the parasitic resistances of a BJT are summarized in Table A4.3.   The remaining three groups of the Gummel-Poon parameters, dealing with nonideal transistor effects, transistor time response, and with the temperature dependencies, are summarized in Tables A4.4, A4.5, and A4.6, respectively.   A few the Gummel-Poon parameters in SPICE (such as parameters QCO, RCO, VO, GAMMA, KF, AF of Table A4.4 and parameters given in Table A4.5) are beyond the scope of this book.   These parameters are listed since they may be specified in the SPICE parameters, which may be provided by transistor manufacturers.   Minor variations of the Gummel-Poon model may occur in different versions of SPICE.   The tables given below are sufficient for both AIM-Spice and PSpice[tm].

| SPICE parameter | SPICE parameter name | Unit | SPICE default | Chap. 5 notation | Relevant equation |
|---|---|---|---|---|---|
| IS | Saturation current | A | 1.0e-16 | $I_s$ | (5-2-27) |
| NF | Ideality factor for ideal emitter-base diode | - | 1 | $n_F$ | (5-2-27) |
| BF | Ideal maximum forward beta | | 100 | $\beta_F$ | (5-2-27) |
| NR | Ideality factor for ideal collector-base diode | - | 1 | $n_R$ | (5-2-29) |
| BR | Ideal maximum reverse beta | | 1 | $\beta_R$ | (5-2-29) |
| ISE | Saturation current for leakage emitter-base diode | A | 0 | $I_{se}$ | (5-2-30) |
| NE | Ideality factor for leakage emitter-base diode | - | 1.5 | $n_E$ | (5-2-27) |
| ISS | Substrate saturation current | A | 1.0e-16 | $I_{ss}$ | (5-2-32) |
| NS | Ideality factor for leakage emitter-base diode | - | 1.5 | $n_S$ | (5-2-32) |

**Table A4.1.**  BJT Gummel-Poon SPICE model parameters related to $I$-$V$ characteristics of emitter-base, collector-base, and substrate diodes.

| Spice parameter | Spice parameter name | Unit | Spice default | Chap. 4 notation | Relevant equation |
|---|---|---|---|---|---|
| CJE | Zero-bias base-emitter capacitance | F | 0 | $C_{jo}$ | (4-4-10) |
| CJC | Zero-bias base-collector capacitance | F | 0 | $C_{jo}$ | (4-4-10) |
| VJE | Built-in base-emitter voltage potential | V | 1 | $V_{bi}$ | (4-4-10) |
| VJC | Built-in base-collector voltage | V | 1 | $V_{bi}$ | (4-4-10) |
| ME | Base-emitter grading coefficient | - | 0.5 | $m$ | (4-4-10) |
| MC | Base-collector grading coefficient | - | 0.5 | $m$ | (4-4-10) |
| CJS | Zero-bias substrate capacitance | F | 0 | $C_{jo}$ | (4-4-10) |
| VJS | Built-in substrate voltage | V | 1 | $V_{bi}$ | (4-4-10) |
| MS | Substrate grading coefficient | - | 0.5 | $m$ | (4-4-10) |
| FC | Coefficient for forward-bias depletion capacitance | - | 0.5 | - | - |
| XCJC | Fraction of base-collector capacitance connected internally to base spreading resistance, RB | - | 1 | - | - |

**Table A4.2.** Parameters of the Gummel-Poon SPICE model used to describe the *C-V* depletion characteristics for emitter-base, collector base, and substrate diodes.

| SPICE parameter | SPICE parameter name | Unit | SPICE default | Chap. 5 notation | Relevant equation |
|---|---|---|---|---|---|
| RE | Emitter ohmic resistance | $\Omega$ | 0 | $R_E$ | - |
| RE | Collector ohmic resistance | $\Omega$ | 0 | $R_C$ | - |
| RB | Maximum base resistance | $\Omega$ | 0 | $R_B$ | (5-2-33) |
| RBM | Minimum base resistance | $\Omega$ | RB | $R_{Bm}$ | (5-2-33) |
| IRB | Base current at which base resistance is $(R_B+R_{Bm})/2$ | - | Infinity | $I_{RB}$ | (5-2-33) |

**Table A4.3.** Parasitic resistances in the Gummel-Poon SPICE model.

| SPICE parameter | SPICE parameter name | Unit | SPICE default | Chap. 5 notation | Relevant equation |
|---|---|---|---|---|---|
| VAF | Forward Early voltage | V | Infinity | $V_A$ | (5-2-20) |
| VAR | Reverse Early voltage | V | Infinity | - | (5-2-20) |
| IKF | Forward beta roll-off current | A | Infinity | $I_{KF}$ | (5-2-35) |
| IKR | Forward beta roll-off current | A | Infinity | $I_{KR}$ | (5-2-36) |
| QCO | Epitaxial region charge factor | C | 0 | - | - |
| RCO | Epitaxial region resistance | $\Omega$ | 0 | - | - |
| VO | Carrier mobility knee voltage | V | 10 | - | - |
| GAMMA | Epitaxial region doping factor | - | $1\times10^{-11}$ | - | - |
| KF | Flicker noise coefficient | - | 0 | - | - |
| AF | Flicker noise exponent | - | 1 | - | - |

**Table A4.4.** Parameters of the Gummel-Poon SPICE model dealing with nonideal transistor effects and noise.

| SPICE parameter | SPICE parameter name | Unit | SPICE default | Chap. 5 notation | Relevant equation |
|---|---|---|---|---|---|
| TF | Ideal forward transit time | s | 0 | - | - |
| XTF | Forward transit time bias dependence coefficient | - | 0 | - | - |
| VTF | Parameter characterizing the dependence of TF on the base-collector voltage | V | infinity | - | - |
| ITF | Parameter characterizing the dependence of TF on the collector current | A | 0 | - | - |
| TR | Ideal reverse transit time | s | 0 | - | - |

**Table A4.5.** Parameters of the Gummel-Poon SPICE model dealing with transistor time constants.

| SPICE parameter | SPICE parameter name | Unit | SPICE default | Chap. 5 notation | Relevant equation |
|---|---|---|---|---|---|
| TNOM | Temperature at which parameters are specified | ºC | 27 | $T_o$ (K) | (5-2-28) |
| EG | Energy gap | eV | 1.11 | $E_g$ | (5-2-28) |
| XTI | IS temperature exponent | - | 3 | | (5-2-28) |
| XTB | Beta temperature coefficient | - | 0 | - | - |
| TRE1 | Linear emitter resistance temperature coefficient | ºC$^{-1}$ | 0 | - | |
| TRE2 | Quadratic emitter resistance temperature coefficient | ºC$^{-2}$ | 0 | - | |
| TRB1 | Linear base resistance (RB) temperature coefficient | ºC$^{-1}$ | 0 | - | |
| TRB2 | Quadratic base resistance (RB) temperature coefficient | ºC$^{-2}$ | 0 | - | |
| TRM1 | Linear base resistance (RBM) temperature coefficient | ºC$^{-1}$ | 0 | - | |
| TRM2 | Quadratic base resistance (RBM) temperature coefficient | ºC$^{-2}$ | 0 | - | |
| TRC1 | Linear collector resistance temperature coefficient | ºC$^{-1}$ | 0 | - | |
| TRC2 | Quadratic collector resistance temperature coefficient | ºC$^{-2}$ | 0 | - | |

**Table A4.6.**   Parameters of the Gummel-Poon SPICE model dealing with temperature dependence.  Temperature coefficients for resistances introduced in Table A4.6 are defined for each resistance as follows:

$$R(T)=R(T_{nom})\left[1+T_{c1}(T-T_{nom})+T_{c2}(T-T_{nom})^2\right]$$

| SPICE parameter | SPICE parameter name | Unit | SPICE default | Chap. 5 notation | Ebers-Moll parameter |
|---|---|---|---|---|---|
| IS | Saturation current | A | 1.0e-16 | $I_s$ | $-a_{11}$ |
| BF | Ideal maximum forward beta | | 100 | $\beta_F$ | $-\dfrac{a_{12}}{a_{11}+a_{12}}$ |
| BR | Ideal maximum reverse beta | | 1 | $\beta_R$ | $-\dfrac{a_{12}}{a_{22}+a_{12}}$ |
| ISS | Substrate saturation current | A | 1.0e-16 | $I_{ss}$ | 0 |

**Table A4.7.** SPICE parameters in terms of the parameters of the Ebers-Moll model. (When other SPICE parameters specified in Table A4.1 have the default values, the Gummel-Poon model reduces to the Ebers-Moll model.)

## APPENDIX A5.  DEFAULT SPICE PARAMETERS FOR MOSFET MODELS LEVEL 1, 2, 3, and 6

| Name | Parameter | Units | Default |
|---|---|---|---|
| VTO | Zero-bias threshold voltage | V | 0.0 |
| KP | Transconductance parameter | $A/V^2$ | $2.0 \times 10^{-5}$ |
| GAMMA | Bulk threshold parameter | | 0.0 |
| PHI | Surface potential | V | 0.6 |
| LAMBDA | Channel length modulation (Only MOS1 and MOS2) | 1/V | 0.0 |
| RD | Drain resistance | $\Omega$ | 0.0 |
| RS | Source resistance | $\Omega$ | 0.0 |
| CBD | Zero-bias bulk-drain capacitance | F | 0.0 |
| CBS | Zero-bias bulk-source capacitance | F | 0.0 |
| IS | Bulk junction saturation current | A | $1.0 \times 10^{-14}$ |
| PB | Bulk junction potential | V | 0.8 |
| CGSO | Gate-source overlap capacitance per meter channel width | F/m | 0.0 |
| CGDO | Gate-drain overlap capacitance per meter channel width | F/m | 0.0 |
| CGBO | Gate-bulk overlap capacitance per meter channel width | F/m | 0.0 |
| RSH | Drain and source diffusion sheet resistance | $\Omega/Sq$ | 0.0 |
| CJ | Zero-bias bulk junction bottom capacitance per square meter of junction area | $F/m^2$ | 0.0 |
| MJ | Bulk junction bottom grading coefficient | - | 0.5 |
| CJSW | Zero-bias bulk junction sidewall capacitance per meter of junction perimeter | F/m | 0.0 |
| MJSW | Bulk junction sidewall grading coefficient | - | 0.50 (level 1) 0.33 (level 2) |
| JS | Bulk junction saturation current per unit of junction area | $A/m^2$ | 0 |

| TOX | Oxide thickness | m | $1.0 \times 10^{-7}$ |
|---|---|---|---|
| NSUB | Substrate doping | $1/\text{cm}^3$ | 0.0 |
| NSS | Surface state density | $1/\text{cm}^2$ | 0.0 |
| NFS | Fast surface state density | $1/\text{cm}^2$ | 0.0 |
| TPG | Type of gate material:<br>+1 : opposite of substrate<br>-1 : same as substrate<br>0 : Al gate | - | 1.0 |
| XJ | Metallurgical junction depth | m | 0.0 |
| LD | Lateral diffusion | m | 0.0 |
| U0 | Surface mobility | $\text{cm}^2/\text{Vs}$ | 600 |
| UCRIT | Critical field for mobility degradation (only MOS2) | V/cm | $1.0 \times 10^4$ |
| UEXP | Critical field exponent in mobility degradation (only MOS2) | - | 0.0 |
| UTRA | Transverse field coefficient (deleted for MOS2) | - | 0.0 |
| VMAX | Maximum drift velocity for carriers | m/s | 0.0 |
| NEFF | Total channel charge (fixed and mobile) coefficient (only MOS2) | - | 1.0 |
| KF | Flicker noise coefficient | - | 0.0 |
| AF | Flicker noise exponent | - | 1.0 |
| FC | Coefficient for forward-bias depletion capacitance formula | - | 0.5 |
| DELTA | Width effect on threshold voltage (only MOS2 and MOS3) | - | 0.0 |
| THETA | Mobility modulation (only MOS3) | 1/V | 0.0 |
| ETA | Static feedback (only MOS3) | - | 0.0 |
| KAPPA | Saturation field factor (MOS3 only) | - | 0.2 |
| TNOM | Parameter measurement temperature | °C | 27 |

From K. Lee, M. Shur, T. Fjeldly, and T. Ytterdal, *Semiconductor Modeling for VLSI*, Prentice Hall, 1993, copyright © Prentice Hall, 1993, reproduced by permission of Prentice Hall, Inc., Englewood Cliffs, NJ.

## APPENDIX A6.   PROPERTIES OF SILICON (Si)

| | |
|---|---|
| Atomic number | 14 |
| Atoms/cm$^3$ | $5.02 \times 10^{22}$ |
| Electronic shell configuration | $1s^2\, 2s^2\, 2p^6\, 3s^2\, 3p^2$ |
| Atomic weight | 28.09 |
| Crystal structure | Diamond |
| Breakdown field (V/cm)[a] | $\sim 3.0 \times 10^5$ |
| Density (g/cm$^3$) | 2.329 (at 298 K) |
| Dielectric constant | 11.7 |
| Diffusion constant (cm$^2$/s) (at 300 K)[a] | 37.5 (electrons) 13 (holes) |
| Effective density of states | |
|    in the conduction band (cm$^{-3}$) | $3.22 \times 10^{19}$ (at 300 K) |
|    in the valence band (cm$^{-3}$) | $1.83 \times 10^{19}$ (at 300 K) |
| Effective electron mass (in units of $m_e$) | longitudinal    : 0.92 (at 1.26 K) |
| | transverse    : 0.19 (at 1.26 K) |
| | density of states  : 1.28 (at 600 K) |
| | 1.18 (at 300 K) 1.08 (at 77 K) |
| | 1.026 (at 4.2 K) |
| Effective hole mass (in units of $m_e$) | heavy hole    : 0.537 (at 4.2 K) |
| | heavy hole    : 0.49  (at 300 K) |
| | light hole    : 0.153  (at 4.2 K) |
| | light hole    : 0.16  (at 300  K) |
| | density of states  : 0.591 (at 4.2 K) |
| | 0.62 (at 77 K)   0.81 (at 300 K) |
| Electron affinity (V) | 4.05 |
| Energy gap (eV) | 1.12 (at 300 K)     1.17 (at 77 K) |
| Index of refraction | 3.42 |
| Intrinsic  carrier concentration (cm$^{-3}$) | $1.02 \times 10^{10}$ cm$^{-3}$ (at 300 K) |
| Intrinsic Debye length ($\mu$m) | 24 |
| Intrinsic resistivity  (ohm-cm) | $3.16 \times 10^5$  (at 300 K) |
| Lattice constant (Å) | 5.43107 (at 298.2 K) |
| Melting point ($^o$C) | 1412 |
| Mobility (cm$^2$/V-s) (at 300 K)** | 1450 (electrons)     500 (holes) |
| Optical phonon energy (eV) | 0.063 |
| Specific heat (J/g-$^o$C) | 0.7 |
| Thermal conductivity (W/cm-$^o$C) | 1.31  (at 300 K) |
| Thermal diffusivity (W/cm-$^o$C) | 0.9 |
| Thermal expansion, linear  ($^o$C$^{-1}$) | $2.6 \times 10^{-6}$ (at 300 K) |
| Young's modulus (dyn/cm$^2$) | $1.9 \times 10^{12}$ in [111] direction |

From K. Lee, M. Shur, T. Fjeldly, and T. Ytterdal, *Semiconductor Modeling for VLSI*, copyright
© Prentice Hall, 1993, reproduced by permission of Prentice Hall, Inc., Englewood Cliffs, NJ.
[a] For undoped or low doped material.

## APPENDIX A7.  PROPERTIES OF GALLIUM ARSENIDE (GaAs)

| | |
|---|---|
| Crystal structure | zinc blende |
| Breakdown field (V/cm) | $\sim4.0\times10^5$ |
| Density (g/cm$^3$) | 5.3176 (at 298 K) |
| Dielectric constant ($\kappa_s$) | 12.9 (at 300 K) |
| ($\kappa_o$) | 10.89 (at 300 K) |
| Diffusion constant (cm$^2$/s) (at 300 K) | 207 (electrons) 10 (holes) |
| Effective density of states | |
| in the conduction band (cm$^{-3}$) | $4.7\times10^{17}$ (at 300 K) |
| in the valence band (cm$^{-3}$) | $7.0\times10^{18}$ (at 300 K) |
| Effective electron mass (in units of $m_e$) | 0.067  (0 K)   0.063 (300 K) |
| Effective hole mass (in units of $m_e$) | heavy hole    : 0.51 (at $T < 100$ K) |
| | : 0.50  (at 300 K) |
| | light hole    : 0.084 (at $T < 100$ K) |
| | : 0.076 (at 300 K) |
| | density of states : 0.53 |
| Electron affinity (V) | 4.07 |
| Energy gap (eV) | 1.424 (at 300 K) |
| | 1.507 (at 77 K) |
| | 1.519 (at 0 K) |
| Index of refraction | 3.3 |
| Intrinsic  carrier concentration (cm$^{-3}$) | $2.1\times10^6$ (at 300 K) |
| Intrinsic Debye length ($\mu$m) | 2,250  (at 300 K) |
| Intrinsic resistivity (ohm-cm) | $10^8$  (at 300 K) |
| Lattice constant (Å) | 5.6533 (at 300 K) |
| Melting point ($^o$C) | 1,240 |
| Mobility (cm$^2$/V-s) | 8,500 (electrons at 300 K) |
| | 400   (holes at 300 K) |
| Optical phonon energy (eV) | 0.035 |
| Specific heat (J/g-$^o$C) | 0.35 |
| Thermal conductivity  (W/cm-$^o$C) | 0.46 |
| Thermal diffusivity (W/cm-$^o$C) | 0.44 |
| Thermal expansion, linear  ($^o$C$^{-1}$) | $6.86\times10^{-6}$ (at 300 K) |

## APPENDIX A8.  DOWNLOADING STUDENT VERSION OF AIM-SPICE

AIM-Spice can be downloaded from a bulletin board using anonymous *ftp*.  The address is:

> *sdlsun4.ee.virginia.edu*

You should use the following command to connect:

> *ftp sdlsun4.ee.virginia.edu*

and then give *anonymous* as the user name and your *e-mail address* as the password.  After you have logged on, change to binary transfer mode with the command *bin*.  To download files, you should use the command *get <filename>*.  To list the files, you use the command *ls*.  The command *cd* is used to change directories.

AIM-Spice can be found in the file *aimstud.zip* in the pub directory.  After you copied *aimstud.zip* to a directory on your PC's hard disk drive, you should install AIM-Spice using the following procedure:

1. Run *pkunzip* to unpack *aimstud.zip*.  This will produce a number of files, including *setup.exe*.
2. Start Microsoft Windows.
3. From the File menu in Program Manager select the *Run* command.
4. Specify that you want to run *setup.exe*.

For instance, if you copied *aimstud.zip* to *c:\aimstud*, enter the following *run* command in the text box:  *c:\aimstud\setup.exe*.

This will install AIM-Spice.

You must specify a new directory for the installation (not the directory where you placed *aimstud.zip*.

The *ftp*-site currently contains version 1.6.

A detailed information on AIM-Spice is given in the following book:  K. Lee, M. Shur, T. A. Fjeldly, and T. Ytterdal, *Semiconductor Device Modeling for VLSI*, Prentice Hall, Englewood Cliffs, NJ (1993), ISBN 0-13-805656-0.  A brief description of AIM-Spice features is also available on World Wide Web.  The address of the AIM-Spice Home Page is

> http://fulton.seas.virginia.edu/~ty2n/aimspice.html

## APPENDIX A9.  BRIEF HISTORY OF SEMICONDUCTOR DEVICES

In 1821, the German physicist, Tomas Seebeck, first noticed unusual properties of semiconductor materials such as lead sulfur (PbS).

In 1833, the English physicist, Michael Faraday, reported on the temperature dependence of the conductivity for a new class of materials – semiconductors.

In 1873, the British engineer, W. Smith, discovered that the resistivity of selenium – a semiconductor material – is very sensitive to light.

In 1875, Alexander Graham Bell used this device for a wireless telephone communication system.

In 1907, Round discovered electroluminescence.

In 1914, Köenigsberger published the first review on the properties of semiconductors.

In the early 1930s, Lilienfeld and Heil introduced the concept of a field effect transistor.

In 1948, Bardeen, Brattain, and Shockley discovered a Bipolar Junction Transistor, and the MODERN AGE BEGAN.

In 1954, Chapin, Fuller, and Pearson developed a solar cell.

In 1958, the American engineer, John Kilby, invented the Integrated Circuit (IC).

In 1958, Leo Esaki discovered a tunnel diode (Esaki diode).

In 1960, Kahng and Atalla demonstrated the first working MOSFET.

In 1962, three groups headed by Hall, Nathan, and Quist demonstrated a semiconductor laser.

In 1963, Gunn discovered microwave oscillations in GaAs and InP (Ridley-Watkins-Hilsum-Gunn effect).

In 1963, Wanlass and Sah introduced complementary MOS (CMOS) technology.

# Glossary

| | |
|---|---|
| **Acceptor** | An impurity atom capable of supplying a hole into a valence band. (Ionized acceptors are negatively charged.) |
| **Accumulation layer** | A layer with a higher concentration of charge carriers than in the rest of semiconductor material. |
| **Activation energy** | Energy, $E_a$, in the Boltzmann factor, $\exp(-E_a/k_BT)$, which determines the temperature dependence of a material or device parameter. |
| **Active Matrix Liquid Crystal Display (AMLCD)** | A Liquid Crystal Display where each pixel is driven by its own Thin Film Transistor. |
| **$A_{II}B_{VI}$** | Compound semiconductors, which are composed of element A of group II and element B of group VI of the Periodic Table. |
| **$A_{III}B_{V}$** | Compound semiconductors, which are composed of element A of group III and an element B of group V of the Periodic Table of Elements. |
| **Ambipolar diffusion** | The diffusion of a nearly neutral fluctuation of electrons and holes. |
| **Ambipolar mobility** | The mobility of a nearly neutral fluctuation of electrons and holes. |
| **Auger recombination** | A recombination process during which an electron in the conduction band (or a hole in the valence band) receives energy from a recombining electron-hole pair. |
| **Avalanche Photo Diode** | A photodiode where light-generated electrons or holes cause impact ionization creating more carriers. |
| **Avalanche breakdown** | A breakdown mechanism caused by impact ionization. The generated electrons or holes create more electron-hole pairs via impact ionization in high electric field, and so on. |

| | |
|---|---|
| **Ballistic transport** | Free electron motion uninterrupted by collisions. |
| **Band bending** | The dependence of energy bands on distance. |
| **Band diagram** | The plot of the bottom of the conduction band and the top of the valence band as functions of distance. |
| **Band discontinuity** | The difference in a band's energy at a heterointerface. |
| **Band gap** | The difference between the bottom of the conduction band and the top of the valence band. |
| **Band gap narrowing** | The decrease of the energy gap caused by a very high doping. |
| **Base spreading resistance** | The base resistance seen by a base current. |
| **Beta cutoff frequency** | The frequency at which the common current emitter gain is equal to unity. |
| **Binary compound** | Compound consisting of two elements. |
| **Bohr energy** | The energy of the ground state of a hydrogen atom. |
| **Bohr radius** | Characteristic radius of a hydrogen atom. |
| **Boltzmann distribution function** | The equilibrium distribution function for a gas of particles that obey the laws of classical mechanics. |
| **Boltzmann transport equation** | Equation for the particle distribution function in an electric field. |
| **Bragg reflection** | Reflection caused by constructive interference of reflected waves from crystal planes. |
| **Breakdown** | A sharp increase in the current flowing through a device. |
| **Brillouin zone** | A region in $k$-space (reciprocal space) with boundaries determined by the values of $\mathbf{k}$ that correspond to the Bragg reflections from different crystal planes. |
| **BSIM** | A popular (and a fairly sophisticated) MOSFET model in SPICE. |
| **Built-in electric field** | Electric field created by charges of ionized impurities, typically near the boundary between $n$-type and $p$-type regions or near a metal-semiconductor interface. |
| **Built-in voltage** | Voltage drop created by charges of ionized impurities, typically near the boundary between $n$-type and $p$-type regions or near a metal-semiconductor interface. |

| | |
|---|---|
| **Central valley** | A conduction band minimum at wave vector $k = 0$. |
| **Charge carriers** | Electrons in the conduction band and/or holes in the valence band. |
| **Charge control model** | A model based on the dependence of a device charge on the applied voltage. |
| **Close-packed structure** | A crystal structure consisting of identical spheres with the largest density of the spheres. |
| **CMOS** | Complementary Metal Oxide Semiconductor technology which utilizes both $n$-channel and $p$-channel MOS transistors. |
| **Compensated semiconductor** | A semiconductor containing both acceptor and donor impurities. |
| **Conduction band edge (or conduction band bottom)** | The lowest energy in the conduction band. |
| **Conducting channel** | The conducting region connecting drain and source contacts in a field effect transistor. |
| **Cooper pair** | An electron pair responsible for superconductivity. |
| **Coulomb blockade** | The suppression of tunneling at very small voltages caused by the finite value of the electron charge. |
| **Cutoff frequency** | The frequency at which the current gain drops to unity. |
| **Dangling bonds** | Bonds that are not attached to the nearest neighbors. |
| **De Broglie wavelength** | The wavelength of the wave function describing a free particle. |
| **Degenerate semiconductor** | A semiconductor with the Fermi level in the conduction band or in the valence band. |
| **Density of states** | The number of states per unit volume and per unit energy. |
| **Depletion approximation** | The approximation that assumes that the charge density is equal to that of fully ionized donors or acceptors. |
| **Depletion layer** | The layer where donors or acceptors are fully ionized and the charge of ionized impurities is much larger than the charges of free electrons or holes. |
| **Donor** | Impurity supplying an electron to the conduction band. |
| **Drift velocity** | Carrier velocity caused by electric field. |

| | |
|---|---|
| **Einstein relation** | $D = \mu k_B T/q$ where $D$ is the diffusion coefficient, $\mu$ is the low field mobility, $k_B$ is the Boltzmann constant, $T$ is temperature, and $q$ is the electronic charge. |
| **Electroluminescence** | The emission of light caused by an electric current. |
| **Electron affinity** | The energy difference between the vacuum level and the bottom of the conduction band. |
| **Electron-volt (eV)** | A unit of energy that is equal to the energy of an electron accelerated by 1 V potential. $1\ eV = 1.602 \times 10^{-19}$ J. |
| **Equivalent circuit** | A lumped element circuit, which consists of resistances, capacitances, inductances, current and voltage sources, etc., and which models the device response. |
| **Fermi energy (Fermi level)** | The energy separating energy states that are mostly filled by electrons (below the Fermi level) from the energy states that are mostly empty (above the Fermi level). At the Fermi level, the Fermi-Dirac occupation function is equal to 1/2. |
| **First Brillouin zone** | The region in the reciprocal space ($k$-space) that contains all electronic states. |
| **Gapless semiconductor** | A semiconductor with zero energy gap. |
| **Generation current** | The current caused by the generation of electron-hole pairs in the depletion region under reverse bias. |
| **Geometric magnetoresistance** | The magnetoresistance in short and wide samples. |
| **Ground state** | The lowest energy state of a system. |
| **Heterodimensional devices** | Devices using interfaces between electron gases of different dimensions (such as the interface between a two-dimensional electron gas and a three-dimensional metal). |
| **Heterointerface** | An interface between different semiconductor materials. |
| **Heterojunction Bipolar Transistor** | A bipolar transistor using a heterointerface. |

**Heteropolar bond**

An ionic bond, i. e., a bond such that the electrons responsible for bonding primarily reside on one of the two bonded atoms.

**Heterostructure Field Effect Transistor**

A field effect transistor using a heterointerface.

**Hole**

A fictitious positively charged particle representing the propagation of an electron vacancy in a valence band.

**Hot electron**

An electron in a semiconductor whose energy is substantially higher than the thermal energy for the semiconductor (usually due to the energy acquired from an electric field).

**Impact ionization**

A process of generating electron-hole pairs by energetic electrons or holes.

**Impurity band**

A band of states formed due to the overlap between wave functions of impurities in a highly doped semiconductor.

**Imref**

See Quasi-Fermi level.

**Induced transitions**

Electron transitions between energy states caused by photons with energies equal to the difference between the energy states.

**Injection**

Supply of carriers (electrons or holes) into a semiconductor volume from a contact or a boundary.

**Intervalley transition**

Electron transitions between different minima of the conduction band.

**Intrinsic carrier concentration**

Carrier concentration in an intrinsic semiconductor.

**Intrinsic Fermi level**

The Fermi level of an intrinsic semiconductor.

**Intrinsic semiconductor**

Undoped semiconductor.

**Ion implantation**

Introduction of impurities into a semiconductor bombarding its surface by the energetic impurity ions (usually at elevated temperatures).

**JFET (Junction Field Effect Transistor)**

A field effect transistor with a $p\text{-}n$ junction gate.

**Josephson junction**

A junction consisting of two superconductors (or a normal metal and a superconductor) separated by a thin insulating barrier.

| | |
|---|---|
| **$k$-space** | Reciprocal space, i. e., a space where distances are measured in 1/meter. |
| | |
| **Laser** | A device producing coherent (stimulated) emission of light. |
| **Lattice (or thermal) vibrations** | Vibrations of atoms in a crystal. |
| **Law of the junction** | $pn=n_i^2 \exp\left(\dfrac{qV}{k_BT}\right)$. |
| **Lifetime** | A characteristic recombination time of minority carriers. |
| **Light Emitting Diode** | A light-emitting $p$-$n$ junction diode. |
| **Liquid crystal** | A liquid that consists of molecules that have an elongated shape or a disc shape. |
| **Lorentz force** | A force exerted on moving electrons or holes by a magnetic field. |
| **Low field mobility** | A coefficient of proportionality between the electron (or hole) velocity and the electric field. |
| | |
| **Mass-action law** | $pn=n_i^2$. |
| **Mean free path** | An average length traveled by an electron (or a hole) between collisions. |
| **Meissner-Ochsenfeld effect** | The effect of the expulsion of magnetic field streamlines from a superconductor. |
| **Miller indices** | A set of numbers determining directions or planes in a crystal lattice. |
| **Minority carriers** | Holes in an $n$-type semiconductor, electrons in a $p$-type semiconductor. |
| **Monte Carlo technique** | A computer simulation method using sequences of random numbers. |

**Nondegenerate semiconductor**  A semiconductor with the Fermi level in the energy gap.

**Nonparabolicity**  The deviation of the $E$ versus $k$ relation from a parabolic dependence, such as $E=\dfrac{\hbar^2 k^2}{2m_n}$, where $m_n$ is the effective mass.

**Ohmic contact**  A contact to a semiconductor with a constant (hopefully small) resistance, independent of the electric current.

**One-dimensional electron gas**  Electron gas whose motion is constricted (quantized) in two direction and which is free to move in one direction.

**Pauli exclusion principle**  The principle stating that not more than two electrons can occupy an energy state with the same quantum numbers such as $n$, $l$, and $m$ for an electron in an atom. (These two electrons have different values of spin $S = \pm 1/2$.)

**Phonon**  A quantum of lattice vibrations.

**Photon**  A quantum of electromagnetic radiation.

**Primitive basis vectors**  The shortest linearly independent vectors connecting crystal lattice sites.

**Primitive cell**  The parallelepiped formed by the primitive vectors.

**Quasi-Fermi level**  An energy related to the carrier concentration in a semiconductor under nonequilibrium conditions in the same way as the Fermi level is related to the carrier concentration in a semiconductor under equilibrium conditions.

**Reciprocal space**  See $k$-space.

**Recombination**  The process of destruction of electron-hole pairs.

| | |
|---|---|
| **Saturation velocity** | The carrier velocity in high electric field, which is practically independent of the electric field. |
| **Schottky barrier** | The barrier between a metal and a semiconductor. |
| **Shallow acceptor** | An acceptor with an energy level close to the top of the valence band. |
| **Shallow donor** | A donor with an energy level close to the bottom of the conduction band. |
| **Spin** | A quantum number that is related to an internal electron motion and that can have two values (1/2 and −1/2). |
| **Spontaneous emission** | The photon emission as a result of the system transition from a higher to a lower energy state (not induced by another photon with the same frequency and phase). |
| **Stimulated emission** | The emission of a photon induced by another photon with the same frequency and phase. |
| **Superconductivity** | The phenomenon of zero resistance to an electric current. |
| **Superlattice** | A crystal structure with a superimposed periodic potential. |
| **Thermal velocity** | Typical carrier velocity related to thermal motion. |
| **Thermal voltage** | $V_{th}=k_B T / q$ |
| **Thermionic emission** | The mechanism of the electric current in a Schottky barrier related to the electrons passing over the barrier. |
| **Thermionic-field emission** | The mechanism of the electric current related to the electron tunneling through the barrier at energies above the bottom of the conduction band and above the Fermi level. |
| **Threshold voltage** | The voltage separating the off and on states of a semiconductor device. |
| **Transconductance** | The derivative of the output current with respect to the input voltage. |
| **Transferred electron devices** | Devices using the negative differential resistance related to intervalley transitions. |
| **Tunneling** | Electron penetration through a potential barrier. |
| **Two-dimensional electron gas** | An electron gas with motion restricted (quantized) in one dimension and free to move in two other dimensions. |

**UCCM (Unified Charge Control Model)**    A field effect transistor model that describes subthreshold and above-threshold regimes using one equation.

**Unit cell**    The smallest unit of a crystal that retains its symmetry.

**Vacuum level**    The lowest allowed electron energy level in vacuum surrounding a semiconductor.

**Valence band edge**    The highest energy level in the valence band.

**Wave vector**    A vector whose magnitude is $k = 2\pi/\lambda$ where $\lambda$ is the wavelength and whose direction coincides with the direction of the wave propagation.

**Work function**    The energy difference between the vacuum level and the Fermi level.

# Index